The Theories of Chemistry

The Theories of Chemistry

JAN C.A. BOEYENS
University of Pretoria
Department of Chemistry
Pretoria, South Africa

2003

ELSEVIER

Amsterdam – Boston – Heidelberg – London – New York – Oxford
Paris – San Diego – San Francisco – Singapore – Sydney – Tokyo

ELSEVIER B.V.
Sara Burgerhartstraat 25
P.O. Box 211, 1000 AE
Amsterdam, The Netherlands

ELSEVIER Inc.
525 B Street, Suite 1900
San Diego, CA 92101-4495
USA

ELSEVIER Ltd
The Boulevard, Langford Lane
Kidlington, Oxford OX5 1GB
UK

ELSEVIER Ltd
84 Theobalds Road
London WC1X 8RR
UK

First edition 2003

Library of Congress Cataloging in Publication Data
A catalog record is available from the Library of Congress.

British Library Cataloguing in Publication Data
A catalogue record is available from the British Library.

ISBN: 0-444-51491-0

♾ The paper used in this publication meets the requirements of ANSI/NISO Z39.48-1992 (Permanence of Paper).
Printed in Hungary.

Preface

At the end of the nineteenth century chemistry was at the cutting edge as a theoretically well-founded experimental science. The most advanced and controversial physical theories of the day had their origin in chemical research, which concerned itself with all aspects related to the nature and constitution of matter. The theories of electrons (*sic*), atoms and molecules were the working models of practising chemists. Optical activity, like other forms of spectroscopy in its infancy, was the pursuit of analytical chemistry.

The theoretical chemistry of a hundred years ago comes across as an exciting, vibrant activity, hotly disputed at the laboratory benches of the leading research schools. By comparison, present-day chemistry has very little by way of an innate theory to stimulate the experimentalist. Instead, the necessity of specialization dictates that theoretical pursuits be performed elsewhere and stimulate chemical research by some two-way flow of information. This cross-fertilization however, has dwindled gradually until it finally disappeared during the latter half of the twentieth century. That is why it is not uncommon today, to find synthetic chemists designing new advanced materials such as nano-structures or superconductors, in blissful ignorance of the basic theories that determine the behaviour of these systems.

Symmetry is another example. Its importance is constantly emphasized, but rarely examined beyond the concepts of point-group symmetry, applied to molecules and orbitals. The implications of translational symmetry and continuous symmetry groups in chemical contexts, are simply ignored, and the fundamental significance of conservation (or non-conservation) laws never appreciated. Consequently, chemical thermodynamics remains a sterile recycling of yesterday's formulae. It embodies nothing more challenging than distinguishing between equivalent energy functions, without the guidance of basic theory.

Reaction mechanisms are formulated strictly in terms of time-like events, while the rest of the scientific world marvels at the wonders of non-local correlations and teleportation. Interpretative chemistry still operates in terms of empirical concepts such as electronegativity and molecular shape, commonly said to be without theoretical underpinning. Admittedly, there are those who pontificate against the use of such classical concepts, while remaining singularly unsuccessful themselves to account for the predictive powers of these ideas, or to provide alternatives that are theoretically more soundly based.

Despite a lot of posturing the electron of chemistry is still the electron of Lewis [1], untouched by quantum electrodynamics. The lip service paid to wave mechanics and electron spin does not alter the fact that the curly arrow of chemistry signifies no more than an inductive redistribution of negative charge. A fundamental theory of electrode processes is still lacking. The working theories are exclusively phenomenological and formulated entirely in terms of ionic distributions in the vicinity of electrode interfaces. An early incomplete attempt [2] to develop a quantum-mechanical theory of electrolysis based on electron tunnelling, is still invoked and extensively misunderstood as the basis of charge-transfer. It is clear from too many statements about the nature of electrons[1] that the symbol e is considered sufficient to summarize their important function. The size, spin and mass of the electron never feature in the dynamics of electrochemistry.

Instead of a theory to elucidate the important unsolved problems of chemistry, theoretical chemistry has become synonymous with what is also known as *Quantum Chemistry.* This discipline has patently failed to have any impact on the progress of mainstream chemistry. A new edition of the world's leading Physical Chemistry textbook [4] was published in the year that the Nobel prize was awarded to two quantum chemists, without mentioning either the subject of their work, nor the names of the laureates. Nevertheless, the teaching of chemistry, especially at the introductory level, continues in terms of handwaiving by reference to the same quantum chemistry, that never penetrates the surface of advanced quantum theory.

The natural enabling theories behind chemistry have been left dormant for so long that they are no longer recognized as part of the discipline. It is rarely appreciated that the theories of relativity, quantum phenomena and

[1]From a modern textbook [3]: Electrons are too weak in energy to climb the potential-energy hill at the metal-solution interface. Nevertheless, their quantum qualities permit the electrons to sneak through the barrier in ghostlike fashion provided there are hospitable acceptor states for the electron, *i.e.*, welcoming vacant energy levels in hydrated hydrogen ions.

fields, are as fundamental to chemistry as to physics. Like the elementary particles of physics, the objects of chemistry are best defined in terms of gauge fields, symmetry breaking and the structure of space-time manifolds.

It is futile to hope for physicists to apply their modern views to chemical phenomena, or even expect these concepts to emerge in user-friendly form. It is incumbent on the chemists themselves to incorporate the ideas of advanced quantum mechanics, relativity and field theory into the canon of chemistry. This procedure requires recasting of the formalisms in different terms; into the language of chemistry.

The theme of this book is not a reformulation of theoretical physics, but an attempt to identify the theoretical ideas fundamental to chemistry and recast them in more familiar style. There is no doubt but, that the chemist of this century will have to be familiar with the concepts that appear to be new, even alien, at present. The philosophy that inspires this work is that specialization in science is detrimental in the long term. Specialization may well stimulate productivity in the short term, but this productivity becomes sterile in the absence of innovation, that depends on cross-pollenation, or the adoption of new theoretical models.

The subject area of chemistry is substances and their interaction or evolution, which means a study of matter and time. A clear understanding of time avoids the necessity of statistical arguments when applying mechanics directly to microsystems in chemistry. Delving into the nature of matter brings into focus chemically active substances such as molecules, atoms and electrons, ranging from classical to quantum-mechanical objects. To probe their interaction requires insight into the nature of fields and photons. These issues, approached from a chemical perspective and with emphasis on understanding rather than simulation, will be addressed in a future volume.

The sequence in which to introduce the range of topics presents a problem. To end up with a theory of chemistry based on relativity and quantum mechanics a minimum background in physical chemistry, mechanics and electromagnetism is essential, which in turn requires a knowledge of vectors, complex numbers and differential equations. The selection of material within the preliminary topics is strictly biased by later needs and presented in the usual style of the parent disciplines. Many readers may like to avoid some tedium by treating the introductory material only for reference, as and when required.

The book contains very little original material, but reviews a fair amount of forgotten results that point to new lines of enquiry. Concepts such as *quaternions, Bessel functions, Lie groups, Hamilton-Jacobi theory, solitons, Rydberg atoms, spherical waves* and others, not commonly emphasized in chemical discussion, acquire new importance. To prepare the ground, the book has been designed as a self-contained summary of the essential theoretical background on which to build a modern theory of chemistry.

Contents

Chapter 1

Basic Mathematics

While it is desirable to formulate the theories of physical sciences in terms of the most lucid and simple language, this language often turns out to be mathematics. An equation with its economy of symbols and power to avoid misinterpretation, communicates concepts and ideas more precisely and better than words, provided an agreed mathematical vocabulary exists. In the spirit of this observation, the purpose of this introductory chapter is to review the interpretation of mathematical concepts that feature in the definition of important chemical theories. It is not a substitute for mathematical studies and does not strive to achieve mathematical rigour. It is assumed that the reader is already familiar with algebra, geometry, trigonometry and calculus, but not necessarily with their use in science.

An important objective has been a book that is sufficiently self-contained to allow first reading without consulting too many primary sources. The introductory material should elucidate most mathematical concepts and applications not familiar to the reader. For more detailed assistance one should refer to specialized volumes treating of mathematical methods for scientists, *e.g.* [5, 6, 7, 8, 9]. It may be unavoidable to consult appropriate texts in pure mathematics, of which liberal use has been made here without reference.

1.1 Elementary Vector Algebra

1.1.1 Vectors

Quantities that have both magnitude and direction are called vectors. Examples of vectors in chemistry include things like the dipole moments of molecules, magnetic and electric fields, velocities and angular momenta. It is convenient to represent vectors geometrically and the simplest example is a vector, $\boldsymbol{\rho}$, in two dimensions. It has the magnitude ρ and its direction can be specified by the polar angle φ.

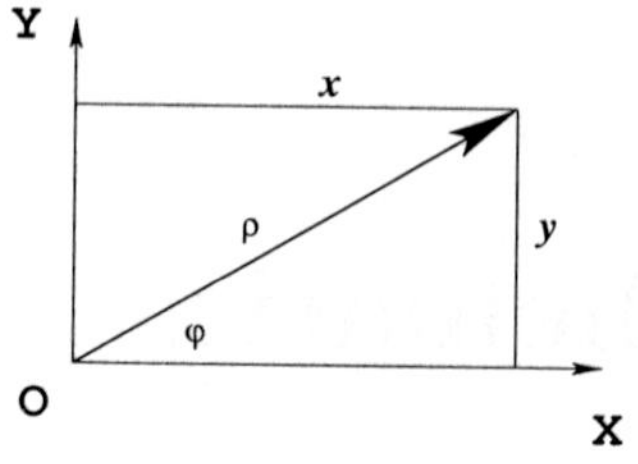

It has two components, (x, y) in cartesian coordinates, and these can be transformed into plane polar coordinates to give

$$x = \rho \cos \varphi \quad , \quad y = \rho \sin \varphi$$

The magnitude of the vector is $\rho = \sqrt{x^2 + y^2}$.

1.1.2 Sum of Vectors

A position vector like $\boldsymbol{\rho}$ above is usually represented as starting at the origin, but it remains unchanged when moved to another position in the plane, as long as its length and direction do not change. Such an operation is used to form the sum of two vectors, **A** and **B**. By moving either **A** or **B** to start from the end point of the second vector, the same vector sum is obtained. It is easy to see how the components add up to

$$C_x = A_x + B_x$$

$$C_y = A_y + B_y$$

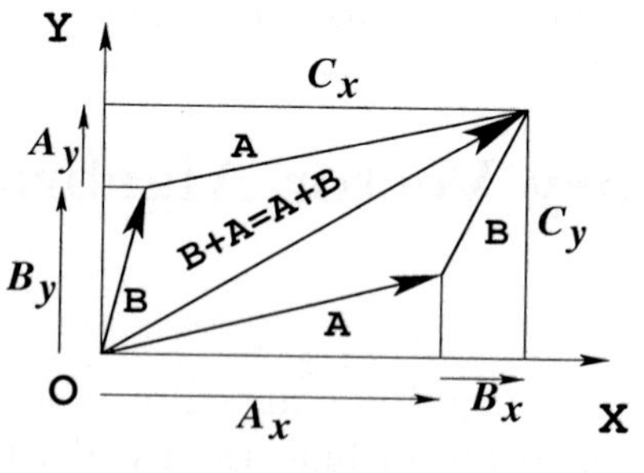

1.1.3 Scalar Product

The product of two vectors $\mathbf{A} \cdot \mathbf{B}$ is the simple product of the magnitudes AB, only if the two vectors are parallel.

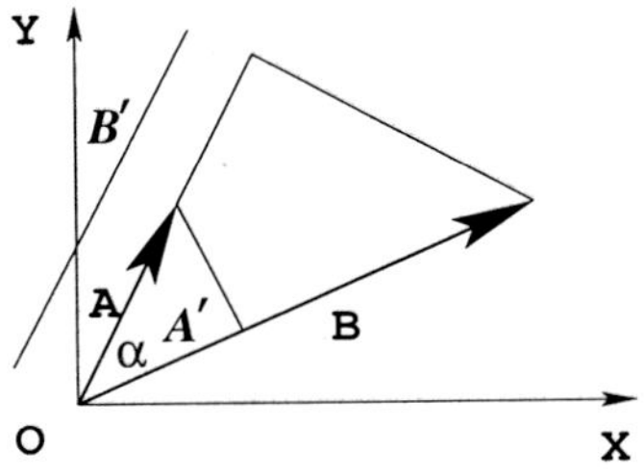

If the angle between the vectors is α, their product, called the scalar or dot product, is defined as $\mathbf{A} \cdot \mathbf{B} = AB \cos\alpha$, which is the product of any of the two vectors in magnitude times the projection of the other (B' or A') on the direction of the first.

$$\cos\alpha = B'/B = A'/A$$

$$A'B = B'A = AB\cos\alpha$$

1.1.4 Three-dimensional Vectors

Another way to represent vectors is by using unit vectors. If $\mathbf{i}$ is a unit vector then any vector $\mathbf{v}$ parallel to $\mathbf{i}$ can be expressed as a multiple of $\mathbf{i}$. If $\mathbf{v}$ has the magnitude v, then $\mathbf{i}v$ is equal to $\mathbf{v}$ in both magnitude and direction. A vector of unit length is now defined in the direction of each of the cartesian axes.

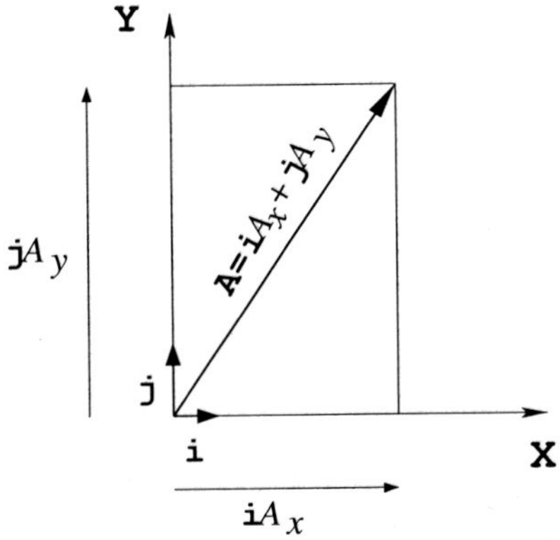

A two-dimensional vector $\mathbf{A}$ is represented as $\mathbf{A} = \mathbf{i}A_x + \mathbf{j}A_y$ and any other vector, $\mathbf{B} = \mathbf{i}B_x + \mathbf{j}B_y$. The scalar product is then written as

$$\begin{aligned}\mathbf{A} \cdot \mathbf{B} &= (\mathbf{i}A_x + \mathbf{j}A_y) \cdot (\mathbf{i}B_x + \mathbf{j}B_y) \\ &= \mathbf{i} \cdot \mathbf{i}A_xB_x + \mathbf{i} \cdot \mathbf{j}A_xB_y + \mathbf{j} \cdot \mathbf{i}A_yB_x + \mathbf{j} \cdot \mathbf{j}A_yB_y\end{aligned}$$

From the definitions of **i** and **j** it follows that

$$\mathbf{i}\cdot\mathbf{i}=\mathbf{j}\cdot\mathbf{j}=1$$

$$\mathbf{i}\cdot\mathbf{j}=\mathbf{j}\cdot\mathbf{i}=0$$

so that $\mathbf{A}\cdot\mathbf{B}=A_xB_x+A_yB_y$.

A position vector in 3 dimensions has components $P(x,y,z)$ in cartesian coordinates, and magnitude of $\mathbf{r}=\mathbf{i}x+\mathbf{j}y+\mathbf{k}z$, following from $r^2=\rho^2+z^2=x^2+y^2+z^2$.

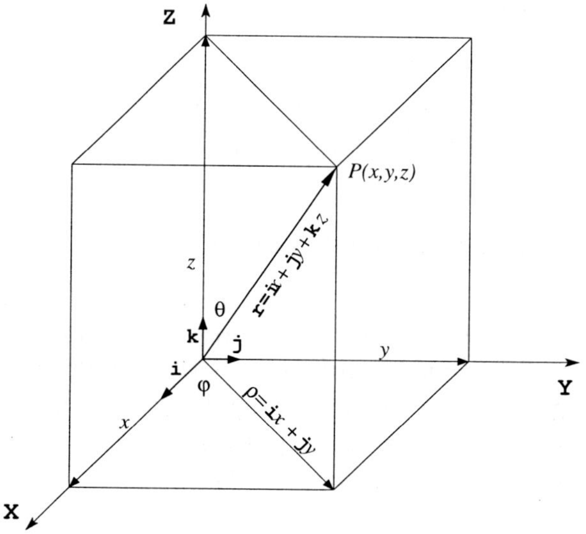

The transformation between cartesian and spherical polar coordinates is

$$x=r\sin\theta\cos\varphi$$

$$y=r\sin\theta\sin\varphi$$

$$z=r\cos\theta$$

The sum and scalar product of two three-dimensional vectors are similar to those quantities in two dimensions, as seen from the following relationships:

$$\mathbf{A}=\mathbf{i}A_x+\mathbf{j}A_y+\mathbf{k}A_z$$

$$\mathbf{B}=\mathbf{i}B_x+\mathbf{j}B_y+\mathbf{k}B_z$$

$$A=\left(A_x^2+A_y^2+A_z^2\right)^{\frac{1}{2}}$$

The scalar product of the two vectors is still given by

$$\mathbf{A} \cdot \mathbf{B} = AB \cos \alpha = A_x B_x + A_y B_y + A_z B_z$$

where α is the angle between the two vectors. The dot product of two unit vectors is equal to the direction cosine relating the two directions.

The value of the dot product is a measure of the coalignment of two vectors and is independent of the coordinate system. The dot product therefore is a true scalar, the simplest *invariant* which can be formed from the two vectors. It provides a useful form for expressing many physical properties: the work done in moving a body equals the dot product of the force and the displacement; the electrical energy density in space is proportional to the dot product of electrical intensity and electrical displacement; quantum mechanical observations are dot products of an operator and a state vector; the invariants of special relativity are the dot products of four-vectors. The invariants involving a set of quantities may be used to establish if these quantities are the components of a vector. For instance, if $\sum A_i B_i$ forms an invariant and B_i are the components of a vector, then A_i must be the components of another vector.

1.1.5 Vector Product

In three dimensions there is another kind of product between two vectors. Consider the same two vectors **A** and **B** together with a third , **C** perpendicular to the plane that contains **A** and **B**.

Now form the scalar products

$$\mathbf{C} \cdot \mathbf{A} = C_x A_x + C_y A_y + C_z A_z$$

$$\mathbf{C} \cdot \mathbf{B} = C_x B_x + C_y B_y + C_z B_z$$

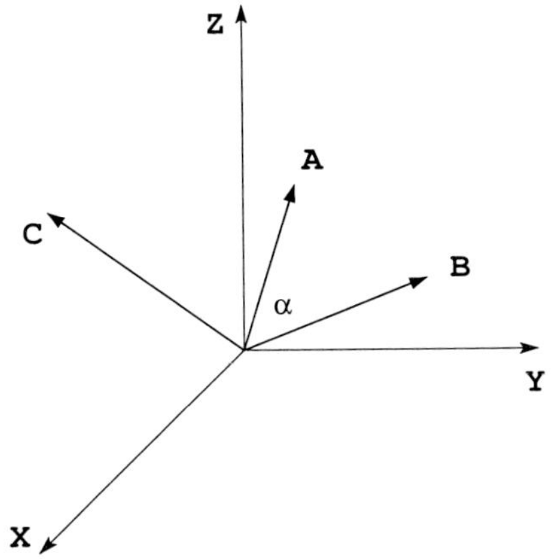

When trying to solve these equations one finds that

$$C_x = m(A_y B_z - A_z B_y)$$

$$C_y = m(A_z B_x - A_x B_z)$$

$$C_z = m(A_x B_y - A_y B_x)$$

where m is some arbitrary constant. If it is defined as $m = +1$, then one has

$$\mathbf{C}^2 = C_x^2 + C_y^2 + C_z^2$$

$$= (A_x^2 + A_y^2 + A_z^2)(B_x^2 + B_y^2 + B_z^2) - (A_x B_x + A_y B_y + A_z B_z)^2$$

$$= (\mathbf{A}^2)(\mathbf{B}^2) - (\mathbf{A} \cdot \mathbf{B})^2$$

Hence

$$\mathbf{C}^2 = (A^2 B^2 - A^2 B^2 \cos^2 \alpha)$$

$$= A^2 B^2 (1 - \cos^2 \alpha) = A^2 B^2 \sin^2 \alpha$$

The vector $\mathbf{C}$ can therefore be considered to be the product of $\mathbf{A}$ and $\mathbf{B}$ with the magnitude $AB \sin \alpha$. This is the basis of defining the vector product, or cross product of two vectors as

$$\mathbf{A} \times \mathbf{B} = AB \sin \alpha$$

By definition $\mathbf{A} \times \mathbf{B} = -\mathbf{B} \times \mathbf{A}$, but it may be shown that

$$\mathbf{A} \times (\mathbf{B} + \mathbf{C}) = \mathbf{A} \times \mathbf{B} + \mathbf{A} \times \mathbf{C}.$$

The vector products of the unit cartesian vectors are seen to obey the following relations

$$\begin{aligned} \mathbf{i} \times \mathbf{i} &= \mathbf{j} \times \mathbf{j} = \mathbf{k} \times \mathbf{k} = 0 \\ \mathbf{i} \times \mathbf{j} &= -\mathbf{j} \times \mathbf{i} = \mathbf{k} \\ \mathbf{j} \times \mathbf{k} &= -\mathbf{k} \times \mathbf{j} = \mathbf{i} \\ \mathbf{k} \times \mathbf{i} &= -\mathbf{i} \times \mathbf{k} = \mathbf{j}. \end{aligned}$$

Further,

$$\begin{aligned} \mathbf{A} \times \mathbf{B} &= (\mathbf{i}A_x + \mathbf{j}A_y + \mathbf{k}A_z) \times (\mathbf{i}B_x + \mathbf{j}B_y + \mathbf{k}B_z) \\ &= \mathbf{i}(A_y B_z - A_z B_y) + \mathbf{j}(A_z B_x - A_x B_z) + \mathbf{k}(A_x B_y - A_y B_z) \\ &= \begin{vmatrix} \mathbf{i} & \mathbf{j} & \mathbf{k} \\ A_x & A_y & A_z \\ B_x & B_y & B_z \end{vmatrix} \end{aligned}$$

Two vectors commonly represented in terms of cross products are: the angular momentum of a particle about some point, equal to the cross product of the momentum vector of the particle and the radius vector from the origin to the particle; and torque, equal to the cross product of the force vector and the vector representing the lever arm.

1.1.6 Three-vector products

The triple scalar product $\boldsymbol{A} \cdot (\boldsymbol{B} \times \boldsymbol{C})$, interpreted geometrically, is the volume of the paralellipiped with sides $\boldsymbol{A}$, $\boldsymbol{B}$, $\boldsymbol{C}$ and represented by the determinant

$$\boldsymbol{A} \cdot (\boldsymbol{B} \times \boldsymbol{C}) = \begin{vmatrix} A_x & A_y & A_z \\ B_x & B_y & B_z \\ C_x & C_y & C_z \end{vmatrix}$$

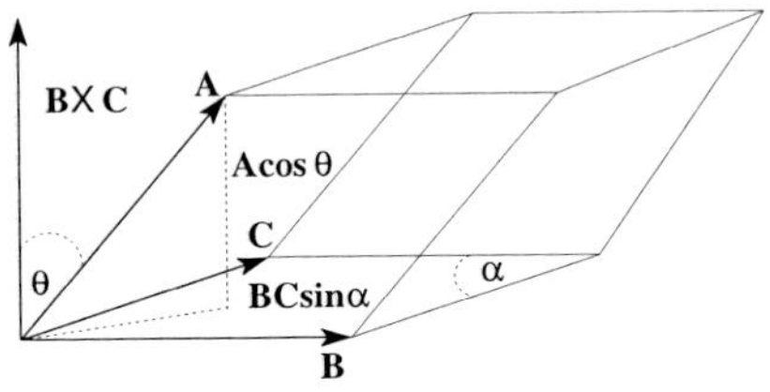

Since a determinant changes sign when two of its rows are interchanged it is readily shown that

$$\begin{aligned} \boldsymbol{A} \cdot (\boldsymbol{B} \times \boldsymbol{C}) &= -\boldsymbol{A} \cdot (\boldsymbol{B} \times \boldsymbol{C}) = -\boldsymbol{B} \cdot (\boldsymbol{A} \times \boldsymbol{C}) = \\ &- \boldsymbol{C} \cdot (\boldsymbol{B} \times \boldsymbol{A}) = \boldsymbol{B} \cdot (\boldsymbol{C} \times \boldsymbol{A}) = \boldsymbol{C} \cdot (\boldsymbol{A} \times \boldsymbol{B}) \end{aligned}$$

Since the order of the scalar product is unimportant, it means that

$$\boldsymbol{A} \cdot (\boldsymbol{B} \times \boldsymbol{C}) = (\boldsymbol{B} \times \boldsymbol{C}) \cdot \boldsymbol{A} = (\boldsymbol{A} \times \boldsymbol{B}) \cdot \boldsymbol{C} = (\boldsymbol{C} \times \boldsymbol{A}) \cdot \boldsymbol{B} \quad (1.1)$$

The vector triple product $\boldsymbol{A} \times (\boldsymbol{B} \times \boldsymbol{C})$ is a vector since it is the product of two vectors, $\boldsymbol{A}$ and $\boldsymbol{B} \times \boldsymbol{C}$. It may be shown by expansion that

$$\boldsymbol{A} \times (\boldsymbol{B} \times \boldsymbol{C}) = \boldsymbol{B}(\boldsymbol{A} \cdot \boldsymbol{C}) - \boldsymbol{C}(\boldsymbol{A} \cdot \boldsymbol{B}) \quad (1.2)$$

which permits decomposition into two scalar products.

1.1.7 Complex Numbers

The most important and fundamental difference between quantum mechanics and classical mechanics is the appearance of complex quantities as an essential ingredient of the former.

A complex number consists of two parts: a real and a so-called imaginary part, $c = a + ib$. The imaginary part always contains the quantity i, which represents the square root of -1, $i = \sqrt{-1}$. The real and imaginary parts of c are often denoted by $a = R(c)$ and $b = I(c)$. All the common rules of ordinary arithmetic apply to complex numbers, which in addition allow extraction of the square root of any negative number. If

$$z = x + iy$$

then the reciprocal of z, called z^{-1}, is given by

$$z^{-1} = \frac{x}{x^2 + y^2} - i\frac{y}{x^2 + y^2}$$

To prove this, form the product

$$z(z^{-1}) = (x + iy)\left(\frac{x}{x^2 + y^2} - i\frac{y}{x^2 + y^2}\right)$$

$$= \frac{1}{x^2 + y^2}(x^2 + ixy - ixy - i^2y^2)$$

$$= \frac{x^2 - i^2y^2}{x^2 + y^2} = 1$$

Since the real and imaginary parts of a complex number are independent of each other, a complex number is always specified in terms of two real numbers, like the coordinates of a point in a plane, or the two components of a two-dimensional vector. In an *Argand diagram* a complex number is represented as a point in the *complex plane* by a *real* and an *imaginary* axis.

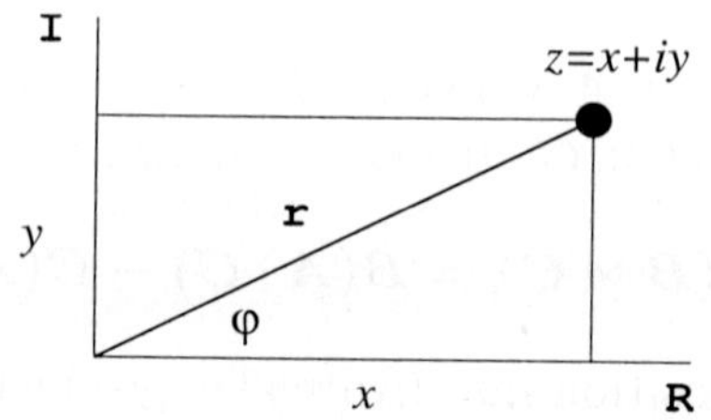

When expressed in polar coordinates, the quantity r is the *magnitude* or *absolute value* or *modulus* and ϕ is the *argument* or *phase* of the complex number. It follows immediately that

$$x + iy = r\cos\phi + ir\sin\phi$$

A standard way to represent trigonometric functions is in terms of infinite series:

$$\sin x = x - \frac{x^3}{3!} + \frac{x^5}{5!} - \frac{x^7}{7!} + \cdots = \sum_{n=0}^{\infty}(-1)^n \frac{x^{2n+1}}{(2n+1)!} \tag{1.3}$$

$$\cos x = 1 - \frac{x^2}{2!} + \frac{x^4}{4!} - \frac{x^6}{6!} + \cdots = \sum_{n=0}^{\infty}(-1)^n \frac{x^{2n}}{(2n)!} \tag{1.4}$$

while the exponential function

$$e^x = 1 + x + \frac{x^2}{2!} + \frac{x^3}{3!} + \cdots = \sum_{n=0}^{\infty} \frac{x^n}{n!}$$

It is left as an exercise to the reader to show that $e^{ix} = \cos x + i\sin x$, which is known as Euler's theorem. In the present notation

$$re^{i\phi} = r\cos\phi + ir\sin\phi = x + iy$$

where e is the base of natural logarithms, $e = 2.7182813\cdots$. Viewed as an operator, this quantity rotates any vector $z = x + iy$ by an angle ϕ and increases its length by a factor r.

In the polar representation, the product of two complex numbers $z_1 = r_1e^{i\phi_1}$ and $z_2 = r_2e^{i\phi_2}$ is $z_1z_2 = r_1r_2e^{i(\phi_1+\phi_2)}$ and their quotient,

$$\frac{z_1}{z_2} = \left(\frac{r_1}{r_2}\right) e^{i(\phi_1-\phi_2)}$$

In all of these formulae, ϕ must be measured in radians.

De Moivre's formula for raising a complex number to a given power is:

$$\left(re^{i\phi}\right)^n = r^n e^{in\phi} = r^n[\cos(n\phi) + i\sin(n\phi)]$$

The magnitude and phase of a complex number $z = x + iy$ are calculated as

$$r = \sqrt{x^2 + y^2}$$

$$\phi = \arctan\left(\frac{y}{x}\right)$$

The *complex conjugate* of $z = x + iy$ is defined as

$$z^* = (x + iy)^* = x - iy$$

as shown in the following Argand diagram

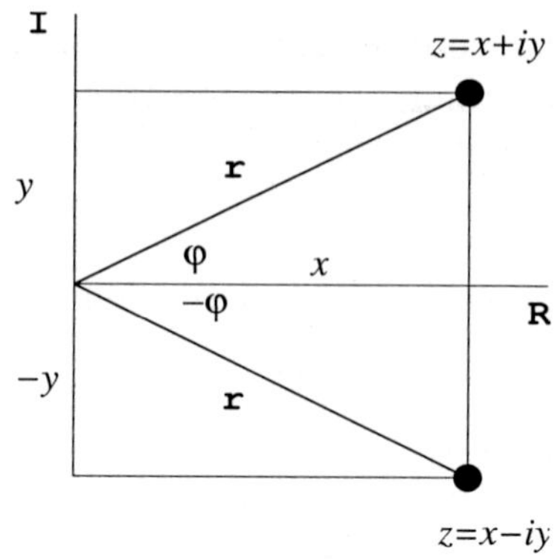

The phase of the complex conjugate[1] is $-\phi$. The magnitude is the same, so

$$\left(re^{i\phi}\right)^* = re^{-i\phi}$$

The magnitude of a complex quantity can be obtained from the product

$$zz^* = \left(re^{i\phi}\right)\left(re^{-i\phi}\right) = r^2$$

so that $r = \sqrt{zz^*}$, where the positive square root is taken. Note that zz^* is always real and nonnegative.

1.1.8 N-dimensional Vectors

In the course of this work there will be occasion to use vectors defined in more than three dimensions. Although an n-dimensional vector is difficult to visualize geometrically the algebraic definition is no more complicated than for three-dimensional vectors.

An n-dimensional vector is considered to have n orthogonal components and the vector is defined by specifying all of these as an array. There are two conventions to define either a column vector, like

$$\mathbf{c} = \begin{pmatrix} c_1 \\ c_2 \\ \vdots \\ c_n \end{pmatrix} = \{c_1, c_2, \cdots, c_n\}$$

or a row vector, like

$$\mathbf{c} = [c_1, c_2, \cdots, c_n]$$

[1]The complex conjugate is obtained by changing the sign in front of every i that occurs.

The scalar product of two n-dimensional vectors is only defined when one vector is a column vector and the other is a row vector, i.e.

$$\begin{aligned}\mathbf{w} \cdot \mathbf{v} &= [w_1, w_2, \cdots, w_n]\{v_1, v_2, \cdots, v_n\} \\ &= w_1 v_1 + w_2 v_2 + \cdots + w_n v_n \\ &= \sum_{i=1}^{n} w_i v_i\end{aligned}$$

The Hermitian conjugate $\mathbf{c}^\dagger$(dagger) of a column vector $\mathbf{c}$, is a row vector, with the components c_i^*. The scalar product of the row vector $\mathbf{w}^\dagger$ and a column vector, $\mathbf{v}$ is

$$\begin{aligned}\mathbf{w}^\dagger \cdot \mathbf{v} &= [w_1^*, w_2^*, \cdots, w_n^*]\{v_1, v_2, \cdots, v_n\} \\ &= \sum_{i=1}^{n} w_i^* v_i\end{aligned}$$

If $\mathbf{w}^\dagger \mathbf{v} = 0$ the column vectors $\mathbf{v}$ and $\mathbf{w}$ (or equivalently the row vectors $\mathbf{v}^\dagger$ and $\mathbf{w}^\dagger$) are orthogonal.

The value of $\mathbf{v}^\dagger \mathbf{v}$ for any vector is always positive, since

$$\mathbf{v}^\dagger \mathbf{v} = \sum_{i=1}^{n} |v_i|^2$$

The length (norm) of the column vector $\mathbf{v}$ is the positive root $\sqrt{\mathbf{v}^\dagger \mathbf{v}}$. A vector is normalized if its length is 1. Two vectors $\boldsymbol{r_i}$ and $\boldsymbol{r_j}$ of an n-dimensional set are said to be *linearly independent* of each other if one is not a constant multiple of the other, *i.e.*, it is impossible to find a scalar c such that $\boldsymbol{r}_i = c\boldsymbol{r}_j$. In simple words, this means that $\boldsymbol{r}_j$ and $\boldsymbol{r}_i$ are not parallel. In general, m vectors constitute a set of linearly independent vectors if and only if the equation

$$\sum_{i=1}^{m} a_i \boldsymbol{r}_i = 0$$

is satisfied only when all the scalars $a_i = 0$ for $1 \le i \le m$. In other words, the set of m vectors is linearly independent if none of them can be expressed as a linear combination of the remaining $m - 1$ vectors. A simple test for the linear independence of a set of vectors is to construct the determinant of their scalar products with each other as

$$\Gamma = \begin{vmatrix} \boldsymbol{r}_1 \cdot \boldsymbol{r}_1 & \boldsymbol{r}_1 \cdot \boldsymbol{r}_2 & \cdots & \boldsymbol{r}_1 \cdot \boldsymbol{r}_m \\ \boldsymbol{r}_2 \cdot \boldsymbol{r}_1 & \boldsymbol{r}_2 \cdot \boldsymbol{r}_2 & \cdots & \boldsymbol{r}_2 \cdot \boldsymbol{r}_m \\ \cdots & \cdots & \cdots & \cdots \\ \boldsymbol{r}_m \cdot \boldsymbol{r}_1 & \boldsymbol{r}_m \cdot \boldsymbol{r}_2 & \cdots & \boldsymbol{r}_m \cdot \boldsymbol{r}_m \end{vmatrix}$$

known as the Gram determinant. If $\Gamma = 0$ (see 1.2.3), it follows that one of the vectors can be expressed as a linear combination of the remaining $m-1$; if $\Gamma \neq 0$, the vectors are linearly independent.

1.1.9 Quaternions

There is similarity between two-dimensional vectors and complex numbers, but also subtle differences. One striking difference is between the product functions of complex numbers and vectors respectively. The product of two complex numbers is

$$\begin{aligned} z_1 z_2 &= (x_1 + iy_1)(x_2 + iy_2) \\ &= (x_1x_2 - y_1y_2) + i(x_1y_2 + x_2y_1), \\ z_3 &= x_3 + iy_3 \end{aligned}$$

or in polar form

$$\begin{aligned} z_1 z_2 &= r_1 r_2 \{\cos(\theta_1 + \theta_2) + i\sin(\theta_1 + \theta_2)\} \\ z_3 &= r_3(\cos\theta_3 + i\sin\theta_3) \end{aligned}$$

The product of two vectors is either a scalar

$$z_3 = \boldsymbol{z_1} \cdot \boldsymbol{z_2} = z_1 z_2 \cos\theta = x_1x_2 + y_1y_2$$

or a vector

$$\boldsymbol{z_3} = \boldsymbol{z_1} \times \boldsymbol{z_2} = z_1 z_2 \sin\theta = y_1x_2 - y_2x_1.$$

The complex product appears to be made up of two terms, not too unlike the scalar and vector products, from which it seems to differ only because of a sign convention.

The extension of vector methods to more dimensions suggests the definition of related *hypercomplex numbers*. When the multiplication of two three-dimensional vectors is performed without defining the mathematical properties of the unit vectors $\boldsymbol{i}, \boldsymbol{j}, \boldsymbol{k}$, the formal result is

$$\begin{aligned} q &= (\boldsymbol{i}x_1 + \boldsymbol{j}y_1 + \boldsymbol{k}z_1)(\boldsymbol{i}x_2 + \boldsymbol{j}y_2 + \boldsymbol{k}z_2) \\ &= \boldsymbol{i}^2x_1x_2 + \boldsymbol{j}^2y_1y_2 + \boldsymbol{k}^2z_1z_2 \\ &+ \boldsymbol{ij}x_1y_2 + \boldsymbol{ji}y_1x_2 + \boldsymbol{ik}x_1z_2 + \boldsymbol{ki}z_1x_2 + \boldsymbol{jk}y_1z_2 + \boldsymbol{kj}z_1y_2 \end{aligned}$$

By introducing the following definitions

$$\begin{aligned} \boldsymbol{i}^2 &= \boldsymbol{j}^2 = \boldsymbol{k}^2 = -1 \\ \boldsymbol{ij} &= \boldsymbol{k} \quad, \boldsymbol{jk} = \boldsymbol{i} \quad, \boldsymbol{ki} = \boldsymbol{j} \\ \boldsymbol{ji} &= -\boldsymbol{k} \quad, \boldsymbol{kj} = -\boldsymbol{i} \quad, \boldsymbol{ik} = -\boldsymbol{j} \end{aligned}$$

a physically meaningful result,

$$\begin{aligned} q = & - (x_1x_2 + y_1y_2 + z_1z_2) + \boldsymbol{i}(y_1z_2 - y_2z_1) \\ & + \boldsymbol{j}(z_1x_2 - z_2x_1) + \boldsymbol{k}(x_1y_2 - x_2y_1) \end{aligned}$$

is obtained. The first term on the right is noticed to be the negative of the scalar product of two vectors and the remaining three terms define the vector product in terms of unit vectors $\boldsymbol{i}, \boldsymbol{j}, \boldsymbol{k}$. The formal multiplication therefore yields the scalar and vector products in one blow. This solution first occured to Sir William Hamilton as a eureka experience on a bridge in Dublin, into which he carved the result

$$\boldsymbol{i}^2 = \boldsymbol{j}^2 = \boldsymbol{k}^2 = \boldsymbol{ijk} = -1.$$

He called the product function a *quaternion.*

In terms of the Hamilton formalism a hypercomplex number of unit norm can now be defined in the form

$$a_0 = a_i e_i$$

where a_0 and the a_i are generalizations of $\sqrt{-1}$. In the case of three e_i the hypercomplex number is called a quaternion and the e_i obey the following rules of multiplication:

$$\begin{aligned} e_i^2 &= -1, \quad i = 1, 2, 3 \\ e_i e_j &= \epsilon_{ijk} e_k \qquad (i \neq j) \end{aligned}$$

in which

$$\epsilon = \begin{Bmatrix} 1 \\ -1 \\ 0 \end{Bmatrix} \quad \text{for} \quad ijk \quad \text{an} \quad \begin{Bmatrix} \text{even} \\ \text{odd} \\ \text{no} \end{Bmatrix} \quad \text{permutation of 1, 2, 3.}$$

These numbers do not obey all of the laws of the algebra of complex numbers. They add like complex numbers, but their multiplication is not commutative. The general rules of multiplication of n-dimensional hypercomplex numbers were investigated by Grassmann who found a number of laws of multiplication, including Hamilton's rule. These methods still await full implimentation in physical theory.

1.2 Determinants and Matrices

1.2.1 Introduction

The first successful formulation of quantum mechanics by Heisenberg was achieved in terms of matrices, and has not diminished in importance since then. Matrices are rectangular or, more important in the present context, square arrays of numbers, manipulated only in terms of the well-defined rules of matrix algebra. These procedures probably developed from techniques, familiar to any schoolboy, for solving a set of linear equations, like

$$a_{11}x_1 + a_{12}x_2 + a_{13}x_3 = b_1$$

$$a_{21}x_1 + a_{22}x_2 + a_{23}x_3 = b_2$$

$$a_{31}x_1 + a_{32}x_2 + a_{33}x_3 = b_3$$

In shorthand notation this is written as

$$\begin{bmatrix} a_{11} & a_{12} & a_{13} \\ a_{21} & a_{22} & a_{23} \\ a_{31} & a_{32} & a_{33} \end{bmatrix} \begin{bmatrix} x_1 \\ x_2 \\ x_3 \end{bmatrix} = \begin{bmatrix} b_1 \\ b_2 \\ b_3 \end{bmatrix}$$

or

$$\mathbf{Ax} = \mathbf{b}$$

where $\mathbf{x}$ and $\mathbf{b}$ are recognized as column vectors and $\mathbf{A}$ is a 3×3 matrix.

1.2.2 Matrix Operations

To reproduce the system of equations it is noted that each *row* of the matrix is multiplied into the *column* vector, adding elementary products together to yield the elements of the product vector, $\mathbf{b}$. This is a special case of matrix multiplication which can be represented symbolically by

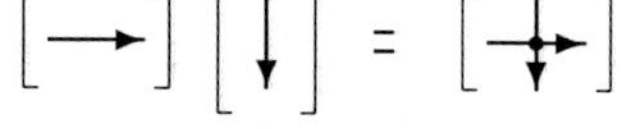

The ith row of the pre-factor is multiplied term-by-term with the jth column of the post-factor and added to yield the ijth element of the product.

The product of two matrices is therefore similar to the scalar product of two vectors. $\mathbf{C}$ is the product of $\mathbf{AB}$, according to

$$c_{ij} = \sum_{k=1}^{n} a_{ik} b_{kj}$$

n is the number of columns in $\mathbf{A}$, and $\mathbf{B}$ is required to have as many rows as $\mathbf{A}$ has columns. $\mathbf{C}$ will have as many rows as $\mathbf{A}$ and as many columns as $\mathbf{B}$.

If two matrices are square, they can be multiplied together in any order. In general, the multiplication is not commutative. That is $\mathbf{AB} \neq \mathbf{BA}$, except in some special cases. It is said that the matrices do not *commute*, and this is the property of major importance in quantum mechanics, where it is common practice to define the *commutator* of two matrices as

$$[\mathbf{A}, \mathbf{B}] = [\mathbf{AB} - \mathbf{BA}]$$

However, matrix multiplication is associative

$$\mathbf{A}(\mathbf{BC}) = (\mathbf{AB})\mathbf{C}$$

and matrix multiplication and addition are distributive

$$\mathbf{A}(\mathbf{B} + \mathbf{C}) = \mathbf{AB} + \mathbf{AC}$$

Two matrices are equal to each other if and only if every element of one is equal to the corresponding element of the other. The two matrices must, of course have the same numbers of rows and of columns.

The sum of two matrices is defined by $\mathbf{C} = \mathbf{A} + \mathbf{B}$ if and only if $c_{ij} = a_{ij} + b_{ij}$ for all i and j.

The product of a matrix and a scalar is defined by $\mathbf{B} = c\mathbf{A}$, if and only if $b_{ij} = ca_{ij}$ for all i and j.

An *identity matrix* is defined to have the property $\mathbf{AI} = \mathbf{IA} = \mathbf{A}$. Since it gives the same result for pre and post multiplication it must be a square matrix, with the form

$$\mathbf{I} = \begin{bmatrix} 1 & 0 & 0 & \cdots & 0 \\ 0 & 1 & 0 & \cdots & 0 \\ \cdots & \cdots & \cdots & \cdots & \cdots \\ 0 & 0 & 0 & \cdots & 1 \end{bmatrix}$$

The nonzero elements are those with both indices the same and are called the *diagonal elements*. The sum of the diagonal elements of a square matrix is called the *trace*

$$\mathrm{Tr}\boldsymbol{A} = \sum_{i=1}^{n} A_{ii}$$

The trace of the product of two or more matrices is independent of the order of multiplication. The proof is simple:

$$\mathrm{Tr}\boldsymbol{AB} = \sum_{i}(AB)_{ii} = \sum_{i}\sum_{j} A_{ij}B_{ji} = \mathrm{Tr}\boldsymbol{BA}$$

If $\boldsymbol{A} \times \boldsymbol{B} = \boldsymbol{C}$, $\text{Tr}\boldsymbol{C} = \text{Tr}\boldsymbol{A} \cdot \text{Tr}\boldsymbol{B}$.

Another important matrix is the *transpose* which is obtained by interchanging rows and columns. It is denoted by $\tilde{\mathbf{A}}$ (tilde) and defined by

$$(\tilde{A})_{ij} = \tilde{a}_{ij} = a_{ji}$$

If a matrix is equal to its transpose, it is said to be a *symmetric* matrix.

If the elements of **A** are complex numbers, the *complex conjugate* of **A** is defined as

$$\mathbf{A}^* = [A_{ij}^*]$$

The transpose of the complex conjugate

$$\mathbf{A}^\dagger = (\tilde{\mathbf{A}}^*) = (\tilde{\mathbf{A}})^* = (A^\dagger)_{ij} = a_{ji}^*$$

is called the *hermitian conjugate.*

A matrix that is equal to its hermitian conjugate is called *hermitian*, and these are the matrices used in matrix mechanics, $\mathbf{A}^\dagger = \mathbf{A}$. A matrix is *antihermitian* if $\mathbf{A}^\dagger = -\mathbf{A}$.

An *orthogonal* matrix is one whose transpose is equal to its *inverse*,

$$\tilde{\mathbf{A}} = \mathbf{A}^{-1} \text{ , } (\mathbf{A} = \tilde{\mathbf{A}}^{-1})$$

A *unitary* matrix is one whose inverse is equal to its hermitian conjugate, $\mathbf{A}^{-1} = \mathbf{A}^\dagger = \tilde{\mathbf{A}}^*$.

1.2.3 Inverse of a Matrix

In order to solve the set of equations it is necessary to find the inverse of **A** (denoted by $\mathbf{A}^{-1}$) and multiply it into $\mathbf{A} = \mathbf{bx}$, noting that $\mathbf{AA}^{-1} = \mathbf{A}^{-1}\mathbf{A} = \mathbf{I}$, the *unit* matrix,

$$\mathbf{I} = \begin{bmatrix} 1 & 0 & 0 \\ 0 & 1 & 0 \\ 0 & 0 & 1 \end{bmatrix}$$

Hence

$$\mathbf{A}^{-1}\mathbf{Ax} = \mathbf{x} = \mathbf{A}^{-1}\mathbf{b}$$

produces the required solutions. The unit matrix is also presented by the *Kronecker delta* function

$$\delta_{ij} = \begin{cases} 1, & \text{for } i = j \\ 0, & \text{for } i \neq j \end{cases}$$

The systematic high-school procedure to solve the set of three equations therefore amounts to the procedure of obtaining the inverse matrix $\mathbf{A}^{-1}$.

To follow this procedure it is useful to define the *determinant* associated with the square matrix of interest, written as

$$\begin{vmatrix} a_{11} & a_{12} & a_{13} \\ a_{21} & a_{22} & a_{23} \\ a_{31} & a_{32} & a_{33} \end{vmatrix} = a_{11}\begin{vmatrix} a_{22} & a_{23} \\ a_{32} & a_{33} \end{vmatrix} - a_{12}\begin{vmatrix} a_{21} & a_{23} \\ a_{31} & a_{33} \end{vmatrix} + a_{13}\begin{vmatrix} a_{21} & a_{22} \\ a_{31} & a_{32} \end{vmatrix}$$

$$= a_{11}a_{22}a_{33} - a_{11}a_{23}a_{33} - a_{12}a_{21}a_{33} + a_{12}a_{23}a_{31} + a_{13}a_{21}a_{32} - a_{13}a_{22}a_{31}$$

The sub-determinants, with appropriate signs, as they appear in the first stage of expansion of the determinant, are examples of *cofactors*. The *minor*, M_{ij} of an element a_{ij} is obtained by deleting the ith row and the jth column from the determinant, whereby the cofactor

$$A_{ij} = (-1)^{i+j}M_{ij}$$

The evaluation of a determinant is a tedious process, which for an $n \times n$ determinant is summarized by

$$|A| = \begin{vmatrix} a_{11} & a_{12} & \cdots & a_{1n} \\ a_{21} & a_{22} & \cdots & a_{2n} \\ \cdots & \cdots & \cdots & \cdots \\ a_{n1} & a_{n2} & \cdots & a_{nn} \end{vmatrix} = \sum_{m=1}^{n} a_{mi}A_{mj}$$

$$\begin{cases} = 0 & \text{for } i = j \\ = |A| & \text{for } i \neq j \end{cases} = |A|\delta_{ij}$$

If the rows and columns of the cofactor matrix are transposed a matrix $\mathbf{A}^{ji}$, called *adjugate* to $\mathbf{A}$ is obtained, such that

$$\mathbf{A} \cdot \mathbf{A}^{ji} = |A| \begin{bmatrix} 1 & 0 & \cdots & 0 \\ 0 & 1 & \cdots & 0 \\ \cdots & \cdots & \cdots & \cdots \\ 0 & 0 & \cdots & 1 \end{bmatrix} = |A|[\mathbf{I}]$$

It follows that

$$\mathbf{A} \times \frac{\mathbf{A}^{ji}}{|A|} = \mathbf{I} = \mathbf{A}\mathbf{A}^{-1}$$

and hence the definition of the inverse matrix as

$$\mathbf{A}^{-1} = \frac{\mathbf{A}^{ji}}{|A|}$$

To calculate the inverse of a matrix by this procedure is equally tedious and probably more work than solving a set of equations by the brute-force high-school technique. However, the procedure is readily converted into computer code and this is now the only way recommended for matrix inversions.

It is noted that the inverse of a matrix only exists if $|A| \neq 0$. Any matrix with $|A| = 0$ is called *singular*.

1.2.4 Linear Homogeneous Equations

It is a common problem to solve a set of *homogeneous* equations of the form $\mathbf{Ax} = 0$. If the matrix is non-singular the only solutions are the trivial ones, $x_1 = x_2 = \cdots = x_n = 0$. It follows that the set of homogeneous equations has non-trivial solutions only if $|A| = 0$. This means that the matrix has no inverse and a new strategy is required in order to get a solution.

Since $\mathbf{A}$ is singular, one of the consequences is that the columns (or rows) of the matrix are linearly dependent. This means that a set of non-zero coefficients c_i, for which an identity relation

$$\begin{pmatrix} a_{11} \\ a_{21} \\ \cdots \\ a_{n1} \end{pmatrix} c_1 + \begin{pmatrix} a_{12} \\ a_{22} \\ \cdots \\ a_{n2} \end{pmatrix} c_2 + \cdots + \begin{pmatrix} a_{1n} \\ a_{2n} \\ \cdots \\ a_{nn} \end{pmatrix} c_n = 0$$

exists between the columns (rows) of the matrix, can be found.

On the other hand if $\mathbf{x}_j (1 \leq j \leq n)$ is a set of n linearly independent column vectors (matrices of order $n \times 1$), then any column matrix (vector) $\mathbf{y}$ can be expressed as a linear combination of the vectors $\mathbf{x}_j$; so that the coefficients c_i exist such that

$$\mathbf{y} = c_1\mathbf{x}_1 + c_2\mathbf{x}_2 + \cdots + c_n\mathbf{x}_n$$

If these n column vectors are assembled into a square matrix

$$\mathbf{X} = \begin{pmatrix} x_{11} & x_{12} & \cdots & x_{1n} \\ x_{21} & x_{22} & \cdots & x_{2n} \\ \cdot & \cdot & \cdots & \cdot \\ \cdot & \cdot & \cdots & \cdot \\ \cdot & \cdot & \cdots & \cdot \\ x_{n1} & x_{n2} & \cdots & x_{nn} \end{pmatrix}$$

then, since the columns of the matrix are linearly independent, the matrix $\mathbf{X}$ is non-singular. It follows that any column matrix $\mathbf{y}$ of order n can be expressed in the form $\mathbf{y} = \mathbf{X}\mathbf{c}$, and the required values of c_j are given by $\mathbf{c} = \mathbf{X}^{-1}\mathbf{y}$.

This result can be generalized into the statement that any arbitrary vector in n dimensions can always be expressed as a linear combination of n basic vectors, provided these are linearly independent. It will be shown that the latent solutions of a singular matrix provide an acceptable set of basis vectors, just like the eigen-solutions of certain differential equations provide an acceptable set of basis functions.

The Latent roots of a Matrix

A problem of particular importance in quantum theory is the calculation of special column vectors $\mathbf{x}$ and the eigenvalues λ associated with a given square matrix $\mathbf{A}$, by the equation

$$\mathbf{A}\mathbf{x} = \lambda\mathbf{x}, \quad \text{or} \quad (\mathbf{A} - \lambda\mathbf{I})\mathbf{x} = 0$$

The solution $\mathbf{x} = 0$ is excluded. There may be several different column vectors $\mathbf{x}$, each with a different value of λ, and each satisfying the equation. The numbers λ are called the *latent roots* or *eigenvalues* of $\mathbf{A}$. The vectors $\mathbf{x}$ are the *latent solutions* or *eigenfunctions* of $\mathbf{A}$. The equation, written out in full, is a homogeneous system of linear equations, and will have a solution, other than $\mathbf{x} = 0$, if and only if

$$\begin{vmatrix} a_{11} - \lambda & a_{12} & \cdots & a_{1n} \\ a_{21} & a_{22} - \lambda & \cdots & a_{2n} \\ \cdots & \cdots & \cdots & \cdots \\ a_{n1} & a_{n2} & \cdots & a_{nn} - \lambda \end{vmatrix} = 0$$

This is an algebraic equation in λ of degree n and will have n roots $\lambda_1, \lambda_2, \cdots, \lambda_n$ (possibly including repeated and complex roots). To each value of λ there will correspond in general a distinct solution $\mathbf{x}$. Let $\mathbf{x}_{(1)}, \cdots, \mathbf{x}_{(n)}$ correspond to $\lambda_1, \lambda_2, \cdots, \lambda_n$ then

$$\left.\begin{array}{rcl} \mathbf{A}\mathbf{x}_{(1)} & = & \lambda_1\mathbf{x}_{(1)} \\ \mathbf{A}\mathbf{x}_{(2)} & = & \lambda_2\mathbf{x}_{(2)} \\ \cdot & \cdot & \cdot \\ \mathbf{A}\mathbf{x}_{(n)} & = & \lambda_n\mathbf{x}_{(n)} \end{array}\right\}$$

These solutions are not unique; for example, if $\mathbf{x}_{(1)}$ is a solution, then $k\mathbf{x}_{(1)}$ is also a solution.

To show that the eigenvectors are linearly independent assume that a linear relationship does exist. Then

$$c_1\mathbf{x}_{(1)} + c_2\mathbf{x}_{(2)} + \cdots + c_k\mathbf{x}_{(k)} = 0$$

If this is multiplied by the matrix $(\mathbf{A} - \lambda_i\mathbf{I})$ terms like

$$(\mathbf{A} - \lambda_i\mathbf{I})\mathbf{x}_{(j)} = \mathbf{A}\mathbf{x}_{(j)} - \lambda_i\mathbf{x}_{(j)} = (\lambda_j - \lambda_i)\mathbf{x}_{(j)}$$

will be generated for all $i \neq j$. A new linear combination, but with $\mathbf{x}_{(i)}$ missing is obtained. Multiplying in succession by $(\mathbf{A} - \lambda_i\mathbf{I})(i = 2 \to k)$ transforms the expression into

$$c_1(\lambda_1 - \lambda_2)(\lambda_1 - \lambda_3)\cdots(\lambda_1 - \lambda_k)\mathbf{x}_{(1)} = 0$$

If all the λ's are different, as assumed, it follows that $c_1 = 0$. In the same way all of the c_i can be shown to be zero. The linear relationship can therefore not exist and it follows that if $\mathbf{A}$ has n distinct eigenvalues the n corresponding eigenfunctions provide a set of basic vectors as discussed before.

Diagonalization of a Matrix

A *similarity transformation* is effected by multiplying a matrix by another matrix and its inverse to produce yet another matrix, according to

$$\mathbf{Q}^{-1}\mathbf{A}\mathbf{Q} = \mathbf{B}$$

Applied to the *characteristic matrix* of $\mathbf{A}$,

$$\mathbf{Q}^{-1}[\mathbf{A} - \lambda\mathbf{I}]\mathbf{Q} = [\mathbf{Q}^{-1}\mathbf{A}\mathbf{Q} - \lambda\mathbf{I}] = [\mathbf{B} - \lambda\mathbf{I}]$$

Moreover, if $\mathbf{B} = \mathbf{Q}^{-1}\mathbf{A}\mathbf{Q}$, then

$$|\mathbf{B}| = |\mathbf{Q}^{-1}\mathbf{A}\mathbf{Q}| = |\mathbf{Q}^{-1}| \cdot |\mathbf{A}| \cdot |\mathbf{Q}| = |\mathbf{Q}^{-1}| \cdot |\mathbf{Q}| \cdot |\mathbf{A}|$$

$$= \frac{\mathbf{1}}{|\mathbf{Q}|} \cdot |\mathbf{Q}| \cdot |\mathbf{A}| = |\mathbf{A}|$$

Therefore $|\mathbf{A} - \lambda\mathbf{I}| = |\mathbf{B} - \lambda\mathbf{I}|$, whereby two matrices related by a similarity transformation have the same eigenvalues.

Now suppose that $\mathbf{B}$ is a diagonal matrix (all off-diagonal elements equal to zero); then the roots of its characteristic equation (eigenvalues) are identical with its diagonal elements. If $\mathbf{A}$ is not a diagonal matrix but is related to $\mathbf{B}$ by a similarity transformation, it follows that it has the same characteristic equation and roots as $\mathbf{B}$. The problem of finding the eigenvalues of **A** is therefore related to reducing the matrix to diagonal form by means of similarity transformations. The aim is to find a matrix **X**, such that $\mathbf{X}^{-1}\mathbf{AX} = \mathbf{\Lambda} = [\lambda_i \delta_{ij}]$.

Suppose that the n eigenvectors of **A** are compacted together as the n columns of a matrix **X**, i.e. $\mathbf{x}_{i(j)} = x_{ij}$ and that the eigenvalues of **A** are $\lambda_1, \lambda_2, \cdots, \lambda_n$. Then

$$\mathbf{AX} = \begin{bmatrix} a_{11} & a_{12} & \cdots & a_{1n} \\ a_{21} & a_{22} & \cdots & a_{2n} \\ \cdots & \cdots & \cdots & \cdots \\ a_{n1} & a_{n2} & \cdots & a_{nn} \end{bmatrix} \begin{bmatrix} x_{11} & x_{12} & \cdots & x_{1n} \\ x_{21} & x_{22} & \cdots & x_{2n} \\ \cdots & \cdots & \cdots & \cdots \\ x_{n1} & x_{n2} & \cdots & x_{nn} \end{bmatrix}$$

$$= \begin{bmatrix} \lambda_1 x_{11} & \lambda_2 x_{12} & \cdots & \lambda_n x_{1n} \\ \lambda_1 x_{21} & \lambda_2 x_{22} & \cdots & \lambda_n x_{2n} \\ \cdots & \cdots & \cdots & \cdots \\ \lambda_1 x_{n1} & \lambda_2 x_{n2} & \cdots & \lambda_n x_{nn} \end{bmatrix}$$

since $\mathbf{Ax}_{(i)} = \lambda_i \mathbf{x}_{(i)}$. It follows that

$$\mathbf{AX} = \begin{bmatrix} x_{11} & x_{12} & \cdots & x_{1n} \\ x_{21} & x_{22} & \cdots & x_{2n} \\ \cdots & \cdots & \cdots & \cdots \\ x_{n1} & x_{n2} & \cdots & x_{nn} \end{bmatrix} \begin{bmatrix} \lambda_1 & 0 & \cdots & 0 \\ 0 & \lambda_2 & \cdots & 0 \\ \cdots & \cdots & \cdots & \cdots \\ 0 & 0 & \cdots & \lambda_n \end{bmatrix}$$

$= \mathbf{XA}'$, where $\mathbf{A}'$ is a diagonal matrix, whose elements are the eigenvalues of **A**.

Finally, since **X** is non-singular its inverse $\mathbf{X}^{-1}$ exists and premultiplication by $\mathbf{X}^{-1}$ yields the desired result

$$\mathbf{X}^{-1}\mathbf{AX} = \mathbf{A}' = [\lambda_i \delta_{ij}]$$

This result is often stated in words that, the trace of a matrix is independent of the representation of the matrix, *i.e.*

$$\mathrm{Tr}\boldsymbol{A} = \mathrm{Tr}\boldsymbol{S}\boldsymbol{S}^{-1}\boldsymbol{A} = \mathrm{Tr}\boldsymbol{S}^{-1}\boldsymbol{A}\boldsymbol{S}$$

1.2.5 Linear Transformations

One important application of matrix algebra is formulating the transformations of points or vectors which define a geometrical entity in space. In ordinary three-dimensional space that involves three axes, any point is located by means of three coordinates measured along these axes. Similarly in n-dimensional vector space a set of n independent vectors is required to *span* the whole space. A linear transformation in general is brought about by an operation such as translation, twisting, rotation, stretching or some other kind of distortion of a vector to produce another vector in the same space. The operation is denoted by $T\boldsymbol{u} = \boldsymbol{v}$ and the space is said to be *invariant* under the action of T. If the outcome is unique for all $\boldsymbol{u}$ and the inverse transformation is also uniquely defined, T is said to be a *one-to-one mapping* of the space onto itself.

An operator such as T can be represented by some matrix $\boldsymbol{A}$, with appropriate elements, chosen in such a way as to mimic the operation of T when multiplied into the vector that represents $\boldsymbol{u}$, *e.g.*

$$\boldsymbol{x} = \boldsymbol{e}_1 x_1 + \boldsymbol{e}_2 x_2 + ... + \boldsymbol{e}_n x_n$$

where the $\boldsymbol{e}_i$ are unit vectors. Hence

$$\sum_{i,k} A_{ki} x_i \boldsymbol{e}_i = \boldsymbol{x}'$$

The transformed vector can also be expressed in terms of the unit vectors as

$$\boldsymbol{x}' = \boldsymbol{e}_1 x_1' + \boldsymbol{e}_2 x_2' + ... + \boldsymbol{e}_n x_n'$$

It follows that the transformation is represented by

$$\boldsymbol{A}\boldsymbol{x} = \boldsymbol{x}'$$

If, in addition $\boldsymbol{x} = \boldsymbol{B}\boldsymbol{x}''$, then $\boldsymbol{AB} = \boldsymbol{C}$ is a matrix which transforms $\boldsymbol{x}''$ directly into $\boldsymbol{x}'$.

If $\boldsymbol{P}$ and $\boldsymbol{Q}$ are non-singular matrices, then $\boldsymbol{A}$ and $\boldsymbol{B}$ are said to be equivalent when $\boldsymbol{B} = \boldsymbol{PAQ}$. An important special case arises when $\boldsymbol{PQ} = \boldsymbol{I}$, when

$$\boldsymbol{B} = \boldsymbol{Q}^{-1}\boldsymbol{A}\boldsymbol{Q}$$

which is known as a similarity transformation. The matrices $\boldsymbol{A}$ and $\boldsymbol{B}$, in this case, are said to be *transforms* of each other. If the matrix elements are complex the transformation is called *unitary*.

The special class of transformation, known as symmetry (or unitary) transformation, preserves the shape of geometrical objects, and in particular the norm (length) of individual vectors. For this class of transformation the *symmetry* operation becomes equivalent to a transformation of the coordinate system. Rotation, translation, reflection and inversion are obvious examples of such transformations. If the discussion is restricted to real vector space the transformations are called *orthogonal*.

The procedure to find symmetry transformation matrices will be demonstrated here for two-dimensional rotation.

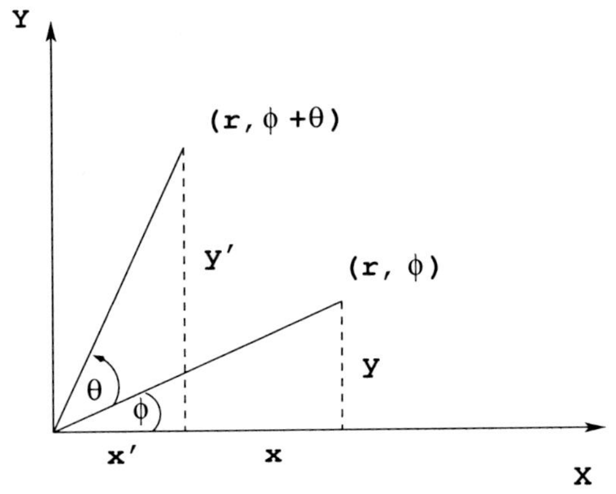

The radial coordinate $\boldsymbol{r}$ remains unchanged during rotation of the vector through an angle θ about the z-axis, while the polar angle ϕ becomes $\phi + \theta$. In terms of cartesian components

$$
\begin{aligned}
x' &= r\cos(\phi + \theta) \\
&= r(\cos\phi\cos\theta - \sin\phi\sin\theta) \\
&= x\cos\theta - y\sin\theta \\
y' &= r\sin(\phi + \theta) \\
&= r(\sin\phi\cos\theta + \cos\phi\sin\theta) \\
&= yy\cos\theta + x\sin\theta
\end{aligned}
$$

Thus

$$\begin{bmatrix} x' \\ y' \end{bmatrix} = \begin{bmatrix} \cos\theta & -\sin\theta \\ \sin\theta & \cos\theta \end{bmatrix} \begin{bmatrix} x \\ y \end{bmatrix}$$

This operation corresponds to a counterclockwise rotation of the vector $\boldsymbol{r}$. Since $\cos\theta = \cos(-\theta)$ and $\sin\theta = -\sin(-\theta)$, the matrix for clockwise rotation is

$$\begin{bmatrix} \cos\theta & \sin\theta \\ -\sin\theta & \cos\theta \end{bmatrix}$$

The matrix equation for clockwise rotation through θ about the z axis becomes

$$\begin{bmatrix} \cos\theta & \sin\theta & 0 \\ -\sin\theta & \cos\theta & 0 \\ 0 & 0 & 1 \end{bmatrix} \begin{bmatrix} x \\ y \\ z \end{bmatrix} = \begin{bmatrix} x' \\ y' \\ z' \end{bmatrix}$$

An improper rotation which has the same effect as a proper rotation but in addition, changes the sign of z, is represented by the matrix

$$\begin{bmatrix} \cos\theta & \sin\theta & 0 \\ -\sin\theta & \cos\theta & 0 \\ 0 & 0 & -1 \end{bmatrix}$$

Other symmetry operations and their matrices are: Reflection in the xy cartesian plane:

$$\begin{bmatrix} 1 & 0 & 0 \\ 0 & 1 & 0 \\ 0 & 0 & -1 \end{bmatrix} \begin{bmatrix} x \\ y \\ z \end{bmatrix} = \begin{bmatrix} x \\ y \\ \bar{z} \end{bmatrix}$$

Inversion through the origin:

$$\begin{bmatrix} -1 & 0 & 0 \\ 0 & -1 & 0 \\ 0 & 0 & -1 \end{bmatrix} \begin{bmatrix} x \\ y \\ z \end{bmatrix} = \begin{bmatrix} \bar{x} \\ \bar{y} \\ \bar{z} \end{bmatrix}$$

The identity operation:

$$\begin{bmatrix} 1 & 0 & 0 \\ 0 & 1 & 0 \\ 0 & 0 & 1 \end{bmatrix} \begin{bmatrix} x \\ y \\ z \end{bmatrix} = \begin{bmatrix} x \\ y \\ z \end{bmatrix}$$

1.2.6 Direct Sums and Products

The direct sum of two square matrices $\boldsymbol{A} \equiv [A_{ij}]$ of order m and $\boldsymbol{B} \equiv [B_{ij}]$ of order n is a square matrix $\boldsymbol{C}$ of order $m+n$ defined by

$$\boldsymbol{C} = \boldsymbol{A} \oplus \boldsymbol{B} = \begin{bmatrix} \boldsymbol{A} & \boldsymbol{0} \\ \boldsymbol{0} & \boldsymbol{B} \end{bmatrix} = \begin{bmatrix} A_{11} & \ldots & A_{1m} & . & & & \\ . & & . & . & & O_1 & \\ A_{m1} & \ldots & A_{mn} & . & & & \\ \ldots & \ldots & \ldots & . & \ldots & \ldots & \ldots \\ & & & . & B_{11} & \ldots & B_{1n} \\ & O_2 & & . & & & . \\ & & & . & B_{n1} & \ldots & B_{nn} \end{bmatrix}$$

where $\boldsymbol{O}_1$ and $\boldsymbol{O}_2$ are null matrices of order $m \times n$ and $n \times m$ respectively.

This idea can easily be extended to more than two matrices to yield a matrix with non-vanishing elements in square blocks along the main diagonal and zeros elsewhere. Such a *block-diagonal matrix* (*e.g.* $\boldsymbol{D} = \boldsymbol{A} \oplus \boldsymbol{B} \oplus \boldsymbol{C}$) has the self-evident important properties:

$$\begin{aligned} \det\boldsymbol{D} &= (\det\boldsymbol{A})(\det\boldsymbol{B})(\det\boldsymbol{C}) \\ \text{trace}\boldsymbol{D} &= \text{trace}\boldsymbol{A} + \text{trace}\boldsymbol{B} + \text{trace}\boldsymbol{C} \end{aligned}$$

The direct product of two matrices is best explained in terms of an example.

The direct product of $\boldsymbol{A} = \begin{bmatrix} a & b \\ c & d \end{bmatrix}$ and $\boldsymbol{B} = \begin{bmatrix} e & f & g \\ h & k & n \\ r & s & t \end{bmatrix}$ is

$$\boldsymbol{C} = \boldsymbol{A} \otimes \boldsymbol{B} = \begin{bmatrix} a\boldsymbol{B} & b\boldsymbol{B} \\ c\boldsymbol{B} & d\boldsymbol{B} \end{bmatrix} = \begin{bmatrix} ae & af & ag & be & bf & bg \\ ah & ak & an & bh & bk & bn \\ ar & as & at & br & bs & bt \\ ce & cf & cg & de & df & dg \\ ch & ck & cn & dh & dk & dn \\ cr & cs & ct & dr & ds & dt \end{bmatrix}$$

The concept can once again be extended to the direct product of more than two matrices.

If $\boldsymbol{A}_1$, $\boldsymbol{A}_2$, $\boldsymbol{B}_1$, and $\boldsymbol{B}_2$ are any matrices whose dimensions are such that the ordinary matrix products $\boldsymbol{A}_1\boldsymbol{A}_2$ and $\boldsymbol{B}_1\boldsymbol{B}_2$ are defined, then the direct product has the property

$$(\boldsymbol{A}_1 \otimes \boldsymbol{B}_1)(\boldsymbol{A}_2 \otimes \boldsymbol{B}_2) = (\boldsymbol{A}_1\boldsymbol{A}_2) \otimes (\boldsymbol{B}_1\boldsymbol{B}_2)$$

Further, if $\boldsymbol{F}$ is the direct product of matrices $\boldsymbol{A}$, $\boldsymbol{B}$, $\boldsymbol{C}$, ... , that is,

$$\boldsymbol{F} = \boldsymbol{A} \otimes \boldsymbol{B} \otimes \boldsymbol{C} \otimes \ \ldots \ , \text{ then}$$

$$\text{trace}\boldsymbol{F} = (\text{trace}\boldsymbol{A})(\text{trace}\boldsymbol{B})(\text{trace}\boldsymbol{C})....$$

The operation of direct product of matrices is both associative and also distributive with respect to matrix addition, and hence finally

$$(\boldsymbol{AB})\otimes(\boldsymbol{AB})\otimes(\boldsymbol{AB}) = (\boldsymbol{AB})\otimes[(\boldsymbol{A} \otimes \boldsymbol{A})(\boldsymbol{B} \otimes \boldsymbol{B})] = (\boldsymbol{A} \otimes \boldsymbol{A} \otimes \boldsymbol{A})(\boldsymbol{B} \otimes \boldsymbol{B} \otimes \boldsymbol{B})$$

i.e.

$$(\boldsymbol{AB})^{[k]} = (\boldsymbol{A})^{[k]}(\boldsymbol{B})^{[k]}$$

where

$$\boldsymbol{A}^{[k]} = \boldsymbol{A} \otimes \boldsymbol{A} ... \otimes \boldsymbol{A} \quad \text{(k times)}.$$

1.3 Vector Fields

In formulating physical problems it is often necessary to associate with every point (x, y, z) of a region R of space some vector $\boldsymbol{a}(x, y, z)$. It is usual to call $\boldsymbol{a}(x, y, z)$ a vector function and to say that a vector field exists in R. If a scalar $\phi(x, y, z)$ is defined at every point of R then a scalar field is said to exist in R.

1.3.1 The Gradient

If the scalar field in R is differentiable the gradient of $\phi(x, y, z)$ is defined as

$$\text{grad}\,\phi \equiv \nabla\phi = \boldsymbol{i}\frac{\partial\phi}{\partial x} + \boldsymbol{j}\frac{\partial\phi}{\partial y} + \boldsymbol{k}\frac{\partial\phi}{\partial z}$$

(The symbol ∇ is pronounced nabla or del.)

Clearly grad ϕ is a vector function whose (x, y, z) components are the first partial derivatives of ϕ. The gradient of a vector function is undefined.

Consider an infinitesimal vector displacement such that

$$\frac{d\boldsymbol{r}}{dt} = \boldsymbol{i}\frac{dx}{dt} + \boldsymbol{j}\frac{dy}{dt} + \boldsymbol{k}\frac{dz}{dt}$$

where t is some parameter. Then

$$\begin{aligned} \nabla\phi \cdot \frac{d\boldsymbol{r}}{dt} &= \left(\boldsymbol{i}\frac{\partial\phi}{\partial x} + \boldsymbol{j}\frac{\partial\phi}{\partial y} + \boldsymbol{k}\frac{\partial\phi}{\partial z}\right) \cdot \left(\boldsymbol{i}\frac{dx}{dt} + \boldsymbol{j}\frac{dy}{dt} + \boldsymbol{k}\frac{dz}{dt}\right) \\ &= \frac{\partial\phi}{\partial x}\frac{dx}{dt} + \frac{\partial\phi}{\partial y}\frac{dy}{dt} + \frac{\partial\phi}{\partial z}\frac{dz}{dt} = \frac{d\phi}{dt} \quad , \end{aligned} \tag{1.5}$$

the total differential coefficient of $\phi(x, y, z)$ with respect to t. Hence considering a surface defined by $\phi(x, y, z) = \text{constant}$, it follows from (5) that $\nabla\phi$ must be perpendicular to dr/dt (since $d\phi/dt = 0$). In other words $\nabla\phi$ is a vector normal to the surface $\phi(x, y, z) = \text{constant}$ at every point. If $\boldsymbol{n}$ is a unit vector normal to this surface in the direction of increasing $\phi(x, y, z)$ and $\partial\phi/\partial n$ is the derivative in this direction then

$$\nabla\phi = \frac{\partial\phi}{\partial n}\boldsymbol{n}.$$

1.3.2 The Laplacian

The ∇ operator may be used to define further quantities, such as $\nabla^2\phi$ (nabla squared ϕ) defined as the scalar product

$$\begin{aligned} \nabla \cdot \nabla &= \left(\boldsymbol{i}\frac{\partial}{\partial x} + \boldsymbol{j}\frac{\partial}{\partial y} + \boldsymbol{k}\frac{\partial}{\partial z}\right) \cdot \left(\boldsymbol{i}\frac{\partial\phi}{\partial x} + \boldsymbol{j}\frac{\partial\phi}{\partial y} + \boldsymbol{k}\frac{\partial\phi}{\partial z}\right) \\ &= \frac{\partial^2\phi}{\partial x^2} + \frac{\partial^2\phi}{\partial y^2} + \frac{\partial^2\phi}{\partial z^2} \end{aligned}$$

which is the three-dimensional Laplacian, so called after Laplace's equation which in this notation becomes

$$\nabla^2\phi = 0. \tag{1.6}$$

1.3.3 The Divergence

The operator $\nabla\cdot$ may be applied to any vector function and the result is called the *divergence*, *e.g.*

$$\begin{aligned}\nabla \cdot \boldsymbol{a}(x,y,z) &= \text{div } \boldsymbol{a} = \left(\boldsymbol{i}\frac{\partial}{\partial x} + \boldsymbol{j}\frac{\partial}{\partial y} + \boldsymbol{k}\frac{\partial}{\partial z}\right) \cdot (\boldsymbol{i}a_x + \boldsymbol{j}a_y + \boldsymbol{k}a_z) \\ &= \frac{\partial a_x}{\partial x} + \frac{\partial a_y}{\partial y} + \frac{\partial a_z}{\partial z}\end{aligned}$$

The divergence operator is the three-dimensional analogue of the differential du of the scalar function $u(x)$ of one variable. The analogue of the derivative is the net outflow integral that describes the flux of a vector field across a surface S

$$\oint \boldsymbol{F} \cdot d\boldsymbol{\sigma} = \left(\frac{\partial F_x}{\partial x} + \frac{\partial F_y}{\partial y} + \frac{\partial F_z}{\partial z}\right) dx\, dy\, dz$$

The flux per unit volume, at the point (x, y, z) defines the divergence

$$\text{div } \boldsymbol{F} = \lim_{vol \to 0} \left[\frac{\oint \boldsymbol{F} \cdot d\boldsymbol{\sigma}}{\text{volume}}\right] = \boldsymbol{\nabla} \cdot \boldsymbol{F}$$

The outflow is therefore described equally well by the total divergence inside S, and hence[2]

$$\int_V \boldsymbol{\nabla} \cdot \boldsymbol{F} d\tau = \oint_S \boldsymbol{F} d\boldsymbol{\sigma}$$

where $\boldsymbol{F} = \rho\boldsymbol{v}$ is the mass flux density, defined in terms of density ρ and velocity of flow $\boldsymbol{v}$.

For a fluid in a region of space where there is neither a source nor a sink, the total mass flow out of the region must balance the rate of decrease of the total mass in the volume. This balance means that

$$\int_V \nabla \cdot (\rho \boldsymbol{v}) d\tau = -\frac{\partial}{\partial t} \int_V \rho\, d\tau$$

and hence the continuity condition

$$\frac{\partial \rho}{\partial t} + \text{div } (\rho \boldsymbol{v}) = 0 \tag{1.7}$$

should hold everywhere in the volume.

[2]This integral relation is known as Gauss's theorem. The most familiar example is in electrostatics.

Unlike grad ϕ div $\boldsymbol{a}$ is seen to be a scalar function. The Laplacian function in terms of this notation is written

$$\nabla^2\phi = \nabla \cdot \nabla\phi = \text{div grad}\,\phi$$

1.3.4 The Curl

Instead of the dot product the vector product of the nabla operator can also be formed to produce a function called curl or rot,

$$\nabla \times \boldsymbol{a} = \text{curl }\boldsymbol{a} = \left(\boldsymbol{i}\frac{\partial}{\partial x} + \boldsymbol{j}\frac{\partial}{\partial y} + \boldsymbol{k}\frac{\partial}{\partial z}\right) \times (\boldsymbol{i}a_x + \boldsymbol{j}a_y + \boldsymbol{k}a_z)$$

Noting that $\boldsymbol{i} \times \boldsymbol{j} = \boldsymbol{k}$, $\boldsymbol{j} \times \boldsymbol{k} = \boldsymbol{i}$, $\boldsymbol{k} \times \boldsymbol{i} = \boldsymbol{j}$ and $\boldsymbol{i} \times \boldsymbol{j} = -\boldsymbol{j} \times \boldsymbol{i}$, *etc.*, it follows that

$$\begin{aligned} \text{curl }\boldsymbol{a} &= \boldsymbol{i}\left(\frac{\partial a_z}{\partial y} - \frac{\partial a_y}{\partial z}\right) + \boldsymbol{j}\left(\frac{\partial a_x}{\partial z} - \frac{\partial a_z}{\partial x}\right) + \boldsymbol{k}\left(\frac{\partial a_y}{\partial x} - \frac{\partial a_x}{\partial y}\right) \\ &= \begin{vmatrix} \boldsymbol{i} & \boldsymbol{j} & \boldsymbol{k} \\ \frac{\partial}{\partial x} & \frac{\partial}{\partial y} & \frac{\partial}{\partial z} \\ a_x & a_y & a_z \end{vmatrix} \end{aligned}$$

The vector curl $\boldsymbol{F}$ is a measure of the vorticity of the field at $P(x, y, z)$ and is related to the net circulation integral per unit area around an element of area at P.

The circulation of a vector field along a closed curve c is defined by the line integral

$$\Gamma_c = \oint_c (F_x dx + F_y dy + F_z dz)\,.$$

The direction of integration decides the sign of the integral, $\oint_c = -\oint_{-c}$.

Consider circulation at constant speed v around a closed rectangular loop on the plane $z =$ constant, *i.e.*

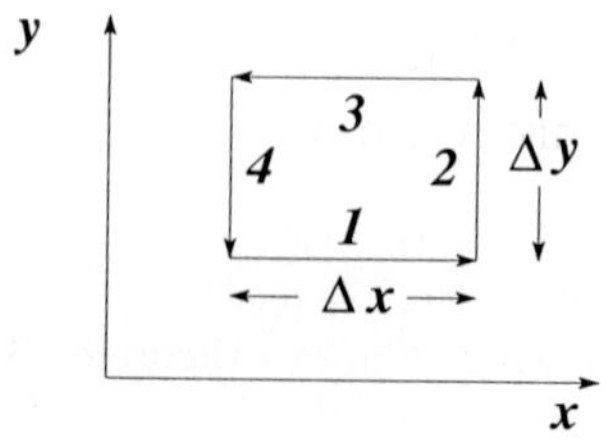

$$\begin{aligned}
\Gamma_z &= \oint (v_x dx + v_y dy) \\
&\simeq \Delta x[v_x(x_1, y) - v_x(x_3, y + \Delta y)] + \Delta y[v_y(x + \Delta x), y_2) - v_y(x, y_4)] \\
&= \Delta x[-\Delta_y v_x] + \Delta y[\Delta_x v_y] \\
&= \Delta x \Delta y \left(-\frac{\Delta_y v_x}{\Delta y} + \frac{\Delta_x v_y}{\Delta x} \right) \\
&\simeq \Delta x \Delta y \left(-\frac{\partial v_x}{\partial y} + \frac{\partial v_y}{\partial x} \right)
\end{aligned}$$

If Γ_z is interpreted as the z-component of a vector, the three-dimensional circulation integral per unit area defines the vector curl $\boldsymbol{v}$.

Since curl $\boldsymbol{a}$ is a vector function one can form its divergence

$$\begin{aligned}
\text{div curl } \boldsymbol{a} &= \nabla \cdot (\nabla \times \boldsymbol{a}) \\
&= \left(\boldsymbol{i}\frac{\partial}{\partial x} + \boldsymbol{j}\frac{\partial}{\partial y} + \boldsymbol{k}\frac{\partial}{\partial z} \right) \cdot \begin{vmatrix} \boldsymbol{i} & \boldsymbol{j} & \boldsymbol{k} \\ \frac{\partial}{\partial x} & \frac{\partial}{\partial y} & \frac{\partial}{\partial z} \\ a_x & a_y & a_z \end{vmatrix} \\
&= \begin{vmatrix} \frac{\partial}{\partial x} & \frac{\partial}{\partial y} & \frac{\partial}{\partial z} \\ \frac{\partial}{\partial x} & \frac{\partial}{\partial y} & \frac{\partial}{\partial z} \\ a_x & a_y & a_z \end{vmatrix} = 0
\end{aligned}$$

since two rows of the determinant are identical. Hence div curl $\boldsymbol{a} = 0$ for all $\boldsymbol{a}$. Conversely if $\boldsymbol{b}$ is a vector function such that div $\boldsymbol{b} = 0$, then it may be inferred that $\boldsymbol{b}$ is the curl of some vector function $\boldsymbol{a}$. Vector functions with identically zero divergence are said to be *solenoidal*.

Likewise, since grad ϕ is a vector function it is allowed to take its curl, *i.e.*

$$\nabla \times \nabla\phi = \text{curl grad}\,\phi = \begin{vmatrix} \boldsymbol{i} & \boldsymbol{j} & \boldsymbol{k} \\ \frac{\partial}{\partial x} & \frac{\partial}{\partial y} & \frac{\partial}{\partial z} \\ \frac{\partial\phi}{\partial x} & \frac{\partial\phi}{\partial y} & \frac{\partial\phi}{\partial z} \end{vmatrix} = 0$$

Hence curl grad $\phi = 0$ for all ϕ. Again, conversely may be inferred that if $\boldsymbol{b}$ is a vector function with identically zero curl, then $\boldsymbol{b}$ must be a gradient of some scalar function. Vector functions with identically zero curl are said to be *irrotational*.

Using the definitions above many vector identities can be derived and these are listed in many sources *e.g.* [10]. Some useful identities are

$$\nabla \cdot (\phi\boldsymbol{a}) = \phi(\nabla \cdot \boldsymbol{a}) + \boldsymbol{a} \cdot \nabla\phi \tag{1.8}$$

$$\nabla \times (\nabla \times \boldsymbol{a}) = \nabla(\nabla \cdot \boldsymbol{a}) - \nabla^2\boldsymbol{a} \tag{1.9}$$

$$\nabla(\boldsymbol{a} \cdot \boldsymbol{b}) = \boldsymbol{a} \times \text{ curl}\,\boldsymbol{b} + (\boldsymbol{a} \cdot \boldsymbol{\nabla})\boldsymbol{b} + \boldsymbol{b} \times \text{ curl } \boldsymbol{a} + (\boldsymbol{b} \cdot \boldsymbol{\nabla})\boldsymbol{a} \tag{1.10}$$

1.3.5 Orthogonal Curvilinear Coordinates

It is often necessary to express div $\boldsymbol{A}$, grad V, curl $\boldsymbol{A}$, and $\nabla^2 V$ in terms of coordinates other than rectangular. Of special importance for problems with spherical symmetry are spherical polar coordinates, already introduced before. Any three-dimensional coordinate system consists of three sets of surfaces that intersect each other. On each surface a certain quantity, the coordinate, is constant. This coordinate has a different value for each surface of the set, and it will be assumed that there is a continuum of surfaces, represented by all possible values of the coordinate. In Cartesian coordinates the surfaces are planes intersecting each other at right angles. The intersection of three of these planes determines a point, and the point is designated by the values of the three coordinates that specify the planes. Similarly, in spherical polar coordinates the surfaces are a set of concentric spheres specified by the value of r, a set of planes which all pass through the polar axis and are specified by the values of φ, and a set of circular cones of which the vertices are all at the origin and which are specified by the values of the variable θ. In these two examples the surfaces intersect each other at right angles, and consequently these coordinate systems are examples of orthogonal coordinates.

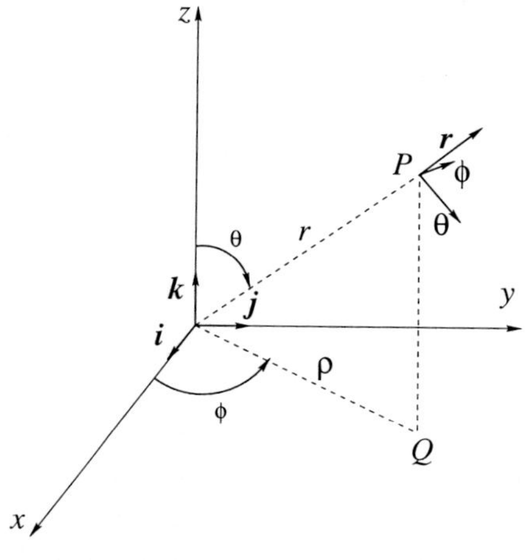

Any three-dimensional orthogonal coordinate system may be specified in terms of the three coordinates q_1, q_2 and q_3. Because of the orthogonality of the coordinate surfaces, it is possible to set up, at any point, an orthogonal set of three unit vectors $\boldsymbol{e}_1$, $\boldsymbol{e}_2$, $\boldsymbol{e}_3$, in the directions of increasing q_1, q_2, q_3, respectively. It is important to select the q_i such that the unit vectors define a right-handed system of axes. The set of three unit vectors defines a Cartesian coordinate system that coincides with the curvilinear system in the immediate neighbourhood of this one point. In the diagram above the polar unit vectors are shown at a point P, together with the cartesian unit vectors and axes.

The differentials of the curvilinear system *i.e.* the orthogonal elements of length, are connected with the differentials of the q_i by the relations

$$ds_1 = h_1 dq_1 \quad ds_2 = h_2 dq_2 \quad ds_3 = h_3 dq_3$$

where the h_i are functions of the coordinates, q_i and vary from point to point. In cylindrical coordinates

$$ds_1 = dr \quad ds_2 = r d\theta \quad ds_3 = dz$$

so that

$$h_1 = 1 \quad h_2 = r \quad h_3 = 1$$

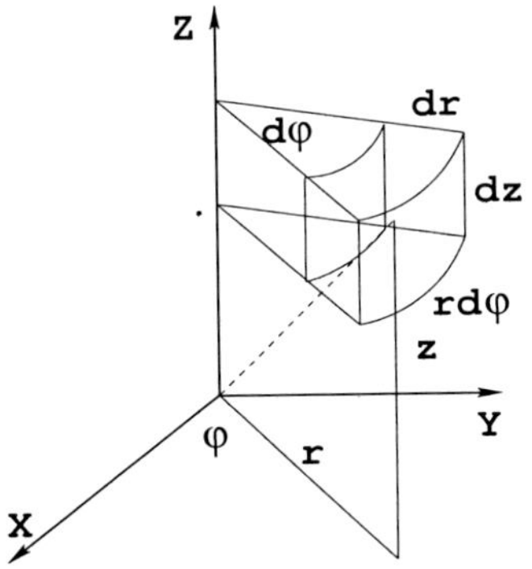

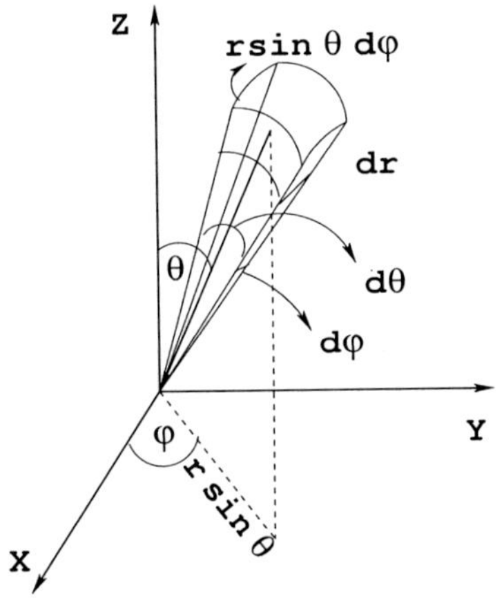

In polar coordinates

$$ds_1 = dr \quad ds_2 = r d\theta \quad ds_3 = r \sin\theta d\varphi$$

$$h_1 = 1 \quad h_2 = r \quad h_3 = r \sin\theta$$

In general, the q_i do not have the dimensions of length, and the h_i are necessary to translate a change in q into a length. The three surface elements are $h_1 h_2 dq_1 dq_2$, $h_1 h_3 dq_1 dq_3$ and $h_2 h_3 dq_2 dq_3$, while the typical volume element is $h_1 h_2 h_3 dq_1 dq_2 dq_3$. An expression for grad w follows immediately as

$$\text{grad } w = \frac{1}{h_1}\frac{dw}{dq_1}\mathbf{e}_1 + \frac{1}{h_2}\frac{dw}{dq_2}\mathbf{e}_2 + \frac{1}{h_3}\frac{dw}{dq_3}\mathbf{e}_3$$

where the $\boldsymbol{e}_i$ are unit vectors in the directions of the elements $h_i dq_i$. Substitution of the proper line elements gives the components of the gradient in cylindrical and in polar coordinates as

$$\text{grad}_r w = \frac{\partial w}{\partial r} \qquad \text{grad}_r w = \frac{\partial w}{\partial r}$$

$$\text{grad}_\varphi w = \frac{1}{r}\frac{\partial w}{\partial \varphi} \qquad \text{grad}_\theta w = \frac{1}{r}\frac{\partial w}{\partial \theta}$$

$$\text{grad}_z w = \frac{\partial w}{\partial z} \qquad \text{grad}_\varphi = \frac{1}{r\sin\theta}\frac{\partial w}{\partial \varphi}$$

To derive an expression for div$\boldsymbol{A}$ in curvilinear coordinates the net outward flow through the surface bounding the volume dv is defined as div$\boldsymbol{A}dv$.

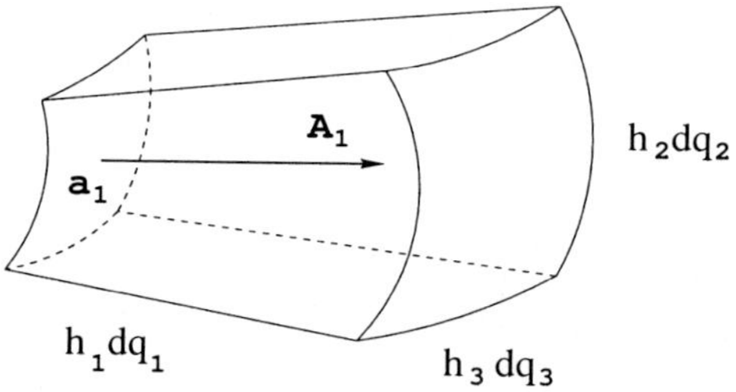

It is important to note that opposite faces of elementary volumes no longer have equal areas. Suppose that the surface on the left has area a_1 and that the component of the vector $\boldsymbol{A}$ in the direction of $h_1 dq_1$ has a value of A_1 at this surface. The outward flow through this surface is $-a_1 A_1$. The contribution from the opposite face will be

$$a_1 A_1 + \frac{\partial}{h_1 \partial q_1}(a_1 A_1) dq_1$$

The net outward flow through the pair of faces therefore is

$$\frac{\partial}{\partial q_1}(a_1 A_1) dw_1 = \frac{\partial}{\partial q_1}(A_1 h_2 h_3) dq_1 dq_2 dq_3$$

Similar contributions from the other two pairs of faces yield the net flow across the entire surface as

$$\left[\frac{\partial}{\partial q_1}(A_1 h_2 h_3) + \frac{\partial}{\partial q_2}(A_2 h_3 h_1) + \frac{\partial}{\partial q_3}(A_3 h_1 h_2)\right] dq_1 dq_2 dq_3$$

Hence

$$\text{div}\boldsymbol{A} = \frac{1}{h_1 h_2 h_3}\left[\frac{\partial}{\partial q_1}(A h_2 h_3) + \frac{\partial}{\partial q_2}(A h_3 h_1) + \frac{\partial}{\partial q_3}(A h_1 h_2)\right]$$

The expression for curl$\boldsymbol{A}$ is obtained in an analogous manner as

$$\text{curl}\boldsymbol{A} = \frac{1}{h_1h_2h_3}\begin{vmatrix} h_1\boldsymbol{e}_1 & h_2\boldsymbol{e}_2 & h_3\boldsymbol{e}_3 \\ \frac{\partial}{\partial q_1} & \frac{\partial}{\partial q_2} & \frac{\partial}{\partial q_3} \\ h_1A_1 & h_2A_2 & h_3A_3 \end{vmatrix}$$

The Laplacian in curvilinear coordinates follows from the definition

$$\begin{aligned} \text{div grad}V &= \nabla^2 V \\ &= \frac{1}{h_1h_2h_3}\left[\frac{\partial}{\partial q_1}\left(\frac{h_2h_3}{h_1}\frac{\partial V}{\partial q_1}\right) + \frac{\partial}{\partial q_2}\left(\frac{h_3h_1}{h_2}\frac{\partial V}{\partial q_2}\right) + \frac{\partial}{\partial q_3}\left(\frac{h_1h_2}{h_3}\frac{\partial V}{\partial q_3}\right)\right] \end{aligned}$$

Appropriate substitutions show that in cylindrical coordinates

$$\nabla^2 V = \frac{1}{r}\left[\frac{\partial}{\partial r}\left(r\frac{\partial V}{\partial r}\right) + \frac{1}{r}\frac{\partial^2 V}{\partial \theta^2} + r\frac{\partial^2}{\partial \varphi^2}\right]$$

and in polar coordinates

$$\begin{aligned} \nabla^2 V &= \text{div grad}\, V \\ &= \frac{1}{r^2}\frac{\partial}{\partial r}\left(r^2\frac{\partial V}{\partial r}\right) + \frac{1}{r^2\sin\theta}\frac{\partial}{\partial\theta}\left(\sin\theta\frac{\partial V}{\partial\theta}\right) + \frac{1}{r^2\sin^2\theta}\frac{\partial^2 V}{\partial\phi^2} \end{aligned} \tag{1.11}$$

Another curvilinear coordinate system of importance in two-centre problems, such as the diatomic molecule, derives from the more general system of confocal elliptical coordinates. The general discussion as represented, for instance by Margenau and Murphy [5], will not be repeated here. Of special interest is the case of prolate spheroidal coordinates. In this system each point lies at the intersection of an ellipsoid, a hyperboloid and and a cylinder, such that

$$h_1^2 = h_2^2 = a^2(\sinh^2 u + \sin^2 v) \quad h_3^2 = a^2(\sinh^2 u \sin^2 v)$$

where

$$R = a \qquad r_A + r_B = 2R\cosh u = 2R\xi \qquad r_A - r_B = 2R\cos v = 2R\eta$$

In terms of these variables the volume element takes the form

$$d\tau = R^3(\xi^2 - \eta^2)d\xi d\eta d\varphi$$

and

$$\begin{aligned} \nabla^2 &= \frac{1}{R^2(\xi^2-\eta^2)}\left[\frac{\partial}{\partial\xi}\left\{(\xi^2-1)\frac{\partial}{\partial\xi}\right\}\frac{\partial}{\partial\eta}\left\{(1-\eta^2)\frac{\partial}{\partial\eta}\right\}\right. \\ &+ \left.\frac{\xi^2-\eta^2}{(\xi^2-1)(1-\eta^2)}\frac{\partial^2}{\partial\varphi^2}\right] \end{aligned} \tag{1.12}$$

1.3.6 Tensor Analysis

In an isotropic medium, vectors such as stress $\boldsymbol{S}$ and strain $\boldsymbol{X}$ are related by vector equations such as, $\boldsymbol{S} = k\boldsymbol{X}$, where $\boldsymbol{S}$ and $\boldsymbol{X}$ have the same direction. If the medium is not isotropic the use of vectors to describe the response may be too restrictive and the scalar k may need to be replaced by a more general operator, capable of changing not only the magnitude of the vector $\boldsymbol{X}$, but also its direction. Such a construct is called a *tensor*.

An important purpose of tensor analysis is to describe any physical or geometrical quantity in a form that remains invariant under a change of coordinate system. The simplest type of invariant is a scalar. The square of the line element ds of a space is an example of a scalar, or a *tensor of rank zero*.

In a space of ν dimensions two coordinate systems may be defined in such a way that the same point has coordinates $(x^1, x^2, \ldots, x^\nu)$ and $\bar{x}^1, \bar{x}^2, \ldots, \bar{x}^\nu)$ in terms of the two systems respectively. The two coordinate systems are related in such a way that a transformation can be effected from one to the other, *i.e.*

$$\bar{x}^j = f^j(x^1, x^2, \ldots, x^\nu) \quad ; \quad x^j = h^j(\bar{x}^1, \bar{x}^2, \ldots, \bar{x}^\nu), \quad j = 1, \nu$$

Suppose that an infinitesimal displacement moves point A (coordinates x^i) to position B (coordinates $x^i + dx^i$). To describe the same displacement in the other coordinate system, it is necessary to differentiate the expression for $\bar{x}^i$, *i.e.*

$$d\bar{x}^i = \sum_{j=1}^{\nu} \frac{\partial h^i}{\partial x^j} dx^j = \sum_{j=1}^{\nu} \frac{\partial \bar{x}^i}{\partial x^j} dx^j, \quad j = 1, 2, \ldots, \nu$$

Any set of quantities A^i that transform in this way, *i.e.*

$$\bar{A}^i = \sum_{j=1}^{\nu} \frac{\partial \bar{x}^i}{\partial x^j} A^j$$

are the contravariant elements of a vector, or a *tensor of first rank*.

To simplify the notation, it is customary to omit the summation sign and sum over indices which are repeated on the same side of the equation. An index which is not repeated is understood to take successively the values $1, 2, \ldots, \nu$, so that there are altogether ν different equations. With these conventions the transformation equations become

$$\bar{A}^i = \frac{\partial \bar{x}^i}{\partial x^j} A^j$$

Consider a function (or scalar field) $\phi(x^i)$ of the point M (coordinates x^i) and defined in the neighbourhood of M. Being a function of a point, the value of ϕ does not change when described in a different coordinate system. By the rules of differentiation

$$\frac{\partial\phi}{\partial\bar{x}^i} = \sum_{j=1}^{\nu} \frac{\partial x^j}{\partial\bar{x}^i}\frac{\partial\phi}{\partial x^j}.$$

It follows that the quantities $A_i = \partial\phi/\partial x^j$ (*e.g.* gradients) transform under a change of coordinate system from x^i to $\bar{x}^i$ like

$$\bar{A}_i = \frac{\partial x^j}{\partial\bar{x}^i}A_j.$$

Any set of quantities that transform according to this prescription are known as the covariant components of a vector, and represented by subscripted indices[3].

These ideas may be extended to define tensors of any rank. There are three varieties of second rank tensors, defined by the transformations

$$\begin{aligned}
\bar{A}^{mn} &= \frac{\partial\bar{x}^m}{\partial x^i}\frac{\partial\bar{x}^n}{\partial x^j}A^{ij}\\
\bar{A}_{mn} &= \frac{\partial x^i}{\partial\bar{x}^m}\frac{\partial x^j}{\partial\bar{x}^n}A_{ij}\\
\bar{A}^m_n &= \frac{\partial\bar{x}^m}{\partial x^i}\frac{\partial x^j}{\partial\bar{x}^n}A^i_j
\end{aligned}$$

They are called contravariant, covariant and mixed tensors, respectively. A useful mixed tensor of the second rank is the Kronecker delta

$$\delta^m_n = 1\,;(m=n), \quad = 0\,;(m\neq n).$$

This is demonstrated as follows. If δ^i_j is a tensor in the coordinate system x^i,

$$\begin{aligned}
\delta^m_n &= \frac{\partial\bar{x}^m}{\partial x^i}\frac{\partial x^j}{\partial\bar{x}^n}\delta^i_j = \frac{\partial\bar{x}^m}{\partial x^i}\frac{\partial x^i}{\partial\bar{x}^n}\\
&= \frac{\partial\bar{x}^m}{\partial\bar{x}^n} = \delta^m_n
\end{aligned}$$

[3]Although it is customary to refer to covariant and contravariant vectors, this may be misleading. Any vector can be described in terms of its contravariant or its covariant components with equal validity. There is no reason other than numerical simplicity for the preference of one set of components over the other.

It follows that δ_n^m has the same components in all coordinate systems.

Tensors of higher rank are defined in the same way, for example, a mixed tensor of rank four is

$$\bar{A}^m_{npq} = \frac{\partial \bar{x}^m}{\partial x^i}\frac{\partial x^j}{\partial \bar{x}^n}\frac{\partial x_k}{\partial \bar{x}^p}\frac{\partial x^h}{\partial \bar{x}^q}A^i_{jkh}.$$

If ν is the number of dimensions of the coordinate systems, then a tensor of order α has α^ν components.

Tensor Algebra

The sum or difference of two or more tensors of the same rank and type is a tensor of the same rank and type, *e.g.*

$$A^{mn} + B^{mn} = C^{mn}.$$

If the components of a tensor satisfy the relation $A^{mn} = A^{nm}$, such a tensor is called symmetric. If $A^{mn} = -A^{nm}$, the tensor is skew-symmetric.

There is an important relationship between vectors and skew-symmetric tensors. Suppose $\boldsymbol{A}$ and $\boldsymbol{B}$ are two vectors in a three-dimensional rectangular coordinate system whose components are connected by

$$\begin{bmatrix} B_x \\ B_y \\ B_z \end{bmatrix} = \begin{bmatrix} a_{11} & a_{12} & a_{13} \\ a_{21} & a_{22} & a_{23} \\ a_{31} & a_{32} & a_{33} \end{bmatrix} \begin{bmatrix} A_x \\ A_y \\ A_z \end{bmatrix}$$

If the coefficients were components of a skew-symmetric tensor, $a_{ij} = -a_{ji}$; $a_{ii} = 0$, then

$$\begin{bmatrix} B_x \\ B_y \\ B_z \end{bmatrix} = \begin{bmatrix} 0 & -a_{21} & a_{13} \\ a_{21} & 0 & -a_{32} \\ -aa_{13} & a_{32} & 0 \end{bmatrix} \begin{bmatrix} A_x \\ A_y \\ A_z \end{bmatrix} = \begin{bmatrix} -a_{21}A_y + a_{13}A_z \\ a_{21}A_x - a_{32}A_z \\ -a_{13}A_x + a_{32}A_y \end{bmatrix}$$

which may be written

$$\begin{bmatrix} -T_zA_y + T_yA_z \\ T_zA_x - T_xA_z \\ -T_yA_x + T_xA_y \end{bmatrix}$$

and recognized as the components of the vector product

$$\boldsymbol{B} = \boldsymbol{T} \times \boldsymbol{A} = \begin{vmatrix} \boldsymbol{i} & \boldsymbol{j} & \boldsymbol{k} \\ T_x & T_y & T_z \\ A_x & A_y & A_z \end{vmatrix}$$

Tensors and matrices are evidently closely related and the components of a tensor can always be written as the elements of a matrix.

Multiplication of A^m by B_n yields a mixed tensor $A^m B_n = C_n^m$, called the outer product, which may be formed with tensors of any rank or type, *e.g.*

$$A_n^m B_{pq} = c_{npq}^m.$$

Let $m = q$ in the mixed tensor $\bar{A}_{npq}^m = A_{jkh}^i$, defined before, and write

$$\begin{aligned}
\bar{B}_{np} &= \bar{A}_{npm}^m = \frac{\partial \bar{x}^m}{\partial x^i}\frac{\partial x^j}{\partial \bar{x}^n}\frac{\partial x^k}{\partial \bar{x}^p}\frac{\partial x^h}{\partial \bar{x}^m}A_{jkh}^i \\
&= \frac{\partial x^j}{\partial \bar{x}^n}\frac{\partial x^k}{\partial \bar{x}^p}\delta_i^h A_{jkh}^i = \frac{x^j}{\partial \bar{x}^n}\frac{\partial x^k}{\partial \bar{x}^p}A_{jki}^i
\end{aligned}$$

This result shows that, by its transformation properties, A_{jki}^i is equivalent to a covariant vector of rank two. This process of summing over a pair of contravariant and covariant indices is called *contraction.* It always reduces the rank of a mixed tensor by two and thus, when applied to a mixed tensor of rank two, the result is a scalar:

$$\bar{A}_m^m = \frac{\partial \bar{x}^m}{\partial x^i}\frac{\partial x^j}{\partial \bar{x}^m} = A_j^i = A_i^i = A_m^m.$$

When two tensors are multiplied together and then contracted the process is known as inner multiplication, thus

$$A^{mn}B_{npq} = C_{pq}^m \quad ; \quad A_m B^m = \text{a scalar}.$$

The last example is clearly equivalent to the scalar product in rectangular coordinates, and l may be viewed as the length of A^m, $l^2 = A^m A_m$.

The angle between two vectors A_m and B_m is defined by

$$\cos\theta = \frac{A_m B^m}{\sqrt{(A_m A^m)(B_m B^m)}}$$

and if A_m and B_m are perpendicular, $A_m B^m = 0$.

The most important second order tensor is the *metric tensor* g, whose components in a Riemann space are defined by the relations

$$g_{ij} = (\boldsymbol{a}_i \boldsymbol{a}_j) \quad ; \quad \boldsymbol{x} = x^i \boldsymbol{a}_i.$$

In terms of this tensor the length of any vector is defined by any of

$$|\boldsymbol{x}|^2 = g_{ij}x^i x^j = g^{ij}x_i x_j = x^i x_i.$$

It is often of value to define a unit vector $\boldsymbol{\lambda}$ in the direction of a vector $\boldsymbol{x}$. Clearly $\lambda_r = x_r/|\boldsymbol{x}|$ will be covariant and its length

$$|\boldsymbol{\lambda}|^2 = g^{mn}\lambda_m\lambda_n = g_{mn}\lambda^m\lambda^n = \lambda^m\lambda_m = 1.$$

The angle between two such unit vectors $\boldsymbol{\lambda}$ and $\boldsymbol{\mu}$ will be

$$\cos\theta = g^{mn}\lambda_m\lambda_n = g_{mn}\lambda^m\lambda^n = \lambda^m\mu_m.$$

1.4 Differential Equations

One of the most useful formulations of physical laws is in terms of differential equations. The main reason for this is that a differential description is, like experimentally obtained knowledge, highly localized. Experimental regularities on which scientific knowledge depends, refer to a limited neighbourhood, since a complete knowledge of the entire universe is not needed in order to make objective predictions of local events. However, doing science in relative ignorance of the actual nature of the universe means that the regularities that can be discovered and quantified are relatively simple relationships. They tend to be smooth functions of space-time and do not involve high-order derivatives or complicated functions of low-order derivatives. As a result, physical laws are often differential equations.

An *ordinary differential equation* has only one variable. Those with more variables are *partial differential equations.* In most applications to be considered here the differential equations are of the homogeneous type. This means that if $y_1(x)$ and $y_2(x)$ are two solutions of the equation

$$\mathcal{L}y(x) = 0$$

then $y(x) = c_1y_1(x) + c_2y_2(x)$, $c_i = constant$ is also a solution.

This is because the linear operator $\mathcal{L}$ has the property

$$\mathcal{L}(c_1y_1 + c_2y_2) = c_1\mathcal{L}y_1 + c_2\mathcal{L}y_2 = 0$$

The complete determination of a solution of a partial differential equation requires the specification of a suitable set of boundary and initial conditions. The boundaries could have a variety of forms depending on the nature of the problem. The role of boundary conditions should become clear from their usage later on.

The solution of differential equations is a specialized pursuit; the precise method is often unique for a specific problem. A common equation with numerous applications will be solved by way of demonstration:

In $\frac{d^2y}{dx^2} + k^2y = 0$, let $\frac{d}{dx} \equiv D$, hence

$$(D^2 + k^2)y = 0 \tag{1.13}$$

which is a special case of a more general family of equations

$$\begin{aligned} (D - p_1)(D - p_2)y &= 0 \\ i.e. \quad D^2y - (p_1 + p_2)Dy + p_1p_2y &= 0 \end{aligned}$$

This form is the same as (13) provided that $p_1 = -p_2$. It follows that $p_1^2 = -k^2$, $p_{1,2} = \pm ik$.

For each term, $(D - p_1)y = 0$, $i.e.$ $\frac{dy}{dx} = p_1y$ or $\frac{dy}{y} = p_1dx$.

Integration gives $\ln y = p_1x + const \quad (= \ln A)$

Thus $\ln \frac{y}{A} = p_1x = \pm ik$

which is $y = A\exp(ikx) + B\exp(-ikx)$

1.4.1 Series Solution of Differential Equations

It is often possible to find a solution of homogeneous differential equations in the form of a power series. According to Frobenius, the power series should have the general form

$$y = \sum_{n=0}^{\infty} A_n x^{m+n}.$$

Such a series is completely described by a knowledge of the initial term, the relations between exponents of x, and the relations between coefficients. To illustrate the procedure equation (13)

$$\left(\frac{d^2}{dx^2} + k^2\right) y = 0$$

is considered once more, assuming a Frobenius solution, $A_0 \neq 0$, for which

$$\begin{aligned} \frac{dy}{dx} &= \sum (m+n)A_n x^{m+n-1} \\ \frac{d^2y}{dx^2} &= \sum (m+n)(m+n-1)A_n x^{m+n-2} \end{aligned}$$

Substitution into (13) gives

$$\sum (m+n)(m+n-1)A_n x^{m+n-2} + k^2 \sum A_n x^{m+n} = 0$$

To be a solution this series must identically equal zero, and hence the coefficient of each power of x therein must be zero. The lowest power of x, namely x^{m-2} occurs in the first term only for $n = 0$. To make this term vanish requires

$$m(m-1)A_0 = 0.$$

This equation, which determines the powers of the Frobenius series is called the *indicial* equation and it has solutions $m = 0$ or $m = 1$. The coefficient of the next power must also vanish:

$$A_1(m+1)m = 0$$

This expression shows that if $m = 0$, A_1 can be non-zero, but for $m = 1$, it is required that $A_1 = 0$.

A general power (x^{m+r}) occurs when $n = r+2$ in the first term and $n = r$ in the second. For the general term to vanish

$$A_{r+2}(m+r+2)(m+r+1) = -k^2 A_r$$

defining the *recursion* formula

$$A_{r+2} = -\frac{k^2}{(m+r+2)(m+r+1)} A_r$$

In this way, all the coefficients A_r of the Frobenius series can be determined step by step. The recursion formula generates two independent series for odd and even values of r. For

$$r = 0 \qquad : A_2 = -\frac{k^2}{(m+2)(m+1)} A_0$$

there are two possibilities:

$$\begin{aligned} m &= 0 \quad : \quad A_2 = -\frac{k^2}{2} A_0 \\ m &= 1 \quad : \quad A_2 = -\frac{k}{3.2} A_0 \end{aligned}$$

$$\begin{aligned} \text{For} \qquad r = 1 \quad &: \quad A_3 = -\frac{k^2}{(m+3)(m+2)} A_1 \\ m = 0 \quad &: \quad A_3 = -\frac{k^2}{3.2} A_1 \\ m = 1 \quad &, \quad \text{no solution,} \quad A_1 = 0. \end{aligned}$$

The most general solution is obtained for $m = 0, \quad r = 0, 1$, in the form of two series:

$$y_{even}(x) = A_0\left(1 - \frac{k^2x^2}{2!} + \frac{k^4x^4}{4!} - \cdots\right) \tag{1.14}$$

$$y_{odd}(x) = A_1 x\left(1 - \frac{k^2x^2}{3!} + \frac{k^4x^4}{5!} - \cdots\right) \tag{1.15}$$

The series (14) and (15) with $A_0 = A_1 = 1$ are recognized as the Taylor expansion about $x = 0$ of the functions, (4),(3)

$$\begin{aligned} y_e(x) &= \cos kx \\ y_o(x) &= \frac{1}{k}\sin kx \end{aligned}$$

which are the linearly independent solutions of (13). Putting $A_0 = 1$, $A_1 = ik$,

$$\begin{aligned} y_e(x) + y_o(x) &= \cos kx + i\sin kx \\ &= \exp(\pm ikx) \end{aligned}$$

as before.

Legendre's Equation

The equation

$$(1 - x^2)\frac{d^2y}{dx^2} - 2x\frac{dy}{dx} + cy = 0 \tag{1.16}$$

can be solved by the same method to demonstrate a different aspect of the Frobenius procedure. The equation (16) has the special feature that the first term disappears for $x = \pm 1$; singular points of the differential equation. It is solved by the same Frobenius series as before, giving

$$\sum A_n(m+n)(m+n-1)x^{m+n-2} - \sum A_n(m+n)(m+m-1)x^{m+n}$$
$$-2\sum A_n(m+n)x^{m+n} + c\sum A_n x^{m+n} = 0$$

The lowest power of x^{m-2} occurs in the first term only for $n = 0$. Hence for this power to vanish

$$A_0 m(m-1)c^{m-2} = 0$$

which yields $m = 0$ or $m = 1$. The next power occurs in the first term for $n = 1$, *i.e.*

$$A_1(m+1)m = 0$$

If $m = 1$ it is required that $A_1 = 0$. A general power x^r occurs for $n = r + 2$ in the first term and for $n = r$ in the others. The recursion formula so defined generates the two series:

$$\begin{aligned} y_e(x) &= A_0\left[1 - \frac{c}{2}x^2 + \frac{c}{2}\left(\frac{c-6}{12}\right)x^4 + \cdots\right] \\ y_o(x) &= A_1\left[x + \frac{2-c}{6}x^3 + \frac{2-c}{6}\left(\frac{12-c}{20}\right)x^5 + \cdots\right] \end{aligned}$$

The Frobenius series can be shown not to converge for$|x| \geq 1$. An important special case occurs when $c = c_\ell \equiv l(l+1)$, $l = 0, 1, 2, \ldots$, any non-negative integer. For each of these special values, one of the two chains of coefficients terminates at A_l because

$$A_{l+2} = A_l\left[\frac{l(l+1) - c_\ell}{(l+2)(l+1)}\right] = 0$$

In particular the even chain terminates when l is even, while the odd chain terminates when l is odd. The corresponding Frobenius solution then simplifies to a polynomial. It is called the Legendre polynomial of degree l, or $P_l(x)$. The modified form of equation (16) becomes

$$(1 - x^2)\frac{d^2y}{dx^2} - 2x\frac{dy}{dx} + l(l+1)y = 0 \qquad (1.17)$$

and is known as Legendre's equation.

Choosing the arbitrary coefficients such that $P_l(x = 1) = 1$ has become standard practice, whereby the Legendre polynomials,

$$\begin{aligned} &P_0(x) = 1 \quad , \quad P_1(x) = x \\ &P_2(x) = \tfrac{1}{2}(3x^2 - 1) \quad , \quad P_3(x) = \tfrac{1}{2}(5x^3 - 3x) \\ &P_4(x) = \tfrac{1}{8}(35x^4 - 30x^2 + 3) \quad , \quad \textit{etc.} \end{aligned}$$

The importance of the special case $c = c_\ell$ lies therein that it avoids the function becoming infinite at $x = \pm 1$, and it is this form that can therefore be used to describe bounded physical properties at these points.

Laguerre's Equation

The equation

$$x\frac{d^2y}{dx^2} + (1 - x)\frac{dy}{dx} + ny = 0$$

is solved by assuming a series solution

$$y = \sum_{k=0}^{\infty} A_k x^{m-k}$$

The indicial equation is $m = n$, the recursion formula is

$$A_{r+1} = \frac{(-1)^r (n!)^2}{[(n-r)!]^2 r!} A_r,$$

and the solution

$$y = \sum \frac{(-1)^r (n!)^2}{[(n-r)!]^2 r!} A_0 x^{n-r}$$

for integral n, is a polynomial of order n, called the nth Laguerre polynomial.

The sth derivative of the nth Laguerre polynomial is the associated Laguerre polynomial of degree $n - s$ and order s,

$$L_n^s(x) = \frac{d^s}{dx^s} L_n(x)$$

By differentiating Laguerre's equation s times the equation satisfied by L_n^s follows as

$$x\frac{d^2}{dx^2} L_n^s(x) + (s + 1 - x)\frac{d}{dx} L_n^s(x) + (n - s) L_n^s(x) = 0 \tag{1.18}$$

Bessel's Equation

Bessel functions are prominent in theoretical chemistry and physics. These functions were first obtained as solutions of Bessel's equation:

$$x^2 \frac{d^2y}{dx^2} + x\frac{dy}{dx} + (x^2 - n^2)y = 0 \tag{1.19}$$

This equation can be solved by the Frobenius method, assuming a series solution

$$y = \sum_{m=0}^{\infty} a_m x^{m+k}$$

to give

$$\sum A_m(m+k)(m+k-1)x^{m+k} + \sum A_m(m+k)x^{m+k} + \sum A_m x^{m+k+2} - n^2 \sum Amx^{m+k} = 0$$

The lowest power occurs for $m = 0$. It has to vanish independently, giving rise to the indicial equation

$$a_0 k(k-1)x^k + A_0 k x^k - n^2 A_0 x^k = 0$$

$$A_0(k^2 - n^2) = 0$$

$$\text{If} \qquad A_0 \neq 0 \quad , \quad k = \pm n$$

A general power x^{r+k} occurs for $r = m - 2$ in the third term and for $r = m$ in the other terms. For this power to vanish independently it is therefore required that

$$A_m[(m+k)(m+k-1) + (m+k) - n^2] + A_{m-2} = 0$$

For $n = k$ this equation defines the recursion formula

$$A_m = -\frac{A_{m-2}}{m(2n+m)} \qquad \text{and the series:}$$

$$y = \sum_{m=0}^{\infty} A_m x^{m+n} \quad , \quad A_1 = 0.$$

This is

$$\begin{aligned} y &= A_0 x^n \left[1 - \frac{x^2}{2(2n+2)} + \frac{x^4}{2.4(2n+2)(2n-4)} - \dots \right. \\ &+ \left. (-1)^r \frac{x^{2r}}{2.4\dots 2r(2n+2)(2n+4)\dots(2n+2r)} + \dots \right] \\ &= \sum_{r=0}^{\infty} \left[(-1)^r \frac{x^{2r} n!}{2^r r! 2^r (n+r)!}\right] A_0 x^n \\ &+ 2^n A_0 \sum \left(\frac{x}{2}\right)^{n+2r} \frac{(-1)^r n!}{r!(n+r)!} \end{aligned}$$

By choosing

$$A_0 = \frac{1}{2^n n!}$$

$$y = J_n(x) = \sum_{r=0}^{\infty} \frac{(-1)^r}{r!(n+r)!} \frac{x}{2}^{n+2r}$$

is known as a Bessel function of order n.

When n is non-integral the factorial function is represented by a Γ (gamma) function. For integer n, $\Gamma(n+1) = n!$

In general

$$\Gamma(x) = \lim_{n\to\infty} \frac{1.2.3\ldots(n-1)}{x(x+1)\ldots(x+n+1)} n^x$$

Thus

$$\Gamma(x+1) = \lim_{n\to\infty} \frac{1.2\ldots(n-1)}{(x+1)(x+2)\ldots(x+n)} n^{x+1}$$

$$\text{whereby} \qquad \Gamma(x+1) = x\Gamma(x)$$

Other important properties of Γ functions, not too difficult to demonstrate, include

$$\Gamma(x)\Gamma(1-x) = \frac{\pi}{\sin(\pi x)}$$

$$\Gamma(x)\Gamma(x+\frac{1}{2}) = 2^{1-2x}\pi^{\frac{1}{2}}\Gamma(2x)$$

$$\Gamma(1) = \lim_{n\to\infty} \frac{n!}{n!} = 1$$

It therefore follows that

$$J_n(x) = \sum_{r=)}^{\infty} \frac{(-1)^r}{\Gamma(r+1)\Gamma(r+n+1)} \left(\frac{x}{2}\right)^{n+2r}$$

Bessel functions have many interesting properties that will be presented here without proof, *e.g.* the recursion formula

$$\frac{d}{dx}\left[x^{-n}J_n(x)\right] = -x^{-n}J_{n+1}(x)$$

i.e.

$$J_n'(x) = \frac{n}{x}J_n(x) - J_{n+1}(x)$$

Using these formulae it can be shown that when n is half an odd integer, *e.g.* $l + \frac{1}{2}$, then $J_n(x)$ takes a particularly simple form and is related to trigonometric functions. By definition, for instance

$$J_{\frac{1}{2}}(x) = \left(\frac{x}{2}\right)^{\frac{1}{2}} \sum_r \frac{(-1)^r x^{2r}}{2^{2r} r! \Gamma(\frac{3}{2}+r)}$$

To evaluate the factor $\Gamma(r+\frac{3}{2})$ the relationship $\Gamma(x+1) = x\Gamma(x)$ is applied r times, by writing the factor in the form $\Gamma(x+1)$ for each step. For the first step, set $x = (2r+1)/2$, so that

$$\begin{aligned} f = \Gamma\left(\frac{2r+3}{2}\right) &= \Gamma(x+1) = x\Gamma(x) = \left(\frac{2r+1}{2}\right)\Gamma(r+\frac{1}{2}) \\ &= \left(\frac{2r+1}{2}\right)\Gamma\left\{\left(\frac{2r-1}{2}\right)+1\right\} \quad , \quad etc. \end{aligned}$$

After r passes

$$\begin{aligned} f &= \left(\frac{2r+1}{2}\right)\cdot\left(\frac{2r-1}{2}\right)\cdots\frac{3}{2}\Gamma\left(\frac{3}{2}\right) \\ &= \frac{(2r+1)!}{2^{2r}r!}\Gamma\left(\frac{3}{2}\right) \end{aligned}$$

Using $\Gamma(1)\Gamma\left(\frac{3}{2}\right) = 2^{-1}\sqrt{\pi}\,\Gamma(2)$

$$f = \frac{(2r+1)!}{s^{2r+1}r!}\sqrt{\pi}$$

When this is substituted into the series for$J_{\frac{1}{2}}(x)$, there results

$$\begin{aligned} J_{\frac{1}{2}}(x) &= \left(\frac{x}{2}\right)^{\frac{1}{2}}\sum_r \frac{2(-1)^r x^{2r}}{(2r+1)!\sqrt{\pi}} = \left(\frac{2}{\pi x}\right)^{\frac{1}{2}}\sum_r \frac{(-1)^r x^{2r+1}}{(2r+1)!} \\ &= \left(\frac{2}{\pi x}\right)^{\frac{1}{2}}\sin x \end{aligned}$$

It follows immediately that

$$J_{\frac{3}{2}}(x) = -x^{\frac{1}{2}}\frac{d}{dx}\left[x^{-\frac{1}{2}}J_{\frac{1}{2}}(x)\right] = \left(\frac{2}{\pi x}\right)^{\frac{1}{2}}\left\{\frac{\sin x}{x} - \cos x\right\}$$

The so-called spherical Bessel functions are now defined as

$$j_n(x) = \sqrt{\frac{\pi}{2x}}J_{n+\frac{1}{2}}(x) \tag{1.20}$$

By using the recursion formula it follows that

$$\begin{aligned} j_0(x) &= \frac{\sin x}{x} \\ j_1(x) &= \frac{\sin x}{x^2} - \frac{\cos x}{x} \\ j_2(x) &= \left(\frac{3}{x^3} - \frac{1}{x}\right)\sin x - \frac{3}{x^2}\cos x \\ j_3(x) &= \left(\frac{15}{x^4} - \frac{6}{x^2}\right)\sin x - \left(\frac{15}{x^3} - \frac{1}{x}\right)\cos x \end{aligned}$$

1.4.2 Separation of Variables

It is often possible to write the solution of a partial differential equation as a sum of terms, each of which is a function in one of the variables only. This procedure is called solution by separation of variables. The one-dimensional wave equation

$$\frac{\partial^2 u(x,t)}{\partial x^2} = \frac{1}{c^2}\frac{\partial^2}{\partial t^2}u(x,t)$$

is a typical example[4]. To solve the equation one tries the possibility of solutions in separable form,

$$u(x,t) = X(x)\cdot T(t)$$

Direct substitution of this product into the wave equation gives

$$\frac{\partial^2 u}{\partial x^2} = T\frac{d^2X}{dx^2} = \frac{1}{c^2}X\frac{d^2T}{dt^2}$$

Thus

$$\frac{1}{X}\frac{d^2X}{dx^2} = \frac{1}{c^2T}\frac{d^2T}{dt^2} = \lambda$$

where the l.h.s. is a function of x only, while the r.h.s. is a function of t only. These two sides must be equal for all values of x and t and this is possible only if they are separately equal to the same constant λ. The result is two separated equations

$$\begin{aligned}\frac{d^2X(x)}{dx^2} &= \lambda X(x)\\ \frac{d^2T(t)}{dt^2} &= c^2\lambda T(t)\end{aligned}$$

The X equation has the general solution

$$X(x) = A\cos(x\sqrt{-\lambda}) + \frac{B}{\sqrt{-\lambda}}\sin(x\sqrt{-\lambda})$$

[4]This equation is second order in time, and therefore remains invariant under time reversal, that is, the transformation $t \to -t$. A movie of a wave propagating to the left, run backwards therefore pictures a wave propagating to the right. In diffusion or heat conduction, the field equation (for concentration or temperature field) is only first order in time. The equation is not invariant under time reversal supporting the observation that diffusion and heat-flow are irreversible processes.

The solution for T is rather similar to X with the constant $c^2\lambda$ replacing λ. Boundary conditions are used to select acceptable solutions from the infinite set. It can happen that one or more of these boundary conditions can be satisfied only when the separation constant takes on some special values. The subset so generated contains only permissible values, or *eigenvalues*, for the problem. The corresponding solutions are called *eigenfunctions*.

A reasonable restriction for a wave is that acceptable solutions should be periodic with period 2π, *i.e.* the solutions 1, $\cos n\pi$, $\sin n\pi$, where n is a positive integer. This condition implies $\lambda_n = -n^2$, $n = 0, 1, 2, \ldots$. For $\lambda = \lambda_n$

$$\begin{aligned} X_n(x)T_n(t) &= a_n \cos nx \cos cnt + b_n \sin nx \cos cnt \\ &+ c_n \cos nx \sin cnt + d_n \sin nx \sin cnt \end{aligned} \tag{1.21}$$

Since the one-dimensional wave equation is linear, the general solution periodic in x with period 2π is the linear superposition

$$u(x,t) = \frac{1}{2}a_0 + \sum_{n=1}^{\infty} X_n(x)T_n(t)$$

of all possible solutions[5]. Equation (21) can be rewritten in the form

$$\begin{aligned} X_n(x)T_n(t) &= A_n \cos n(x-ct) + B_n \sin n(x-ct) \\ &+ C_n \cos n(x+ct) + D_n \sin n(x+ct) \end{aligned}$$

where the linearly independent terms $f(x-ct)$ and $g(x+ct)$ represent waves propagating along the $+x$ and $-x$ direction, respectively. A general solution in complex notation is

$$u(x,t) = a \exp in(x-ct) \tag{1.22}$$

as before.

1.4.3 Special Functions

The solutions of differential equations often define series of related functions that can be obtained from simple generating functions or formulae. As an example consider the Legendre polynomials

$$P_n(x) = \sum_{r=0}^{L} \frac{(-1)^r(2n-2r)!}{2^n r!(n-2r)!(n-r)!} x^{n-2r}$$

[5]The function is called a double Fourier series.

where $L = (n-1)/2$ for n odd and $L = n/2$ for n even. According to the binomial theorem

$$\begin{aligned}(x^2-1)^n &= \sum_{r=0}^{n} {}^nC_r(-1)^r\left(x^2\right)^{n-r} \\ &= \sum \frac{(-1)^r n!}{r!(n-r)!}\left(x^2\right)^{n-r}\end{aligned}$$

and by direct differentiation follows that

$$\left(\frac{d}{dx}\right)^n \left(x^{2n-2r}\right) = \frac{(2n-2r)!}{(n-2r)!}x^{n-2r} \quad \text{, for} \quad r \leq \frac{n}{2}.$$

Combining these results leads to the Rodriques formula for Legendre polynomials:

$$P_n(x) = \frac{1}{2^n n!}\left(\frac{d}{dx}\right)^n \left(x^2-1\right)^n$$

The Legendre polynomials are one example of a family of polynomials, said to be orthogonal on the closed interval $[a,b]$ with respect to a weight function $w(x)$ if

$$\int_a^b w(x) f_m(x) f_n(x) dx = h_n \delta_{mn}.$$

Further examples are shown in the following table.

$f_n(x)$	Name	a	b	$w(x)$	h_n
$P_n(x)$	Legendre	-1	1	1	$2/(2n+1)$
$L_n(x)$	Laguerre	0	∞	e^{-x}	1
$H_n(x)$	Hermite	$-\infty$	∞	e^{-x^2}	$\sqrt{\pi}\,2^n n!$

The first four terms for each series are given in the next table

n	0	1	2	3
$P_n(x)$	1	x	$\frac{1}{2}(3x^2-1)$	$\frac{1}{2}(5x^3-3x)$
$L_n(x)$	1	$-x+1$	x^2-4x+2	$-x^3+9x^2-18x+6$
$H_n(x)$	1	$2x$	$4x^2-2$	$8x^3-12x$

Each of these has a Rodriques formula,

$$f_n(x) = \frac{1}{a_n w(x)}\left(\frac{d}{dx}\right)^n \left\{w(x)[s(x)]^n\right\}.$$

$f_n(x)$	$w(x)$	$s(x)$	a_n
$P_n(x)$	1	$1-x^2$	$(-1)^n 2^n n!$
$L_n(x)$	e^{-x}	x	1
$H_n(x)$	e^{-x^2}	1	$(-1)^n$

Formulae of this type are especially useful when looking for solutions of more complicated differential equations related to the simpler classical ones. Two important examples are the associated Legendre and associated Laguerre (18) equations, respectively

$$\left[\left(1-x^2\right)\frac{d^2}{dx^2} - 2x\frac{d}{dx} + \left\{l(l+1) - \frac{m^2}{1-x^2}\right\}\right] y = 0 \qquad (1.23)$$

$$\left[x\frac{d^2}{dx^2} + (s+1-x)\frac{d}{dx} + (n-s)\right] y = 0 \qquad (1.24)$$

The relationship between (23) and Legendre's equation can be demonstrated in two steps. First, substitute

$$y = \left(1-x^2\right)^{\frac{m}{2}} z$$

into (23), transforming it into

$$\left(1-x^2\right)\frac{d^2z}{dx^2} - 2(m+1)\frac{dz}{dx} + \left[l(l+1) - m(m+1)\right] z = 0 \qquad (1.25)$$

The next step is to differentiate Legendre's equation n times. Using the Leibniz [6] theorem one gets

$$\left(1-x^2\right) D^{m+2}w - 2(m+1)xD^{m+1}w + [l(l+1) - m(m+1]D^m w = 0 \quad (1.26)$$

where $D = \frac{d}{dx}$ and w solves Legendre's equation. Equation (26) is identical to (25) for $z = D^m w$. The solution of (23) follows immediately as

$$P_l^m(x) = \left(1-x^2\right)^{\frac{m}{2}} P_l(x) \quad , \quad 0 \leq m \leq l.$$

Substitution into the Rodriques formula gives the associated Legendre functions in the form

$$P_l^m(x) = \frac{1}{2^n n!}\left(1-x^2\right)^{\frac{m}{2}}\left(\frac{d}{dx}\right)^{m+n}\left(x^2-1\right)^n$$

For m an integer the $P_l^m(x)$ are polynomials. The polynomials $P_l^0(x)$ are identical with the Legendre polynomials. The first few associated Legendre polynomials for $x = \cos\theta$ (a common form of Legendre's equation) are:

$$\begin{aligned}
P_1^1 &= \left(1-x^2\right)^{\frac{1}{2}} = \sin\theta \\
P_2^1 &= 3\left(1-x^2\right)^{\frac{1}{2}} x = 3\sin\theta\cos\theta \\
P_3^1 &= \frac{3}{2}\left(1-x^2\right)^{\frac{1}{2}}\left(5x^2-1\right) = \frac{1}{2}\sin\theta(15\cos^2\theta-3) \\
P_2^2 &= 3\left(1-x^2\right) = 3\sin^2\theta \\
P_3^2 &= 15\left(1-x^2\right)x = 15\sin^2\theta\cos\theta
\end{aligned}$$

To normalize a function such as $P_l^m(\cos\theta)$ it is necessary to equate the integral $\int_{-1}^{1}[P_l^m(\cos\theta)]^2dx = 1$. Starting from the Rodriques formula and integrating by parts it can be shown that the normalized associated Legendre functions are

$$\Theta(\theta) = \sqrt{\frac{(2l+1)(l-m)!}{2(n+m)!}}P_l^m(\cos\theta)$$

From its Rodriques formula the next Laguerre polynomial is

$$L_4(x) = x^4 - 16x^3 + 72x^2 - 96x + 24.$$

The associated Laguerre polynomials derived by differentiation are:

$$\begin{aligned}
L_1^1(x) &= -1 \\
L_2^1(x) &= 2x-4 \\
L_2^2(x) &= 2! \\
L_3^1(x) &= -3x^2+18x-18 \\
L_3^2(x) &= -6x+18 \\
L_3^3(x) &= -(3!) \\
L_4^1(x) &= 4x^3-48x^2+144x-96 \qquad (1.27) \\
L_4^2(x) &= 12x^2-96x+144 \\
L_4^3(x) &= 24x-96 \\
L_4^4(x) &= 4!
\end{aligned}$$

To demonstrate that Laguerre polynomials in the interval $0 \leq x \leq \infty$ may be obtained from the generating function

$$U(u,x) = \sum_{r=0}^{\infty}\frac{L_r(x)}{r!}u^r = \frac{\exp[-xu/(i-u)]}{1-u}$$

this function is differentiated with respect to both u and x. In the first instance

$$\frac{\partial U}{\partial u} = \sum \frac{L_r(x)}{(r-1)!}u^{r-1} = \frac{\exp[-xu/(1-u)]}{(1-u)^3}(1-u-x)$$

i.e.

$$(1-u)^2 \sum \frac{L_r(x)}{(r-1)!}u^{r-1} = (1-u-x)\sum \frac{L_r(x)}{r!}u^r$$

or

$$\sum \frac{L_r(x)}{(r-1)!}u^{r-1} - 2\sum \frac{L_r(x)}{(r-1)!}u^r + \sum \frac{L_r(x)}{(r-1)!}u^{r+1}$$
$$= \sum \frac{L_r(x)}{r!}u^r - \sum \frac{L_r(x)}{r!}u^{r+1} - x\sum \frac{L_r(x)}{r!}u^r$$

Equating coefficients of similar powers (*e.g.* u^r) on the l.h.s. and the r.h.s. gives the recursion formula

$$\frac{L_{r+1}(x)}{r!} - 2\frac{L_r(x)}{(r-1)!} + \frac{L_{r-1}(x)}{(r-2)!} = \frac{L_r(x)}{r!} - \frac{L_{r-1}(x)}{(r-1)!} - x\frac{L_r(x)}{r!}$$

i.e.

$$L_{r+1}(x) + (x-1-2r)L_r(x) + r^2 L_{r-1}(x) = 0 \tag{1.28}$$

Similarly

$$\frac{\partial U}{\partial x} = \sum \frac{u^r}{r!}\frac{\partial L_r(x)}{\partial x} = -\frac{u}{1-u}\sum \frac{L_r(x)}{r!}u^r$$

or

$$\frac{\partial L_r(x)}{\partial x} - r\frac{\partial L_{r-1}(x)}{\partial x} + rL_{r-1}(x) = 0 \tag{1.29}$$

Equivalent forms of (29) are

$$\frac{\partial L_{r+1}(x)}{\partial x} - (r+1)\frac{\partial L_r(x)}{\partial x} + (r+1)L_r(x) = 0$$
$$\frac{\partial L_{r+2}(x)}{\partial x} - (r+2)\frac{\partial L_{r+1}(x)}{\partial x} + (r+2)L_{r+1}(x) = 0$$

differentiating to expressions

$$\frac{\partial^2 L_{r+1}(x)}{\partial x^2} = (r+1)\left[\frac{\partial^2 L_r(x)}{\partial x^2} - \frac{\partial L_r(x)}{\partial x}\right] = 0$$

etc., which can be used to turn (28) into an ordinary differential equation. First replace r by $(r+1)$ in (28) and differentiate twice, such that

$$\frac{\partial^2 L_{r+2}(x)}{\partial x^2} + (x-3-2r)\frac{\partial^2 L_{r+1}(x)}{\partial x^2} + (r+1)^2\frac{\partial^2 L_r(x)}{\partial x^2}$$

Substituting from the previous expressions transforms this equation into one in $L_r(x)$ alone:

$$x\frac{d^2L_r(x)}{dx^2}+(1-x)\frac{dL_r(x)}{dx}+rL_r(x)=0$$

which is the differential equation of the rth Laguerre polynomial.

Since $L_r^s(x)=\left(\frac{d}{dx}\right)^s L_r(x)$ the generating function for the associated Laguerre polynomials follows as

$$U_s(x,u)=\sum_{r=s}^{\infty}\frac{L_r^s(x)}{r!}u^r=\frac{d^s}{dx^s}\left[\frac{\exp[-xu/(1-u)]}{1-u}\right]=(-1)^s\left[\frac{\exp[-xu/(1-u)]}{(1-u)^{s+1}}\right]u^s$$

To normalize radial hydrogenic eigenfunctions it will be necessary to evalute the integral

$$N=\int_0^\infty e^{-x}x^{s+1}\left[L_r^s(x)\right]^2dx$$

which can be considered as a product of two generating functions,

$$\begin{aligned}I &= \int_0^\infty e^{-x}x^{s+1}U_s(x,u)V_s(x,v)dx\\ &= \sum_{r=s}^{\infty}\sum_{t=s}^{\infty}\frac{u^rv^t}{r!t!}\int_0^\infty e^{-x}x^{s+1}L_r^s(x)L_t^s(x)dx\end{aligned}$$

and also

$$I=\frac{(uv)^s}{(1-u)^{s+1}(1-v)^{s+1}}\int_0^\infty x^{s+1}e^{-x\left(1+\frac{u}{1-u}+\frac{v}{1-v}\right)}dx$$

Using

$$\int_0^\infty x^{s+1}e^{-ax}=\frac{(s+1)!}{a^{s+2}}$$

$$I=\frac{(s+1)!(uv)^s(1-u)(1-v)}{(1-uv)^{s+2}}$$

The factor $(1-uv)^{-(s+2)}$ is expanded by the binomial theorem[6] to give

$$I=(s+1)!(1-u-v+uv)\sum_{k=0}^{\infty}\frac{(s+k+1)!}{k!(s+1)!}(uv)^{s+k} \qquad (1.30)$$

---

[6]

$$(1-a)^{-n}=1+na+\frac{n(n-1)}{2!}a^2+\ldots+\frac{(n+k-1)!}{(n-1)!k!}a^k+\ldots$$

Also, for $r = t$

$$I = \int_0^\infty e^{-x} x^{s+1} \sum \frac{\left[L_r^s(x)\right]^2}{(r!)^2} (uv)^r \, dx$$

The normalization integral is given by $(r!)^2$ times the coefficient of $(uv)^r$ in (30), *i.e.*

$$\begin{aligned} N &= (r!)^2(s+1)! \left[\frac{(r+1)!}{(r-s)!(s+1)!} + \frac{r!}{(r-s-1)!(s+1)!} \right] \\ &= (r!)^3 \left[\frac{r+1}{(r-s)!} + \frac{r-s}{(r-s)!} \right] \\ &= \frac{(r!)^3(2r-s+1)}{(r-s)!} \end{aligned} \tag{1.31}$$

Chapter 2

Group Theory

2.1 Introduction

Group theory, although a topic of pure mathematics has had such a powerful influence on the physical sciences that many features thereof have been adapted especially to emphasize scientific rather than mathematical aspects. For the same reason it has become traditional for physical scientists to present group theory as a consummate mixture of science and maths. The same bias will be evident in this chapter despite a conscious effort to present the predominantly physical aspects at their more appropriate places in the text.

Even mathematicians often find it hard to present group theory in totally abstract terms.

There are several texts that deal with group theory for non-mathematicians [11, 12, 13, 14, 15].

A mathematical group, by definition, consists of a set of distinct elements $G = \{E, A, B, C, D, \ldots\}$, endowed with a law of composition (such as multiplication, addition, or some other operation), such that the following properties are satisfied:

(a) The product or composition of any two elements A and B of G under the group law results in an element which also belongs to G. Thus

$$A \circ B \equiv AB = C \in G$$

$$B \circ A = D \in G$$

This property is known as the *closure* property of the group. The group is said to be closed under the given law of composition.

(b) There exists an identity element $E \in G$ such that (for all) $\forall A$ (in) $\in G$, $EA = AE = A$.

(c) For any element $A \in G$, there exists an unique element $B \in G$ such that $AB = BA = E$. B is called the inverse of A, and *vice versa*, $(B = A^{-1})$.

(d) The law of composition of the group elements is associative, *i.e.*, for any $A, B, C \in G$, $A(BC) = (AB)C$. The number of elements in a group is called its *order*. A group containing a finite number of elements is called a *finite group*; a group containing an infinite number of elements is called an *infinite group*. An infinite group may be either *discrete* or *continuous*. If the number of elements of a group is denumerably infinite (such as the number of all integers), the group is discrete. If the number of elements in a group is non-denumerably infinite (such as the number of all real numbers), the group is continuous.

Examples of a group are:
(i) The group of order two consisting of the real numbers 1 and -1, with ordinary multiplication as the law of composition.
(ii) The group of order four consisting of the complex numbers $1, i, -1, -i$, under multiplication.
(iii) The discrete infinite group of all real integers (positive, negative and zero) under addition as the law of combination. The identity element of this group is zero.
(iv) The set of all real numbers under addition. This is a continuous group with 0 as the identity element. The inverse of a number b is its negative $-b$.
(v) The set of all positive (non-zero) real numbers under multiplication. The identity element is 1 and the inverse of x is its reciprocal $1/x$.
(vi) The smallest non-empty set containing only unity is a group of order one under multiplication.
(vii) The set of two matrices $\begin{bmatrix} 1 & 0 \\ 0 & 1 \end{bmatrix}$ and $\begin{bmatrix} -1 & 0 \\ 0 & -1 \end{bmatrix}$ under matrix multiplication.
(viii) The set of all non-singular matrices of order n (n a positive integer) under matrix multiplication.
The product of group elements is not necessarily commutative and in general $AB \neq BA$. If all the elements of a group commute with each other, it is said to be an Abelian group.

Of particular importance in the physical sciences is the fact that the symmetry operations of any symmetrical system constitute a group under the operators that effect symmetry transformations, such as rotations or reflections. A symmetry transformation is an operation that leaves a physical system invariant. Thus any rotation of a circle about the perpendicular axis through its centre is a symmetry transformation for the circle. The permutation of any two identical atoms in a molecule is a symmetry transformation for the molecule.

It is noted that two successive symmetry transformations of a system leave that system invariant. The product of the two operations is therefore also a symmetry operation of the system. The set of symmetry transformations is therefore closed under the law of successive transformations. An identity transformation that leaves the system unchanged clearly belongs to the set. It is not difficult to see that any given symmetry transformation has an inverse that also belongs to the set. Since successive transformations of the set obey the associative law it finally follows that the set constitutes a group.

2.2 The Multiplication Table

Because of the group property of closure all operations of any group can be presented in the form of a multiplication table that contains all elements of the group, *e.g.*

G	E	A	B	C
E	E	A	B	C
A	A	E	C	B
B	B	C	E	A
C	C	B	A	E

Groups with similar multiplication tables have the same structure and are said to be *isomorphic*. There exists an *isomorphism* between two groups $G = \{E, A, B, C, ...\}$ and $G' = \{E', A', B', C', ...\}$ of the same order, if there is a one-to-one correspondence between the elements of G and G'. A multiplication such as $AB = C$ in one group then implies that $A'B' = C'$ in G'.

It is noticed from the multiplication table that each element of the group occurs once and only once in each row or column. The arrangement of elements in a row (column) is different from that in any other row (column)[1]. An important consequence of this arrangement is that if f is any function of the group elements, then

$$\sum_{A \in G} f(A) = \sum_{A \in G} f(AB)$$

[1]To prove the general validity of these statements, suppose that some element D occurs twice in some row or column corresponding to the element A. This means that there exist two elements, B and C say, such that $BA = D$ and $CA = D$. Multiplication from the right yields $B = DA^{-1}$, $C = DA^{-1}$, showing that $B = C$, contrary to the assumption that the group elements are distinct.

where B is an element of the finite group G and the summation runs over all the group elements.

2.3 Conjugate Elements and Classes

If a relation of the type

$$A^{-1}BA = C$$

exists between the elements A, B, C of a group, the elements B and C are said to be *conjugate*. The operation is called a *similarity transformation* of B by A. It is clear that $ACA^{-1} = B$. It is easy to show that if B is conjugate to C and B is also conjugate to D, the C and D are conjugate elements; or B, C and D are all conjugate to each other. It follows almost immediately that the elements of a group can be split into sets such that all the elements of a set are conjugate to each other, but no two elements belonging to different sets are conjugate to each other. These sets of elements are known as the conjugate classes, or simply the classes, of a group. The identity element E always constitutes a class by itself in any group, since, for any elements A of the group, $A^{-1}EA = E$.

2.4 Homomorphism

There often occurs a one-to-many correspondence between two groups. Suppose that $G = \{E, A, B, C, ...\}$ is a group of order g and that the group $G' = \{E_1, E_2, ..., E_n, A_1, A_2, A_3, ..., A_n\}$ is of order ng[2]. Suppose it is possible to split the elements of G' into g sets (E_i), (A_i), *etc.*, each containing n elements such that if $AB = C$ in G, then the product of A_i and B_i $(1 \le i, j \le n)$ of G', belongs to the set $(C_k) \equiv (C_1, C_2, ..., C_n)$ of G', *i.e.* $A_iB_j = C_k$, $1 \le k \le n$. This correspondence may be denoted by

$$\begin{aligned} E \leftrightarrow (E_i) &= (E_1, E_2, ..., E_n) \\ A \leftrightarrow (A_i) &= (A_1, A_2, ..., A_n) \end{aligned}$$

The groups G and G' are said to be *homomorphic* to each other with a one-to-n correspondence, or that there is a *homomorphism* between G and G'.

[2]Only one element, E_1 say, is the identity of G'.

2.5 Some examples of symmetry groups

To demonstrate group theory in action and its close relationship to the intuitive notion of symmetry, some elementary groups of importance in solid-state science will be considered.

(i) The Cyclic Group C_n

This is a symmetry group that describes the rotations of a regular polygon with n directed sides, *e.g.*

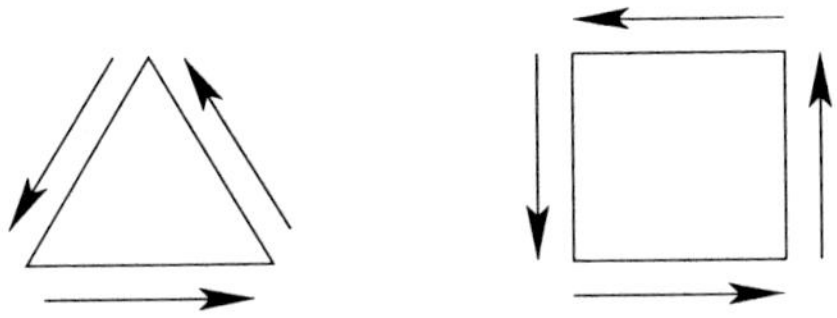

The elements of the group are rotations through an angle $2\pi j/n$ $(j = 0, 1, 2, ..., n-1)$ about an axis through the centre and normal to the plane of the figure. Calling these rotations C_n^j it is clear that a rotation C_n^j is equivalent to repeating the smallest non-trivial rotation C_n^1 j times, *i.e.*

$$C_n^j = jC_n^1 = jc$$

where c is a rotation through $2\pi/n$. Note that $c^n = e$, the identity element, which is equivalent to a rotation through 2π.

Mathematically this group is known as a cyclic group of order n, generated by the element c, and consisting of the elements $\{e, c, c^2, ..., c^{n-1}\}$. The multiplication table of C_3 is

g_1/g_2	e	c	$b(=c^2)$
e	e	c	b
c	c	b	e
b	b	e	c

Each element appears once only in each row and in each column.

(ii) The Dihedral Group D_n

This is the symmetry group of rotations of a regular polygon with n undirected sides.

The group D_3, for instance, contains all the same elements as C_3 together with three additional elements

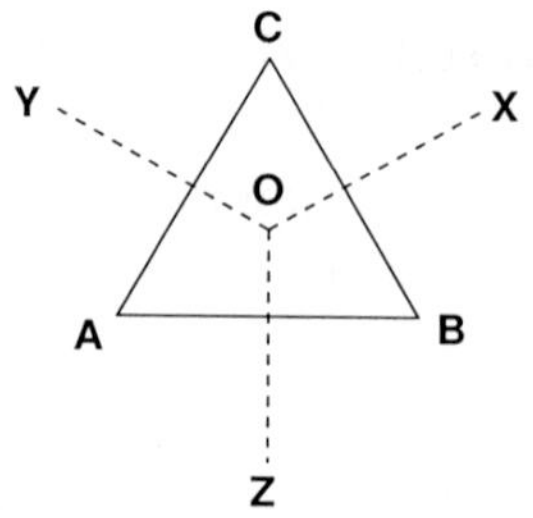

corresponding to rotations of π about the axes OX, OY and OZ. These *two-fold* rotations b_1, b_2, b_3 (say) satisfy the condition $b_i = e$. An anticlockwise rotation c of $2\pi/3$ in the plane of the paper, rotates the three *two-fold axes* into each other, *e.g.*

$$c(AO) = BO$$

This equality implies that b_1 and b_2 are related by *conjugation* with c:

$$b_2 = cb_1c^{-1}$$

That is, the effect of b_2 can be reproduced by first performing the inverse of c, namely a clockwise rotation through $2\pi/3$, then a π rotation about OY (considered as fixed in space) and finally undoing the first rotation[3]. There is another relationship between b_2 and b_1, *i.e.*

$$b_2 = b_1c$$

whereby

$$cb_1c^{-1} = b_1c$$

Multiplying on the right by c gives

$$cb_1 = b_1c^2$$

By dropping the subscript on b_1 the group is seen to be *generated* by the elements b and c in the sense that every element can be represented as the product of b's and c's:

$$\begin{aligned} b_1 &= b \\ b_2 &= bc \\ b_3 &= bc^{-1} = bc^2 \\ c^3 &= b^2 = e \end{aligned}$$

[3]This sequence of operations is a common routine with the Rubik cube.

The six independent elements of $D_3 = gp\{c, b\}$ are $\{e, c, c^2, b, bc, bc^2\}$

A square with undirected sides is transformed into itself by the rotations of the group D_4.

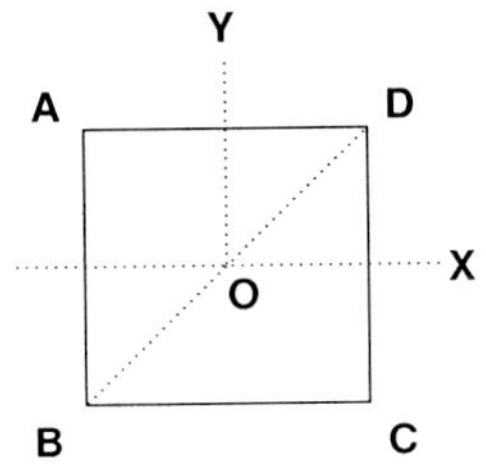

If c is an anticlockwise rotation through $\pi/2$, satisfying $c^4 = e$, about an axis through the centre O perpendicular to the plane of the figure, and $b = b_1$, is the two-fold rotation about the axis OX, other two-fold axes can be identified as b_2 about AC, b_3 about OY and b_4 about BD, *i.e.*

$$\begin{aligned} b_2 &= bc \\ b_3 &= bc^2 \\ b_4 &= bc^3 \end{aligned}$$

D_4 therefore has parameters $\mathrm{gp}\{c, b\}$ and elements $\{e, c, c^2, c^3, b, bc, bc^2, bc^3\}$.

The generalization to larger values of n now becomes clear: D_n is the group of symmetry rotations of a regular polygon with n sides, is generated by an n-fold rotation about an axis through the centre and perpendicular to the plane of the figure and any one of the 2-fold rotations about axes in the plane, *i.e.*

$$D_n = \mathrm{gp}\{c, b\} \qquad c^n = b^2 = (bc)^2 = e$$

The group is of order $2n$, and the $2n$ elements can be written as

$$\{e, c, c^2, ..., c^{n-1}, b, bc, bc^2, ..., bc^{n-1}\}$$

Note that the rotations c^s comprising C_n form a *subgroup* of D_n.

(iii) The Permutation Group S_n

This group describes the permutation of n objects, or their labels, and is of order $n!$. Multiplication of two permutations is defined as successive application. A permutation P under which the object with label $i (i = 1, ..., n)$ is changed into that with label p_i can be written as

$$P = \begin{pmatrix} 1 & 2 & \cdots & n \\ p_1 & p_2 & \cdots & p_n \end{pmatrix}$$

The order of the columns is irrelevant. Multiplication, or successive application of permutations is not commutative, $PQ \neq QP$, for example:

$$\begin{pmatrix} 1 & 2 & 3 \\ 1 & 3 & 2 \end{pmatrix}\begin{pmatrix} 1 & 2 & 3 \\ 3 & 1 & 2 \end{pmatrix} = \begin{pmatrix} 1 & 2 & 3 \\ 3 & 2 & 1 \end{pmatrix}$$

whereas

$$\begin{pmatrix} 1 & 2 & 3 \\ 3 & 1 & 2 \end{pmatrix}\begin{pmatrix} 1 & 2 & 3 \\ 1 & 3 & 2 \end{pmatrix} = \begin{pmatrix} 1 & 2 & 3 \\ 2 & 1 & 3 \end{pmatrix}$$

The identity element is

$$e = \begin{pmatrix} 1 & 2 & \ldots & n \\ 1 & 2 & \ldots & n \end{pmatrix}$$

and the inverse of P is

$$P^{-1} = \begin{pmatrix} p_1 & p_2 & \ldots & p_n \\ 1 & 2 & \ldots & n \end{pmatrix}$$

An alternative way of writing permutations is in *cycle notation.* To achieve this the permutation is followed, starting from label 1 until the sequence returns to 1, and recording all labels along the way. Should there be further labels, unaccounted for in cycle 1, the process is continued, starting from the lowest unaccounted label, creating additional cycles until all labels are accounted for. By this procedure the permutation is decomposed into a number of disjoint cycles (corresponding to the classes of the group). As an example the permutation

$$\begin{pmatrix} 1 & 2 & 3 & 4 & 5 & 6 \\ 2 & 4 & 6 & 1 & 5 & 3 \end{pmatrix}$$

decomposes into the disjoint cycles $1 \to 2 \to 4 \to 1$, $3 \to 6 \to 3$, and $5 \to 5$. By reordering the columns the permutation can be written

$$\left(\begin{array}{ccc|cc|c} 1 & 2 & 4 & 3 & 6 & 5 \\ 2 & 4 & 1 & 6 & 3 & 5 \end{array}\right) = \begin{pmatrix} 1 & 2 & 4 \\ 2 & 4 & 1 \end{pmatrix}\begin{pmatrix} 3 & 6 \\ 6 & 3 \end{pmatrix}\begin{pmatrix} 5 \\ 5 \end{pmatrix}$$

In cycle notation the bottom row is superfluous since all the information is encoded in the order of the labels in the top row. The last example is therefore written as $(1\ 2\ 4)(3\ 6)$. One-cycles are omitted, assuming that all labels not showing transform into themselves. Returning to the example illustrating $PQ \neq QP$, in cycle notation it reads $(1\ 3) \neq (1\ 2)$.

A general permutation can always be written as the product of disjoint cycles. Since they operate on different indices these cycles commute, and so the order in which they appear is immaterial.

The group S_2 consists of an identity permutation written as () and the *transposition* $1 \leftrightarrow 2$, or in cycle notation (1 2). As a group it is identical to C_2.

S_3 is of order 6. It consists of the identity, three 2-cycles and two 3-cycles:

$$S_3 = \{(); \ (1\ 2),\ (1\ 3),\ (2\ 3); \ (1\ 2\ 3),\ (1\ 3\ 2)\}$$

It is isomorphic to D_3. In fact, referring to the figure of all the rotations of D_3 they can be considered as permutations of the three vertices A, B, C. Relabelling these as 1, 2, 3 respectively c can be identified with the permutation (1 2 3) and b with the permutation (2 3). Then

$$bc = (2\ 3)(1\ 2\ 3) = (1\ 3) = b_2$$

which confirms the result obtained geometrically. Continuing in this way establishes the complete correspondence

$$\{(),\ (2\ 3),\ (3\ 1),\ (1\ 2); \ (1\ 2\ 3),\ (1\ 3\ 2)\} \leftrightarrow \{e, b, bc, bc^2; c, c^2\}$$

2.5.1 Cayley's theorem

This important theorem states that every finite group of order N is isomorhic to a subgroup of S_n. The theorem arises from the observation that for a finite group G, multiplication of all the elements $\{a_i\}$ by a given element g simply permutes them, *i.e.*

$$\{ga_1, ga_2, ga_3, ..., ga_n\} = \{a_{\pi_1}, a_{\pi_2}, ..., a_{\pi_n}\}$$

so that to each element g of the group there corresponds a permutation $\Pi(g)$ of S_n:

$$g \to \Pi(g) = \begin{pmatrix} 1 & 2 & \dots & n \\ \pi_1 & \pi_2 & \dots & \pi_n \end{pmatrix}$$

This is a true permutation as each index appears once and once only in the bottom row. To show that the n elements ga_i are all distinct suppose that $ga_j = ga_k$ and multiply on the left by g^{-1} to give $a_j = a_k$. However, this is impossible since the elements $\{a_i\}$ are all distinct. Moreover, the correspondence is invertible, *i.e.* a given permutation $\Pi(g)$ cannot arise from any other group element g', since $ga_i = g'a_i \Rightarrow g = g'$ on multiplication from the right by a_i^{-1}. These relationships imply a 1:1 correspondence $g \leftrightarrow \Pi(g)$, which respects the group structures of G and S_n. Thus if $g_1 \leftrightarrow \Pi(g_1)$ and $g_2 \leftrightarrow \Pi(g_2)$ then $g_1g_2 \leftrightarrow \Pi(g_1)\Pi(g_2)$, the permutation resulting from successive left multiplications by g_2 and g_1. There are n distinct permutations

$\Pi(g_i)$, and in general they represent only a small subset of the $n!$ permutations of S_n. It is a subset, closed under the composition law of S_n, thus forming a subgroup thereof. This correspondence is summarized by stating that the two groups are isomorhic.

2.6 Field and Space

The two concepts, *field* and *space* are familiar from a study of ordinary three-dimensional vectors and may be considered as special cases of mathematically much wider concepts that carry the same names.

In mathematical terms a field F is defined as a set with elements $(a, b, c, ...)$ for which two binary operations are defined and which satisfies the following conditions:
(i) F is an abelian group under the first operation (called addition) with an identity element, 0 (zero);
(ii) the non-zero elements of F constitute an abelian group under multiplication, with identity element 1 (unity).

Familiar fields are the set of real numbers $\mathbb{R}$, the set of complex numbers $\mathbb{C}$, and the set of all rational numbers $\mathbb{Q}$. The elements of a field are called *scalars*. A set L of elements $(u, v, w, ...)$ is called a vector space[4] over a field F if the following conditions are fulfilled:
(i) An operation of addition is defined in L, denoted by $+$, such that L is an abelian group under addition. The identity element of this this group will be denoted by $\mathbf{0}$.
(ii) Any scalar of the field F and any element of L can be combined by an operation called scalar multiplication to give an element of L such that for every $u, v \in L$ and $a, b \in F$, one has

$$\begin{aligned} a(u+v) &= (au+av) \in L \\ (a+b)u &= (au+bu) \in L \\ a(bu) &= (a.b)u \\ 1u = u \quad &, \quad 0u = \mathbf{0} \end{aligned}$$

Note that 0 is an element of the field F, whereas $\mathbf{0}$ is the *null* element of L. Henceforth the two types of zero will no longer be distinguished. The elements of a vector space are called *vectors*. The multiplication of two elements of a vector space is not necessarily defined.

[4]The names *vector space, linear vector space* and *linear space* are synonyms.

Examples of vector space are:
(i) The familiar three-dimensional space of position vectors over the field of real numbers:- the set of all position vectors together with the operations of ordinary vector addition and multiplication of a scalar by a vector.
(ii) The set of all n-tuplets of numbers such as $u \equiv (u_1, u_2, u_3, ..., u_n)$ over a field to which the scalars u_j belong. Thus, the set of all n-tuplets of complex numbers is a vector space over $\mathbb{C}$; the set of all n-tuplets of real numbers is a vector space over $\mathbb{R}$, *etc.* Two elements u and $w \equiv (w_1, w_2, ..., w_n)$ of this set are said to be equal if and only if $u_i = w_i \quad \forall 1 \leq j \leq n$. This is denoted as $u = w$. The addition of two vectors u and $v = (v_1, v_2, ..., v_n)$ of this space and the scalar multiplication are defined by

$$(u_1, u_2, ..., u_n) + (v_1, v_2, ...v_n) = (u_1 + v_1, u_2 + v_2, ..., u_n + v_n)$$

$$c(u_1, u_2, ..., u_n) = (cu_1, cu_2, ..., cu_n)$$

Moreover, if all $u_j = 0$ for $1 \leq j \leq n$, it is said that $u = 0$.

A vector field defined over the field of real (complex, *etc.*) numbers, is called a real (complex, *etc.*) vector space.

2.6.1 Unitary Space

A vector space L defined over a field F is further called an *inner-product space* or *unitary space* if its elements satisfy one more condition:
With every pair of elements $u, v \in L$, there is associated a unique number belonging to the field F - denoted by (u, v) or $u.v$, called the inner product or scalar product of u and v, for which the following properties hold:

$$\begin{aligned}(u, v) &= (v, u)^* \\ (au, bv) &= a^* b(u, v) \\ (w, au + bv) &= a(w, u) + b(w, v)\end{aligned}$$

where the asterisk denotes the complex conjugate.

The linear space of all n-tuplets of complex numbers becomes an inner-product space if the scalar product of the two elements u and v is defined as the complex number given by

$$(u, v) = \sum_{j=1}^{n} u_j^* v_j$$

The ordinary three-dimensional space of position vectors is also an inner product space with the familiar rule for taking the scalar product of two vectors.

The inner product of an element with itself is

$$(u, u) = \sum_{j=1}^{n} |u_j|^2 = ||u||^2,$$

called the *norm* of the vector u. In familiar language this corresponds to the length of a vector. The norm has the following important properties
(i)$||u|| \geq 0$, and $||u|| = 0$ if, and only if $u = 0$.
(ii)$||u + v|| \leq ||u|| + ||v||$; this is the so-called triangle inequality.
(iii)$||au|| = |a| \, ||u||$.

2.6.2 Hilbert Space

The distance between two vectors (or points, or functions) x and y defined in terms of an inner product is

$$|x - y|^2 = (x - y, x - y)$$

A sequence of vectors $x_n (n = 1, 2, ..., \infty)$ in L is said to converge to the vector x in L if

$$\lim_{n \to \infty} |x_n - x| = 0 \tag{2.1}$$

i.e., if for any $\epsilon > 0$ there is an integer $n(\epsilon)$ such that $|x_m - x| < \epsilon$ for $m > n(\epsilon)$. The sequence of vectors x_n is said to *converge in itself* or to be a *fundamental sequence* if

$$\lim_{m,n \to \infty} |x_m - x_n| = 0 \tag{2.2}$$

The space L is said to be complete if every fundamental sequence converges to a vector in L, *i.e.* if eq. (2) implies the existence of a vector x in L for which (1) holds.

A complete unitary space is called a *Hilbert space.* The unitary spaces of finite dimension are necessarily complete. For reasons of completeness the vector space of all n-tuplets of rational numbers is not a Hilbert space, since it is not complete. For instance, it is possible to define a sequence of rational numbers that approaches the irrational number $\sqrt{2}$ as a limit. The set of all rational numbers therefore does not define a Hilbert space. Similar arguments apply to the set of all n-tuplets of rational numbers.

In an n-dimensional Hilbert space, L_n the set of n independent vectors define a complete set in L_n. This set is called a *basis*, and the vectors are called the basis vectors. Basis vectors can be chosen in an infinite number of ways. Any vector u in L_n can be expanded in terms of a complete set of basis vectors, *i.e.*,

$$\boldsymbol{u} = \sum_{i=1}^{n} u_i \boldsymbol{r}_i \tag{2.3}$$

where u_i is the component of $\boldsymbol{u}$ along r_i. The space L_n is said to be fully spanned by the basis vectors. The scalars u_i are also called the Fourier coefficients of $\boldsymbol{u}$ and the expansion (3) is called the Fourier expansion of $\boldsymbol{u}$. The vector of unit norm is called a unit vector or a normalized vector. It is customary to choose a basis consisting of the unit vectors e_i in L_n. If the e_i are the orthonormal basis vectors, $(e_i, e_j) = \delta_{ij}$, the Kronecker delta.

The scalar product of two vectors $u = \sum_{i=1}^{n} u_i e_i$ and $v = \sum_{i=1}^{n} v_i e_i$ is now found to be

$$(u, v) = (v, u)^* = \sum_{i=1}^{n} u_i^* v_i$$

Also

$$||u||^2 \equiv (u, u) = \sum_{i=1}^{n} |u_i|^2.$$

Transformation of the unit vectors e_i (keeping everything else fixed) results in a new set of basis vectors e_i':

$$e_i \to e_i' = Te_i = \sum_{j=1}^{n} e_j T_{ji}, \quad 1 \leq i \leq n, \tag{2.4}$$

where T_{ji} is a scalar denoting the component of e_i' along e_j. (Compare 1.2.5).

Transformations that take one orthonormal set of basis vectors into another orthonormal set are called *unitary transformations*; the operators associated with them are called unitary operators. This definition preserves the norms and scalar products of vectors in L_n. The transformation (4) is in fact a set of linear equations

$$\begin{aligned} T(e_1, e_2, ..., e_n) &= (e_1', e_2', ..., e_n') \\ &= (e_1, e_2, ..., e_n) \begin{bmatrix} T_{11} & T_{12} & ... & T_{1n} \\ T_{21} & T_{22} & ... & T_{2n} \\ ... & ... & ... & ... \\ T_{n1} & T_{n2} & ... & T_{nn} \end{bmatrix} \end{aligned}$$

The square matrix $[T_{ij}]$[5]of order n is called a representation of the operator

[5]The matrix $[T_{ij}]$ should not be confused with the operator T although the same symbol is commonly used for an operator and a matrix representing it.

T in the basis (e_i).

The scalar product of a vector e'_i on the transformed basis with another vector e_k on the original basis e, is

$$(e_k, e'_i) = (e_k, Te_i) = (e_k, \sum_{j=1}^{n} e_j T_{ji})$$

or $(e_k, Te_i) = T_{ki}$, using the orthonormality property. This T_{ki} is called the *matrix element* of the operator T between the basis vectors e_k and e_i. It means that the operator T when applied on e_i, has a projection T_{ki} along the vector e_k.

Operators

In this section the symbols $\phi_n(x)$ are used for the orthonormal basis functions of a Hilbert space L, which may be finite or infinite, and x stands for the variables on which the functions of L may depend. An operator defined on L has the action $Tf(x) = g(x)$ where $g \in L$. The action of T on a basis function $\phi_n(x)$ is described by

$$T\phi_n(x) = \phi'_n(x) = \sum_m \phi_m(x) T_{mn}, \quad n, m = 1, 2, ...$$

where the transformed $\phi'(x)$ is expanded in a linear combination of the original basis functions. The expansion represents a set of linear equations

$$\begin{aligned}(\phi'_1, \phi'_2, ..., \phi'_n, ...) &= T(\phi_1, \phi_2, ..., \phi_n, ...) \\ &= (\phi_1, \phi_2, ..., \phi_n, ...) \begin{bmatrix} T_{11} & T_{12} & ... & T_{1n} & ... \\ T_{21} & T_{22} & ... & T_{2n} & ... \\ ... & ... & ... & ... & ... \\ T_{n1} & T_{n2} & ... & T_{nn} & ... \\ ... & ... & ... & ... & ... \end{bmatrix}\end{aligned}$$

The matrix $[T_{ij}]$ is the representation of the operator T with the basis $\{\phi_n\}$. As before, a matrix element of T is given by

$$\begin{aligned}T_{mn} = (\phi_m, \phi'_n) &= (\phi_m, T\phi_n) \\ &= S\phi^*_m(x) T\phi_n(x)\end{aligned}$$

where S denotes summation over discrete variables and integration over the continuous variables of the set x on which the ϕ's depend. In terms of row vectors such as

$$\Phi \equiv (\phi_1, \phi_2, ..., \phi_n, ...), \quad \Phi' = \Phi T.$$

An operator T is *linear* if for every f and g in L,

$$T(cf + dg) = cTf + dTg$$

where c and d are scalars of the field over which L is defined. T is *antilinear* if

$$T(cf + dg) = c^*Tf + d^*Tg$$

for every f and g in L. An example is the operator for complex conjugation.

An operator T is *hermitian* if

$$(f, Tg) = (Tf, g), \text{ for every } f, g \in L.$$

Let $f = \sum_n a_n\phi_n$ and $g = \sum_n b_n\phi_n$. For hermitian T

$$\sum_{n,m} a_m^* b_n T_{nm} = \sum_{n,m} a_m^* b_n T_{mn}^*$$

and since this must be true for all scalars a_m and b_n, it follows that

$$T_{nm} = T_{mn}^*$$

an alternative definition of a hermitian matrix: a matrix that equals its own hermitian conjugate,

$$T = T^\dagger = (\tilde{T})^* = (\widetilde{T^*})$$

where T represents the transposed matrix. T is a unitary operator if

$$(Tf, Tg) = (f, g)\,\forall f, g \in L.$$

If T is a unitary operator, then

$$TT^\dagger = T^\dagger T = E$$

where E is the identity operator. If the scalars of L are real, this reduces to

$$T\tilde{T} = \tilde{T}T = E,$$

in which case T is called an *orthogonal operator*.

2.6.3 The Eigenvalue Problem

The operation of T on a basis function,

$$T\phi_n = \sum_m \phi_m T_{mn}$$

is not uniquely defined by the basis set $\{\phi_n\}$ and the simplest choice of an orthonormal basis set is such that the sum on the right has only one non-vanishing term:

$$T\psi_n = T_{nn}\psi_n \equiv t_n\psi_n \tag{2.5}$$

which defines the scalars t_n. A non-zero vector ψ_n that satisfies (5) is called an *eigenvector* or an eigenfunction of T corresponding to the eigenvalue t_n. The problem of finding the eigenvalues and eigenfunctions of an operator acting on Hilbert space is referred to as the eigenvalue problem.

The eigenvalues need not all be distinct, that is, two or more eigenvectors may correspond to the same eigenvalue: in this case, such eigenvectors are said to be *degenerate.* The multiplicity of an eigenvalue is defined as the number of independent eigenvectors which have this same eigenvalue.

In the vector space L defined over the field of real numbers, every operator acting on L does not necessarily have eigenvalues and eigenvectors. Thus for the operation of $\pi/2$ rotation on a two-dimensional vector space of (real) position vectors, the operator has no eigenvectors since there is no non-zero vector in this space which transforms into a real multiple of itself. However, if L is a vector space over the field of complex numbers, every operator on L has eigenvectors. The number of eigenvalues is equal to the dimension of the space L. The set of eigenvalues of an operator is called its *spectrum.*

Consider a finite dimensional Hilbert space L_n of T, defined over the field of complex numbers, so that T has exactly n eigenvalues. The distinct eigenvalues are $t_1, t_2, ..., t_m$, for $m < n$. If the eigenvalue of t_i is k_i-fold degenerate, there are k_i independent eigenvectors of T in L_n which have the same eigenvalue t_i. These eigenvectors constitute the basis for a k_i-dimensional subspace M_i of L_n; M_i is called the *eigenspace* of T corresponding to t_i. Any vector of M_i is an eigenvector of T with the eigenvalue t_i. There are eigenspaces $M_1, M_2, ..., M_m$ corresponding to the eigenvalues $t_1, t_2, ..., t_m$ respectively. If T is a hermitian or a unitary operator, then these subspaces are pairwise orthogonal. Two spaces are orthogonal if every vector of one space is orthogonal to every vector of the other space.

Any vector $u \in L_n$ can now be expressed uniquely in the form

$$u = u_1 + u_2 + ... + u_i + ... + u_m$$

where u_i is in M_i. The u_i's are pairwise orthogonal. The operation of T on u gives

$$Tu = Tu_1 + Tu_2 + ... + Tu_m = t_1u_1 + t_2u_2 + ... + t_mu_m$$

This operation uniquely determines the action of T on any vector of the Hilbert space L_n. The result can be expressed in terms of m *projection* operators defined on the eigenspaces M_i such that the action of P_i on u gives the projection of u on M_i, or

$$P_iu = u_i$$

Hence,

$$Tu = t_1P_1u + t_2P_2u + ... + t_mP_mu$$

for every $u \in L_n$. Therefore

$$T = t_1P_1 + t_2P_2 + ... + t_mP_m.$$

This expression is known as the *spectral resolution* of T.

2.7 Representation of Groups

A set of operators $\{A, B, ...\}$ in a vector space L forms a group if the operators satisfy the group postulates. Hence the product of the operators A and B means the single operator C such that

$$C\boldsymbol{x} = A(B\boldsymbol{x}) \quad \forall \boldsymbol{x} \in L.$$

The identity of the group is the unit operator which leaves all vectors in L unchanged. All operators in the group have inverses.

Mapping L on another space L', using the operator S creates an isomorphic group of operators in the space L' which are the transforms of $A, B, ...$ by the operator S:

$$A' = SAS^{-1} \quad , \quad B' = SBS^{-1} \quad , ...$$

If the arbitrary group G is mapped homomorphically on a group of operators $D(G)$ in the vector space L, the operator group $D(G)$ is said to be a *representation* of the group G in the representation space L. If the dimensionality of L is n, the representation is said to have degree n (or is an *n-dimensional representation*). The operator corresponding to the element R of G may be denoted by $D(R)$. If R and S are elements of the group G, then

$$\begin{aligned} D(RS) &= D(R)D(S) \\ D(R^{-1}) &= [D(R)]^{-1} \\ D(E) &= 1 \end{aligned} \tag{2.6}$$

A *linear representation* is a representation in terms of linear operators.

In an n-dimensional space L, the linear operators of the representation can be described by their matrix representatives. This procedure produces a homomorphic mapping of the group G on a group of $n \times n$ matrices $D(G)$, *i.e.*, a *matrix representation* of the group G. From equations (6) it follows that the matrices are non-singular, and that

$$\begin{aligned} D_{ij}(E) &= \delta_{ij} \quad , \quad i,j = 1,2,...,n; \\ D_{ij}(RS) &= \sum_k D_{ik}(R)D_{kj}(S) = D_{ik}(R)D_{ik}(S) \end{aligned}$$

2.7.1 Equivalent representations

By changing the basis in the n-dimensional space L, the matrices $D(R)$ will be replaced by their transforms by some matrix C. The matrices $D'(R) = CD(R)C^{-1}$ also provide a representation of the group G, which is equivalent to the representation $D(R)$. It should be clear that equivalent representations have the same structure, even though the matrices look different. What is needed to avoid any possible ambiguity are appropriate aspects of $D(R)$ which remain invariant under a change of coordinate axes. One such invariant is easily defined in terms of the diagonal elements of the matrix, as

$$\begin{aligned} \sum_i [CD(R)C^{-1}]_{ii} &= \sum_{ikl} C_{ik}D_{kl}(R)C^{-1}_{li} \\ &= \sum_{kl} \delta_{kl}D_{kl}(R) = \sum_k D_{kk}(R) \end{aligned}$$

Thus the sum of the diagonal elements, or trace, of a matrix $D(R)$ is invariant under a transformation of the coordinate axes. When dealing with group representations the trace $\sum_i D_{ii}(R)$ is called the *character* of R in the representation D and is denoted by

$$\chi(R) = \sum_i D_{ii}(R)$$

2.7.2 Reducible Representations

Consider the vector space L_n which is used to generate a representation of the group G. For every element A of G and every vector $\phi \in L$, $A\phi$ also belongs to L_n. The vector space is said to be invariant under the transformations of G.

A vector space L_m is said to be a subspace of another vector space L_n if every vector of L_m is also contained in L_n. L_m is called a *proper subspace* of L_n if the vectors of L_m do not exhaust the space L_n.

The vector space L_n, which is invariant under G, may contain a proper subspace which is also invariant under G. In such a case, L_m is an invariant subspace of L_n under G, and the space L_n is said to be *reducible* under G.

Let $\{T(E), T(A), T(B), ...\}$ be a representation of G in L_n. Then, if L_n has an invariant subspace L_m $(m < n)$ under G, the matrices of the representation have the form

$$T(A) = \begin{bmatrix} D^{(1)}(A) & | & \mathbf{0} \\ \cdots & \cdots & \cdots \\ X(A) & | & D^{(2)}(A) \end{bmatrix}, \quad etc. \ ,$$

where $D^{(1)}(A)$ and $D^{(2)}(A)$ are square matrices of order m and $n - m$ respectively, $X(A)$ is of order $(n - m) \times m$ and $\mathbf{0}$ is a null matrix of order $m \times (n - m)$.

To demonstrate this proposition the unit basis vectors $\{\phi_1, \phi_2, ..., \phi_n\}$ are arranged in a matrix, placing the m rows in L_m first, *i.e.*

$$M = \begin{bmatrix} 1 & 0 & 0 & \cdots & \cdots & \cdot & \cdots & \cdots \\ 0 & 1 & 0 & \cdots & \cdots & \cdot & \cdots & \cdots \\ \cdots & \cdots & \cdots & \cdots & \cdots & \cdot & \cdots & \cdots \\ 0 & 0 & 0 & \cdots & 1(mm) & \cdot & 0 & \cdots \\ \cdot & \cdot & \cdot & \cdot & \cdot & \cdot & \cdot & \cdot \\ & & & & & \cdot & & \\ & & & & & \cdot & & \\ & & & & & \cdot & & \end{bmatrix}$$

Since L_m is invariant under G, any operator $A \in G$ transforms each vector in L_m into another vector in L_m. Hence, the operation AM results in a matrix of the same form as $T(A)$. It should be clear that the two sets of matrices $D^{(1)}$ and $D^{(2)}$ give two new representations of dimensions m and $n - m$ respectively for the group G. For $D^{(1)}$ there exists a set of basis vectors $\{\phi_1, \phi_2, ..., \phi_m\}$ for the representation, and another set $\{\phi_{m+1}, ..., \phi_n\}$ for $D^{(2)}$. The representation T is said to be reducible. It follows that the reducibility of a representation is linked to the existence of a proper invariant subspace in the full space. Only the subspace of the first m components is invariant:

$$\begin{bmatrix} D^{(1)}(A) & \cdot & 0 \\ \cdots & \vdots & \cdots \\ X(A) & \cdot & D^{(2)}(A) \end{bmatrix} \begin{bmatrix} x_{<m} \\ \cdots \\ x_{>m} \end{bmatrix} = \begin{bmatrix} D^{(1)}(A)x_{<m} \\ \cdots \\ X(A)x_{<m} + D^{(1)}(A)x_{>m} \end{bmatrix},$$

while the complementary $(n - m)$ dimensional subspace is not invariant. If it is possible to find a basis in which all the matrices of the representation assume the form of $T(A)$, but with $X(A) = 0$, *i.e.*

$$D(A) = \begin{bmatrix} D^{(1)}(A) & & \cdot & & 0 \\ \cdots & \cdots & \vdots & \cdots & \cdots \\ 0 & & \cdot & & D^{(2)}(A) \end{bmatrix}.$$

the representation is *fully reducible.* In this case both the m and the $m - n$ dimensional subspaces are invariant. This means that the basis functions $\phi_1, ...\phi_m$ transform among themselves, and the basis functions $\phi_{m+1}, ..., \phi_n$ transform among themselves, and the transformations of the group do not couple the two sets of functions. The space L is decomposed into the direct sum of $L^{(1)}$ and $L^{(2)}$,

$$L = L^{(1)} \oplus L^{(2)}$$

and the representation D is the sum of $D^{(1)}$ and $D^{(2)}$. If the matrices of a representation are unitary the matrix $A(X) = 0$ and reducibility implies full reducibility. This is always the case for finite groups.

If the L_m and L_{n-m} subspaces contain further invariant (proper) subspaces within them the process of reduction can be carried on until no further unitary transformation can be found to further reduce the matrices of the representation. The final form of the matrices of the representation Γ may look like

$$\Gamma(A) = \begin{bmatrix} \Gamma^{(1)}(A) & \vdots & & & \\ \cdots\ \cdots\ \cdots & \cdots & \cdots\ \cdots & & \\ & \vdots & \Gamma^{(2)}(A) & \vdots & 0 \\ & & \cdots\ \cdots & & \\ & & & \ddots & \\ & & & & \cdots\ \cdots \\ 0 & & & \vdots & \Gamma^{(s)}(A) \end{bmatrix}$$

etc. with all the matrices of Γ having the same reduced structure.

When such a complete reduction has been achieved, the component representations $\Gamma^{(1)}, \Gamma^{(2)}, ..., \Gamma^{(s)}$ are called the *irreducible representations* of the group G and the representation Γ is said to be *fully reduced.* An irreducible representation may occur more than once in the reduction of a reducible representation Γ. Symbolically

$$\begin{aligned} \Gamma &= a_1\Gamma^{(1)} \oplus a_2\Gamma^{(2)} \oplus ... \oplus a_s\Gamma^{(s)} \\ &= \sum_i a_i\Gamma^{(i)} \end{aligned}$$

2.7.3 Orthogonality

There are two theorems of fundamental importance, known as Schur's lemmas, which are useful in the study of the irreducible representations of a group.

Lemma 1

If $\Gamma^{(i)}$ is an irreducible representation of a group G and if a matrix P commutes with all the matrices of $\Gamma^{(i)}$, then P must be a constant matrix, that is, $P = cE$ where c is a scalar.

Proof

Assume that A is an element of the group G and that

$$\Gamma^{(i)}(A)P = P\Gamma^{(i)}(A) \quad \forall A \in G \tag{2.7}$$

If the dimension of $\Gamma^{(i)}$ is n, P is a square matrix of order n. Let x_j be the eigenvectors of P with eigenvalues c_j. Then $Px_j = c_j x_j$. Multiply both sides from the left by $\Gamma^{(i)}(A)$ to give

$$\Gamma^{(i)}(A)Px_j = \Gamma^{(i)}(A)c_j x_j$$

$$\text{or} \qquad P\Gamma^{(i)}(A)x_j = c_j\Gamma^{(i)}(A)x_j \text{ , using (7)}$$

This means that $\Gamma^{(i)}(A)x_j$, $\forall A \in G$, are eigenvectors of P with the same eigenvalue c_j. Let there be m such independent eigenvectors in P having the same eigenvalue c_j. Recall that the eigenvectors belonging to to an eigenvalue generate a subspace L_m which is invariant under G. Now if L_m is a proper subspace of L_n, that is, if L_m is not the same as L_n, then L_n has an invariant subspace and the representation $\Gamma^{(i)}$ must be reducible, which contradicts the assumption. Therefore $L_m = L_n$, making all the eigenvectors of P equal to each other, $c_j = c$ (say), giving $P = cE$.

Lemma 2

If $\Gamma^{(i)}$ and $\Gamma^{(j)}$ are two irreducible representations of dimensions l_i and l_j respectively of a group G and if a matrix M (of order $l_i \times l_j$) satisfies the relation

$$\Gamma^{(i)}(A)M = M\Gamma^{(j)}(A) \quad \forall A \in G \tag{2.8}$$

then either (a) $M = \boldsymbol{O}$, the null matrix, or (b) det$M \neq 0$, in which case $\Gamma^{(i)}$ and $\Gamma^{(j)}$ are equivalent representations[6].

Proof:

Take the hermitian conjugate of both sides of (8) to give

$$M^\dagger\Gamma^{(i)\dagger}(A) = \Gamma^{(j)\dagger}(A)M^\dagger \quad \forall A \in G$$

$$\text{or} \qquad M^\dagger\Gamma^{(i)}(A^{-1}) = \Gamma^{(j)}(A^{-1})M^\dagger \quad \forall A \in G$$

Multiply from the right by M:

$$M^\dagger\Gamma^{(i)}(A^{-1})M = \Gamma^{(j)}(A^{-1})M^\dagger M \quad \forall A \in G$$

or by using (8)

$$M^\dagger M\Gamma^{(j)}(A^{-1}) = \Gamma^{(j)}(A^{-1})M^\dagger M \quad \forall A \in G.$$

[6]It is noted that two representations can be equivalent only if their dimensions are equal. Hence if $l_i \neq l_j$ only case (a) applies.

Thus the matrix $M^\dagger M$ commutes with $\Gamma^{(j)}(A^{-1})$ $\forall A \in G$ and therefore, by lemma 1, must be a constant matrix:

$$M^\dagger M = cE \tag{2.9}$$

In the first instance, consider the case $l_i = l_j = n$ (say). Then from (9)

$$\det(M^\dagger M) = \det(M^\dagger)\det(M) = c^n$$

If $c \neq 0$, then $\det(M^\dagger) \neq 0$ (because $\det M^\dagger = (\det M)^*$); therefore M^{-1} exists and from (8) follows that

$$\Gamma^{(j)}(A) = M^{-1}\Gamma^{(i)}(A)M \quad \forall A \in G$$

showing that $\Gamma^{(i)}$ and $\Gamma^{(j)}$ are equivalent representations. If $c = 0$, then taking the $(i,i)^{\text{th}}$ element of (8) shows that

$$\sum_k M^\dagger_{ik} M_{ki} = 0$$

$$\text{or} \qquad \sum_k M^*_{ki} M_{ki} = \sum_k |M_{ki}|^2 = 0$$

which is possible only if $M_{ki} = 0$ for $1 \leq k \leq n$. But, i is arbitrary between 1 and n, hence $M = \boldsymbol{O}$.

The proof when $l_i \neq l_j$ $(l_i < l_j)$ (say) consists of supplementing the matrix M by adding $l_j - l_i$ rows of zeros to give the new matrix

$$M' = \begin{bmatrix} M_{(l_i)} \\ \cdots\cdots \\ O_{(l_j - l_i)} \end{bmatrix}$$

and taking M' through the same steps as for M in the first instance.

The Orthogonality Theorem

Construct a matrix

$$M = \frac{1}{g}\sum_{A \in G} \Gamma^{(i)}(A) X \Gamma^{(j)}(A^{-1}) \tag{2.10}$$

where $\Gamma^{(i)}$ and $\Gamma^{(j)}$ are two irreducible representations of dimensions l_i and l_j respectively of a group G of order g, and X is an arbitrary matrix of order $(l_i - l_j)$ independent of the group elements.

Multiply both sides of (10) from the left by $\Gamma^{(i)}(B)$, where $B \in G$, to give

$$\begin{aligned}
\Gamma^{(i)}(B)M &= \frac{\Gamma^{(i)}(B)}{g}\sum_{A\in G}\Gamma^{(i)}(A)X\Gamma^{(j)}(A^{-1}) \\
&= \frac{1}{g}\sum_{A\in G}\Gamma^{(i)}(BA)X\Gamma^{(j)}(A^{-1}) \\
&= \frac{1}{g}\sum_{A\in G}\Gamma^{(i)}(C)X\Gamma^{(j)}(C^{-1}B)\ ,\ \text{where } BA = C \\
&= \frac{1}{g}\sum_{C\in G}\Gamma^{(i)}(C)X\Gamma^{(j)}(C^{-1})\Gamma^{(j)}(B) \\
&= M\Gamma^{(i)} \quad \forall B \in G.
\end{aligned}$$

Therefore by Schur's second lemma it follows that $M = \boldsymbol{O}$. Taking the $(k,s)^{\text{th}}$ element of (10), gives

$$\sum_{A\in G}\sum_{p,q}\Gamma^{(i)}_{kp}(A)X_{pq}\Gamma^{(j)}_{qs}(A^{-1}) = 0 \tag{2.11}$$

It is now convenient to choose the arbitrary matrix X to be a matrix with all zero elements except the (m,n)th element which is taken to be unity, *i.e.* $X_{pq} = \delta_{pm}\delta_{qn}$. From (11) follows that

$$\sum_{A\in G}\Gamma^{(i)}_{km}(A)\Gamma^{(j)}_{ns}(A^{-1}) = 0$$

or

$$\sum_{A\in G}\Gamma^{(i)}_{km}\Gamma^{(j)^*}(A) = 0 \tag{2.12}$$

for all $1 \leq k,\ m \leq l_i$, $1 \leq n,\ s \leq l_j$. This means that the product of the (k,m)th element of the irreducible representation $\Gamma^{(i)}$, summed over the group elements, equals zero.

Next construct a matrix N by replacing $\Gamma^{(j)}$ in (11) by $\Gamma^{(i)}$, that is

$$N = \frac{1}{g}\sum_{a\in G}\Gamma^{(i)}(A)X\Gamma^{(i)}(A^{-1}) \tag{2.13}$$

By the same procedure as before it follows that

$$\Gamma^{(i)}(A)N = N\Gamma^{(i)}(A) \quad \forall A \in G.$$

From Schur's first lemma it follows that N must be a constant matrix, $N = aE$ (say), where E is the unit matrix of order l_i. Again, taking the (k,s)th element of (13) one gets

$$\frac{1}{g}\sum_{A\in G}\sum_{p,q}\Gamma^{(i)}_{kp}(A)X_{pq}\Gamma^{(i)}_{qs}(A^{-1}) = a\delta_{ks}$$

As before, take $X_{pq} = \delta_{pm}\delta_{qn}$, to give

$$\frac{1}{g}\sum_{A\in G}\Gamma^{(i)}_{km}\Gamma^{(i)}_{ns}(A^{-1}) = a\delta_{ks} \tag{2.14}$$

To interpret (14) it is necessary to identify the scalar a. For this purpose, take the traces of the matrices on both sides of (13):

$$\text{trace}N = al_i = \frac{1}{g}\sum_{k=1}^{l_i}\sum_{A\in G}\sum_{p,q}\Gamma^{(i)}_{kp}(A)X_{pq}\Gamma^{(i)}_{qk}(A^{-1})$$

or

$$\begin{aligned} al_ig &= \sum_{p,q}X_{pq}\sum_{A\in G}\sum_{k}\Gamma^{(i)}_{qk}(A^{-1})\Gamma^{(i)}_{kp}(A) \\ &= \sum_{p,q}X_{p,q}\sum_{A\in G}\Gamma^{(i)}_{qp}(E) \\ &= g\sum_{p,q}X_{pq}\delta_{pq} = g\text{trace}X \end{aligned}$$

or

$$a = (\text{trace}X)/l_i$$

But, by choice of X, trace$X = 0$, unless $m = n$, in which case trace$X = 1$. In short, trace$X = \delta_{mn}$. Hence, from (14) it follows that

$$\sum_{A\in G}\Gamma^{(i)}_{km}(A)\Gamma^{(i)}_{ns}(A^{-1}) = \left(\frac{g}{l_i}\right)\delta_{ks}\delta_{mn} \quad \forall 1 \le k,m,n,s \le l_i. \tag{2.15}$$

Combining (12) with (15) into a single equation gives

$$\sum_{A\in G}\Gamma^{(i)}_{km}(A)\Gamma^{(j)}_{ns}(A^{-1}) = \left(\frac{g}{l_i}\right)\delta_{ij}\delta_{ks}\delta_{mn}$$

or

$$\sum_{A\in G}\Gamma^{(i)}_{km}(A)\Gamma^{(j)*}_{sn}(A) = \left(\frac{g}{l_i}\right)\delta_{ij}\delta_{ks}\delta_{mn} \tag{2.16}$$

This equation (16) is known as the great orthogonality theorem for the irreducible representations of a group and occupies a central position in the theory of group representations.

2.7.4 Orthogonal Relationships

Each irreducible representation of a group consists of a set of square matrices of order l_i. The set of matrix elements with the same index, grouped together, one from each matrix in the set, constitutes a vector in g-dimensional space. The great orthogonality theorem (16) states that all these vectors are mutually orthogonal and that each of them is normalized so that the square of its length is equal to g/l_i. This interpretation becomes more obvious when (16) is unpacked into separate expressions:

$$\sum_A \Gamma_i(A)_{mn}\Gamma_j(A)_{mn} = \begin{cases} 0 & \text{if } i \neq j \\ g/l_i & \text{if } i = j \end{cases} \tag{2.17}$$

$$\sum_A \Gamma_i(A)_{mn}\Gamma_i(A)_{rs} = 0 \quad \text{if} \quad m \neq r \quad \text{and/or} \quad n \neq s. \tag{2.18}$$

For practical applications these relationships lead to a number of important rules about irreducible representations and their characters:
(i) *The sum of the squares of the dimensions of the irreducible representations of a group is equal to the order of the group, i.e.*

$$\sum_i l_i^2 = g \tag{2.19}$$

Proof:
A matrix of order l has l^2 elements. Each irreducible representation Γ_i must therefore contribute l_i^2 g-dimensional vectors. The orthogonality theorem requires that the total set of $\sum l_i^2$ vectors must be mutually orthogonal. Since there can be no more than g orthogonal vectors in g-dimensional space, the sum $\sum l_i^2$ cannot exceed g. For a complete set (19) is implied. Since the character of an identity matrix always equals the order of the representation it further follows that

$$\sum_i [\chi_i(E)]^2 = g$$

(ii) *The sum of the squares of the characters in any irreducible representation, equals g, i.e.*

$$\sum_A [\chi_i(A)]^2 = g$$

Proof:
From (16) it follows that

$$\sum_A \Gamma_i(A)_{mm}\Gamma_i(A)_{rr} = \frac{g}{l_i}\delta_{mr}$$

Summing the l.h.s. over m and r gives

$$\begin{aligned}\sum_r \sum_m \sum_A \Gamma_i(A)_{mm}\Gamma_i(A)_{rr} &= \sum_A \left[\sum_m \Gamma_i(A)_{mm} \sum_r \Gamma_i(A)_{rr}\right] \\ &= \sum_A [\chi_i(A)]^2\end{aligned}$$

while summing the r.h.s. over m and r gives

$$\frac{g}{l_i} \sum_r \sum_m \delta_{mr} = \frac{g}{l_i} l_i = g$$

(iii) *The vectors whose components are the characters of two different irreducible representations are orthogonal, i.e.*

$$\sum_A \chi_i(A)\chi_j(A) = 0 \quad \text{for} \quad i \neq j.$$

Proof:
Let $m = n$ in (17), *i.e.*

$$\sum_A \Gamma_i(A)_{mm}\Gamma_j(A)_{mm} = 0 \quad \text{if} \quad i \neq j$$

Now

$$\begin{aligned}\sum_A \chi_i(A)\chi_j(A) &= \sum_A \left[\sum_m \Gamma_i(A)_{mm} \sum_m \Gamma_j(A)_{mm}\right] \\ &= \sum_m \left[\sum_A \Gamma_i(A)_{mm}\Gamma_j(A)_{mm}\right] = 0.\end{aligned}$$

(iv) *In any given representation the characters of all matrices belonging to operations in the same class are identical.*

Proof:
Since all elements in the same class are conjugate to one another, all matrices belonging to elements in the same class in a representation must be conjugate, *e.g.*

$$R = A^{-1}PA$$

Hence

$$\begin{aligned}\chi(R) &= \chi(A^{-1}PA) = \chi(A^{-1}P)A \\ &= \chi A(A^{-1}P) = \chi(AA^{-1})P \\ &= \chi(P)\end{aligned}$$

(v) *The number of irreducible representations of a group is equal to the number of classes in the group.*

Proof:

Since

$$\sum_A [\chi_i(A)]^2 = g$$

$$\sum_A \chi_i(A)\chi_j(A) = g\delta_{ij} \tag{2.20}$$

Let h_p be the number of elements in class p from a total of k classes. Equation (20) then becomes

$$\sum_{p=1}^{k} \chi_i(A_p)\chi_j(A_p)h_p = g\delta_{ij}$$

where A_p refers to any one of the operations in class p. This equation implies that the k quantities, $\chi_l(A_p)$ in each representation Γ_l behave like the components of a k-dimensional vector and these k-vectors are mutually orthogonal. For a complete set condition (v) is satisfied.

(vi) *For a finite or compact group a given irreducible representation may occur more than once in the decomposition of a reducible representation.* In the general case the decomposition is written

$$D = \sum_{\oplus} a_\nu D^\nu$$

where the non-negative integer a_ν denotes the number of times D^ν appears in the decomposition. To find a_ν the trace is taken on both sides. For a general group it follows that

$$\chi(g) = \sum_\nu a_\nu \chi^\nu g$$

This equation is multiplied by $\chi^\mu(g^{-1})$ and summed over (g) to give

$$\sum_g \chi^\mu(g^{-1})\chi(g) = \sum_\nu a_\nu \sum_g \chi^\mu(g^{-1})\chi^\nu(g)$$

Using (20) in the form

$$\sum_g \chi^\mu(g)\chi^{\nu *}(g) = g\delta^{\mu\nu}$$

the r.h.s. is just

$$\sum_\nu a_\nu g\delta^{\mu\nu}(g^{-1}) = ga_\mu$$

Thus

$$\begin{aligned} a_\mu &= \frac{1}{g}\sum_g \chi(g)\chi^\mu(g^{-1}) \\ &\equiv \langle \chi^\mu, \chi \rangle \end{aligned}$$

where $\langle \chi, \phi \rangle = \frac{1}{g}\sum_g \chi(g)\phi(g^{-1})$ is defined as the scalar product of two characters χ and ϕ.

Group characters and the rules that specify their their properties find application in the analysis of molecular spectra and conformation. Many useful examples are given in [13].

2.7.5 Direct Product Representations

Vector spaces which occur in physical applications are often direct products of smaller vector spaces that correspond to different degrees of freedom of the physical system (*e.g.* translations and rotations of a rigid body, or orbital and spin motion of a particle such as an electron). The characterization of such a situation depends on the relationship between the representations of a symmetry group realized on the product space and those defined on the component spaces.

Let U and V be inner product spaces with orthonormal bases $\{\boldsymbol{u}_i; i = 1, ..., n_u\}$ and $\{\boldsymbol{v}_i; i = 1, ..., n_v\}$. The direct product $W = U \times V$ consists of all linear combinations of the orthonormal basis vectors $\{\boldsymbol{w}_k; k = (i,j); i = 1, ..., n_u; j = 1, ...n_v\}$ where $\boldsymbol{w}_k = (\boldsymbol{u}_i \cdot \boldsymbol{v}_j)$.

Let G be a symmetry group of a physical system (*e.g.* rotations in space), and W be the direct product space of physical solutions consisting of two sets of degrees of freedom U, V (*e.g.* orbital motion of two different particles). Suppose $D^\mu(G)$ and $D^\nu(G)$ are the representations of G on U and V respectively. Then the operators $D^{\mu\times\nu}(g) = D^\mu(g) \times D^\nu(g)$ on W with $g \in G$ also form a representation of $D^\mu(G)$ (on U) and $D^\nu(G)$ (on V). The representation $D^{\mu\times\nu}(G)$ defined on W is called the *direct product representation* of $D^\mu(G)$ (on U) and $D^\nu(G)$ (on V). It is easy to verify that the group characters of the direct product representation $D^{\mu\times\nu}$ are equal to the product of the characters of the two representations $D^\mu(G)$ and $D^\nu(G)$,

$$\chi^{\mu\times\nu} = \chi^\mu \chi^\nu$$

The product representation $D^{\mu\times\nu}$ is in general reducible. It can be decomposed into its irreducible components

$$D^{\mu\times\nu} = D^\mu \otimes D^\nu = \sum_\oplus a_\sigma D^\sigma$$

known as the *Clebsch-Gordon series.*

The character of $D^{\mu\times\nu}(g)$ is by definition

$$\chi^{\mu\times\nu}(g) = D^{\mu\times\nu}_{AA}(g) = D^{\mu}_{aa}(g)D^{\nu}_{cc}(g) = \chi^{\mu}(g)\chi^{\nu}(g)$$

That is, the character of the product representation is just the product of the characters. Hence the *Clebsch-Gordon coefficients* are

$$a_\sigma = \langle\chi^\sigma, \chi^\mu\chi^\nu\rangle.$$

2.8 Continuous Groups (Lie Groups)

The theory of discrete infinite groups closely parallels that of finite groups, but for infinite continuous groups there are several important differences.

The elements of a continuous group can be characterized by a set of real parameters $a_1, a_2, ..., a_n$, at least one of which varies continuously over a certain interval. The choice of parameters should be restricted to the minimum number required to characterize all elements of the group. If the number of parameters is finite, the continuous group is called *finite* and the number of parameters defines the *order* of the continuous group.

The group of all rotations about an axis is a continuous group of order 1, whose parameter may be chosen to be the angle of rotation, θ taking values in the interval $[-\pi, \pi]$. A group like this, where the domain of variation of parameters is finite, is called a *closed* group.

For an r-parameter group, continuity is expressed in terms of distances in the parameter space. Two group elements $R(a)$ and $R(a')$, called *images* of the element a, are defined to be near to each other if the distance

$$\left[\sum_{i=1}^{r}(a_i - a_i')^2\right]^2 < \epsilon \qquad (2.21)$$

where ϵ is small.

Consider a group whose elements can be put in one-to-one correspondence with the points of a subset of an n-dimensional real inner-product space S_n, as parameter space. Let x_1 and x_2 be any two elements of a group G with images $P(x_1)$ and $P(x_2)$ in S_n. If it is possible to connect $P(x_1)$ and $P(x_2)$ by one or more paths lying entirely within the parameter space, the parameter space is said to be *connected*; otherwise it is disconnected.

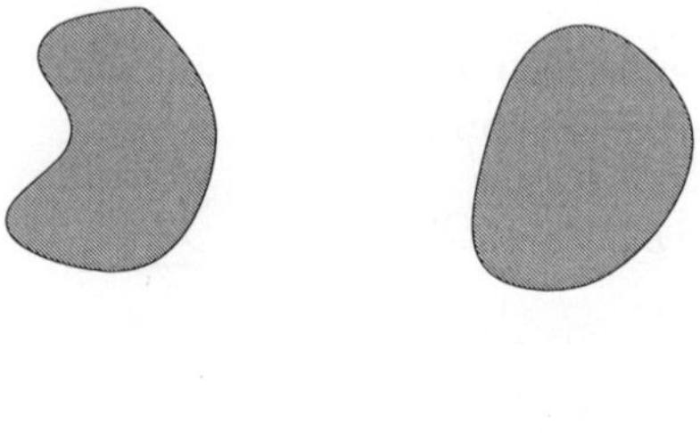

As an example, the group of rotations about an axis is a connected group. The property of connectedness is not the same as the continuous nature of a group. A continuous group, for instance the rotation-inversion group in three dimensions may be disconnected. The parameter space of a continuous disconnected group consists of two or more disjoint subsets such that each subset is a connected space, but where it is impossible to go continuously from a point in one subset to a point in another without going outside the parameter space.

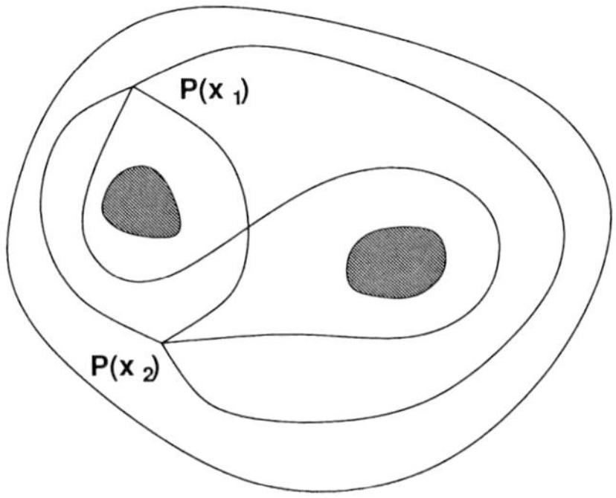

A continuous connected group may be *simply connected* or *multiply connected*, depending on the topology of the parameter space. A subset of the euclidean space S_n is said to be *k-fold* connected if there are precisely k distinct paths connecting any two points of the subset which cannot be brought into each other by continuous deformation without going outside the subset. A schematic of four-fold connected space is shown in the lower diagram.

A topological group is said to be *compact* if its parameter space is closed and bounded.

The requirements that the elements $R(a)$ form a continuous group are the same as for finite groups. First there must be a set of parameter values a° such that

$$R(a^\circ)R(a) = R(a)R(a^\circ) = R(a)$$

for all a. $R(a^\circ)$ is the identity element of the group. It is common practice to choose $a^\circ = 0$.

Next, for any a there is an $\bar{a}$ such that

$$R(\bar{a})R(a) = R(a)R(\bar{a}) = R(0)$$

$R(\bar{a})$ is the element inverse to $R(a)$:

$$R(\bar{a}) = [R(a)]^{-1}$$

The product of two elements of the set must also belong to the set. Given parameters a and b it is possible to find c such that

$$R(c) = R(b)R(a).$$

The parameters c are real functions of the real parameters a and b:

$$c_k = \phi_k(a_1, ..., a_r; b_1, ..., b_r),\ k = 1, ..., r, \tag{2.22}$$

or, symbolically

$$c = \phi(a; b)$$

So far the requirements are the same as for finite or denumerable groups. If, in addition, it is now stipulated that the parameters of a product be analytic functions of the parameters of the factors[7] and that the $\bar{a}$ be analytic functions of the a, the group is known as an r-parameter Lie group[8]. It is convenient to choose the parameters of a Lie group such that the image of the identity element E is the origin of the parameter space, *i.e.* $E \equiv x(0, 0, ..., 0)$.

The *generators* of a Lie group are defined by considering elements infinitesimally close to the identity element. The operator $T(a)x \rightarrow x'$ takes variables of space from their initial values x to final values x' as a function of the parameter a. The gradual shift of the space variables as the parameters vary continuously from their initial values $a = 0$ may be used to introduce the concept of infinitesimal transformation associated with an infinitesimal operator P. Since the transformation with parameter a takes x to x' the neighbouring parameter value $a + da$ will take the variables x to $x' + dx'$, since x is an analytical function of a. However, some parameter value δa very close to zero (*i.e.* the identity) may also be found to take x' to $x' + dx'$. Two alternative paths from x to $x' + dx'$ therefore exist, symbolized by

$$[T(a + da)]x = [T(a) + \frac{dT(a)}{da}da]x \rightarrow x' + dx'$$

$$[T(a + da)]x = [T(da)T(a)]x \rightarrow x' + dx'$$

[7] *i.e.* the function in (18) should possess derivatives of all orders with respect to both arguments.

[8] The r parameters are all essential. As an example consider the transformations

$$x' = x + a + b$$

The two parameters a and b are not essential for their sum can be represented by a continuous variable $c = a + b$, giving all transformations of the group, which hence is a one-parameter group.

Hence

$$\frac{dT(a)}{da} = \frac{T(a)\{T(da) - 1\}}{da}$$

which defines the operator P, called the generator of the Lie group, as

$$P = \left.\frac{dT}{da}\right|_{a=0} = \lim_{a\to 0} \frac{\{T(a) - 1\}}{a}$$

The composition of a one-parameter Lie group can now be described in terms of a canonical parameter t so that

$$R(t_1 + t_2) = R(t_1)R(t_2) \quad , \quad R(0) = E$$

The element $R(t)$ is described by the same symbol as the operator.

The operator equation is differentiated with respect to t_1, followed by setting $t_1 = 0$ and $t_2 = t$, to give

$$\begin{aligned} \frac{dR(t)}{dt} &= \left.\frac{dR}{dt}\right|_{t=0} \cdot R(t) \\ &= PR(t) \end{aligned}$$

in terms of the infinitesimal operator P. Under the boundary condition $R(0) = E$, the differential equation has the unique solution

$$R(t) = e^{Pt}$$

A representation of the Lie group will be unitary if the operators $R(t)$ are unitary:

$$(R(t)\psi, R(t)\phi) = (\psi, \phi)$$

Differentiating with respect to t and setting $t = 0$, give

$$(P\psi, \phi) + (\psi, P\phi) = 0 \quad \text{for all } \psi, \phi.$$

Hence $P + P^\dagger = 0$.

Thus unitary operators for the group are associated with anti-Hermitian operators for the Lie algebra. Replacing $P \leftarrow iP$, gives $P = P^\dagger$,

$$R(t) = \exp(iPt)$$

and

$$P = \lim_{a\to 0} \left\{ \frac{1}{ia} [T(a) - 1] \right\} \tag{2.23}$$

2.8.1 Representations of a Lie group

Suppose that a set of matrices $\Gamma(x)$ generate a representation of the Lie group G. Γ is a *continuous* representation of G if

$$\Gamma(x) \to \Gamma(x') \text{ as } x \to x'.$$

The group Γ of matrices is homomorphic to G and the matrices Γ can be characterized by the same parameters as used for characterizing the elements of G. It is implied that as the values of the parameters change continuously from those for x to those for x' in the parameter space, the corresponding matrix $\Gamma(x)$ goes continuously to $\Gamma(x')$.

By analogy with finite groups the representations of compact continuous groups have the following important properties:
(a) Any representation has an equivalent representation whose matrices are unitary.
(b) Any unitary representation is completely reducible.
(c) Any unitary representation is finite dimensional.

2.8.2 Important Lie Groups

The Axial Rotation Group $O(2)$

Consider the set of rotations of a circle about an axis normal to the plane of the circle and passing through its centre. Each element of this set is characterized by one parameter which may be chosen to be the angle of rotation ϕ which varies in the interval $[0, 2\pi]$. This is a one-parameter, continuous, connected, abelian, compact Lie group, known as the *axial rotation group*, denoted by $O(2)$.

Since rotations by ϕ and $\phi + 2\pi n$ (n an integer) are identical, the parameter space is the subset $[0, 2\pi]$ of the real line and the group is doubly connected because there are two paths connecting any two group elements which cannot be brought into each other by continuous deformation without going outside the space, as shown in the next diagram. If the element of the group is denoted by $T(\phi)$, the law of composition is

$$T(\phi)T(\theta) = t(\theta)T(\phi) = \begin{cases} T(\phi+\theta) & \text{if } \phi+\theta < 2\pi \\ T(\phi+\theta-2\pi) & \text{if } \phi+\theta > 2\pi \end{cases} \tag{2.24}$$

The identity element is $T(0)$ and the inverse of $T(\phi)$ is $T(2\pi - \phi)$).

The transformations of a cartesian coordinate system (x, y) in the plane of the circle can be used to generate a representation of the group. The operation of an element $T(\phi)$ on (x, y) is given by

$$T(\phi)(x,y) \equiv (x',y') = (x,y)\begin{bmatrix} \cos\phi & \sin\phi \\ -\sin\phi & \cos\phi \end{bmatrix}$$

The transformation matrix is orthogonal of order 2. With every element $T(\phi)$ of the group can be associated a 2×2 orthogonal matrix with determinant $+1$ and the correspondence is one-to-one. The set of all orthogonal matrices of order 2 having determinant $+1$ is a group isomorphic to $O(2)$ and therefore provides a two-dimensional representation for it. The matrix group is also denoted by the symbol $O(2)$.

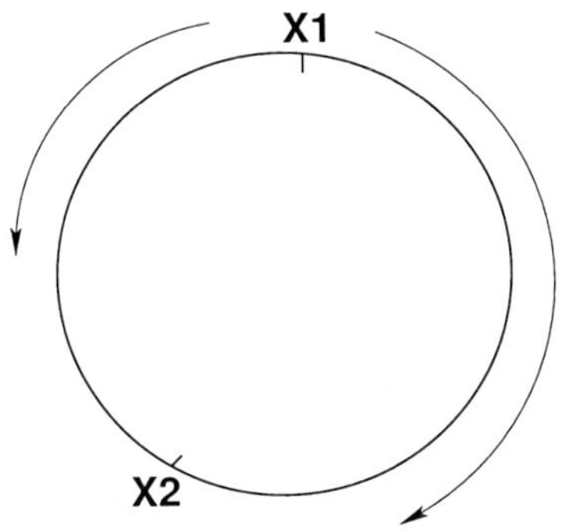

Since $O(2)$ is abelian, all its irreducible representations must be one-dimensional. To identify these irreducible representations, the product rule (20) is used, noting that the only numbers (1×1 matrices) that satisfy the rule are of the form

$$\chi(\phi) = \exp(c\phi)$$

where c is a number and $\chi(\phi)$ is the character of $T(\phi)$. Since $T(2\pi) = e$, the identity, and e must be represented by unity in one-dimensional representation, it follows that $\exp(2\pi c) = 1$, giving $c = im$, where m is an integer, or

$$\chi^{(m)}(\phi) = \exp(im\phi) \tag{2.25}$$

For every integral value of m there is an irreducible representation of $O(2)$, given by (23). The orthogonality theorem for characters in this case becomes

$$\int_0^{2\pi} \chi^{(m)*}(\phi)\chi^{(m')}(\phi)d\phi = 2\pi\delta_{mm'}.$$

The three-dimensional rotation group $O(3)$

The set of all orthogonal transformations in a three-dimensional real vector space (*i.e.* a space defined over the field of real numbers) constitutes a group denoted by $O_I(3)$. Alternatively it may be defined as the group of all 3×3 orthogonal matrices. The two groups are isomorphic.

If R is an orthogonal matrix, it satisfies the equation $R\tilde{R} = \tilde{R}R = E$. Taking the determinants of both sides of the equation and noting that $\det R = \det\tilde{R}$, it follows that $(\det R)^2 = 1$, or $\det R = \pm 1$. The matrices of $O_I(3)$ therefore divide into two sets with determinants $= \pm 1$ respectively.

The group of all real orthogonal matrices of order 3 and determinant $+1$ will be denoted by $O(3)$. Such matrices correspond to pure rotation or *proper* rotation of the coordinate system. An orthogonal matrix with determinant -1 corresponds to the product of pure rotation and inversion. Such transformations are called *improper* rotations. The matrix corresponding to inversion is the negative of the unit matrix

$$= \begin{bmatrix} -1 & 0 & 0 \\ 0 & -1 & 0 \\ 0 & 0 & -1 \end{bmatrix}$$

The inversion and the identity constitute a group of order 2. Since inversion commutes with all rotations it follows that

$$O_I(3) = O(3) \otimes (E, J).$$

The group (E, J) has only two one-dimensional irreducible representations. The representations of $O_I(3)$ can therefore be obtained from those of $O(3)$ as direct products. The group $O_I(3)$ is called the three-dimensional *rotation-inversion* group. It is isomorphic with the crystallographic space group $P\bar{1}$.

The parameters of $O(3)$ can be chosen in various ways. In a cartesian coordinate system (x, y, z) two parameters are required to fix the direction of the rotation axis $\boldsymbol{u}$, *e.g.* the angular polar coordinates θ, ϕ, and another, ψ to specify the angle of rotation.

A more covenient method is to express rotations in terms of the Euler angles. A rotation through the Euler angles (α, β, γ), denoted by $R(\alpha, \beta, \gamma)$ consists of three successive rotations: a rotation through α about the z-axis, followed by a rotation through β about the new y axis, followed by a rotation through γ about the transformed z axis. Thus,

$$R(\alpha, \beta, \gamma) = R(\gamma, 0, 0)R(0, \beta, 0)R(\alpha, 0, 0)$$

To find the transformation matrix first consider the element $R(\alpha, 0, 0)$ with matrix of transformation

$$\begin{bmatrix} \cos\phi & \sin\phi & 0 \\ -\sin\phi & \cos\phi & 0 \\ 0 & 0 & 1 \end{bmatrix}$$

Writing down similar matrices for $R(0, \beta, 0)$ and $R0, 0, \gamma)$ and taking the product as indicated

$$R(\alpha, \beta, \gamma) = \begin{bmatrix} \cos\alpha\cos\beta\cos\gamma & \sin\alpha\cos\beta\cos\gamma + \cos\alpha\sin\gamma & \sin\beta\cos\gamma \\ -\cos\alpha\cos\beta\sin\gamma & -\sin\alpha\cos\beta\sin\gamma + \cos\alpha\cos\gamma & \sin\beta\sin\gamma \\ -\cos\alpha\sin\beta & -\sin\alpha\sin\beta & \cos\beta \end{bmatrix}$$

This is an orthogonal matrix with determinant +1 and gives the general element of the matrix group $O(3)$.

The full rotation-inversion group $O_I(3)$ has four parameters which may be taken to be $(\alpha, \beta, \gamma, d)$ where α, β, γ are the parameters of $O(3)$ and d denotes the determinant of an element and can take values ± 1. The parameter space of $O_I(3)$ thus consists of two disconnected regions. It therefore is a four-parameter continuous compact group which is, however, not connected. It is also not a Lie group because one of its parameters is discrete.

To find the irreducible representations of $O(3)$ it is necessary to find a set of basis functions which transform into their linear combinations on operating with the elements of $O(3)$. The set of $2l + 1$ spherical harmonics $Y_l^m(\theta, \phi)$, where $l = 0, 1, 2...$ and $-l \leq m \leq l$ transforms in this way on rotating the coordinate system.

Now consider two operations $R_{\boldsymbol{u}}(\alpha)$ and $R_{\boldsymbol{v}}(\alpha)$, assumed to denote rotations through the same angle α about two distinct axes $\boldsymbol{u}$ and $\boldsymbol{v}$. Since there exists in $O(3)$ an operation which can bring the axis $\boldsymbol{u}$ into the axis $\boldsymbol{v}$, $R_{\boldsymbol{u}}(\alpha)$ and $R_{\boldsymbol{v}}(\alpha)$ must belong to the same class, from which follows that in $O(3)$, rotations through a given angle about all axes belong to a class. It is therefore not necessary to know the transformation properties of the spherical harmonics under all rotations. In any representation, the characters of the elements of $O(3)$ depend only on the angle of rotation and not the axis of rotation. The axis of rotation may be chosen as the z axis. The operation of $R_z(\alpha)$ on a spherical harmonic $Y_l^m(\theta, \phi)$ is then known to be

$$R_z(\alpha) Y_l^m(\theta, \phi) = Y_l^m(\theta, \phi - \alpha) = \exp(-im\alpha) Y_l^m(\theta, \phi)$$

The matrix representing $R(\alpha)$ is therefore diagonal and the character follows as

$$\chi^{(l)}(\alpha) = \sum_{m=-l}^{l} e^{im\alpha}.$$

Every rotation of $O(3)$ is characterized by a vector of length equal to the angle of rotation and directed along the direction of rotation. The end-points of all such vectors thus fill a sphere of radius π. Every element of $O(3)$, except those denoting rotations through π, has associated with it a unique point inside the sphere. However, since rotations through π and $-\pi$ about any axis denote the same element, all diametrically opposite points on the surface of the sphere must be identified, *i.e.* these diametrically opposed points must be considered the same. This procedure has important topological consequences for the connectedness of the group $O(3)$.

Consider the elements R_1 and R_2 of $O(3)$. There are two distinct paths connecting the images of R_1 and R_2 in the parameter space:

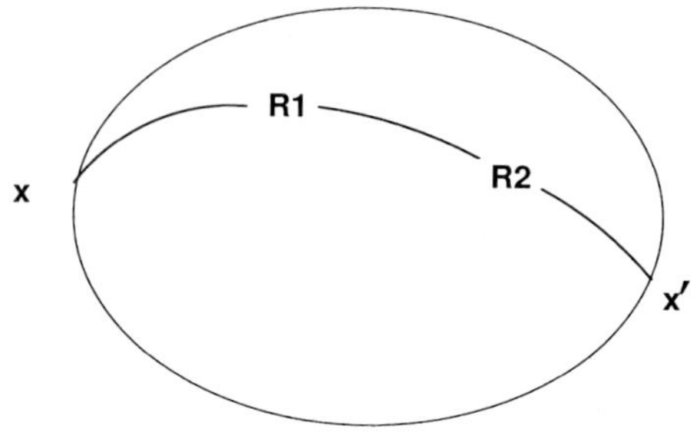

A direct path from R_1 to R_2 and the path from R_1 to x on the surface, cotinuing from the diametrically opposed x' to R_2. The group $O(3)$ is therefore doubly connected.

The group O_n

It should be clear that the set of all real orthogonal matrices of order n with determinants $+1$ constitutes a group. This group is denoted by $O(n)$ and is a continuous, connected, compact, $n(n-1)/2$ parameter[9] Lie group. It can be thought of as the set of all proper rotations in a real n-dimensional vector space. If $x_1, x_2, ..., x_n$ are the orthonormal basis vectors in this space, a transformation of $O(n)$ leaves the quadratic form $\sum_{i=1}^{n} x_i^2$ invariant.

The Special Unitary Group $SU(2)$

Let u and v be a pair of vectors in a two-dimensional vector space defined over the field of complex numbers. A rotation in this space transforms u and

[9] An orthogonal matrix of order n has $n(n-1)/2$ independent elements.

v into their linear combinations

$$u' = au + bv \quad , \quad cu + dv$$

or

$$[u', v'] = [u, v] \begin{bmatrix} a & c \\ b & d \end{bmatrix}$$

where a, b, c and d are complex numbers and hence the transformation matrix involves 8 parameters. By looking only at the rotations that leave the quadratic form

$$uu^* + vv^* = |u|^2 + |v|^2$$

invariant, it turns out that the transformation matrix must be unitary. In other words, the stipulation that

$$|u'|^2 + |v'|^2 = |u|^2 + |v|^2$$

leads to the conditions

$$aa^* + cc^* = 1 \quad , \quad bb^* + dd^* = 1 \quad , \quad ab^* + cd^* = 0$$

Since the scalars are complex, this last condition is equivalent to two conditions. The number of parameters is therefore reduced from 8 to 4. The most general unitary matrix of order two involving four real parameters can be expressed in the form

$$\begin{bmatrix} \cos\theta e^{i\alpha} & \sin\theta e^{i\gamma} \\ -\sin\theta e^{i(\beta-\gamma)} & \cos\theta e^{i(\beta-\alpha)} \end{bmatrix}$$

whose determinant is $\exp(i\beta)$.

The set of all such transformations constitutes the group $U(2)$ which is isomorphic to the group of all unitary matrices of order 2. It is a 4 parameter, continuous, connected, compact, Lie group. The subgroup of $U(2)$ which contains all the unitary matrices of order two with determinant $+1$, is the set of matrices whose general element is

$$\begin{bmatrix} a & -b^* \\ b & -a^* \end{bmatrix} \quad \text{with } aa^* + bb^* = 1.$$

It is known as the *unitary unimodular* group, or the *special unitary group* denoted by $SU(2)$. Because of the extra condition on the determinant, $SU(2)$ is a three-parameter group.

2.8.3 Homomorphism of $SU(2)$ with $O(3)$

For transformations under $SU(2)$ it is required that

$$\begin{array}{lcl} aa^* + bb^* = 1 & , & cc^* + dd^* = 1 \\ a^*c + b^*d = 0 & , & ad - bc = 1 \end{array} \tag{2.26}$$

Solving the last equation for d and substituting into the others give

$$d = -\frac{a^*c}{b^*} \; , \; \frac{-aa^*c}{b^*} - bc = 1 \; , \; c = -b^* \; , \; d = a^*$$

This solution leads to the group $SU(2)$ of transformations

$$\begin{aligned} u' &= au + bv \quad (aa^* + bb^* = 1) \\ v' &= -b^*u + a^*v \end{aligned} \tag{2.27}$$

with three independent real parameters.

The matrices (27) provide one representation of $SU(2)$. Other representations can be constructed by taking symmetric product representations with itself. The transformations of the symmetric products $u^2, uv, v^2 (\equiv x_1, x_2, x_3)$ according to (27) are

$$\begin{aligned} x_1' &= a^2x_1 + 2abx_2 + b^2x_3 \\ x_2' &= -ab^*x_1 + (aa^* - bb^*)x_2 + a^*bx_3 \\ x_3' &= b^{*2}x_1 - 2a^*b^*x_2 + a^{*2}x_3 \end{aligned}$$

By defining new variables $x_1 \pm x_3, x_2$ and setting $x = \frac{1}{2}(x_1 - x_3)$, $y = \frac{1}{2i}(x_1 + x_3)$, $z = x_2$ it can be shown [11] that

$$\begin{aligned} x' &= \frac{1}{2}(a^2 - b^{*2} - b^2 + a^{*2})x - \frac{i}{2}(a^2 - b^{*2} + b^2 - a^{*2})y + (ab + a^*b^*)z \\ y' &= -\frac{i}{2}(a^2 + b^{*2} - b^2 - a^{*2})x + \frac{1}{2}(a^2 + b^{*2} + b^2 + a^{*2})y - i(ab - a^*b^*)z \\ z' &= -(a^*b + ab^*)x + i(a^*b - ab^*)y + (aa^* - bb^*)z \end{aligned}$$

These expressions have real coefficients. By squaring and using (26) the expressions reduce to the invariant

$$x'^2 + y'^2 + z'^2 = x^2 + y^2 + z^2$$

They therefore represent a pure rotation of (x, y, z). In fact, it can be shown that all rotations are associated with unitary transformations in this way.

The rotations are characterized by the Euler angles α, β, γ. Choosing a unitary transformation with $a = \exp i\alpha/2$, $b = 0$, (26) reduces to

$$\begin{aligned} x' &= x\cos\alpha - y\sin\alpha \\ y' &= x\sin\alpha + y\cos\alpha \\ z' &= z \end{aligned}$$

i.e., to the rotation $R(\alpha, 0, 0)$ through angle α about the z-axis:

$$\begin{bmatrix} e^{i\alpha/2} & 0 \\ 0 & e^{-i\alpha/2} \end{bmatrix} \to \begin{bmatrix} \cos\alpha & -\sin\alpha & 0 \\ \sin\alpha & \cos\alpha & 0 \\ 0 & 0 & 1 \end{bmatrix}$$

Choosing $a = \cos\beta/2$, $b = \sin\beta/2$, equation (26) reduces to

$$\begin{aligned} x' &= x\cos\beta + z\sin\beta \\ y' &= y \\ z' &= -x\sin\beta + z\cos\beta \end{aligned}$$

i.e. to the rotation $R(0, \beta, 0)$ through angle β about the y-axis.

Thus the general rotation

$$R(\alpha, \beta, \gamma) = R(0, 0, \gamma)R(0, \beta, 0)R(\alpha, 0, 0)$$

can be associated with the unitary transformation

$$\begin{bmatrix} e^{i\gamma/2} & 0 \\ 0 & e - i\gamma/2 \end{bmatrix} \begin{bmatrix} \cos\beta/2 & \sin\beta/2 \\ -\sin\beta/2 & \cos\beta/2 \end{bmatrix} \begin{bmatrix} e^{i\alpha/2} & 0 \\ 0 & e^{-i\alpha/2} \end{bmatrix}$$

or

$$\begin{bmatrix} \cos(\beta/2)e^{(i/2)(\alpha+\gamma)} & \sin(\beta/2)e^{(i/2)(\gamma-\alpha)} \\ -\sin(\beta/2)e^{(i/2)(\alpha-\gamma)} & \cos(\beta/2)e^{-(i/2)(\alpha+\gamma)} \end{bmatrix} \to R(\alpha, \beta, \gamma) \qquad (2.28)$$

This result amounts to a 1 to 2 homomorphic mapping of the unitary group $SU(2)$ onto the rotation group. From (28) it follows that the two unitary matrices

$$I = \begin{bmatrix} 1 & 0 \\ 0 & 1 \end{bmatrix} \quad \text{and} \quad -I = \begin{bmatrix} -1 & 0 \\ 0 & -1 \end{bmatrix}$$

are mapped into the identity element of the rotation group. These two elements therefore form an invariant subgroup of $SU(2)$, and their products with any element of $SU(2)$ are mapped into the same element of the rotation group. So each element of the rotation group is associated with a pair of elements of the unitary group which differ merely in a change of sign of all their coefficients. The group manifold of $SU(2)$ is in one-to-one correspondence with the points

$$|a|^2 + |b|^2 = 1$$

of the surface of a sphere in a four-dimensional space. Since the sphere is simply connected the group manifold of $SU(2)$ is simply connected. $SU(2)$ has only single-valued representations and is the universal covering of $O^+(3)$.

Summary

The most commonly encountered Lie groups, using slightly different notation, are [16]:

(i) The orthogonal groups O(n). These are the $n \times n$ matrices with real entries such that the inverse is the transpose

$$O(n) = \{A : \tilde{A} = A^{-1}\}$$

(ii) The special orthogonal groups SO(n).
These are the subgroups of O(n) with determinant one

$$SO(n) = \{A : \tilde{A} = A^{-1} \text{ and } \det A = 1\}$$

(iii) The unitary groups U(n).
These are the $n \times n$ matrices with complex entries such that the inverse is the conjugate transpose. That is

$$U(n) = \{A : \widetilde{A^\dagger} = A^{-1}\}$$

(iv) The special unitary groups SU(n).
These are subgroups of U(n) with determinant one. That is

$$SU(n) = \{A : \widetilde{A^\dagger} = A^{-1} \text{ and } \det A = 1\}$$

(v) The symplectic group Sp(n).
The group Sp(n) consists of $n \times n$ matrices with quaternion entries such that the inverse is the conjugate transpose

$$Sp(n) = \{A : \widetilde{A^\dagger} = A^{-1}\}$$

Conjugation of a quaternion is given by

$$\overline{w + ix + jy + kz} = w - ix - jy - kz$$

Chapter 3

Particles and Waves

3.1 Introduction

Chemistry is the science that deals with the interaction between different forms of matter to bring about a change in the nature of the interacting matter. According to the nuclear model of the atom it is evident that chemical interaction is due to interacting electron clouds. In order to understand chemical interactions it is therefore necessary to understand, in the first place the behaviour of electrons.

It is not obvious how a stable nuclear atom can be rationalized in terms of a static charge distribution in a Coulomb field, where the force between two charges is given by

$$F = \frac{q_1 q_2}{4\pi\varepsilon_o r^2}$$

It is necessary to postulate a dynamic charge distribution as in the well-known, but unrealistic planetary model of the atom. A stable electronic orbit can only be maintained by a constantly accelerated electron, which according to the principles of electrodynamics constitutes a source of radiation. The stability of the atom can simply not be accounted for in terms of classical mechanics. A radically different description of electronic behaviour is required. As a matter of fact, a radically different system of mechanics is required to describe electronic motion correctly and this is where a theoretical understanding of chemistry must start.

Classical mechanics which correctly describes the behaviour of macroscopic particles like bullets or space craft is not derived from more basic principles. It derives from the three laws of motion proposed by Newton. The only justification for this model is the fact that a logical mathematical development of a mechanical system, based on these laws, is fully consistent with observation, at both terrestrial and cosmic levels, with two important exceptions:

(i) at velocities approaching the speed of light;
(ii) for sub-atomic particles like electrons.

These exceptions do not invalidate the system of mechanics. It simply shows the need for refinement of the system in order to deal correctly with these esoteric cases.

Particles moving at velocities approaching the speed of light are better described by relativistic mechanics. At moderate velocities, this mechanics which is based on the postulate that the velocity of light, c cannot be exceeded, reduces to the more familiar system of Newtonian mechanics. In the same way one expects the mechanics that describes the motion of sub-atomic particles reduces to the familiar form of mechanics for more massive particles.

The reason why classical mechanics is inadequate at the sub-atomic level is because it ignores quantum effects. Not only matter, but also radiation is discontinuous at the sub-microscopic level. Matter occurs as separate atoms and radiation as small packets, quanta or photons. At this level the two descriptions become identical - matter behaves like radiation and radiation behaves like matter. This is appropriate since matter and energy are interconvertible according to the relativistic relationship, $E = mc^2$. Wave and particle theories must therefore merge at the quantum level.

Macroscopic descriptions of matter and radiation are adequate without taking the discontinuous nature of matter and/or radiation into account. However, when dealing with particles approaching the size of elementary quanta, the quantum effects become increasingly important and must be taken into account explicitly in the mechanical description of these particles. Unlike relativistic mechanics, quantum mechanics cannot be used to describe macroscopic events. There is a fundamental difference between classical and non-classical, or quantum, phenomena and the two systems are complimentary rather than alternatives.

An electron is a particle of such small dimensions that its mechanical behaviour can be correctly described only by medium of quantum mechanics. The aim of this discussion is therefore to introduce quantum mechanics and to investigate the mechanical behaviour of electrons in situations with implications for chemistry. Since quantum and classical mechanics yield the same results in the so-called classical limit, the more complicated quantum system can be developed by analogy with the simpler classical system.

The short introduction that follows is no more than a brief summary of some aspects of classical mechanics described in greater detail elsewhere

[17, 5].

3.2 Review of Classical Dynamics

The laws of classical dynamics were first formulated by Newton. The first law states that any particle will persist in its state of uniform unaccelerated motion unless it is acted upon by a force. Using the notation x_i, y_i, z_i for the cartesian coordinates of the ith point particle, of mass m_i, Newton's equations for n point particles are

$$\left.\begin{array}{rcl} m_i\frac{d^2x_i}{dt^2} = m_i\ddot{x}_i & = & X_i \\ m_i\ddot{y}_i & = & Y_i \\ m_i\ddot{z}_i & = & Z_i \end{array}\right\} \quad i = 1, 2, \cdots, n$$

where X_i, Y_i and Z_i are the three components of force acting on the ith particle.

This can be cast into a more useful form by introducing the concept of kinetic energy

$$T_i = \frac{1}{2}m_i(\dot{x}_i^2 + \dot{y}_i^2 + \dot{z}_i^2) = \frac{1}{2}m_i v_i^2$$

$$T = \frac{1}{2}\sum_{i=1}^{n} m_i(\dot{x}_i^2 + \dot{y}_i^2 + \dot{z}_i^2)$$

Note that $\partial T/\partial \dot{x}_i = m_i\dot{x}_i$ (momentum)

$$\frac{d}{dt}\left(\frac{\partial T}{\partial \dot{x}_i}\right) = m_i\ddot{x}_i = X_i$$

with similar expressions for Y_i and Z_i.

Expressed in words, the rate of change of momentum represents a force. This is a statement of Newton's second law. The third law states that any action is balanced by a reaction, which implies conservation of momentum.

For conservative systems, it is possible to define another quantity, the potential energy V, which is a function of the coordinates x_i, y_i, z_i of all particles ($i = 1 \longrightarrow n$). The force components acting on each particle are equal to the negative partial derivatives of the potential energy with respect to the coordinates

$$\left.\begin{array}{rcl} X_i & = & -\partial V/\partial x_i \\ Y_i & = & -\partial V/\partial y_i \\ Z_i & = & -\partial V/\partial z_i \end{array}\right\} \quad i = 1, 2, \cdots, n$$

This formalism applies to mechanical, gravitational and electrostatic forces, e.g.

$$V_e = -\frac{q_1 q_2}{4\pi\varepsilon_o r}$$

With these definitions Newton's equations become

$$\left.\begin{aligned} \frac{d}{dt}\left(\frac{\partial T}{\partial \dot{x}_i}\right) + \frac{\partial V}{\partial x_i} &= 0 \\ \frac{d}{dt}\left(\frac{\partial T}{\partial \dot{y}_i}\right) + \frac{\partial V}{\partial y_i} &= 0 \\ \frac{d}{dt}\left(\frac{\partial T}{\partial \dot{z}_i}\right) + \frac{\partial V}{\partial z_i} &= 0 \end{aligned}\right\} \quad i = 1, 2, \cdots, n$$

These equations are restricted to cartesian coordinates and for many problems in atomic physics it is more appropriate to use different coordinate systems. It is therefore important to note that these equations can be turned into a form independent of coordinate system by the introduction of a new function.

3.3 Hamilton's Principle

The *Lagrangian* function is defined as the difference between the kinetic and potential energies

$$L = L(x_1, y_1, z_1, x_2, y_2, z_2, \cdots, \dot{x}_1, \cdots, \dot{z}_n) = T - V$$

and is also referred to as the kinetic potential. Since T is a function of velocities only and V is a function of coordinates only

$$\frac{\partial L}{\partial \dot{x}_i} = \frac{\partial T}{\partial \dot{x}_i} \quad \text{and} \quad \frac{\partial L}{\partial x_i} = -\frac{\partial V}{\partial x_i}$$

The equations of motion therefore take the form

$$\frac{d}{dt}\left(\frac{\partial L}{\partial \dot{x}_i}\right) - \frac{\partial L}{\partial x_i} = 0 \quad \text{, i.e.} \quad \frac{d}{dt}(p_x) = \frac{\partial L}{\partial x_i}$$

with similar expressions for y_i and z_i.

The Lagrangian equations contain nothing more than the original Newtonian equations, but have the advantage that the coordinates may be of any kind whatever. This is of particular importance when analyzing phenomena in which the motion of material particles is not observed directly, such as problems in electricity. Parameters, other than particle positions (*e.g.* currents) are observed here, although the behaviour of the system is really controlled by the motion of particles (*e.g.* electrons) which remain intrinsically concealed. The values assumed by the descriptive parameters, called *generalized coordinates*, must then be connected with the position coordinates of the particles. This formulation gains added importance in quantum mechanics where particle position is an ill-defined parameter.

It is assumed that the rectangular coordinates of a set of particles may be defined in terms of $3n$ generalized coordinates, *i.e.*

$$x_j = x_j(q_1 \cdots q_n) \quad ; \quad y_j = y_j(q_1 \cdots q_n) \quad ; \quad z_j = z_j(q_1 \cdots q_n)$$

Similarly the velocity components of the jth particle will be

$$\dot{x}_j = \sum_{k=1}^{n} \frac{\partial x_j}{\partial q_k}\dot{q}_k \tag{3.1}$$

etc., where the $\dot{q}_k$ are generalized velocities. Now the kinetic energy of the system will by assumption be the total kinetic energy of the particles, *viz.*,

$$T = \frac{1}{2}\sum m_j(\dot{x}_j^2 + \dot{y}_j^2 + \dot{z}_j^2)$$

where the summation is over all particles in the system. Substituting from (1) this expression may be reduced to

$$T = \sum a_{ik}\dot{q}_i\dot{q}_k$$

with double summation over $i, k = 1 \cdots n$ and the a_{ik} are functions of the q's.

To develop a system of mechanics from here without the introduction of any other concepts, apart from energy, some general principle that predicts the course of a mechanical change is required. This could be like the Maupertuis principle of least action or Fermat's principle of least time. It means that the actual path of the change will have an extreme value (*e.g.* minimum) of either action or time, compared to all other possible paths. Based on considerations like these Hamilton formulated the principle that the action integral

$$S = \int_{t_o}^{t_1} L(x, \dot{x})dt \tag{3.2}$$

of the Lagrangian assumes an extreme value. Since

$$\int_{t_o}^{t_1} Ldt = \int_{t_o}^{t_1} Tdt - \int_{t_o}^{t_1} Vdt$$

Hamilton's principle is equivalent to the statement that for the actual motion the average kinetic energy approaches the average potential energy as closely as possible. If the potential energy is a function of position only the equivalent mathematical statement is

$$\delta \int_{t_o}^{t_1} \{T(q,\dot{q}) - V(q)\} dt = 0$$

i.e.

$$\int_{t_o}^{t_1} \left\{ \sum_{j=1}^{n} \left(\frac{\partial L}{\partial q_j} \delta q_j + \frac{\partial L}{\partial \dot{q}_j} \frac{d}{dt}(\delta q_j) \right) \right\} dt = 0 \tag{3.3}$$

Integration by parts[1] of the second term in the integral yields

$$\int_{t_o}^{t_1} \sum_{j=1}^{n} \frac{\partial L}{\partial \dot{q}_j} \frac{d}{dt}(\delta q_j) dt = \sum_{j=1}^{n} \frac{\partial L}{\partial \dot{q}_j} \delta q_j \Bigg]_{t_o}^{t_1} - \int_{t_o}^{t_1} \sum_{j=1}^{n} \frac{d}{dt}\left(\frac{\partial L}{\partial \dot{q}_j}\right) \delta q_j dt$$

Since $\delta q_j = 0$ for all j at $t = t_o$ and at $t = t_1$ the integrated part vanishes and (3) reduces to

$$\int_{t_o}^{t_1} \sum_{j=1}^{n} \left\{ \frac{\partial L}{\partial q_j} - \frac{d}{dt}\left(\frac{\partial L}{\partial \dot{q}_j}\right) \right\} \delta q_j dt = 0$$

Since δq_j is arbitrary in the interval from t_o to t_1 the term in curly brackets should vanish, leading to the set of equations

$$\frac{d}{dt}\left(\frac{\partial L}{\partial \dot{q}_j}\right) - \frac{\partial L}{\partial q_j} = 0 \quad , \qquad j = 1, 2, 3, \cdots, 3n$$

It is necessary only to write the potential and the kinetic energies in the desired coordinates, to obtain the equations of motion by simple differentiation.

3.3.1 Lagrangian Density

The definition of a Lagrangian function can be generalized to a system with many particles and eventually also to a field that represents a continuously

[1]The expression for integrating by parts is

$$\int u dv = uv - \int v du.$$

varying energy density. For a separation distance a between the equilibrium positions of two neighbouring interacting particles and a Lagrangian density of $\mathcal{L}_i$ per unit length, the Lagrangian becomes

$$L = \sum_i^N a\mathcal{L}_i$$

Passing from a discrete system to a continuous system with infinite degrees of freedom and infinitesimal separation distance,

$$L = \int \mathcal{L} d^3x$$

In terms of a four-dimensional field, ϕ the action integral becomes

$$S = \int \mathcal{L}(\phi, d_\mu\phi) d^4x,$$

where $d_\mu\phi = \partial\phi/\partial x^\mu$, $\mu = 1, 4$.

In field theory the Lagrangian density is referred to simply as the Lagrangian, using the same symbol L. The variation of L is written explicitly as

$$\delta L = \frac{\partial L}{\partial \phi}\delta\phi + \frac{\partial L}{\partial_\mu \phi}\delta(\partial_{\mu)}\phi$$

The coordinates are not varied; δL is due entirely to the variation of the field quantities. The variation can be rewritten in the form

$$\delta L = \left(\frac{\partial L}{\partial \phi} - \partial_\mu \frac{\partial L}{\partial_\mu \phi}\right)\delta\phi + \partial_\mu\left(\frac{\partial L}{\partial_\mu \phi}\delta\phi\right)$$

The volume integral of the last term, when transformed into a 3D surface integral by the theorem of Gauss, is required to vanish in the boundary surface [18]. The condition for stationary action thereby becomes

$$\frac{\partial L}{\partial \phi} - \frac{\partial}{\partial x^\mu}\left[\frac{\partial L}{\partial(\partial_\mu \phi)}\right] = 0$$

known as the Euler-Lagrange equation for ϕ.

3.3.2 The Hamiltonian Function

The Lagrangian equations can be turned into another useful form involving generalized coordinates and momenta and by defining the *Hamiltonian* function

$$H = \sum_{k=1}^{3n} p_k\dot{q}_k - L(q_k, \dot{q}_k)$$

where $p_k = m_k\dot{x}_k$, which, since V is restricted to be a function of the coordinates only, can be written as

$$p_k = \frac{\partial T}{\partial \dot{x}_k} = \frac{\partial L}{\partial \dot{x}_k}\ ,\ k = 1, 2, \cdots, 3n$$

From Lagrange's equations it therefore follows that

$$\dot{p}_k = \frac{\partial L}{\partial q_k}\ ,\ k = 1, 2, \cdots, 3n$$

It follows directly that

$$\left.\begin{array}{lllll} \frac{\partial H}{\partial p_k} & = & \dot{q}_i & & \\ \frac{\partial H}{\partial q_k} & = & -\frac{\partial L}{\partial q_k} & = & -\dot{p}_k \end{array}\right\}\ k = 1, 2, \cdots, 3n$$

These are the Hamiltonian or canonical form of the equations of motion. The advantage of the Hamiltonian equations over the Lagrangian is that they contain $6n$ partial differential equations of the first order rather than $3n$ of the second order.

The Hamiltonian function, H is a constant of the motion, since

$$\frac{dH}{dt} = \sum_j \left(\frac{\partial H}{\partial q_j}\right)\left(\frac{\partial q_i}{\partial t}\right) + \left(\frac{\partial H}{\partial p_j}\right)\left(\frac{\partial p_j}{\partial t}\right)$$

$$= \sum_j \left(\frac{\partial H}{\partial q_j}\right)\left(\frac{\partial H}{\partial p_k}\right) - \left(\frac{\partial H}{\partial p_j}\right)\left(\frac{\partial H}{\partial q_j}\right) = 0$$

Since the kinetic energy is a homogeneous quadratic function of velocities it may be formulated as

$$T = \sum_{i,j=1}^{3n} a_{ij}\dot{q}_i\dot{q}_j$$

where the a_{ij} may be functions of the coordinates. It follows that

$$p_k = \frac{\partial L}{\partial \dot{q}_k} = \frac{\partial T}{\partial \dot{q}_k} = \sum_j a_{kj}\dot{q}_i + \sum_i a_{ik}\dot{q}_i$$

and hence that

$$\sum_k p_k\dot{q}_k = \sum_{k,j} a_{kj}\dot{q}_i\dot{q}_k + \sum_{i,k} a_{ik}\dot{q}_i\dot{q}_k = 2T$$

The Hamiltonian becomes

$$H = \sum_k p_k \dot{q}_k - L = 2T - (T - V) = T + V = E \ ,$$

the total energy of the system. For a particle of mass m

$$T = \frac{1}{2} m \boldsymbol{v}^2 = \frac{(m\boldsymbol{v})^2}{2m} = \frac{\boldsymbol{p}^2}{2m} \text{ and } H = \frac{\boldsymbol{p}^2}{2m} + V$$

3.4 Hamilton-Jacobi Theory

Another situation of interest concerns a time-dependent Hamiltonian, i.e. which transforms from $(q_0, \boldsymbol{p}_0)$ at t_0 to $(q, \boldsymbol{p})$ at t.

$$q = q(q_0, \boldsymbol{p}_0, t)$$

$$p = \boldsymbol{p}(q_0, \boldsymbol{p}_0, t)$$

The transformed Hamiltonian can be written as

$$K = H + \frac{\partial F}{\partial t}$$

where F is some generating function. The canonical equations in K are

$$\frac{\partial K}{\partial P_i} = \dot{Q}_i$$

$$-\frac{\partial K}{\partial Q_i} = \dot{P}_i$$

If the same equation is used to describe the inverse transformation, the new variables (starting parameters) are constant in time ($\dot{Q}_i = \dot{P}_i = 0$) which requires $K = 0$, and

$$H(q, \boldsymbol{p}, t) + \frac{\partial F}{\partial t} = 0$$

$$\left[\frac{dF(q, \boldsymbol{P}, t)}{dt} = \sum \frac{\partial F}{\partial q_i}\dot{q}_i + \sum \frac{\partial F}{\partial P_i}\dot{P}_i + \frac{\partial F}{\partial t}\right]$$

$$\sum(p_i \dot{q}_i) - L + \frac{dF}{dt} - \sum\left(\frac{\partial F}{\partial q_i}\dot{q}_i\right) - \sum\left(\frac{\partial F}{\partial P_i}\dot{P}_i\right) = 0$$

Since $\dot{\boldsymbol{P}} = 0$, and since the coefficient of each term is required to vanish independently, it follows, *inter alia* , that $p_i = \frac{\partial F}{\partial q_i}$, $\frac{dF}{dt} = L$.

The equation

$$H(q_i, \frac{\partial F}{\partial q_i}, t) + \frac{\partial F}{\partial t} = 0$$

is known as the *Hamilton-Jacobi* (HJ) equation, and its solution S is known as *Hamilton's Principal Function.*

Since $$H = \frac{\boldsymbol{p}^2}{2m} + V$$ it follows that

$$\frac{1}{2m}\left(\frac{\partial S}{\partial q}\right)^2 + V + \frac{\partial S}{\partial t} = 0 \tag{3.4}$$

which is one common form of the HJ equation.

The principal function S is the same quantity as the one defined by Hamilton's principle as in equation (2). It has the advantage of dealing with the scalars T and V only, whereas transformations of force components using Newton's laws deal with more complicated vector quantities.

The complete time derivative of the action, S

$$\frac{dS}{dt} = \frac{\partial S}{\partial t} + \frac{\partial S}{\partial q}\frac{\partial q}{\partial t} = L$$

from which it can be shown that

$$-\frac{\partial S}{\partial t} = H \ ,$$

which is the Hamiltonian of the system. In integrated form it follows that $S = -Et+W$, where $W(p,q)$ is known as Hamilton's *characteristic function.* It now becomes apparent why S has been called the action of the system. It represents the energy transferred to the system over a period of time. Substituted into the HJ equation it is seen to be the same as Hamilton's principal function, and the HJ equation can be written in the form

$$\left(\frac{\partial S}{\partial q}\right)^2 = 2m(E-V) \tag{3.5}$$

At any given instant the equation $S(\mathbf{x}, t) = const.$ defines a surface in Euclidean space. As t varies the surface traces out a volume. At each point of the moving surface the gradient, ∇S is orthogonal to the surface. In the case of an external scalar potential the particle trajectories associated with S are given by the solutions $m\dot{\mathbf{x}} = \nabla S$. It follows that the mechanical paths of a moving point are perpendicular to the surface $S = c$ for all $\mathbf{x}$ and t. A family of trajectories is therefore obtained by constructing the normals to a set of surfaces, each orbit being distinguished by its starting point $\mathbf{x_o}$. For a single starting point the moving surface describes a spherical wavefront. This intimate connection between wave and particle formalisms is of special interest and has been used to forge a link between geometrical and wave optics. By analogy to this it was also used by Schrödinger in his original formulation of a wave equation for quantum particles.

3.5 Potential Theory

The concept of potential energy in mechanics is one example of a scalar field, defined by a simple number that represents a single function of space and time. Other examples include the displacement of a string or a membrane from equilibrium; the density, pressure and temperature of a fluid; electromagnetic, electrochemical, gravitational and chemical potentials. All of these fields have the property of *invariance* under a transformation of space coordinates. The numerical value of the field at a point is the same, no matter how or in what form the coordinates of the point are expressed.

The surfaces defined by ϕ = constant, where ϕ is the scalar field, and resembling contour lines on a topographic map, may be called *isotimic* surfaces. In potential theory they are referred to as equipotentials; in heat conduction they are isothermals, *etc.* They form a family of non-intersecting surfaces. The gradient of the scalar field measures the rapidity with which the field changes as a function of position. The most rapid change occurs along a family of lines normal to the isotimic surfaces.

An extremely important property of a scalar field is expressed in terms of its second derivatives. In the sense that a force

$$X = -\frac{\partial V}{\partial x} \quad \text{and hence} \quad \frac{\partial X}{\partial x} = -\frac{\partial^2 V}{\partial x^2},$$

Laplace's equation, $\nabla^2 V = 0$, in any number of dimensions, describes a system of balanced forces in a potential field. The equation is satisfied by a variety of functions, such as

(i) Gravitational potential in regions not occupied by attracting matter;
(ii) Electrostatic potential in a uniform dielectric;
(iii) Magnetic potential in free space, in the theory of magnetostatics;
(iv) Electric potential in the theory of steady flow of electric currents in solid coductors;
(v) Temperature, in the theory of thermal equilibrium in solids;
(vi) Velocity potential at points of a homogeneous liquid in hydrodynamic flow;

(vii) Chemical potential during phase transition or at reaction equilibrium;
(viii) Electrochemical steady states.

When a disturbance occurs in any of these systems an extra term is required to describe the resulting propagation, *e.g.* waves

$$\nabla^2 V = \frac{1}{c^2}\frac{d^2V}{dt^2}$$

or heat flow $$\nabla^2 V = \frac{1}{k}\frac{dV}{dt}.$$

In other cases the scalar field is affected by a source function $q(x, y, z)$ according to

$$\nabla^2 V = -q$$

known as Poisson's equation.

The solutions of Laplace's equation are known as *harmonic functions.* In one dimension the equation

$$\frac{d^2V}{dx^2} = 0$$

has the general solution $V = mx + b$, the formula of a straight line. The two arbitrary constants are fixed by the boundary conditions of any specific case. This simple solution has two obvious properties, less obvious, but more important for solutions in more dimensions. Firstly, $V(x)$ is the average of $V(x+R)$ and $V(x-R)$ for any R. The value of V at x is the average of the values of V on the left and the right of x. Secondly, there are no local maxima or minima and extreme values of V only occur at the end points[2]. A physical example of a system that obeys the one-dimensional Laplace equation is a rubber band stretched between two pins along a flat surface.

In two dimensions Laplace's equation becomes

$$\frac{\partial^2 V}{\partial x^2} + \frac{\partial^2 V}{\partial y^2} = 0$$

As a general physical example one may consider a flexible membrane like a thin rubber sheet or a soap film stretched over an irregular support, such as a distorted ring.

[2]This is not a mere consequence of the fact that the second derivative is zero. The second derivative of $y = x^4$ vanishes at $x = 0$ where y has a minimum.

The z-coordinates of points in the resulting surface projected on an $x-y$ Cartesian grid satisfy the 2D equation. The average value of V, *i.e.* z, on a circle of radius R will now be equal to V at the central point x, y

$$V(x,y) = \frac{1}{2\pi R}\oint V dl.$$

Again, V has no local maxima or minima and all extrema occur at the boundaries. Geometrically, just as the straight line is the shortest distance between two points, so a harmonic function in two dimensions minimizes the surface area fitted to the given boundary line.

Laplace's equation in three dimensions solved by separation of variables

$$V(r,\theta,\phi) = R(r)\cdot y(\theta)\cdot z(\phi)$$

and substituting $\cos\theta = x$, reduces to

$$\frac{d}{dr}\left(r^2\frac{dR}{dr}\right) - l(l+1)R = 0$$

$$(1-x^2)\frac{d^2y}{dx^2} - 2x\frac{dy}{dx} + \left[l(l+1) - \frac{m^2}{1-x^2}\right]y = 0$$

$$\frac{d^2z}{d\phi^2} + m^2 z = 0$$

where $l(l+1)$ and m^2 are separation constants. The solution of the radial equation that remains finite at infinity is

$$R(r) = \frac{A}{r^{l+1}}$$

as can be verified by direct substitution. The second equation is the associated Legendre equation (1.23) which has well-behaved solutions $P_l^m(\cos\theta)$ only if l is a positive integer and $l \geq |m|$. For any l there are $2l+1$ different modes, $P_l^m(\cos\theta)z_m(\phi)$ known as surface spherical harmonics. The general solution therefore is

$$\begin{aligned} V = \frac{A_0}{r} &+ \frac{1}{r^2}(\text{sum of three } l_1 \text{ modes}) \\ &+ \frac{1}{r^3}(\text{sum of five } l_2 \text{ modes}) \\ &+ \frac{1}{r^4}(\text{sum of seven } l_3 \text{ modes}) + \ldots \end{aligned}$$

In electrostatics the first term represents the contribution to the field due to the net charge, and higher terms represent dipole, quadrupole, octupole, *etc.*

contributions. The most celebrated solution however, was Laplace's analysis of the tidal wave motion of a flooded planet[3]. For the purpose of this work the appearance of surface spherical harmonics in quantum-mechanical problems is more important. These applications will be examined in more detail later on. However, it is important to note that the two special properties of harmonic functions are satisfied in three dimensions.

As before, the value of V at point P, now is the average value of V over a spherical surface of radius R, centred at P. As a consequence V can have no local maxima or minima; the extreme values of V must occur at the boundaries. By way of illustration the average potential over a spherical surface surrounding a point P, distant r from a charge q, is used to calculate a well-known result.

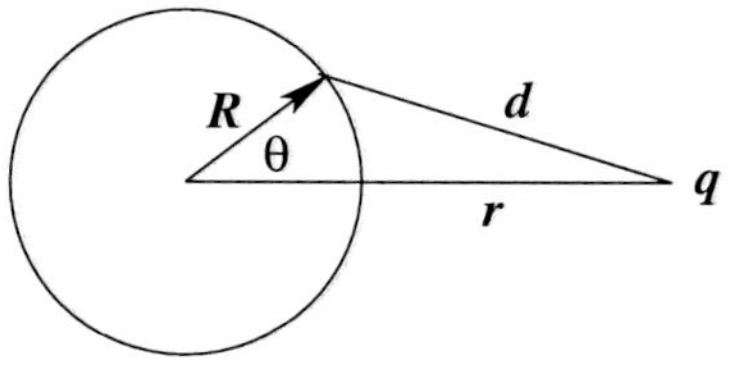

$$V_{av} = \frac{1}{4\pi R^2}\int_{\text{sphere}} V\,da$$

For any point on the surface $V = q/(4\pi\epsilon_0 d)$ and

$$\begin{aligned}
d^2 &= r^2 + R^2 - 2rR\cos\theta \\
V_{av} &= \frac{1}{4\pi R^2}\cdot\frac{q}{4\pi\epsilon_0}\int_S [r^2 + R^2 - 2rR\cos\theta]^{-1/2} R^2 \sin\theta d\theta d\phi \\
&= \frac{1}{4\pi}\cdot\frac{q}{4\pi\epsilon_0}\int_0^{2\pi} d\phi \int_0^{\pi} d\theta [r^2 + R^2 - 2rR\cos\theta]^{-1/2}\sin\theta \\
&= \frac{q}{8\pi\epsilon_0}\left[\frac{\sin\theta}{rR\sin\theta}(r^2 + R^2 - 2rR\cos\theta)^{1/2}\right]_0^{\pi} \\
&= \frac{q}{4\pi\epsilon_0}
\end{aligned}$$

which is recognized as the potential due to q at the centre of the sphere.

[3]It is interesting to note that the action of the moon on the oceans is such as to excite vibrations approximating the l_2 modes with a phase difference of a quarter cycle between them. This is why the period of the tides is 12 hours although the earth rotates under the moon with a period of 24 hours.

3.6 Wave Motion

An intuitive analysis of motion reveals two fundamental components of linear and periodic motion respectively. Linear motion only depends on the connectivity or *affine* structure of space and proceeds along its straight lines or geodesics. It may include accelerated motion by suitable modification of the differential affine geometry, but not rotational motion. The distorting effects experienced by a rotating object seem to depend directly on its relationship to the rest of the matter in the universe, as first envisioned by Mach [19]. Since it is highly unlikely that any actual physical object has exactly zero net external force acting upon it, pure linear motion seems to be equally unlikely. Whereas the combination of linear and rotational motion corresponds to the notion of wave motion, the previous conclusion seems to indicate that all motion is wavelike. Some common phenomena readily recognised as wave motion [20] include sound or acoustic waves, the electromagnetic waves of sight and radio communication and gravity induced waves in the ocean.

Because of its oscillatory component wave motion requires a related, but more complicated description than linear motion. The methods of particle mechanics use vectors to describe displacements, velocities and other quantities of motion in terms of orthogonal unit vectors, *e.g.*

$$\boldsymbol{F} = \boldsymbol{i}F_x + \boldsymbol{j}F_j + \boldsymbol{k}F_z$$

Vectors in N dimensions are said to be orthogonal when

$$\boldsymbol{A} \cdot \boldsymbol{B} = \sum_{i=1}^{N} A_i B_i = 0.$$

If, in a vector space of an infinite number of dimensions the components A_i and B_i become continuously distributed and everywhere dense, i is no longer a denumerable index but a continuous variable (x) and the scalar product turns into an overlap integral $\int A(x)B(x)dx$. If it is zero the functions A and B are said to be orthogonal. This type of function is more suitable for describing wave motion.

For the sake of mathematical simplicity wave motion was introduced before (1.3.2) as a one-dimensional problem. One of the most obvious features of waves, however, is their tendency to spread in all directions from the source of a disturbance. To represent such a wave in three dimensions requires two parameters, t and r, the distance from the source to any point in the medium, in a form like $f(r - vt)$, in which v is the velocity at which the disturbance spreads, and the phase is $r - vt$. As the disturbance spreads it increases in volume and the amplitude should diminish to ensure that the total disturbance does not effectively grow spontaneously. A more appropriate wave function would therefore be[4] $(1/r)f(r \pm vt)$. At any given instant t_0 the disturbance has the value $(1/r_0)f(r_0 - vt_0)$ at all points on the surface of the sphere radius r_0 about the source as centre. This surface in which the disturbance has constant phase is called the *wavefront.* The curves normal to the surface at every point of the wavefront are called *rays.*

There is another simple type of wavefront in three dimensions where the wave propagation is restricted to one direction. The disturbance at a given instant is now the same at all points of a plane perpendicular to this direction. This is called a *plane wave.* For both spherical and plane waves the propagation is described as

$$g[\phi(x, y, z)]f[\phi(x, y, z) - vt].$$

At any arbitrary instant $t = t_0$, the argument of the f-function is constant over the surface

$$\phi(x, y, z) = vt_0 + C,$$

where C is an arbitrary constant. Hence the f function will itself be constant over this surface. The same is true for $g(\phi)$. The surface $\phi(x, y, z)$ is the wavefront. For a spherical wave $\phi(x, y, z) = \sqrt{x^2 + y^2 + z^2}$, reduces to a sphere. For a general plane wave $\phi(x, y, z) = \alpha x + \beta y + \gamma z$, where α/D, β/D, γ/D are the direction cosines of the normal to the plane, $D = \sqrt{\alpha^2 + \beta^2 + \gamma^2}$. In a special case like $\alpha = \beta = 0, \gamma = 1$ the propagation takes place in a plane wave in which the wavefronts are perpendicular to z. A ray describes the direction of propagation at any point.

---

[4]In polar coordinates the radial part of ∇^2 reduces to $\frac{\partial^2}{\partial r^2} + \frac{2}{r}\frac{\partial}{\partial r}$ and the wave equation reads

$$\frac{v^2}{r}\frac{\partial^2(rU)}{\partial r^2} - \frac{\partial^2 U}{\partial t^2} = 0$$

The substitution $\eta = r \pm vt$, $rU = P$ converts it into $v^2\, d^2P/d\eta^2 - v^2\, d^2P/d\eta^2 = 0$, hence

$$U = \frac{1}{r}[f_1(r + vt) + f_2(r - vt)]$$

which represents a pair of travelling waves, moving into and out from the origin respectively. The factor $1/r$ accounts for the attenuation of the spherical wave as it moves out from its source.

3.6.1 Harmonic Waves

The simplest form of a moving wave disturbance is a pure sine or cosine function, periodic in both space and time.

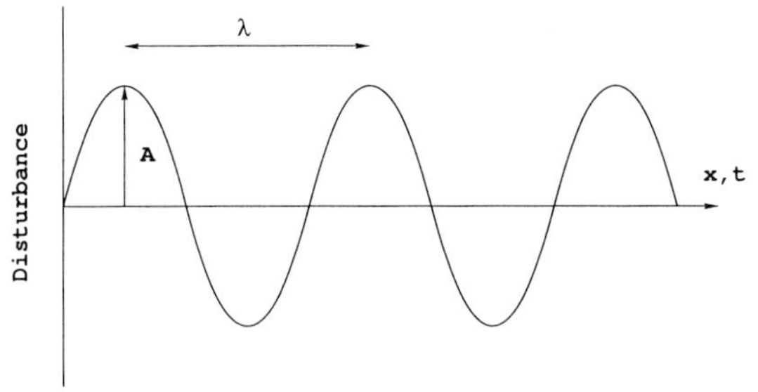

Periodicity in space means that it repeats at regular intervals, known as the wavelength, λ. Periodicity in time means that it moves past a fixed point at a steady rate characterised by the period τ, which counts the crests passing per unit time. By definition, the velocity $v = \lambda/\tau$. It is custom to use the reciprocals of wavelength $1/\lambda = (k/2\pi)$ or $\bar{\nu}$, known as the wavenumber (k = wave vector) and $1/\tau = \nu$, the frequency, or angular frequency $\omega = 2\pi\nu$. Since a sine or cosine (harmonic) wave repeats at intervals of 2π, it can be described in terms of the function

$$U = A\sin\left(\frac{x}{\lambda}\cdot 2\pi - \frac{t}{\tau}\cdot 2\pi\right) = A\sin(kx - \omega t) = S(x)\sin\omega t$$

For a stationary standing wave $S(x) = A\sin(kx + \delta)$, where A and δ are arbitrary constants, and $S(x)$ is a general solution of the space part of the one-dimensional wave equation

$$\frac{d^2S}{dx^2} + \frac{\omega^2}{v^2}S = 0$$

i.e.

$$\frac{d^2S}{dx^2} + k^2S = 0$$

an equation describing harmonic motion.

For a standing wave on a string of length l it is necessary to impose the boundary conditions $S(0) = 0$, $S(l) = 0$. A trivial solution would be $A = 0$, but this means $S = 0$, everywhere. Adjustment of δ can never satisfy both boundary conditions that hence require $\delta = 0$ together with the special values of $k = n\pi/l$, (integral l), known as eigenvalues. To each eigenvalue there corresponds an eigenfunction $S_n = A_n \sin n\pi x/l$. The eigenvalues of $\lambda = 2l/n$ are the well-known wavelenghts of standing waves on a vibrating string.

3.6.2 Fourier Series

The eigenfunctions of the vibrating string can be shown to have the property of orthogonality, as anticipated, *i.e.*

$$\int_0^l S_n(x)S_m(x)dx = c_n\delta_{nm}$$

This is verified by substitution of the S functions which shows that

$$c_n = A_n^2\int_0^l \sin^2\frac{n\pi}{l}xdx = A_n^2\frac{l}{n\pi}\int_0^\pi \sin^2 udu = \frac{lA_n}{2} \tag{3.6}$$

For $c_n = 1$, or $A_n = \sqrt{2/l}$ the eigenfunctions, like the unit vectors of particle mechanics, are normalized. Thus, for $S_n = \sqrt{2/l}\sin n\pi x/l$, the functions are said to be orthonormal and the normalization condition reads

$$\int_0^l S_n(x)S_m(x)dx\ \delta_{nm}$$

In addition to orthogonality the eigenfunctions also have the important property of completeness, first formulated by Fourier. This property implies that any arbitrary function $f(x)$ that satisfies the same boundary conditions as the functions of the set, can be expressed as a Fourier series

$$f(x) = \sum_{n=1}^\infty a_nS_n(x) \tag{3.7}$$

the a_n being constant coefficients. The coefficients in (7) are characterised by the product integral

$$\int_0^l S_m(x)f(x)dx = \sum_{n=1}^\infty a_n\int_0^l S_m(x)S_n(x)dx$$

The integral on the right vanishes for all $n \neq m$ and (6) reduces to

$$\begin{aligned} a_m\int_0^l \sin^2(m\pi x/l)dx &= \frac{l}{2}a_m \\ &= \int_0^l f(x)\sin(n\pi x/l)dx \end{aligned}$$

Each Fourier coefficient therefore has the form

$$a_n = \frac{2}{l}\int_0^l f(x)\sin\left(\frac{n\pi x}{l}\right)dx$$

and together they define a function

$$g(k) = \frac{2}{l}\int_0^l f(x)S_k(x)dx$$

If in a more general formulation the variable $\pi x/l$ is replaced by x, the range changes to $0, \pi$ and the Fourier series becomes

$$f(x) = \sum_{n=1}^{\infty} a_n S(x)$$

where

$$a_n = \frac{2}{\pi}\int_0^{\pi} f(x)S_n(x)dx$$

Without repeating the work, a similar result is obtained for $S(x) = \cos nx$. Combination of sine and cosine series leads to the even more general exponential form, such that

$$f(x) = \sum_{-\infty}^{\infty} a_n e^{in\pi x/l} \quad , \quad a_n = \frac{1}{2l}\int_{-l}^{l} f(x)e^{-in\pi x/l}dx$$

3.6.3 Fourier Transforms

The complex exponential Fourier series has an important limiting form [21] when $l \to \infty$. If $n\pi/l \equiv k$, remains large (ranging from $-\infty$ to ∞) and $a_n \to 0$ (proportional to $1/l$), but

$$g(k) \equiv \lim_{\substack{l\to\infty \\ a_n\to 0}} \left(\frac{l}{\pi}a_n\right) = \frac{1}{2\pi}\int_{-\infty}^{\infty} f(x)e^{-ikx}dx = \text{finite},$$

then

$$\begin{aligned} f(x) &= \sum_{n=-\infty}^{\infty} a_n e^{ikx} \quad , \quad k = \frac{n\pi}{l} \\ &= \lim_{\substack{l\to\infty \\ a_n\to 0}} \sum_{n=-\infty}^{\infty} \frac{\pi}{l} g(k)e^{ikx} \end{aligned}$$

Since k is proportional to n the sum over n steps of $\Delta n = 1$ can be written as a sum over k,

$$\sum_n \left(\frac{\pi}{l}\right) = \left(\sum_{\Delta k} \frac{l}{\pi}\right)\frac{\pi}{l} = \sum_{\Delta k}$$

However, $\Delta k = (\pi/l)\Delta n$ becomes infinitesimally small when l becomes large and the sum over Δk becomes an integral

$$\lim_{\substack{l\to\infty \\ a_n\to 0}} \sum_n \frac{\pi}{l} = \int dk$$

and

$$f(x) = \int_{-\infty}^{\infty} g(k)e^{ikx}dk \tag{3.8}$$

The function $g(k)$ is called the Fourier transform of $f(x)$ and (8) the Fourier inversion formula.

The Fourier transform and its inverse are similar in structure. The symmetry can be made explicit by replacing $g(k)$ with $(1/\sqrt{2\pi})g(k)$, *i.e.*

$$g(k) = \frac{1}{\sqrt{2\pi}}\int_{-\infty}^{\infty} e^{-ikx}f(x)dx$$

$$f(x) = \frac{1}{\sqrt{2\pi}}\int_{-\infty}^{\infty} e^{ikx}g(k)dk.$$

The roles of $f(x)$ and $g(k)$ can be interchanged under the substitutions $x \leftrightarrow k$ and $i \to -i$, known as a *conjugate transformation*. The variables x and k form a *conjugate pair* of variables. The conjugate pair x and k are joined together by the fact that either $f(x)$ or $g(k)$ contains all the information about the function. This is an important feature of wave functions to be highlighted later on.

An important example of Fourier transformation is the box function

$$f(x) = \begin{cases} a & \text{if } |x| < \alpha \\ 0 & \text{if } |x| > \alpha \end{cases}$$

with its Fourier transform

$$g(k) = \frac{\alpha}{\sqrt{2\pi}}\int_{-\alpha}^{\alpha} ae^{-ikx}dx = \frac{a\alpha}{\sqrt{2\pi}}\left[\frac{e^{-ikx}}{-ik}\right]_{-\alpha}^{\alpha}$$

$$= \frac{a\alpha}{\sqrt{2\pi}}\frac{2\sin k\alpha}{k}$$

For the special case $a = \sqrt{2\pi}/2\alpha$,

$$g(k) = \frac{\sin k\alpha}{k\alpha} = j_0(k\alpha) \quad , \tag{3.9}$$

a spherical Bessel function (1.20).

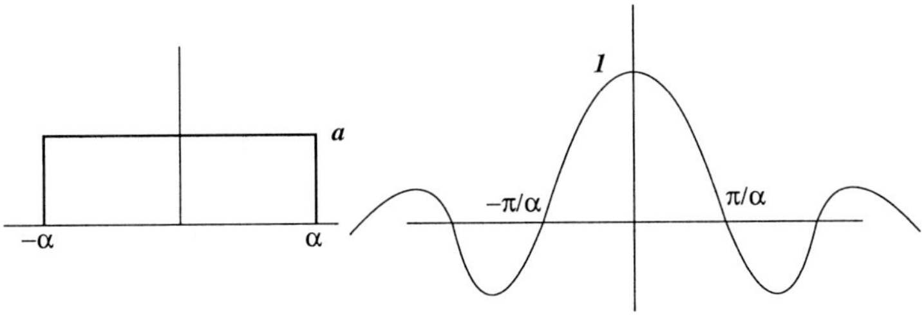

As α approaches zero, $f(x)$ becomes ∞ at $x = 0$, when it is called a "unit impulse", or a delta function. Its transform $g(k) = 1$. Of equal importance is the Fourier transform of the Gaussian function

$$f(x) = Ne^{-\frac{1}{2}cx^2} \tag{3.10}$$

i.e.

$$g(k) = \frac{N}{\sqrt{2\pi}} \int_{-\infty}^{\infty} e^{-ikx} e^{-\frac{1}{2}cx^2} dx$$

It is calculated by completing the square in the exponent:

$$-\frac{1}{2}cx^2 - ikx = -\frac{c}{2}\left(x + \frac{ik}{c}\right)^2 - \frac{k^2}{2c} = -\frac{1}{2}cy^2 - \frac{k^2}{2c}.$$

so that

$$g(k) = \frac{Ne^{-k^2/2c}}{\sqrt{2\pi}} \int_{-\infty}^{\infty} e^{-\frac{1}{2}cy^2} dy$$

Using[5]

$$\int_{-\infty}^{\infty} e^{-ax^2} dx = \sqrt{\frac{\pi}{a}}$$

5

Write $\quad I^2 = \int_{-\infty}^{\infty} e^{-ax^2} dx \int_{-\infty}^{\infty} e^{-ay^2} dy = \int_{-\infty}^{\infty}\int_{-\infty}^{\infty} e^{-a(x^2+y^2)} dx\, dy$

Convert to polar coordinates r, θ, noting that $dx\, dy$ becomes $r\, dr\, d\theta$,

$$I^2 = \int_0^{\infty} r\, dr \int_0^{2\pi} d\theta\, e^{-ar^2} = 2\pi \int_0^{\infty} re^{-ar^2} dr = \frac{\pi}{a},$$

using standard integrals [22].

$$g(k) = \frac{Ne^{-k^2/2c}}{\sqrt{c}}.$$

It follows that the Gaussian function transforms into a Gaussian but with a falloff constant c^{-1} that is inversely proportional to c in $f(x)$. This means that a narrow $f(x)$ gives rise to a broad $g(k)$ and *vice versa*, as shown below, for large c.

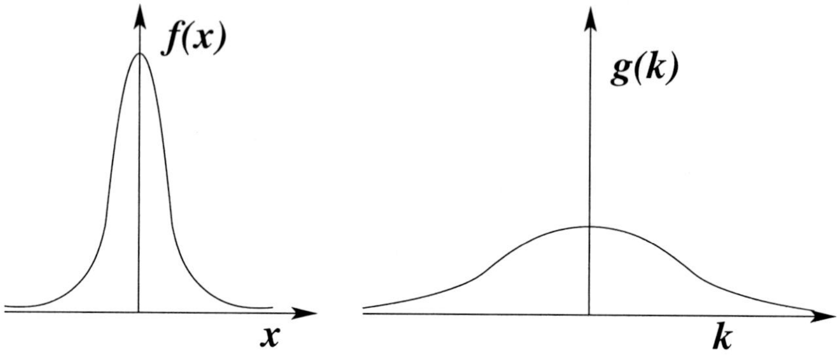

This reciprocal relationship is a general property of Fourier transforms. The δ-function of the previous paragraph and its transform demonstrate the same reciprocity. To characterise this property more precisely a bracket function, called a *scalar* or *inner* (dot) product, is used to define an overlap integral

$$(f_1, f_2) = \int_{-\infty}^{\infty} f_1^*(x) f_2(x) dx.$$

If g_i is the Fourier transform of f_i this expression becomes

$$\begin{aligned}
(f_1, f_2) &= \int_{-\infty}^{\infty} \left(\frac{1}{\sqrt{2\pi}} \int_{-\infty}^{\infty} e^{ikx} g_1(k) dk \right)^* \left(\frac{1}{\sqrt{2\pi}} \int_{-\infty}^{\infty} e^{ik'x} g_2(k') dk' \right) dx \\
&= \int_{-\infty}^{\infty} g_1^*(k) g_2(k') \left(\frac{1}{2\pi} \int_{-\infty}^{\infty} dx\, e^{-i(k-k')x} \right) dk\, dk' \\
&= \int_{-\infty}^{\infty} g_1^*(k) g_2(k) dk = (g_1, g_2)
\end{aligned}$$

The inner product can be calculated using either f or g, or $\phi = f(x)$ or $g(k)$ depending on the integration variable x or k. More general integrals can also be represented in this notation, *e.g.*

$$(\phi_1, A(x)\phi_2) = \int_{-\infty}^{\infty} f_1^*(x) A(x) f_2(x) dx$$

Some inner products of the Gaussian function are:

$$\begin{aligned}(\phi,\phi) &= (f,f) = N^2\int_{-\infty}^{\infty} e^{-cx^2}\,dx = N^2\sqrt{\pi/c}\\(\phi,x\phi) &= N^2\int_{-\infty}^{\infty} e^{-cx^2}x\,dx = 0\\(\phi,x^2\phi) &= N^2\int_{-\infty}^{\infty} x^2e^{-cx^2}\,dx = \frac{N^2}{2c}\sqrt{\pi/c}\end{aligned}$$

The average value of x in the system ϕ is called the *expectation value*

$$\langle x\rangle \equiv (\phi,x\phi)/(\phi,\phi) = \bar{x}$$

while its *uncertainty* or *dispersion* Δx is defined by

$$\begin{aligned}(\Delta x)^2 &= (\phi,[x-\bar{x}]^2\phi)/(\phi,\phi)\\&= (\phi,x^2\phi)/(\phi,\phi)-\bar{x}^2\end{aligned}$$

Hence the Gaussian function (10) has $\bar{x}=0$ and$(\Delta x)^2 = 1/c$. By the same reasoning, also for the same function $\bar{k}=0,\quad (\Delta k)^2 = c/2$. Thus the product of uncertainties has a unique value $\Delta x\Delta k = \frac{1}{2}$, independent of c. This result means that the conjugate variables in ϕ cannot simultaneously be known precisely. The better one of these is known, the poorer is the other defined. Complete knowledge of k (say) is possible only when there is complete ignorance of x. This logic makes sense when x is the position of a wave and k the wave vector. A wave with a unique value of k is infinitely long to ensure that it oscillates in the same way everywhere. An infinitely long wave does not have a definite position since x can be anywhere along its length.

No other function has a smaller uncertainty product than the Gaussian[6], so in general[7] $\Delta x\Delta k \geq \frac{1}{2}$.

[6]This is related to the Schwarz inequality in vector algebra,

$$A^2B^2 \geq (\boldsymbol{A}\cdot\boldsymbol{B}) = (AB\cos\theta)^2$$

where $A^2 = \boldsymbol{A}\cdot\boldsymbol{A}$, and can be demonstrated [21] by substituting $\boldsymbol{A} = (x-\bar{x})\phi,\quad \boldsymbol{B} = (k-\bar{k})\phi$.

[7]For matter waves $\hbar k$ is the particle momentum and the uncertainty relation $\Delta x\Delta p \geq \hbar/2$, known as the Heisenberg uncertainty principle.

3.6.4 Wave Packets

The real part of the general solution (1.22) to the one-dimensional wave equation is

$$\begin{aligned} u_r &= a\cos k(x-ct) \\ &= a\cos(kx-\omega t) \end{aligned}$$

Suppose that two harmonic waves of equal amplitude and almost equal frequencies are combined into a total disturbance [23]

$$z = a\cos(k_1x-\omega_1 t) + a\cos(k_2-\omega_2 t)$$

From the identity

$$\cos\alpha+\cos\beta = 2\cos\frac{1}{2}(\alpha+\beta)\cos\frac{1}{2}(\alpha-\beta)$$

follows that

$$z = 2a\cos\left(\frac{k_1+k_2}{2}x-\frac{\omega_1+\omega_2}{2}t\right)\cos\left(\frac{k_1-k_2}{2}x-\frac{\omega_1-\omega_2}{2}t\right)$$

The first cosine factor represents a wave like the originals with average wavelength and frequency and moving with velocity $(\omega_1+\omega_2)/(k_1+k_2)$. In the case of electromagnetic waves $\omega_1 = ck_1$ and $\omega_2 = ck_2$, so that $v = c(k_1+k_2)/(k_1+k_2) = c$, the original velocity. The composite wave however, has an amplitude that varies within a profile defined by the factor

$$z = 2a\cos\left(\frac{k_1-k_2}{2}x-\frac{\omega_1-\omega_2}{2}t\right)$$

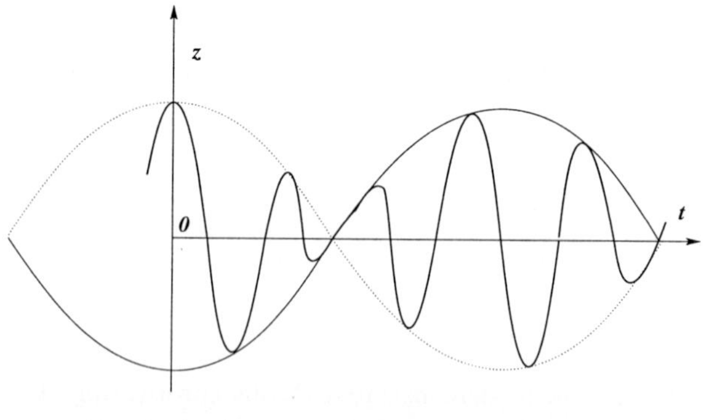

Since the two waves have the same velocity the composite wave moves without change in shape. If the constituent waves are dispersive, *i.e.* have different velocities ($\omega_1/k_1 \neq \omega_2/k_2$) the profile moves with velocity $(\omega_1-\omega_2)/(k_1-k_2)$, different from the velocity $(\omega_1+\omega_2)/(k_1+k_2)$ of the faster oscillating part. The individual waves therefore move through the profile and the wave packet disperses in time as shown in the diagram. This happens whenever the constituent waves have non-constant phase velocities, v_ϕ. The velocity of the profile, called the wave packet, is known as the group velocity, v_g. Where the components are not too different,

$$v_\phi = \omega/k \quad \text{and} \quad v_g = \frac{\omega_1-\omega_2}{k_1-k_2} = \frac{d\omega}{dk}.$$

In terms of wavelength[8]

$$v_g = \frac{d\nu}{d(1/\lambda)} = -\lambda^2\frac{d\nu}{d\lambda}.$$

In general wave packets consist, not of two, but of many components of the type $\exp i(kx - \omega t)$ where k can have all possible values. It may also be assumed that the wave velocity depends on the frequency, *i.e.* $\omega = f(k)$. For an amplitude $a(k)$ the total disturbance is[9]

$$\phi(x,t) = \int_{-\infty}^{\infty} a(k)e^{i(kx-\omega t)}dk$$

[8] For matter waves wavelength and frequency are assumed to relate to particle momentum and energy according to $\lambda = h/p$ and $\nu = E/h$. The proportionality factor is Planck's constant h. Hence

$$v_g = \left(-\frac{h^2}{p^2}\right)\left(\frac{dE}{h}\right)\left(-\frac{p^2}{h\,dp}\right) = \frac{dE}{dp} = \frac{d(p^2/2m)}{dp} = \frac{p}{m} = v,$$

the classical particle velocity. The wave mechanical description of a particle therefore corresponds to the classical in those cases where the internal structure and dimensions of the wave packet can be ignored.

[9] The Fourier integral

$$\phi(x,t) = \int_{-\infty}^{\infty} a(k)e^{i(\mathbf{k}.\mathbf{r}-\omega t)}dk$$

provides a good representation of a particle with wave properties since a quantum mechanical particle cannot have momentum and position both well defined at the same time. The minimum size of a wave packet in any direction is one wavelength. Therefore in the x-direction the uncertainty in position $\Delta x \sim \lambda$. The corresponding uncertainty in momentum $\Delta p = \Delta(h/\lambda)$ and hence $\Delta x.\Delta p \sim h$, as required by the uncertainty principle.

3.6.5 Solitons

The general solution of the linear one-dimensional wave equation describes two waves that move in opposite directions without change in shape or mutual interaction. The two components therefore move apart in finite time and, for most purposes, it is sufficient to consider only one of the components. More specifically, the discussion may be restricted to solutions[10] of

$$\frac{\partial u}{\partial t} + c\frac{\partial u}{\partial x} = 0 \tag{3.11}$$

i.e.

$$u(x,t) = f(x - ct).$$

If the time parameter is redefined as t/c, equation (11) becomes

$$\frac{\partial u}{\partial t} + \frac{\partial u}{\partial x} = 0,$$

as if $c = 1$, and $u(x,t) = f(x - ct)$.

There are many features of real waves, such as dispersion and nonlinearity, that cannot be described directly in terms of the general wave equation. A brief discussion of such effects is needed to understand solitons.

The simplest known dispersive wave equation is

$$\frac{\partial u}{\partial t} + \frac{\partial u}{\partial x} + \frac{\partial^3 u}{\partial x^3} = 0 \tag{3.12}$$

The harmonic wave solution

$$u(x,t) = e^{i(kx-\omega t)}$$

is a solution of (12) if $\omega = k - k^3$, which is the dispersion relation that determines $\omega(k)$ for given k. From

$$kx - \omega t = k[x - (1 - k^2)t]$$

it follows that (12) describes waves that propagate at velocity $c = \omega/k = 1-k^2$, which is a function of k. This condition is characteristic of a dispersive

[10] Note that

$$\left(\frac{\partial}{\partial t} \mp c\frac{\partial}{\partial x}\right)\left(\frac{\partial}{\partial t} \pm c\frac{\partial}{\partial x}\right) u \equiv \left(\frac{\partial^2}{\partial t^2} - c^2\frac{\partial^2}{\partial x^2}\right) u = 0.$$

wave defined by waves of different wave number propagating at different velocities. Thus a single wave profile which can be represented by the sum of many harmonic components may change its shape as time evolves by virtue of the different velocities of the components, as in

$$u(x,t) = \int_{-\infty}^{\infty} A(k)e^{i[kx-\omega(k)t]}dk$$

where $A(k)$ is essentially the Fourier transform of $u(x,0)$. The overall effect is to produce a wave profile which changes its shape as it moves; in fact, since different components move at different velocities the profile will necessarily spread out or disperse.

The velocity c, of an individual wave component is called the phase velocity. There is also another velocity. It can be argued that the major contribution to a wave packet originates from the neighbourhood $k_0(x,t)$ found by setting

$$\frac{d}{dk}[kx - \omega(k)t] = 0$$

i.e.

$$\left[\frac{d\omega(k)}{dk}\right]_{k=k_0} = \frac{x}{t}.$$

This result is used to define the group velocity

$$c_g = \frac{d\omega(k)}{dk} = 1 - 3k^2$$

which represents the velocity of a wave packet. In wave mechanics, group velocity is often identified with particle velocity and the quantum particle with a wave packet. Because of dispersion however, a wave packet cannot be a serious model of a stable particle, unless the dispersion is suppressed by some other factor, such as nonlinearity.

Most wave equations *e.g.* (11) are valid only for sufficiently small amplitudes. At larger amplitudes deviations from linearity may occur as described by a non-linear partial differential equation of the type

$$\frac{\partial u}{\partial t} + (1+u)\frac{\partial u}{\partial x} = 0 \tag{3.13}$$

Rearranged into the form

$$1 + u = -\frac{\partial u}{\partial t}\frac{\partial x}{\partial u}$$

it follows that for constant u

$$1 + u = -\frac{\partial x}{\partial t}$$

and hence that u is constant on the characteristic lines

$$x = (1 + u)t + \text{constant}$$

Thus the general solution of (13) is

$$u(x,t) = f[x - (1 + u)t]$$

where f is an arbitrary function.

For a given wave profile the foregoing solution does not remain single-valued indefinitely but, due to a concentration of the pulse on the leading side of the wave, develops into a shape that resembles a wave that breaks. Recall that the effect of dispersion is to produce a wave which constantly changes shape, but now in the opposite sense, such that it causes the wave to spread out rather than to steepen. When the two effects are balanced, as shown schematically in figure 1, a wave of permanent form is produced.

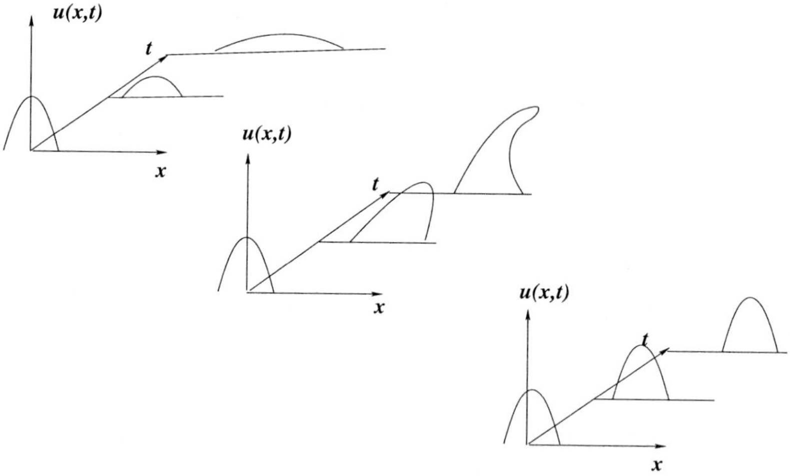

Figure 1: *The frame on the left shows the development of a dispersive wave and the frame in the middle that of a nonlinear wave. When these effects are balanced as in the frame on the right a soliton is formed.*

This balance between nonlinearity and dissipation gives rise to what is called a solitary wave, and is correctly described by the Korteweg-de Vries (KdV) equation

$$\frac{\partial u}{\partial t} + (1 + u)\frac{\partial u}{\partial x} + \frac{\partial^3 u}{\partial x^3} = 0.$$

A common form of the KdV equation obtained by a series of transformations found convenient in engineering applications is

$$\frac{\partial u}{\partial t} - 6u\frac{\partial u}{\partial x} + \frac{\partial^3 u}{\partial x^3} = 0.$$

An analytic solution of the KdV equation can be written as

$$u(x,t) = -\alpha^2/2\text{sech}^2[\alpha/2(x - \alpha^2 t - x_1]$$

Where x_1 is a phase shift, the maximum amplitude is given by $\alpha^2/2$ and the velocity is α^2, that is twice the amplitude.

Solitary waves, especially in shallow water, have been studied for many years[24]. They have the interesting property of interacting with other solitary waves and to separate afterwards as if there had been no interaction at all. This persistence of the wave led to the name *soliton*, to emphasize the particle-like character of these waves which seem to retain their identities in a collision.

Whereas the profile in linear wave equations is usually arbitrary it is important to note that a nonlinear equation will normally describe a restricted class of profiles which ensure persistence of solitons as $t \to \infty$. Any theory of ordered structures starts from the assumption that there exist localized states of nonlinear fields and that these states are stable and robust. A one-dimensional soliton is an example of such a stable structure. Rather than identify elementary particles with simple wave packets, a much better assumption is therefore to regard them as solitons. Although no general formulations of stable two or higher dimensional soliton solutions in non-linear field models are known at present, the conceptual construct is sufficiently well founded to anticipate the future development of standing-wave soliton models of elementary particles.

3.6.6 The Eikonal Equation

Since all waves are not necessarily harmonic and wave velocity need not be constant, the more general formulation of a wave function

$$U = A(x,y,z)e^{i[\omega t - k\phi(x,y,z)]}$$

has variable amplitude A and phase $k\phi$. Substitution of this trial solution into the wave equation

$$\nabla^2 U = \frac{1}{v^2}\frac{d^2 U}{dt^2}$$

leads to

$$\nabla^2 A - 2ik\nabla A \cdot \nabla\phi - ikA\nabla^2\phi - k^2 A(\nabla\phi)^2 + \frac{\omega^2}{v^2}A = 0.$$

After separation of real and imaginary terms[11]

$$\nabla^2 A - k^2 A(\nabla\phi)^2 + \frac{\omega^2}{v^2} = 0 \tag{3.14}$$

$$2\nabla A \cdot \nabla\phi + A\nabla^2\phi = 0$$

To find a physical interpretation of (14) recall that $v = v(x, y, z)$ need not be a constant. Set the vacuum velocity $v_0 = \omega/k$. For velocity v in another medium the index of refraction $n = v_0/v$, in terms of which (14) becomes

$$\frac{\nabla^2 A}{A} - k^2\left[(\nabla\phi)^2 - n^2\right] = 0$$

This equation describes the relationship between geometric and wave optics. The geometric description is valid when the wavelength is small compared to the dimension of any change in the medium [17], *i.e.* n does not change greatly over distances of the order of λ. In this case the term containing $k^2 = 4\pi^2/\lambda^2$ dominates. To satisfy the equation it is therefore required that

$$(\nabla\phi)^2 = n^2 \quad , \tag{3.15}$$

which is known as the eikonal equation of geometrical optics [25]. The mathematical similarity between (15) and the HJ equation (5)

$$(\nabla S)^2 = 2m[E - V(x, y, z)]$$

is unmistakable. They become identical by interpreting ϕ as proportional to S and equating

$$n^2 = \frac{2m(E - V)}{p^2}$$

where p has the dimensions of momentum. The similarity between HJ and eikonal equations suggests the possibility that wave propagation can be interpreted in terms of particle motion or to describe the motion of a particle in terms of wave motion. This observation is the basis of wave mechanics.

[11]For a plane harmonic wave $\phi = \alpha x + \beta y + \gamma z$, where α, β, γ are direction cosines and hence $(\nabla\phi)^2 = \alpha^2 + \beta^2 + \gamma^2 = 1$, $\nabla^2\phi = 0$, $\nabla^2 A = 0$, $\nabla A \cdot \nabla\phi = 0$. This result shows that U is an acceptable solution providing $k = \omega/v$, a known relationship.

Chapter 4

Space and Time

4.1 Introduction

The fundamental concepts on which the theory of chemistry must ultimately be based are fields in space and time. However, many centuries of speculative cosmology has complicated the discussion of cosmic events by confusing the terminology around space and time so much that a discussion of these concepts should properly start by clearing up some ambiguities. It is proposed that the large-scale cosmic structure into which all objects in the universe are embedded could be called *world space*, without specifying its substance, geometry, topology or dimensionality. It is the entity equated by the flat-earther with infinite three-dimensional euclidean space; it is the Minkowski space-time of special relativity; five-dimensional space-time of the Kaluza-Klein unified field; the primordial void of big-bang cosmology; the twenty six dimensions of superstring theory; the vacuum of chemistry; or the chaos of creation science. It is the stage on which the cosmic drama unfolds. The primary substance that pervades world space is a matter of conjecture. Some call it space and others call it aether, or vacuum. In quantum field theory it is a boiling sea of virtual particles. In the view of Maxwell, Kelvin and others it is an array of vortices. Sakharov [26] and others [27] describe it as solidly filled up by elementary units of matter and anti-matter. Others define it as wave motion. Occam's razor demands that whatever the substance of world space, it be the source of all elementary units of matter and energy. It is the ultimate unknown beyond which conjecture becomes fruitless and where physics begins.

It is proposed to use the terms world space and *aether* synonymously and to reserve *space* and *vacuum* for empty three-dimensional space. Curvature of world space may refer to any of its unknown number of dimensions. It is recommended that any discussion of cosmic phenomena should commence with a clear statement of the assumed model of world space.

Of all the mysteries of Nature time is the oldest and most daunting. It has been analyzed from many angles, mostly from a philosophical rather than a scientific point of view. These studies have produced a number of related descriptions, including definitions of psychological, biological, geological and mathematical time [28]. Despite the fact that time intervals can be measured with stupendous accuracy there is no physical model of time. This anomalous situation probably means that the real essence and origin of the concept time is not understood at all.

An elaborate recent effort to establish the physical nature of time was published by Penrose [29]. It is based on the shibboleth popularized by Eddington [30], that the flow of time is dictated by the second law of thermodynamics. In order to fix the direction of time flow Penrose proposed to look for a mechanism that ensures minimum entropy at the beginning of time. However, when the argument is pursued to its logical conclusion it transpires that entropy also has a maximum at this same minimum point. Despite its complexity the Penrose argument therefore fails and adds nothing to Eddington's construction, called time's arrow that aims to link the one-way property of time to elements of randomness and organization, only recognized by the second law of thermodynamics. All other laws of Nature which control the behaviour of single units are indifferent to the direction in which time flows. With the exception of the second law of thermodynamics all the laws of physics are said to be time reversible.

There is no need to pursue Eddington's argument any further. There is simply no mechanism forthcoming to suggest how the laws of probability that describe the state of randomness of a developing system can possibly dictate asymmetric time flow. Statistics may well predict the outcome of shuffling a pack of cards, but it cannot initiate or drive the shuffling process by itself. Therein lies the weakness of the statistical definition of entropy. It certainly provides an adequate book-keeping account of entropy changes during an unfolding event, but remains silent about the engine that drives the process. What possible effect probability can have on the flow of time is even more mysterious. The only contribution that the second law makes to elucidate the nature of time is to record that spontaneous change leads to an increase of entropy (with time).

An alternative to many minds is to consider time as an illusion. The parameter that serves to describe change is just that, and has no independent meaning of its own. This assertion still begs the question of what creates the universal illusion called time. There is no escape from the conclusion that something out there is responsible for the perception that time flows and no denying the fact that measurement of time intervals points at a steady flow. Tied in with the fundamental link between matter, energy and space there must be an involvement of the time factor.

4.2 The Electromagnetic Field

Electric and magnetic effects have been observed since ancient times without suspecting a close relationship between the two phenomena, and certainly not inferring any close connection with visible light. The modern view is that the three effects are different aspects of a single concept, known as the electromagnetic field, which in turn is a manifestation of interactions involving the elementary entities called electrons and photons.

A stationary electron or unit of negative electricity is recognized as the source of an electrostatic field while magnetic dipoles give rise to a magnetostatic field. The electrostatic field is caused by, and causes forces on, static charges; magnetic field is caused by, and causes force on, moving charges. The electric field is determined by a scalar potential, $\mathbf{E} = -\nabla\phi$, like the scalar wave potential for compressional waves in elastic media. A vector potential that determines the magnetic field resembles the vector wave potential for shear waves. The magnetic induction $\boldsymbol{B}$ determines the force on a current element, or charge moving with velocity $\boldsymbol{v}$,

$$\boldsymbol{F} = q(\boldsymbol{v} \times \boldsymbol{B}).$$

If there is also an electric field present, the electric force $q\boldsymbol{E}$ must be taken into account as well. The complete force equation for a charged mass point, also known as the Lorentz force, is

$$\boldsymbol{F} = q(\boldsymbol{E} + \boldsymbol{v} \times \boldsymbol{B}).$$

For more details any text on electromagnetism, such as [10] should be consulted.

A reciprocal relationship between electricity and magnetism is inferred from the observed magnetic effects of a changing electric current (acceleration of electrons) or the electric effects of a rotating magnet. The propagation of electric and magnetic effects through a medium is found to depend on respectively the permittivity ϵ and permeability μ of the medium, formulated as $\mathbf{D} = \epsilon\mathbf{E}$ and $\mathbf{B} = \mu\mathbf{H}$. When the medium is the vacuum, or free space, these quantities assume the characteristic values of ϵ_0 and μ_0.

4.2.1 Units

To ensure that the mechanical effects of electromagnetic activity are described in a consistent manner it is necessary to find the appropriate units in terms of which to measure the absolute values of the constants ϵ_0 and μ_0. In specific applications either or both of these constants can arbitrarily be set to unity or another arbitrary value, in addition to choosing convenient mechanical units of distance, mass and time. These procedures give rise to a bewildering variety of conventions which unfortunately persists in the scientific literature, although a system of absolute units was proposed by Giorgi in 1901. It has subsequently been officially adopted by the international scientific community as the recommended (SI) system of units. This system needs four basic units of measurement for its full expression, *i.e.* measuring distance in meters, mass in kilograms, time in seconds and electric charge in coulombs. In this system the constants

$$\begin{aligned} \epsilon_0 &= 8.854 \times 10^{-12} \quad \text{farad m}^{-2} \\ \mu_0 &= 4\pi \times 10^{-7} \quad \text{henry m}^{-1} \end{aligned}$$

and their product $\epsilon_0\mu_0 = c^{-2}$, where c has the dimensions of a velocity, $c = 3 \times 10^8$ m s^{-1} is commonly known as the velocity of light.

The inverse-square Coulomb law of attraction between two charges, q_1 and q_2 at a distance r apart in vacuum, expressed in SI units takes the form

$$F = \frac{1}{4\pi\epsilon_0}\frac{q_1 q_2}{r^2} \text{ newton}$$

The energy in the electromagnetic field is

$$w = \frac{1}{2}\int \left(\epsilon \boldsymbol{E}^2 + \mu \boldsymbol{H}^2\right) dv$$

4.2.2 The Maxwell Equations

The synthesis of ideas described above was first achieved by Maxwell and is described by two equations that summarize the essential experimental observations pertaining to the reciprocally related electromagnetic effects[1]. In SI units and vector notation these equations, valid at every ordinary point

[1] More exact details are provided in any of many textbooks, *e.g.* ref. [31]

in space, derive from the known laws that describe electric and magnetic fields.

$$\boldsymbol{\nabla}\times\boldsymbol{E}+\frac{\partial\boldsymbol{B}}{\partial t}=0\quad\text{(Faraday-Lenz law.)}\tag{4.1}$$

$$\boldsymbol{\nabla}\times\boldsymbol{H}=\boldsymbol{J}\quad\text{(Ampére's law)}$$

It is noticed that taking the divergence of this last equation leads to conflict with the continuity equation (eqn. 1.7) for electric charge,

$$\frac{\partial\rho}{\partial t}+\boldsymbol{\nabla}.\boldsymbol{J}=0$$

since $\boldsymbol{\nabla}.(\boldsymbol{\nabla}\times\boldsymbol{H})=0$, identically. This condition can only apply in situations where the charge density is constant in time. For the general case, Maxwell therefore modified Ampére's law to read

$$\boldsymbol{\nabla}\times\boldsymbol{H}=\boldsymbol{J}+\frac{\partial\boldsymbol{D}}{\partial t}\tag{4.2}$$

which is consistent with the continuity condition. Apart from the electric and magnetic field intensities $\boldsymbol{E}$ and $\boldsymbol{H}$ the magnetic induction $\boldsymbol{B}$ and the electric displacement $\boldsymbol{D}$ are defined in terms of the magnetic and electric polarization vectors

$$\begin{aligned}\boldsymbol{M}&=\frac{1}{\mu_0}\boldsymbol{B}-\boldsymbol{H}\\ \boldsymbol{P}&=\boldsymbol{D}-\epsilon_0\boldsymbol{E}\end{aligned}$$

$\boldsymbol{E}$ and $\boldsymbol{B}$ are the fundamental force vectors, while $\boldsymbol{D}$ and $\boldsymbol{H}$ are derived vectors associated with the state of matter. $\boldsymbol{J}$ is the vector current density. The Maxwell equations in terms of $\boldsymbol{E}$ and $\boldsymbol{B}$ are

$$\begin{aligned}\boldsymbol{\nabla}\times\boldsymbol{E}&+\frac{\partial\boldsymbol{B}}{\partial t}=0\\ \boldsymbol{\nabla}\times\boldsymbol{B}&-\epsilon_0\mu_0\frac{\partial\,\boldsymbol{E}}{\partial t}=\mu_0(\boldsymbol{J}+\frac{\partial\boldsymbol{P}}{\partial t}+\boldsymbol{\nabla}\times\boldsymbol{M})\end{aligned}$$

By interpreting the term in brackets as the total current density the inhomogeneous Maxwell equation (2) is also written as

$$\boldsymbol{\nabla}\times\boldsymbol{B}-\epsilon_0\mu_0\frac{\partial\boldsymbol{E}}{\partial t}=\mu_0\boldsymbol{J}$$

It is common practice to add two divergence equations

$$\boldsymbol{\nabla} \cdot \boldsymbol{B} = 0 \tag{4.3}$$

$$\boldsymbol{\nabla} \cdot \boldsymbol{D} = \rho \quad \text{(Gauss's law)} \tag{4.4}$$

in order to define the electromagnetic field, although these are not independent equations if conservation of energy is assumed. ρ defines electric charge density at a point. Equation (4) in terms of $\boldsymbol{E}$,

$$\boldsymbol{\nabla} \cdot \boldsymbol{E} = \frac{1}{\epsilon_0}(\rho - \nabla \boldsymbol{P})$$

is also written as

$$\boldsymbol{\nabla} \cdot \boldsymbol{E} = \frac{1}{\epsilon_0}\rho$$

The polarization vectors vanish in free space, so that in the absence of charge and matter $\boldsymbol{D} = \epsilon_0 \boldsymbol{E}$, $\boldsymbol{H} = \frac{1}{\mu_0}\boldsymbol{B}$ and the Maxwell equations are:

$$\begin{aligned}
\boldsymbol{\nabla} \times \boldsymbol{E} &= -\frac{\partial \boldsymbol{B}}{\partial t} \\
\boldsymbol{\nabla} \times \boldsymbol{B} &= \epsilon_0 \mu_0 \frac{\partial \boldsymbol{E}}{\partial t} = \frac{1}{c^2}\frac{\partial \boldsymbol{E}}{\partial t} \\
\boldsymbol{\nabla} \cdot \boldsymbol{B} &= 0 \\
\boldsymbol{\nabla} \cdot \boldsymbol{E} &= 0
\end{aligned}$$

4.2.3 Electromagnetic Potentials

Since the condition $\boldsymbol{\nabla} \cdot \boldsymbol{B} = 0$ is generally valid the vector $\boldsymbol{B}$ is solenoidal (source free) and can therefore be expressed as the curl of another vector $\boldsymbol{A_0}$

$$\boldsymbol{B} = \boldsymbol{\nabla} \times \boldsymbol{A_0} \tag{4.5}$$

However, $\boldsymbol{A_0}$ is not uniquely defined since $\boldsymbol{B}$ could also be equated to the curl of another vector $\boldsymbol{A}$ such that

$$\boldsymbol{A} = \boldsymbol{A_0} - \nabla\psi \tag{4.6}$$

and ψ is an arbitrary scalar function of position. From (1) and (5) it then follows that

$$\boldsymbol{\nabla} \times \left(\boldsymbol{E} + \frac{\partial \boldsymbol{A_0}}{\partial t}\right) = 0 \tag{4.7}$$

$$\boldsymbol{\nabla} \times \left(\boldsymbol{E} + \frac{\partial \boldsymbol{A}}{\partial t}\right) = 0 \tag{4.8}$$

Thus the fields of the vectors $(\boldsymbol{E}+\frac{\partial \boldsymbol{A_0}}{\partial t})$ and $(\boldsymbol{E}+\frac{\partial \boldsymbol{A}}{\partial t})$ are irrotational and equal to the gradients of two scalar functions ϕ_0 and ϕ,

$$\begin{aligned} \boldsymbol{E} &= -\nabla\phi_0 - \frac{\partial \boldsymbol{A_0}}{\partial t} \\ \boldsymbol{E} &= -\nabla\phi - \frac{\partial \boldsymbol{A}}{\partial t} \end{aligned} \tag{4.9}$$

The functions ϕ and ϕ_0 are obviously related by

$$\phi = \phi_0 + \frac{\partial \psi}{\partial t} \tag{4.10}$$

The functions $\boldsymbol{A}$ are vector potentials of the field, and the ϕ are scalar potentials from which the field can be derived through (5) and (9). An infinite number of potentials leading to the same field can be constructed from (6) and (10). Using (5) and (9) the Lorentz force defined in terms of potentials becomes

$$\boldsymbol{F} = q\left[-\nabla\phi - \frac{\partial \boldsymbol{A}}{\partial t} + \boldsymbol{v} \times (\boldsymbol{\nabla} \times \boldsymbol{A})\right]. \tag{4.11}$$

It can be shown by use of the vector identity

$$\boldsymbol{\nabla} \times (\boldsymbol{\nabla} \times \boldsymbol{A}) = \boldsymbol{\nabla}(\boldsymbol{\nabla} \cdot \boldsymbol{A}) - \nabla^2 \boldsymbol{A} \tag{4.12}$$

that for a homogeneous isotropic medium in which ϵ and μ are independent of field intensity a general solution of the Maxwell field equations can be obtained in terms of potential functions, *i.e.*

From the definitions $\boldsymbol{B}= \mu\boldsymbol{H}$ and $\boldsymbol{D} = \epsilon\boldsymbol{E}$ and (2) follows

$$\begin{aligned} \frac{1}{\mu}(\boldsymbol{\nabla} \times \boldsymbol{B}) - \epsilon\frac{\partial \boldsymbol{E}}{\partial t} &= \boldsymbol{J} \\ \boldsymbol{\nabla} \times \boldsymbol{B} - \mu\epsilon\frac{\partial \boldsymbol{E}}{\partial t} &= \mu\boldsymbol{J} \\ \boldsymbol{\nabla} \times (\boldsymbol{\nabla} \times \boldsymbol{A}) &= \mu\epsilon\frac{\partial}{\partial t}\left(-\nabla\phi - \frac{\partial \boldsymbol{A}}{\partial t}\right) + \mu\boldsymbol{J} \\ \boldsymbol{\nabla}(\boldsymbol{\nabla} \cdot \boldsymbol{A}) - \nabla^2\boldsymbol{A} &= -\mu\epsilon\left(\nabla\frac{\partial\phi}{\partial t} + \frac{\partial^2 \boldsymbol{A}}{\partial t^2}\right) + \mu\boldsymbol{J} \\ -\nabla^2\boldsymbol{A} + \mu\epsilon\frac{\partial^2 \boldsymbol{A}}{\partial t^2} &= -\boldsymbol{\nabla}(\boldsymbol{\nabla} \cdot \boldsymbol{A}) - \mu\epsilon\nabla\frac{\partial\phi}{\partial t} + \mu\boldsymbol{J} \\ &= -\nabla\left(\boldsymbol{\nabla} \cdot \boldsymbol{A} + \mu\epsilon\frac{\partial\phi}{\partial t}\right) + \mu\boldsymbol{J} \end{aligned}$$

Since the divergence of $\boldsymbol{A}$ is still undefined it can be chosen so as to render the last expression in brackets equal to zero, and hence

$$\boldsymbol{\nabla} \cdot \boldsymbol{A} = -\mu\epsilon \frac{\partial \phi}{\partial t} \tag{4.13}$$

Equation (13) is known as the Lorentz gauge[2] condition. From (13), (9) and (4) follows

$$\begin{aligned} \frac{\partial}{\partial t}(\nabla \cdot \boldsymbol{A}) &= \nabla \cdot \left(\frac{\partial \boldsymbol{A}}{\partial t}\right) = -\mu\epsilon \frac{\partial^2 \phi}{\partial t^2} \\ &= \nabla \cdot (-\boldsymbol{E} - \nabla\phi) \\ &= \nabla \cdot (-\frac{1}{\epsilon}\boldsymbol{D} - \nabla\phi) \\ &= -\frac{1}{\epsilon}(\boldsymbol{\nabla} \cdot \boldsymbol{D}) - \boldsymbol{\nabla} \cdot \boldsymbol{\nabla}\phi \\ &= -\frac{\rho}{\epsilon} - \nabla^2\phi \end{aligned}$$

[2]The electromagnetic field is a well-known example of a *gauge theory.* A gauge transformation of the form

$$\psi \rightarrow \psi' = \psi e^{i\alpha}$$

is said to be local if α has space-time dependence. Taking a derivative of ψ then generates an extra term and to compensate for this an additional field is introduced by the transformation $A_\mu \rightarrow A'_\mu = A_\mu + \partial_\mu \alpha$.

Equation (3) is automatically satisfied by $\boldsymbol{B} = \boldsymbol{\nabla} \times \boldsymbol{A}$, which when substituted into (1) directly gives (7),

$$\nabla \times \left(\boldsymbol{E} + \frac{\partial \boldsymbol{A}}{\partial t}\right) = 0$$

Using the identity $\nabla \times (\nabla\varphi) = 0$, (7) is seen to be satisfied by

$$\boldsymbol{E} = \frac{\partial \boldsymbol{A}}{\partial t} - \nabla\varphi$$

The electromagnetic potentials $\boldsymbol{A}$ and φ are not determined uniquely by the expressions for $\boldsymbol{B}$ and $\boldsymbol{E}$. Thus the magnetic field is not affected by the addition of a scalar gradient to $\boldsymbol{A}$

$$\boldsymbol{A} \rightarrow \boldsymbol{A'} + \nabla\Omega$$

and the electric field is unchanged if, at the same time, φ is transformed according to

$$\varphi \rightarrow \varphi' = \varphi - \frac{\partial \Omega}{\partial t}$$

Together, these two equations constitute a compensating gauge transformation.

resulting in the two equations

$$\nabla^2\phi \quad - \quad \mu\epsilon\frac{\partial^2\phi}{\partial t^2} = -\frac{\rho}{\epsilon} \tag{4.14}$$

$$\nabla^2\boldsymbol{A} \quad - \quad \mu\epsilon\frac{\partial^2\boldsymbol{A}}{\partial t^2} = -\mu\boldsymbol{J} \tag{4.15}$$

These equations (14) and (15) determine the scalar and vector potentials in terms of ρ and $\boldsymbol{J}$. When ρ and $\boldsymbol{J}$ are zero, these equations become wave equations with wave velocity $c = \sqrt{1/\mu\epsilon}$. That is, $\boldsymbol{A}$ and ϕ are solutions of decoupled equations, where they are related by the wave operator

$$\left(\frac{1}{c^2}\frac{\partial^2}{\partial t^2} - \nabla^2\right) = \Box^2$$

to their respective sources $\mu\boldsymbol{J}$ and ρ/ϵ.

Invariance of the fields with respect to changes in potential is known as gauge invariance. It is used to simplify Maxwell's equations in regions where there is no free charge. In this case ψ itself is a solution of the wave equation, so that it can be adjusted to cancel ϕ and eliminate the scalar potential. This means that in (13) $\boldsymbol{\nabla}\cdot\boldsymbol{A}=0$ and, as before

$$\nabla^2\boldsymbol{A} = \mu\epsilon\frac{\partial^2\boldsymbol{A}}{\partial t^2}$$

and

$$\boldsymbol{E} = -\frac{\partial\boldsymbol{A}}{\partial t}$$

Complex Field Vectors and Potentials

The equations satisfied by the fields and potentials may be reduced to a more compact form by the construction of a complex vector whose real and imaginary parts are formed from the vectors defining the magnetic and electric fields.

In a homogeneous isotropic medium in which $\boldsymbol{D}=\epsilon\boldsymbol{E}$ and $\boldsymbol{B}=\mu\boldsymbol{H}$, a complex field vector is defined as

$$\boldsymbol{Q} = \boldsymbol{B} + i\sqrt{\epsilon\mu}\,\boldsymbol{E} \tag{4.16}$$

The Maxwell equations (1), (2) and (4) reduce to

$$\boldsymbol{\nabla}\times\boldsymbol{Q} \quad + i\sqrt{\epsilon\mu}\frac{\partial\boldsymbol{Q}}{\partial t} = \mu\boldsymbol{J} \tag{4.17}$$

$$\boldsymbol{\nabla}\cdot\boldsymbol{Q} \quad = i\sqrt{\frac{\mu}{\epsilon}}\rho \tag{4.18}$$

The vector operation $\boldsymbol{\nabla}\cdot\boldsymbol{Q}$ may be eliminated by use of the identity (12) to yield

$$\boldsymbol{\nabla\nabla}\cdot\boldsymbol{Q}-\boldsymbol{\nabla}^2\boldsymbol{Q}+i\sqrt{\epsilon\mu}\left(\boldsymbol{\nabla}\times\frac{\partial\boldsymbol{Q}}{\partial t}\right)=\mu(\boldsymbol{\nabla}\times\boldsymbol{J})$$

Resubstituting the values of $\boldsymbol{\nabla}\times\boldsymbol{Q}$ and $\boldsymbol{\nabla}\cdot\boldsymbol{Q}$ from (17) and (18) yields

$$\nabla^2\boldsymbol{Q}-\epsilon\mu\frac{\partial^2\boldsymbol{Q}}{\partial t^2}=-\mu\left(\boldsymbol{\nabla}\times\boldsymbol{J}-i\sqrt{\epsilon\mu}\frac{\partial\boldsymbol{J}}{\partial t}-i\frac{1}{\sqrt{\epsilon\mu}}\nabla\rho\right)$$

When this equation is resolved into its real and imaginary parts one obtains the equations individually satisfied by the vectors $\boldsymbol{E}$ and $\boldsymbol{H}$

$$\nabla^2\boldsymbol{H}-\epsilon\mu\frac{\partial^2\boldsymbol{H}}{\partial t^2}=\boldsymbol{\nabla}\times\boldsymbol{J} \tag{4.19}$$

$$\nabla^2\boldsymbol{E}-\epsilon\mu\frac{\partial^2\boldsymbol{E}}{\partial t^2}=\mu\frac{\partial\boldsymbol{J}}{\partial t}+\frac{1}{\epsilon}\nabla\rho \tag{4.20}$$

In a stationary field the net force transmitted across a closed surface bounding a region containing neither charge nor current is zero. If, however, the field is variable this need not be the case. Dimensional analysis shows that the quantity $\boldsymbol{g}=\frac{1}{c^2}\boldsymbol{E}\times\boldsymbol{H}$ has dimensions of momentum per unit volume. The identity

$$\int_\Sigma\boldsymbol{S}\cdot\boldsymbol{n}da=\frac{\partial}{\partial t}\int_V\boldsymbol{g}dv$$

can be interpreted on the hypothesis that there is associated with an electromagnetic field a momentum distributed with a density $\boldsymbol{g}$. The total momentum of the field contained within V is $\int_V\boldsymbol{g}dv$ (kg m s^{-1}). A direct consequence of this hypothesis is that Newton's third law and the principle of conservation of momentum are strictly valid only when the momentum of an electromagnetic field is taken into account along with that of the matter which reproduced it. There appears to be associated with an electric field an inertial property similar to that of ponderable matter.

The electromagnetic field in free space is described by the pair of coupled equations

$$\nabla^2\boldsymbol{H}=\frac{1}{c^2}\frac{\partial^2\boldsymbol{H}}{\partial t^2}\qquad\nabla^2\boldsymbol{E}=\frac{1}{c^2}\frac{\partial^2\boldsymbol{E}}{\partial t^2}$$

which together define the field vectors. The electric and magnetic vectors of the field are orthogonal to each other and describe waves moving in the direction of the Poynting vector $\boldsymbol{S}=\boldsymbol{E}\times\boldsymbol{H}$, which may be interpreted as the intensity of energy flow at a point in the field, measured in watt m^{-2}.

4.2.4 Electromagnetic Waves

The propagation of an electromagnetic field through space is conveniently described in terms of two mutually perpendicular field vectors, harmonically vibrating perpendicular to their common direction of propagation. The propagation of the two vectors, one electric and one magnetic, through space, is regulated by the product of electric permittivity and magnetic permeability of the vacuum. This product is a natural constant, which means that any electromagnetic disturbance proceeds at a fixed velocity of $c = 1/\sqrt{\epsilon_0\mu_0}$ through the vacuum. This conclusion is a result of Maxwell's theory and not an independent postulate of special relativity, as is often claimed.

The simplest description of wave motion (3.6.1) is in terms of a sine function,

$$\Phi = A\sin\left(\frac{z}{\lambda}\cdot 2\pi - \frac{t}{\tau}\cdot 2\pi\right) = A\sin 2\pi(\bar{\nu}z - \nu t)$$

An equally valid description is provided by the cosine, with the only difference that it is out of phase with the sine function by $\pi/2$. The most general wave function follows as a sum of the form

$$\Phi = A\cos 2\pi(\bar{\nu}x - \nu t) + B\sin 2\pi(\bar{\nu}x - \nu t)$$

or in complex notation

$$\Phi = Ae^{\pm 2\pi i(\bar{\nu}x - \nu t)}$$

which follows from the relationship $e^{i\theta} = \cos\theta + \sin\theta$, for phase angle θ. The angular frequency $\omega = 2\pi\nu$ and the wave vector $k = 2\pi/\lambda$.

Dielectric Phenomena

The velocity of electromagnetic waves through any material other than the vacuum is $(\epsilon\mu)^{-\frac{1}{2}} = v$ and the ratio $n = c/v$ is called the index of refraction of that material. It follows that $n = \sqrt{\epsilon\mu/\epsilon_0\mu_0}$ and, since the ratio $\mu/\mu_0 \simeq 1$, except for ferromagnetic materials, the index of refraction is commonly defined as the square root of the dielectric constant, ϵ/ϵ_0. Since the frequency of the field is not affected by the medium, refraction can be described equally well as a change of the wavelength of light passing between different transparent media.

It is commonly found that the index of refraction is not a constant for a given material, but that it changes with the frequency of the light beam. The reason for this is that absorption of light by the medium affects the index of refraction and to understand this phenomenon it is necessary to examine the absorption process more closely. The details, first recognized by Maxwell are discussed by several authors[32, 33]. Light is absorbed in a medium by a resonance process, which means that electric dipoles in the medium has a vibrational frequency that matches the frequency of the incident light. Light at frequencies below the resonance value induces oscillation of the dipoles in phase with the applied electric field of the wave. Since the velocity of vibration is zero at the extremes of the motion and reaches a maximum at the equilibrium point, it can be demonstrated by simple graphics that when the oscillations of field and medium are in phase, the field works with the motion for only half a cycle and opposes the natural motion during the other half, leading to no effective transfer of energy. As the frequency of the field increases towards the resonant frequency the oscillators in the medium fall behind the field and the same graphic analysis shows that a phase lag of $\pi/2$ at resonance causes the field to drive the oscillation all the time. When the frequency of the applied field is too high, the oscillator cannot follow the field and again the field accelerates the vibrations for only part of the time and decelerates it for the rest.

Since the absorption process is $\pi/2$ out of phase with the electromagnetic field its effect on the index of refraction will also lag in the same way behind the simple dielectric effect. The total index of refraction is therefore formulated as $\eta = n + ik$, where k is the absorptive contribution.

Polarization

If the foregoing created the impression that the electric and magnetic vectors of a propagating electromagnetic field are confined to vibrate in fixed planes, that was unintentional. A field like that would be plane polarized and to create that requires a special device known as a polarizer. Ordinary unpolarized light consists of an array of plane waves that are randomly oriented with respect to a plane perpendicular to the directions of propagation.

Any plane polarized wave may be represented by a vector whose amplitude is given by

$$\boldsymbol{E} = \boldsymbol{E}_0 \cos 2\pi(\bar{\nu}z - \nu t) = \boldsymbol{E}_0 \cos\theta \tag{4.21}$$

If the plane of polarization is defined as xz, the electric vector can be represented in terms of the unit vector $\boldsymbol{i}$ in the direction of x as $\boldsymbol{E} = E_0\boldsymbol{i}\cos\theta$. This plane polarized wave may be decomposed into two circularly polarized components propagating in the same direction. A simple description of these waves are given by

$$\boldsymbol{E_r} = E_0(\boldsymbol{i}\cos\theta_r - \boldsymbol{j}\sin\theta_r)$$

$$\boldsymbol{E}_l = E_0(\boldsymbol{i}\cos\theta_l - \boldsymbol{j}\sin\theta_l)$$

for left and right circularly polarized waves respectively. When the phase angles $\theta_r = \theta_l = \theta$, the sum of the two waves is simply $\boldsymbol{E} = E_0\boldsymbol{i}\cos\theta$.

The meaning of circular polarization is explained in terms of Figure 1, showing the vector sum $\boldsymbol{i}\cos\theta + \boldsymbol{j}\sin\theta$ for a half wave.

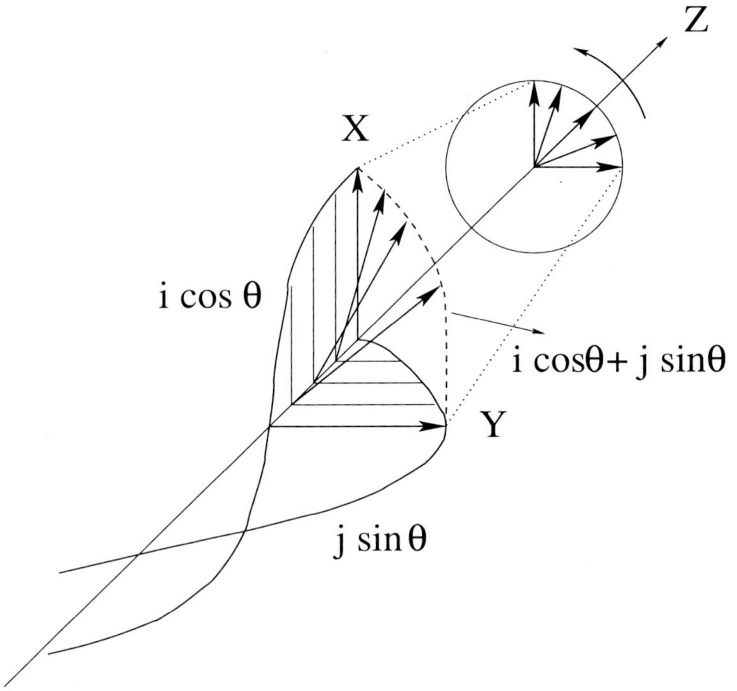

Figure 1: *Graphical decomposition of the plane wave $\boldsymbol{i}\cos\theta$ into clockwise and counterclockwise rotating components. Construction and projection of only the left circulating component $\boldsymbol{i}\cos\theta + \boldsymbol{j}\sin\theta$ is shown.*

The composite vector is seen to spiral around the z-axis and in projection moves anti-clockwise in a circle around the z-axis. The other component which is the mirror image of the first, performs a clockwise circular motion in projection along z. The decomposition into circularly polarized components can also be formulated in complex notation , $\boldsymbol{E}_{r,l} = \boldsymbol{E}_0 e^{\pm i\theta}$.

Optical Activity Certain media are optically active, which means that they interact with plane polarized light by changing the direction of polarization. A simple explanation of this effect is that optically active material interacts differently with left and right circularly polarized light respectively, for instance by exhibiting different indices of refraction with respect to these two types of polarized light. If this is the case one or the other of the circular components of plane polarized light would lag behind in that medium.

Rewriting (21) in the form

$$\boldsymbol{E} = \boldsymbol{E}_0 \cos 2\pi\nu(\frac{nz}{c} - t)$$

using $n = c/v$, it follows that at a time t after entering the active medium the phases θ_r and θ_l will no longer be the same, *i.e.*

$$\begin{aligned}
\theta_r &= 2\pi\nu\left(\frac{n_r z}{c} - t\right) \\
\theta_l &= 2\pi\nu\left(\frac{n_l z}{c} - t\right) \\
\theta_l - \theta_r &= \frac{2\pi\nu z}{c}(n_l - n_r)
\end{aligned}$$

and the sum of the components

$$\begin{aligned}
\boldsymbol{E} = \boldsymbol{E_r} + \boldsymbol{E_l} &= E_0\boldsymbol{i}\left(\cos\theta_r + \cos\theta_l\right) - E_0\boldsymbol{j}\left(\sin\theta_r - \sin\theta_l\right) \\
&= 2E_0\cos\left(\frac{\theta_r + \theta_l}{2}\right)\left[\boldsymbol{i}\cos\left(\frac{\theta_r - \theta_l}{2}\right) - \boldsymbol{j}\sin\left(\frac{\theta_r - \theta_l}{2}\right)\right]
\end{aligned}$$

Since $\boldsymbol{E}$ now has both x and y components it is no longer parallel to the x-axis, having rotated by an angle

$$\delta = \frac{\theta_l - \theta_r}{2} = \frac{\pi\nu z}{c}(n_l - n_r) = \frac{\pi z}{\lambda}(n_l - n_r) \ .$$

δ is the angle (in radians) by which the plane-polarized wave has been rotated in traversing the optically active medium of thickness z.

If the optically active medium is not transparent at the wavelength of the incident radiation, the transmitted intensity may be further reduced by an absorptive contribution to the index of refraction. Because of preferential absorption of either the left or the right circularly polarized component, the emerging beam would no longer be the sum of equal amplitudes and trace out an ellipse with ellipticity $\psi = \frac{\pi z}{\lambda}(k_l - k_r)$. Practical details of the measurement and chemical applications of optical activity are discussed by Charney[34].

4.3 Special Relativity

4.3.1 Introduction

Maxwell's theory of the electromagnetic field represents the reduction of several apparently disparate aspects of physics, to a common basis. It shows how electric, magnetic and optical effects are all shaped by the characteristics of empty space and suggests that mechanics probably reveals another aspect of the same principle. The single most important factor in the equation is the constancy of $c = 1/\sqrt{\epsilon_0\mu_0}$, which is a property of empty space. The constant, c has the dimensions of a velocity and represents the natural rate of progression of an electromagnetic disturbance through space, according to the general equation of wave motion

$$\nabla^2\Phi = \frac{1}{c^2}\frac{\partial^2\Phi}{\partial t^2} \tag{4.22}$$

The disturbance carries momentum and energy, but by a mechanism that seems to be inconsistent with the accepted ideas of classical dynamics. According to Newton's second law the force exerted on a particle of mass m at vector position $\boldsymbol{r}$, is $\boldsymbol{F} = m\ddot{\boldsymbol{r}}$. Seen from another frame of reference, moving at a constant velocity $\boldsymbol{v}$, relative to the first, the position of the particle is perceived to be

$$\boldsymbol{r}' = \boldsymbol{r} - \boldsymbol{v}t \tag{4.23}$$

It follows immediately that the observed force, as viewed by two observers remains constant,

$$\boldsymbol{F}' = m\ddot{\boldsymbol{r}}' = m\ddot{\boldsymbol{r}} = \boldsymbol{F}$$

This invariance defines the principle of (Galilean) relativity, once thought to be universally valid. However, the situation for electromagnetic waves is different and the form of equation (22) is destroyed[3] under a transformation

[3]For relative motion along z the Galilean transformation is $x' = x$, $y' = y$, $z' = z \pm vt$, $t' = t$. To allow for the equivalence of relative motions of $\pm v$, it is sufficient to define the variable $\Phi = \Phi(z,t)_\pm$, *i.e.*

$$\frac{\partial\Phi}{\partial z'} = \left[\frac{\partial\Phi}{\partial z}\cdot\frac{\partial z}{\partial z'}\right]_\pm + \left[\frac{\partial\Phi}{\partial t}\cdot\frac{\partial t}{\partial z'}\right]_+ - \left[\frac{\partial\Phi}{\partial t}\cdot\frac{\partial t}{\partial z'}\right]_-$$

$$\text{where}\quad \frac{\partial z}{\partial z'} = \left(\frac{c}{c\pm v}\right), \quad \frac{\partial t}{\partial z'} = \left(\frac{1}{c\pm v}\right)$$

$$\frac{\partial^2\Phi}{\partial(z')^2} = \frac{\partial^2\Phi}{\partial z^2}\left(\frac{\partial z}{\partial z'}\right)_+\left(\frac{\partial z}{\partial z'}\right)_- + \frac{\partial^2\Phi}{\partial z'\partial t'}\cdot\left[\left(\frac{\partial t}{\partial z'}\right)_+ - \left(\frac{\partial t}{\partial z'}\right)_-\right]$$

After rearrangement

$$\frac{\partial^2\Phi}{\partial z^2} = \left[\frac{(c-v)(c+v)}{c^2}\right]\frac{\partial^2\Phi}{\partial(z')^2} - \frac{\partial^2\Phi}{\partial z'\partial t'}\left[\left(\frac{1}{c+v}\right) - \left(\frac{1}{c-v}\right)\right]\frac{(c-v)(c+v)}{c^2}$$

such as (23). For uniform motion in one dimension (z say) the equation becomes

$$\frac{\partial^2\Phi}{\partial(x')^2}+\frac{\partial^2\Phi}{\partial(y')^2}+\left(1-\frac{v^2}{c^2}\right)\frac{\partial^2\Phi}{\partial(z')^2}+\frac{2v}{c^2}\cdot\frac{\partial^2\Phi}{\partial z'\partial t'}=\frac{1}{c^2}\frac{\partial^2\Phi}{\partial(t')^2}$$

The problem relates directly to the constancy of c, which implies that the velocity of light is independent of both the motion of its source and the direction of propagation, a condition that cannot hold in more than one Newtonian inertial frame if the Galilean principle of relativity applies. Since there is no evidence that the laws of physics are not identical in all inertial frames of reference the only conclusion is that the prescription for Galilean transformations needs modification to be consistent, not only with simple mechanics, but also with electromagnetic effects.

A practical test of electromagnetic transformations, known as the experiment of Michelson and Morley has become virtually synonomous with the theory of special relativity in many minds. In simple terms the experiment was set up to test the Galilean proposition that light should go at speeds of $c\pm v$ in, and against the direction of the earth's orbital motion at linear speed v. The optical measurement was designed to look for interference effects due to different velocities of light beams reflected parallel to and perpendicular to the orbital direction. After the first measurement the entire experimental assembly is turned through ninety degrees and the measurement repeated. This precaution eleminates spurious results due to accidental differences in path lengths. The experiment is equivalent to comparison of times taken by identical power boats to cover equal distances during return journeys on flowing water, parallel and perpendicular to the flow, as shown in figure 2.

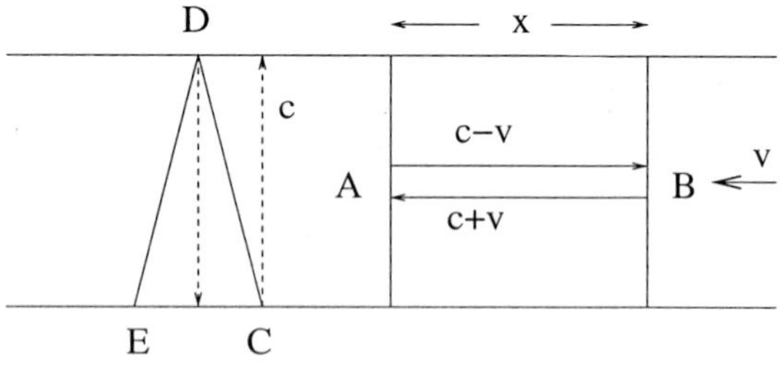

---

i.e. $\frac{\partial^2\Phi}{\partial z^2}=\left(1-\frac{v^2}{c^2}\right)\frac{\partial^2\Phi}{\partial(z')^2}+\frac{2v}{c^2}\cdot\frac{\partial^2\Phi}{\partial z'\partial t'}$

Figure 2: *Two identical power boats covering the same distance $2x$ (CDE and ABA measured at right angles) in times of $T_\perp = \frac{2x}{c}$ and $T_\| = \frac{2x}{c}\frac{1}{1-v^2/c^2}$ respectively.* The device travelling at velocity c perpendicular to the stream covers the projected distance $2x$ in time $T_\perp = 2x/c$. The other device travels two unequal legs. Going against the flow (velocity v) it needs more time to cover the distance x than going with the stream. Going down stream it therefore has insufficient time to recover the ground lost while going against the stream. Simple arithmetic gives the total time $T_\| = t_{up} + t_{down} = \frac{x}{c+v} + \frac{x}{c-v} = T_\perp/(1-\beta^2)$, $\beta = v/c$. The well known null result of the experiment confirmed that electromagnetic radiation does not obey galilean transformation theory.

4.3.2 The Lorentz Transformation

A particularly simple, though elegant way to arrive at the appropriate transformation that leaves (22) invariant[4] is due to Born[35]. Consider two inertial systems S and S' in relative motion along z. The origin of S' with respect to S at time t' has the coordinate $z = vt$, while in S' $z' = 0$. To ensure that the two expressions are consistent it is necessary to stipulate that

$$z - vt = \alpha z' \ .$$

Since the two systems are equivalent the same argument, but with the sign of v inverted, with reference to the origin of S relative to S', then requires that $z' + vt$ is proportional to z, with the same proportionality constant, *i.e.*

$$z' + vt' = \alpha z.$$

Combination of the two expressions leads to

$$vt' = \alpha z - z' = \alpha z - \left(\frac{z - vt}{\alpha}\right) = \frac{1}{\alpha}\{(\alpha^2 - 1)z + vt\}$$

also

$$\alpha t' = \left(\frac{\alpha^2 - 1}{v}\right) z + t$$

The constant α is now chosen so that the velocity of light is constant. A uniform velocity in the two systems is represented respectively by $u = z/t$,

[4]The differentiation procedure of footnote 1 uses the same approach.

and $u' = z'/t'$. It follows directly that

$$\begin{aligned} u' = \frac{z'}{t'} &= \frac{\alpha z'}{\alpha t'} = \frac{z - vt}{\left(\frac{\alpha^2-1}{v}\right) z + t} = \frac{\frac{z}{t} - v}{\left(\frac{\alpha^2-1}{v}\right)\frac{z}{t} + 1} \\ &= \frac{u - v}{\left(\frac{\alpha^2-1}{v}\right) u + 1} \end{aligned}$$

Setting $u = u' = c$, one has

$$c = \frac{c - v}{\left(\frac{\alpha^2-1}{v}\right) c + 1} \quad \text{, which reduces to}$$

$$\alpha^2 - 1 = -\frac{v^2}{c^2} = -\beta^2 \quad \text{, or} \quad \alpha = \sqrt{1-\beta^2} = 1/\gamma$$

The last step is introduced to conform with the conventional definition of γ. The required transformation is

$$z' = \frac{z - vt}{\sqrt{1 - \frac{v^2}{c^2}}} \quad , \quad t' = \frac{t - \frac{vz}{c^2}}{\sqrt{1 - \frac{v^2}{c^2}}} \tag{4.24}$$

It is now found that (22) is indeed invariant under (24), which is known as the Lorentz[5] transformation of Special Relativity. It is important to note that in the limit $v/c \to 0$ the Lorentz formulae reduce to the Galilean transformation, suggesting that Lorentzian (relativistic) effects only become significant at relative velocities that approach c. The condition $t' = t$ which

[5] It was pointed out by Lorentz[36] that the same transformation equations (24) were obtained for (22) by Voigt (Göttinger Nachrichte, 1887, p.41) in an analysis of the Doppler effect.

Now $\frac{\partial z'}{\partial z} = \frac{\partial t'}{\partial t} = \frac{1}{\gamma}\left(\frac{c}{c \pm v}\right)$; $\frac{\partial t}{\partial z'} = \left(\frac{1}{\gamma(c \pm v)}\right)$

$$\begin{aligned} \frac{\partial\phi}{\partial z'} &= \frac{\partial\phi}{\partial z}\frac{\partial z}{\partial z'} + \frac{\partial\phi}{\partial t}\frac{\partial t}{\partial z'} = \frac{c}{\gamma(c \pm v)}\left[\frac{\partial\phi}{\partial z} + \frac{1}{c}\frac{\partial\phi}{\partial t}\right] \\ \frac{\partial^2\phi}{\partial(z')^2} &= \left[\frac{c^2}{\gamma^2(c-v)(c+v)}\right]\frac{\partial^2\phi}{\partial z^2} + \frac{1}{c}\frac{\partial^2\phi}{\partial z'\partial t'} = \frac{\partial^2\phi}{\partial z^2} + \frac{1}{c}\frac{\partial^2\phi}{\partial z'\partial t'} \\ \frac{\partial^2\phi}{\partial(t')^2} &= \frac{\partial^2\phi}{\partial t^2} + c\frac{\partial^2\phi}{\partial z'\partial t'} \end{aligned}$$

$$\text{Hence} \qquad \frac{\partial^2\phi}{\partial(z')^2} - \frac{1}{c^2}\frac{\partial^2\phi}{\partial(t')^2} = \frac{\partial^2\phi}{\partial z^2} - \frac{1}{c^2}\frac{\partial^2\phi}{\partial t^2}$$

is implied by the Galilean transformation means that an absolute standard of simultaneity exists, upon which all inertial observers agree, and furthermore that the distance between two simultaneous events is also the same for all such observers. This distance can be considered to have absolute meaning independent of any observer and may be thought of as a property of space itself. The time interval between two events likewise acquires an absolute meaning. The Lorentz transformation however, preserves neither simultaneity nor distance.

The most important new feature of the Lorentz transformation, absent from the Galilean scheme, is this interdependence of space and time dimensions. At velocities approaching c it is no longer possible to consider the cartesian coordinates of three-dimensional space as being independent of time and the three-dimensional line element $d\sigma = \sqrt{x^2 + y^2 + z^2}$ is no longer invariant within the new relativity. Suppose a point source located at the origin emits a light wave at time $t = 0$. The equation of the wave front is that of a sphere, radius r, such that

$$r^2 = x^2 + y^2 + z^2 = (ct)^2$$

The transformed wave front observed in a primed system at relative velocity v_z, follows from (24):

$$\begin{aligned}(x')^2 + (y')^2 + (z')^2 - (ct')^2 &= x^2 + y^2 + \frac{1}{\gamma^2}\left[(z - vt)^2 - \left(ct - \frac{vz}{c^2}\right)\right] \\ &= x^2 + y^2 + z^2 - (ct)^2\end{aligned}$$

which is also a sphere expanding with speed c. Since the Lorentz transformation involves transformation of time intervals in addition to space intervals, it is convenient to introduce the concept of an *event*, which is specified by the four coordinates x, y, z, ict, which locate it in space and time, also called *Minkowski space*. The invariant element is readily identified as the interval s, defined by

$$\begin{aligned} s^2 &= x^2 + y^2 + z^2 + (ict)^2 \\ \text{or} \quad ds^2 &= dx^2 + dy^2 + dz^2 + d(ict)^2 \\ &= \sum_{\mu=1}^{4} dx_\mu^2 \, , \\ \text{or} \quad \mu &= 0, 1, 2, 3, \quad \text{as preferred by many authors.} \end{aligned} \tag{4.25}$$

The interval ds is also known as the *proper time* between two nearby events ($\boldsymbol{r}, t$) and ($\boldsymbol{r} + d\boldsymbol{r}, t + dt$). This formulation does not imply that any one of the axes is transformable into any other by a rotation, since the complex nature of the time axis differentiates it from the others. The quantity ds is known as the invariant interval between two events, and again, its invariance gives it an absolute meaning independent of any observer.

As a particle moves in ordinary space its corresponding point in 4-space describes a path known as the *world line.* The vector dx_μ represents the change in the position 4-vector for differential motion along a world line. The absolute magnitude of the 4-vector is described by the dot product with itself and defines the invariant *world scalar*

$$(d\tau)^2 = -\frac{1}{c^2}\sum_{\mu=1}^{4} dx_\mu \cdot dx_\mu \equiv -\frac{1}{c^2} dx_\mu dx_\mu$$

This expression demonstrates use of the Einstein summation convention [6]. The significance of τ is made clear by examining a particle momentarily at rest in a Lorentz system. The components of the vector, transformed $dx'_\mu = (0, 0, 0, icdt')$ and $d\tau^2 = -(1/c^2)dx'_\mu dx'_\mu = (dt')^2$. Thus $d\tau$ is the time interval on a clock travelling with the particle and is therefore referred to as the interval of the particle's *proper time* or *world time.* The relationship between $d\tau$ and an interval of time as measured in a given Lorentz system can be derived directly by expanding the equation

$$(d\tau)^2 = -\frac{1}{c^2}\left[(dx)^2 + (dy)^2 + (dz)^2 - c^2(dt)^2\right]$$

or

$$d\tau = dt\sqrt{1 - \frac{1}{c^2(dt)^2}\left[(dx)^2 + (dy)^2 + (dz)^2\right]}$$

which is equivalent to the relation[7]

$$\frac{d\tau}{\sqrt{1-\beta^2}} = dt$$

The equation says that a time interval measured in the rest system is always longer than the corresponding time interval observed in a system in which the particle is not at rest. This is an example of *time dilation.*

[6] The repeated appearance of an index (j say) indicates that the lhs in an expression like $x'_i = a_{ij}x_j \quad (i = 1, 2, 3)$ is a sum over the dummy index j for all possible values (here $j = 1, 2, 3$). Whenever an index occurs two or more times in a term, it is implied, without any further symbols, that the terms are to be summed over all possible values of the index.

[7] β is here used interchangeably for both the relative velocity (in units of c) between the observer's system and the instantaneous rest system, and for the observed velocity of the particle.

Four-vectors for which the square of the magnitude is greater than or equal to zero are called *space-like*; when the squares of the magnitudes are negative they are known as *time-like* vectors. Since these characteristics arise from the dot products of the vectors with reference to themselves, which are world scalars, the designations are invariant under Lorentz transformation[17]. A space-like 4-vector can always be transformed so that its fourth component vanishes. On the other hand, a time-like four-vector must always have a fourth component, but it can be transformed so that the first three vanish. The difference between two world points can be either space-like or time-like. Let $\boldsymbol{X}_\mu$ be the difference vector

$$\boldsymbol{X}_\mu = x_{1\mu} - x_{2\mu}$$

between events 1 and 2. The magnitude of $\boldsymbol{X}_\mu$ is given by

$$\boldsymbol{X}_\mu \cdot \boldsymbol{X}_\mu = |\boldsymbol{r}_1 - \boldsymbol{r}_2|^2 - c^2(t_1 - t_2)^2$$

Thus, $\boldsymbol{X}_\mu$ is space-like if two world points are separated such that $|\boldsymbol{r}_1 - \boldsymbol{r}_2|^2 > c^2(t_1 - t_2)^2$, while while it is time-like if $|\boldsymbol{r}_1 - \boldsymbol{r}_2|^2 < c^2(t_1 - t_2)^2$.

The condition for a time-like difference vector is equivalent to stating that it is possible to bridge the distance between the two events by a light signal, while if the points are separated by a space-like difference vector, they cannot be connected by any wave travelling with the speed c. If the spatial difference vector $|\boldsymbol{r}_1 - \boldsymbol{r}_2|$ is along the z axis, such that $|\boldsymbol{r}_1 - \boldsymbol{r}_2| = z_1 - z_2$, under a Lorentz transformation with velocity v parallel to the z axis, the fourth component of $\boldsymbol{X}_\mu$ transforms as

$$c(t_1 - t_2) = \frac{c(t_1 - t_2) - \frac{v}{c}(z_1 - z_2)}{\sqrt{1 - \beta^2}}$$

If $\boldsymbol{X}_\mu$ is space-like and the events are designated such that $t_2 > t_1$, then $c(t_1 - t_2) < z_1 - z_2$, and it is therefore possible to find a velocity $v < c$ such that $ic(t_1' - t_2') \equiv \boldsymbol{X}_4'$ vanishes. Physically the vanishing of $\boldsymbol{X}_4'$ means that if the distance between two events is space-like, then one can always find a Lorentz system in which the two events have the same time coordinate in the selected frame. On the other hand, for time-like separations between events one cannot find a Lorentz transformation that will make them simultaneous, or change the order of the time sequence of the two events. The concepts "future" and "past" are invariant and causality is preserved. That the sequence of events with space-like separations can be reversed does not violate causality. As an example it is noted that no influence eminating from earth can affect an object one light-year away within the next year.

For the special case $|\boldsymbol{r}_1 - \boldsymbol{r}_2|^2 = c^2(t_1 - t_2)^2$, *i.e.* for a relative velocity of c, the world vector $\boldsymbol{X}_\mu = 0$. The surprising conclusion is that two world points receding at the relative speed of light remain in physical contact, indicating that c should not be interpreted as a velocity at all. Transmission of signals along the worldline of a photon does not correspond to the type of motion normally associated with massive particles and represents a situation in which time and space coordinates coincide.

Four-vectors in Minkowski Space

The four-velocity $\boldsymbol{u}_\nu$ is defined as the rate of change of the position vector of a particle with respect to its proper time

$$\boldsymbol{u}_\nu = \frac{dx_\nu}{d\tau}$$

with space and time components of $u_i = v_i/\sqrt{1-\beta^2}$ and $u_4 = ic/\sqrt{1-\beta^2}$. The world velocity, defined by analogy with the world point, has a constant magnitude given by the sum

$$\boldsymbol{u}_\nu \cdot \boldsymbol{u}_\nu = \frac{1}{1-\beta^2} - \frac{c^2}{1-\beta^2} = -c^2$$

and thus, is also time-like. In the same way a 4-vector force (Minkowski force) is defined by

$$\boldsymbol{K}_\nu = \frac{d}{d\tau}(\boldsymbol{u}_\nu) \tag{4.26}$$

At small velocities this should reduce to

$$F_i = \frac{d}{dt}(mv_i)$$

From the relation between τ and t, and the definition of the world velocity, the spatial components of (26) can be written

$$\frac{d}{dt}\left(\frac{mv_i}{\sqrt{1-\beta^2}}\right) = K_i\sqrt{1-\beta^2}$$

by defining the relationship between Minkowski and ordinary force components $K_i = F_i/\sqrt{1-\beta^2}$.

Four-momentum is defined by

$$\boldsymbol{p}_\nu = m\boldsymbol{u}_\nu \quad , \quad \frac{d\boldsymbol{p}_\nu}{dt} = \boldsymbol{K}_\nu \quad ,$$

with spatial part, $p_i = \frac{m_0 v_i}{\sqrt{1-\beta^2}}$, where m_0 is the mass measured in the frame at rest with the particle (*i.e.* rest frame or proper frame) and m_0 is called the rest mass or proper mass. The time-like part of $\boldsymbol{K}_\nu$ is obtained from the dot product of (26) with the world velocity

$$\boldsymbol{u}_\nu \cdot \frac{d}{d\tau}(m\boldsymbol{u}_\nu) = \frac{d}{d\tau}\left(\frac{m}{2}\boldsymbol{u}_\nu \cdot \boldsymbol{u}_\nu\right) = \boldsymbol{K}_\nu \boldsymbol{u}_\nu \tag{4.27}$$

Since $\boldsymbol{u}_\nu \cdot \boldsymbol{u}_\nu = -c^2$ and m is also constant the lhs of (27) vanishes, leaving[8]

$$\boldsymbol{K}_\nu \boldsymbol{u}_\nu \equiv \frac{\boldsymbol{F} \cdot \boldsymbol{v}}{\sqrt{1-\beta^2}} + \frac{icK_4}{\sqrt{1-\beta^2}} = 0$$

The fourth component of the Minkowski force therefore is

$$K_4 = \frac{i}{c} \frac{\boldsymbol{F} \cdot \boldsymbol{v}}{\sqrt{1-\beta^2}}$$

and the corresponding fourth component according to (26) appears as

$$\frac{d}{dt} \frac{m_0 c^2}{\sqrt{1-\beta^2}} = \boldsymbol{F} \cdot \boldsymbol{v} = \frac{dT}{dt}$$

When integrating the expression for kinetic energy (T) between time limits t_1 can be taken to correspond to the time at which the particle was at rest, ($\beta = 0$), *i.e.*

$$\begin{aligned} \int_{t_1}^{t_2} \frac{dT}{dt} dt &= T_2 - T_1 = \frac{m_0 c^2}{\sqrt{1-\beta^2}} - m_0 c^2 \\ &= mc^2 - m_0 c^2 \end{aligned} \tag{4.28}$$

where $mc^2 = E$ is defined as the total energy.

In terms of (28) the fourth component of the momentum may be expressed as

$$p_4 = \frac{m_0 c}{\sqrt{1-\beta^2}} = mc = \frac{E}{c} \quad .$$

Hence, the four-vector momentum may be written

$$\boldsymbol{P}_\nu = p_i + \frac{iE}{c}$$

[8] $\boldsymbol{F} \cdot \boldsymbol{v}$ is the work done on a particle by the force per unit time and is equal to the time rate of change of the kinetic energy and $F_j = \dot{p}_j$.

Momentum and energy are therefore linked together in a manner like that which joins the concepts of space and time[9]. The world momentum scalar

$$\begin{aligned} \boldsymbol{P}_\nu \cdot \boldsymbol{P}_\nu &= m_0^2 \boldsymbol{u}_\nu \cdot \boldsymbol{u}_\nu = -m_0^2 c^2 \\ &= p^2 - \frac{E^2}{c^2} \end{aligned}$$

and hence $E^2 = p^2c^2 + m_0^2c^4$.

The Lorentz transformation is an orthogonal transformation in the four dimensions of Minkowski space. The condition of constant c is equivalent to the requirement that the magnitude of the 4-vector $\boldsymbol{s}$ be held invariant under the transformation. In matrix notation

$$\boldsymbol{x}' = \boldsymbol{L}\boldsymbol{x}$$

with $L_{\mu\nu}$ as the general element[10]. For the familiar example of Lorentz transformation with relative velocity along z only, the matrix $\boldsymbol{L}$ takes the simple form

$$\boldsymbol{L} = \begin{pmatrix} 1 & 0 & 0 & 0 \\ 0 & 1 & 0 & 0 \\ 0 & 0 & \gamma & i\beta\gamma \\ 0 & 0 & -i\beta\gamma & \gamma \end{pmatrix} \tag{4.29}$$

Looking at (29) it is seen that the 2×2 submatrix of the third and fourth coordinates resembles the corresponding submatrix in the rotation of a three-dimensional coordinate system

$$\boldsymbol{R} = \begin{pmatrix} \cos\phi & \sin\phi \\ -\sin\phi & \cos\phi \end{pmatrix}$$

Equation (29) is therefore equivalent to rotation in the x_3x_4 plane of Minkowski space through an imaginary angle ϕ, such that

$$\cos\phi = 1/\sqrt{1-\beta^2}, \quad \sin\phi = i\beta/\sqrt{1-\beta^2}. \tag{4.30}$$

If the transformation is formulated in terms of the real angle $\psi = i\phi$, for which $\cosh\psi = 1/\sqrt{1-\beta^2}$, $\sinh\psi = \beta/\sqrt{1-\beta^2}$ the submatrix in (29)

[9]An important corollary of this analogy implies that the conservation of momentum is a consequence of the isotropy of space, whereas energy conservation is dictated by time-inversion symmetry.

[10]It is convention to use Greek letters for a subscript that runs over all four coordinates, and Roman letters for subscripts only over the space components.

can be written,

$$\begin{pmatrix} \cosh\psi & i\sinh\psi \\ -i\sinh\psi & \cosh\psi \end{pmatrix}$$

Since two successive Lorentz transformations for relative velocities β and β' along z_3 are rotations in the same plane, the rotation angles simply add. From (30) $\tan\phi = i\beta$, and since

$$\tan\phi'' = \tan(\phi+\phi') = \frac{\tan\phi + \tan\phi'}{1-\tan\phi\tan\phi'}$$

the overall relative speed

$$\beta'' = \frac{\beta+\beta'}{1+\beta\beta'} .$$

This formula defines the Einstein addition law for parallel velocities. It shows that, no matter how closely β and β' approach unity β'' can never exceed unity. In this sense c can be considered to be the ultimate speed allowed by special relativity.

The invariance of the general wave equation (22) is a special consequence of a more general four-vector invariance involving the four-gradient defined as

$$\Box = \frac{\partial}{\partial x} + \frac{\partial}{\partial y} + \frac{\partial}{\partial z} + \frac{1}{ic}\frac{\partial}{\partial t}$$

The second derivative function[11] is Lorentz invariant. The wave equation in this terminology is

$$\Box^2\phi = 0 \quad .$$

Electromagnetism

It is to be expected that the equations relating electromagnetic fields and potentials to the charge current, should bear some resemblance to the Lorentz transformation. Stating that the equations for $\boldsymbol{A}$ and ϕ are Lorentz invariant, means that they should have the same form for any observer, irrespective of relative velocity, as long as it's constant. This will be the case if the quantity $(A_x, A_y, A_z, i\phi/c) = \boldsymbol{V}$ is a Minkowski four-vector. Easiest would be to show that the dot product of $\boldsymbol{V}$ with another four-vector, *e.g.* the four-gradient, is Lorentz invariant, *i.e.* to show that

$$\Box\cdot\boldsymbol{V} = \nabla\cdot\boldsymbol{A} + \frac{1}{c^2}\frac{\partial\phi}{\partial t} \tag{4.31}$$

---

[11] The operator $\Box^2 = \nabla^2 - \frac{1}{c^2}\frac{\partial^2}{\partial t^2}$ is called the d'Alembertian.

is a constant. The rhs of (31) is just the equation that relates the vector and scalar potentials of the electromagnetic field

$$\nabla \cdot \boldsymbol{A} + \mu\epsilon \frac{\partial \phi}{\partial t} = 0$$

which means that the four-potentials transform together as a Minkowski four-vector.

By defining a four-vector with the components of the current-density vector $\boldsymbol{J}$ and charge density, *i.e.*

$$\boldsymbol{I} = (J_x, J_y, J_z, ic\rho)$$

the same procedure shows that

$$\Box \cdot \boldsymbol{I} = \nabla \cdot \boldsymbol{J} + \frac{\partial \rho}{\partial t}(= 0)$$

where the equality in parentheses completes the electromagnetic continuity equation and proves the invariance of $\boldsymbol{I}$.

The equation

$$\begin{aligned} \Box^2 \boldsymbol{V} &= \nabla^2 \boldsymbol{V} - \frac{1}{c^2}\frac{\partial^2 \boldsymbol{V}}{\partial t^2} \\ &= \nabla^2 \boldsymbol{A} + \frac{i}{c}\nabla^2 \phi - \frac{1}{c^2}\left(\frac{\partial^2 \boldsymbol{A}}{\partial t^2} + \frac{i}{c}\frac{\partial^2 \phi}{\partial t^2}\right) \\ &= \left(\nabla^2 \boldsymbol{A} - \frac{1}{c^2}\frac{\partial \boldsymbol{A}}{\partial t^2}\right) + \frac{i}{c}\left(\nabla^2 \phi - \frac{1}{c^2}\frac{\partial^2 \phi}{\partial t^2}\right) \end{aligned}$$

is readily simplified by substituting for the terms in brackets from the wave equations (14,15) for the potentials, *i.e.*

$$\begin{aligned} \Box^2 \boldsymbol{V} &= -\mu \boldsymbol{J} - \frac{i}{c}\frac{\rho}{\epsilon} \\ &= -\mu\left(\boldsymbol{J} + ic\rho\right) \\ &= -\mu \boldsymbol{I} \end{aligned}$$

The relativistic invariance of the electromagnetic field is conveniently expressed in tensor notation. Factorized in Minkowski space the Maxwell equations (1) and (3) are represented by

$$\begin{aligned}
0+\frac{\partial E_3}{\partial x_2}-\frac{\partial E_2}{\partial x_3}+ic\frac{\partial B_1}{\partial x_4} &= 0\\
-\frac{\partial E_3}{\partial x_1}+0+\frac{\partial E_1}{\partial x_3}+ic\frac{\partial B_2}{\partial x_4} &= 0\\
\frac{\partial E_2}{\partial x_1}-\frac{\partial E_1}{\partial x_2}+0+ic\frac{\partial B_3}{\partial x_4} &= 0\\
-\frac{\partial B_1}{\partial x_1}-\frac{\partial B_2}{\partial x_2}-\frac{\partial B_3}{\partial x_3}+0 &= 0
\end{aligned}$$

After division of the first three equations by ic the set of equations can be reduced to a simpler form in terms of an antisymmetric tensor defined as

$$G_{jk}=\begin{pmatrix} 0 & -\frac{i}{c}E_3 & \frac{i}{c}E_2 & B_1\\ \frac{i}{c}E_3 & 0 & -\frac{i}{c}E_1 & B_2\\ -\frac{i}{c}E_2 & \frac{i}{c}E_1 & 0 & B_3\\ -B_1 & -B_2 & -B_3 & 0 \end{pmatrix}$$

Alternatively, starting from Maxwell's equations (2) and (4) and writing $x_4 = ict$, an equivalent set of equations is obtained in the form

$$\begin{aligned}
0+\frac{\partial B_3}{\partial x_2}-\frac{\partial B_2}{\partial x_3}-\frac{i}{c}\frac{\partial E_1}{\partial x_4} &= \mu_0 J_1\\
-\frac{\partial B_3}{\partial x_1}+0+\frac{\partial b_1}{\partial x_3}-\frac{i}{c}\frac{\partial E_2}{\partial x_4} &= \mu_0 J_2\\
\frac{\partial B_2}{\partial x_1}-\frac{\partial B_1}{\partial x_2}+0-\frac{i}{c}\frac{\partial E_3}{\partial x_4} &= \mu_0 J_3\\
\frac{i}{c}\frac{\partial E_1}{\partial x_1}+\frac{i}{c}\frac{\partial E_2}{\partial x_2}+\frac{i}{c}\frac{\partial E_3}{\partial x_3}+0=\frac{i\rho}{c\epsilon_0} &\equiv \mu_0 J_4
\end{aligned}$$

which defines the antisymmetric tensor

$$F_{\mu\nu}=\begin{pmatrix} 0 & B_3 & -B_2 & -\frac{i}{c}E_1\\ -B_3 & 0 & B_1 & -\frac{i}{c}E_2\\ B_2 & -B_1 & 0 & -\frac{i}{c}E_3\\ \frac{i}{c}E_1 & \frac{i}{c}E_2 & \frac{i}{c}E_3 & 0 \end{pmatrix}$$

In the tensor formulation the difference between electric and magnetic fields disappears. What one observer interprets as an electric process another may regard as magnetic, although the actual particle motions that they predict will be identical[12] [37].

The set of equations that defines F_{jk} is readily shown to be contained in the formula

$$\frac{\partial F_{ij}}{\partial x_k}+\frac{\partial F_{ki}}{\partial x_j}+\frac{\partial F_{jk}}{\partial x_i}=0 \qquad (4.32)$$

where i,j,k are any three of the four numbers 1,2,3,4.

Equation (32) is satisfied identically by the field tensor

$$F_{\mu\nu}=\frac{\partial V_\nu}{\partial x_\mu}-\frac{\partial V_\mu}{\partial x_\nu} \quad (\mu,\nu=1,2,3,4) \qquad (4.33)$$

defined in terms of the Minkowski four vector $\boldsymbol{V}$ of vector (5) and scalar (9) potentials. Tensor components according to (9) and (5) are

$$\begin{aligned}-\frac{i}{c}E_\mu &= \frac{\partial}{\partial x_j}\left(\frac{i}{c}\phi\right)-\frac{\partial A_j}{\partial x_4}\\ B_\mu &= \frac{\partial A_k}{\partial x_j}-\frac{\partial A_j}{\partial x_k}\end{aligned}$$

Tensor formalism can be used to generate all aspects of the electromagnetic field. The inhomogeneous Maxwell equation (2) in tensor form

$$\frac{\partial F_{jk}}{\partial x_k}=\mu_0 J^k \qquad (4.34)$$

---

[12] An electric current in a conducting wire can be modelled in terms of two strings of negative and positive charges ($-\lambda$ and $+\lambda$) moving at velocity v to the left and to the right respectively. The net current to the right is $I=2\lambda v$. A test charge q at a distance r from the conductor and moving at velocity u in the same direction as the positive string will experience no electric force due to the current, since the two lines of charge cancel. However, observed from a frame that moves with the test charge the situation appears different. In this frame the relative velocities of positive and negative strings will be observed as

$$w_\pm=\frac{v\mp u}{1\mp vu/c^2}$$

Since $w_- > w_+$ the spacing between negative charges in the string will appear to be contracted relative to the spacing between positive charges and the wire appears to be carrying a net negative charge, λ_{net}. This charge produces an electric field $E=\lambda_{net}/2\pi\epsilon_0 r$ at q and exerts an electric force on the test charge. To account for this force an observer in the stationary frame identifies it as a magnetic interaction.

becomes

$$\frac{\partial}{\partial x_j}\left(\frac{\partial A_k}{\partial x_k}\right) - \frac{\partial}{\partial x_k}\left(\frac{\partial A_j}{\partial x_k}\right) = \mu_0 J^k$$

Since F_{jk} is not affected by the substitution[13]

$$A'_\mu \to A_\mu + \frac{\partial \alpha}{\partial x_\mu}$$

this gauge freedom (compare eqn. 6) allows adjustment of the divergence of $\boldsymbol{A}$. It is noted that the Lorentz condition (13)

$$\boldsymbol{\nabla} \cdot \boldsymbol{A} = -\frac{1}{c^2}\frac{\partial \phi}{\partial x_\mu}$$

in tensor notation becomes

$$\left(\frac{\partial A_k}{\partial x_k}\right) = 0$$

and (34) reduces to

$$\Box^2 A_j = -\mu_0 J^j \tag{4.35}$$

where

$$\Box^2 = \frac{\partial}{\partial x_j}\frac{\partial}{\partial x^j} = \nabla^2 - \frac{1}{c^2}\frac{\partial^2}{\partial t^2}$$

Equation (35) is probably the simplest way of expressing Maxwell's equations.

An invariant scalar density can be constructed from the field tensor as the double sum

$$\begin{aligned} F_{\mu\nu}F^{\mu\nu} &= 2(B_1^2 + B_2^2 + B_3^2) - \frac{2}{c^2}(E_1^2 + E_2^2 + E_3^2) \\ &= 2\mu_0\left(\frac{1}{\mu_0}B^2 - \epsilon_0 E^2\right) \end{aligned}$$

[13] Writing $A_\mu(x) \to A_\mu + \partial\alpha/\partial\mu$, one has

$$\begin{aligned} F_{\mu\nu} &= \frac{\partial}{\partial x_\mu}\left(A_\nu + \frac{\partial \alpha}{\partial x_\nu}\right) - \frac{\partial}{\partial x_\nu}\left(A_\mu + \frac{\partial \alpha}{\partial x_\mu}\right) \\ &= \frac{\partial A_\nu}{\partial x_\mu} + \frac{\partial^2 \alpha}{\partial x_\nu \partial x_\mu} - \frac{\partial A_\mu}{\partial x_\nu} - \frac{\partial^2 \alpha}{\partial x_\mu \partial x_\nu} = F_{\mu\nu} \end{aligned}$$

Written in the form

$$-\frac{1}{4}\frac{1}{\mu_0}F_{\mu\nu}F^{\mu\nu} = \frac{1}{2}\left(\epsilon_0 E^2 - \frac{1}{\mu_0}B^2\right)$$

the r.h.s. is recognized as the difference in electric and magnetic field energies. This difference is of the same form as the Lagrangian function $L = T - U$, which shows that particle motion arises from exchanges of its kinetic and potential energies. By analogy, the field Lagrangian indicates that redistributions of a field, isolated from sources, arise from exchanges between electric and magnetic energy distributions.

A covariant Lagrangian density (*i.e.* an equation that looks the same in all Lorentz frames) will be obtained if the Lagrangian density is defined in terms of a relativistic scalar density, as in

$$L = -\frac{1}{4}F_{\mu\nu}F^{\mu\nu} + \mu_0 J^\mu A_\mu$$

where

$$F_{\mu\nu} = \frac{\partial A_\nu}{\partial x_\mu} - \frac{\partial A_\mu}{\partial x_\nu}$$

and substituted into the Euler-Lagrange equation (2.3.1). By regarding each component of A_μ as an independent field [38]

$$\begin{aligned}
\frac{\partial}{\partial x_\nu}\frac{\partial L}{\partial(\partial A_\mu/\partial x_\nu)} &= -\frac{1}{4}\frac{\partial}{\partial x_\nu}\left\{\frac{\partial}{\partial(\partial A_\mu/\partial x_\nu)}\left[\left(\frac{\partial A_\sigma}{\partial x_\lambda} - \frac{\partial A_\lambda}{\partial x_\sigma}\right)\left(\frac{\partial A_\sigma}{\partial x_\lambda} - \frac{\partial A_\lambda}{\partial x_\sigma}\right)\right]\right\} \\
&= -\frac{1}{4}\frac{\partial}{\partial x_\nu}\left[\frac{\partial}{\partial(\partial A_\mu/\partial x_\nu)}\left(2\frac{\partial A_\sigma}{\partial x_\lambda}\frac{\partial A_\sigma}{\partial x_\lambda} - 2\frac{\partial A_\sigma}{\partial x_\lambda}\frac{\partial A_\lambda}{\partial x_\sigma}\right)\right] \\
&= -\frac{1}{4}\frac{\partial}{\partial x_\nu}\left(4\frac{\partial A_\mu}{\partial x_\nu} - 4\frac{\partial A_\nu}{\partial x_\mu}\right) \\
&= -\frac{\partial}{\partial x_\nu}F_{\nu\mu} \\
\frac{\partial L}{\partial A_\mu} &= \mu_0 J^\mu
\end{aligned}$$

Hence the EL equation

$$\frac{\partial L}{\partial A_\mu} - \frac{\partial}{\partial x_\nu}\left[\frac{\partial L}{\partial(\partial_\nu A_\mu)}\right] = \mu_0 J^\mu - \frac{\partial}{\partial x_\nu}F_{\nu\mu} = 0$$

gives the Maxwell equation (34),

$$\frac{\partial F_{\mu\nu}}{\partial x_\nu} = \mu_0 J^\mu$$

4.3.3 Physical Interpretation of Special Relativity

The generation of invariants in the Lorentz transformation of four-vectors has been interpreted to mean that the transformation is equivalent to a rotation. The most general rotation of a four-vector, defined as the quaternion $q = w + ix + jy + kz$ is given by[39]

$$q' = e^{\theta(i\alpha+j\beta+k\gamma)}qe^{-\phi(i\lambda+j\mu+k\nu)}$$

where the symbols in the expression $\alpha^2+\beta^2+\gamma^2 = \lambda^2+\mu^2+\nu^2 = 1$, represent direction cosines of the rotation axes. When both θ and ϕ are imaginary angles, the transformation corresponds to a Lorentz transformation of a four-vector of the type $\sigma(x, y, z, ict)$. The form of the equations suggests that the proper times for two observers moving with respect to each other are related by an imaginary rotation in space-time, the amount of rotation being related to the relative velocity of the observers.

In order to give a physical interpretation of special relativity it is necessary to understand the implications of the Lorentz rotation. Within Galilean relativity the three-dimensional line element of euclidean space ($r^2 = \boldsymbol{r} \cdot \boldsymbol{r}$) is an invariant and the transformation corresponds to a rotation in three-dimensional space. The fact that this line element is not Lorentz invariant shows that world space has more dimensions than three. When rotated in four-dimensional space the physical invariance of the line element is either masked by the appearance of a fourth coordinate in its definition, or else destroyed if the four-space is not euclidean. An illustration of the second possibility is the geographical surface of the earth, which appears to be euclidean at short range, although on a larger scale it is known to curve out of the euclidean plane.

In a curved space the unit vectors of one coordinate system change scale when transformed to another, *i.e.*

$$\boldsymbol{e'}_\nu = h_\nu \boldsymbol{e}_\nu$$

Transformation from one coordinate system x_ν with scale factors h_ν, to another, x'_ν with scale factors h'_ν is represented by

$$F'_{\mu\nu} = \sum_\mu \gamma_{\mu\nu} F_\mu$$

where

$$\left(\frac{h_\mu}{h'_\nu}\right)\left(\frac{\partial x_\mu}{\partial x'_\nu}\right) = \gamma_{\mu\nu} = \left(\frac{h'_\nu}{h_\mu}\right)\left(\frac{\partial x'_\nu}{\partial x_\mu}\right)$$

If the vector $\boldsymbol{F}$ is now written in terms of the scaled unit vectors

$$\boldsymbol{F} = \sum_\nu f^\nu \boldsymbol{e}_\nu \quad ; \quad f^\nu = \frac{F_\nu}{h_\nu}$$

the components transform into

$$(f^\nu)' = \sum_\mu f^\mu \frac{\partial x'_\nu}{\partial x_\mu} = \sum_\mu f^\mu \left(\frac{h_\mu}{h'_\nu}\right) \frac{\partial x_\mu}{\partial x'_\nu}$$

The quantities f^ν are the *contravariant* components of a vector in the coordinate system x_i. They give an actual vector only when multiplied by the unit vector $\boldsymbol{e'}_\nu = h_\nu \boldsymbol{e}_\nu$. If the unit vectors along the coordinate lines have a scale inverse to the coordinate, $\boldsymbol{e'}_\nu = \boldsymbol{e}_\nu / h_\nu$ so that

$$\boldsymbol{F} = \sum_\nu f_\nu \boldsymbol{e}_\nu \quad ; \quad f_\nu = h_\nu F_\nu$$

the transformation is

$$(f_\nu)' = \sum_\mu f_\mu \frac{\partial x_\mu}{\partial x'_\nu} = \sum_\mu f_\mu \left(\frac{h'_\nu}{h_\mu}\right) \frac{\partial x'_\nu}{\partial x_\mu}$$

The quantities f_μ are called the *covariant* components of a vector in the coordinate system $\boldsymbol{x}$[14].

The quantities b_{ij}, b^{ij}, and b^i_j are respectively called the components of covariant, contravariant or *mixed tensors* of the second order, if they transform according to the formulae

$$b'_{ij} = \sum_{\mu,\nu} b_{\mu\nu} \frac{\partial x_\mu}{\partial x'_i} \frac{\partial x_\nu}{\partial x'_j}$$

$$(b')^{ij} = \sum_{\mu,\nu} b^{\mu\nu} \frac{\partial x'_i}{\partial x_\mu} \frac{\partial x'_j}{\partial x_\nu}$$

$$(b')^i_j = \sum_{\mu,\nu} b^\mu_\nu \frac{\partial x'_i}{\partial x_\mu} \frac{\partial x_\nu}{\partial x'_j}$$

The products of the components of two covariant vectors, taken in pairs, form the components of a covariant tensor. If the vectors are contravariant,

[14]The fractional coordinates in crystallography, referred to direct and reciprocal cells respectively, are examples of covariant and contravariant vector components.

the resulting tensor is contravariant. If A_i and B_j are the components of two ordinary vectors, then $(h_i/h_j)A_iB_j = C_i^j$ are the components of a mixed tensor. The quantity $\sum_\mu b_\mu^\mu$ called a *contracted* tensor, formed from a mixed tensor, does not change its value when transformed, since

$$\sum_\nu (b')_\nu^\nu = \sum_{\mu,\kappa,\nu} b_\kappa^\mu \frac{\partial x'_\kappa}{\partial x_\mu}\frac{\partial x_\kappa}{\partial x'_\nu} = \sum_{\kappa,/mu} b_\kappa^\mu \frac{\partial x_\kappa}{\partial x_\mu} = \sum_\mu b_\mu^\mu \tag{4.36}$$

Such a quantity has here been called an *invariant* or a scalar. The scalar product of two vectors is a contracted tensor, $\sum_\nu A_\nu B_\nu = \sum_\nu (h_\nu/h_\nu)\, A_\nu B_\nu$, and is, therefore an invariant.

4.4 General Relativity

It is commonly accepted that the special and general theories of relativity refer to different aspects of reality; the former being a simplification of the general theory in the special case of unaccelerated uniform non-rotational motion. The major item of simplification is thought to be in the metric tensor of general relativity that reduces to an orthogonal diagonal form in the special case. This assumption is the complicating one that should be examined more closely. A simple comparison of the two theories is given by Dixon [40].

Let (μ) and (ν) be two coordinate systems in the same region of space-time. Each of the variables x^μ $(\mu = 1, 4)$ can be considered as a function of the four coordinates x^ν, *e.g.* the partial derivatives

$$\frac{\partial x^\mu}{\partial x^\nu} \equiv A_\nu^\mu(x)$$

define an array which has indices associated with two different coordinate systems.

Now, one of the fundamental results of special relativity is the invariance of the quadratic form

$$\Phi = \delta x^2 + \delta y^2 + \delta z^2 - c^2\delta t^2.$$

In terms of the diagonal matrix

$$\eta = \begin{pmatrix} 1 & 0 & 0 & 0 \\ 0 & 1 & 0 & 0 \\ 0 & 0 & 1 & 0 \\ 0 & 0 & 0 & -c^2 \end{pmatrix} \tag{4.37}$$

and the coordinate system (α), defined by $x^1 = x$, $x^2 = y$, $x^3 = z$, $x^4 = ict$, the quadratic form may be written

$$\Phi = \sum_{\mu,\nu=1}^{4} \eta_{\mu\nu}\delta x^{\nu}\delta x^{\mu}.$$

To obtain the result that describes the invariance in the gravitational case space-time is assumed to be curved as described by a function $x(\lambda)$ at any point z. Since

$$\frac{dx^{\mu}}{d\lambda} = \sum_{\nu=1}^{4} \frac{\partial x^{\mu}}{\partial x^{\nu}}\frac{\partial x^{\nu}}{\partial \lambda} = \sum_{\nu=1}^{4} A^{\mu}_{\nu}\frac{dx^{\nu}}{d\lambda}$$

the derivatives $\frac{dx^{\nu}}{d\lambda}$ transform under a change of coordinates in exactly the same way as coordinate differences x^{ν}. The quantity Φ is therefore defined at z by

$$\Phi = \sum_{\mu,\nu=1}^{4} \eta_{\mu\nu}\frac{dx^{\mu}}{d\lambda}\frac{dx^{\nu}}{d\lambda} \qquad (4.38)$$

and has the same value in all natural coordinate systems of all reference frames that are inertial at z. It has a constant value that depends on the curve, but is independent of the choice of reference axes. It therefore is a geometric property of space-time itself. The geometry of space-time is described by the metric tensor $\eta_{\mu\nu}$.

If the matrix of the metric tensor can be diagonalized it is independent of position and the corresponding geometry is said to be flat. This is the case of special relativity with the diagonal matrix $\eta = \mathrm{diag}(1,1,1,-c^2)$ defined for the coordinate system $x^1 = x$, $x^2 = y$, $x^3 = x$, $x^4 = t$,

$$\Phi = \sum_{\mu,\nu=1}^{4} \eta_{\mu\nu}\delta x^{\mu}\delta x^{\nu}.$$

The previous result (38)which is valid only at z can be generalized to include all points in a reference frame. Let (a) be an arbitrary coordinate system, such that

$$A^{\nu}_{a} = \frac{dx^{\nu}}{dx^{a}}.$$

Substituting the identity

$$\frac{dx^{\nu}}{d\lambda} = \sum_{a} A^{\nu}_{a}\frac{dx^{a}}{d\lambda}$$

into (37), it follows that

$$\Phi(z) = \sum_{a,b} g_{ab}(z) \frac{dx^a}{d\lambda} \frac{dx^b}{d\lambda}\bigg|_{x=z} \tag{4.39}$$

where
$$g_{ab}(z) = \sum \eta_{ab} A_a^{\nu}(z) A_b^{\mu}(z) \tag{4.40}$$

It is obvious that $g_{ab}(z)$ is independent both of the choice of inertial frame at z, with its corresponding natural coordinate system (ν), and the choice of curve $x(\lambda)$. The elements of g are known as the components of the metric tensor in this coordinate system. Expression (39) is the required generalization that allows evaluation of Φ at all points in terms of $g_{ab}(x)$ and the curve $x(\lambda)$.

The matrix g_{ab} is symmetric and non-singular and has other purely algebraic properties that can be demonstrated by first defining the matrices

$$G = (g_{ab}) \quad , \quad H = (\eta_{ab}) \quad , \quad A = (A_a^{\nu}).$$

In A ν labels rows and a columns. Then (40) can be written as

$$G = \tilde{A}HA \,, \text{ so that } \quad H = \tilde{B}GB \quad \text{where} \quad B = A^{-1}.$$

It is known that any real symmetric $(k \times k)$ matrix G is reducible by a congruence transformation of the form $G \to \tilde{B}GB$, where B is non-singular, to a diagonal form in which all diagonal elements are either 1, 0 or -1. The numbers of positive (p), zero (z) and negative (n) diagonal elements are invariants of the matrix. The numbers $(p+n)$ and $(p-n)$ are known respectively as the *rank* and *signature* of the matrix. When G is non-singular its rank is necessarily k. It follows from (37) that the matrix (g_{ab}) of metric tensor components must have signature 2.

From the foregoing it follows that when gravitation is taken into consideration, the geometry of space-time is described by a metric tensor that associates with every coordinate system a position-dependent matrix (g_{ab}) of components which is symmetric, non-singular and has signature 2. If there exists a coordinate system in which the g_{ab} are independent of position, then the linear coordinate transformation can make $g_{ab} = \eta_{ab}$ throughout its domain, in the notation of (37). This case is commonly assumed [40] to represent special relativity, physically characterized by the absence of gravitation. The corresponding geometry is said to be flat. When it is not possible to make the g_{ab} constant by a suitable choice of coordinates, the geometry is said to be curved. It is said that space-time is curved by a gravitational field.

As the presence of gravity (mass) imparts a variable curvature on space the simple Minkowski form of a 4-dimensional line element is replaced by the more general form

$$ds^2 = \sum_{\mu,\nu=1}^{4} g_{\mu\nu} dx^\mu dx^\nu$$

where the $g_{\mu\nu}$'s are no longer constant. Inertial or cartesian coordinate systems no longer exist in this construct.

The curved space may be considered as covered by local metric neighbourhoods and regions of overlap where the transformation law of general relativity applies. It is called a Riemannian *manifold.* Vectors transplanted to neighbouring points may change their orientation and the intuitive idea of a straight line must be replaced by that of *geodesic*, which may be regarded as a shortest or straightest curve in that region. A manifold on which vector transplantation is described by a law of the form

$$d\xi^i = \Gamma^i_{mj} dx^m \xi^j \tag{4.41}$$

(in which the summation rule operates) is called an *affine* (or linear) space and the Γ coefficients are called the affine connections [41]. To ensure that the length of a vector is not affected by the transplantation the metric requirement that the scalar product of two vectors be invariant is introduced. This condition defines an unique connection, compatible with a given metric tensor g, and its components in any coordinate system (x) are given by

$$\Gamma^i_{jk} = -\frac{1}{2} g^{im} \left(\frac{\partial g_{mk}}{\partial x^i} + \frac{\partial g_{mj}}{\partial x^k} - \frac{\partial g_{jk}}{\partial x^m} \right) .$$

The defining equations for a geodesic in Riemann space becomes

$$\frac{d^2 x^i}{ds^2} - \Gamma^i_{\alpha\beta} \frac{dx^\alpha}{ds} \frac{dx^\beta}{ds} = 0$$

where s is the arc length of the geodesic.

An affine manifold is said to be *flat* or *Euclidean* at a point p, if a coordinate system in which the functions Γ^i_{jk} all vanish, can be found around p. For a cartesian system the geodesics become

$$\frac{d^2 x^i}{ds^2} = 0 \quad , \quad \textit{i.e.} \text{ straight lines.}$$

To specify the directions of two different vectors at nearby points it is necessary to define *tangent* vectors (tangent space) at these points. Important tangent spaces are the light cones. In Minkowski space these all have the same shape and orientation, but not in general relativistic curved space, where the light cones may tilt, expand or contract as one moves from one tangent space to another [19].

The final objective is an equation that relates a geometrical object representing the curvature of space-time to a geometrical object representing the source of the gravitational field. The condition that all affine connections must vanish at a euclidean point, defines a tensor [41]

$$R^{\alpha}_{\mu\beta\nu} = \frac{\partial\Gamma^{\alpha}_{\mu\nu}}{\partial x_{\beta}} - \frac{\partial\Gamma^{\alpha}_{\mu\beta}}{\partial x_{\nu}} + \Gamma^{\gamma}_{\mu\nu}\Gamma^{\alpha}_{\gamma\beta} - \Gamma^{\gamma}_{\mu\beta}\Gamma^{\alpha}_{\gamma\nu}$$

known as the *Riemann curvature tensor*, which can be contracted to the symmetrical tensor $R_{\mu\nu}$ with 10 independent components. Since this tensor has non-zero divergence it cannot feature in a conservation law (of mass and energy), but this can be achieved by definition of the *Ricci tensor*

$$G_{\mu\nu} = R_{\mu\nu} - \frac{1}{2}g_{\mu\nu}R$$

where the doubly contracted tensor R is called the Riemann scalar.

The Ricci tensor that represents the geometry of space is next equated with the so-called energy-momentum (stress) tensor of the matter field that defines the influence of matter and field energy

$$T^{\mu\nu} = \rho_0(x)u^{\mu}(x)u^{\nu}(x)$$

in terms of a scalar density field ρ_0 and a four-vector field of flow.

This procedure leads to the Einstein gravitational field equations, one form of which, without cosmological term, is

$$\begin{aligned} G_{\mu\nu} = R_{\mu\nu} - \frac{1}{2}g_{\mu\nu}R &= -\frac{8\pi\kappa}{c^2}T_{\mu\nu} \\ &= kT_{\mu\nu} \end{aligned} \tag{4.42}$$

where k is Newton's gravitation constant, or equivalently

$$R_{\mu\nu} = -\frac{8\pi\kappa}{c^2}\left(T_{\mu\nu} - \frac{1}{2}g_{\mu\nu}T\right) \tag{4.43}$$

where the Laue scalar T is the doubly contracted stress tensor.

The important result is the obvious symmetry between $T_{\mu\nu}$ and $R_{\mu\nu}$ as shown in (42) and (43). Both of these tensors vanish in empty euclidean space and a reciprocal relationship between them is inferred: The presence of matter causes space to curl up and curvature of space generates matter.

The field equations of general relativity are rarely used without simplifying assumptions. The most common application treats of a mass, sufficiently distant from other masses, so as to move uniformly in a straight line. All applications of special relativity are of this type, in order to stay in Minkowski space-time. A body that moves inertially (or at rest) is thus assumed to have four-dimensionally straight world lines from which they deviate only under acceleration or rotation. The well-known Minkowski diagram of special relativity is a graphical representation of this assumption and therefore refers to a highly idealized situation, only realized in isolated free fall or improbable regions of deep intergalactic space.

In the real world the stress tensor never vanishes and so requires a non-vanishing curvature tensor under all circumstances. Alternatively, the concept of mass is strictly undefined in flat Minkowski space-time. Any mass point in Minkowski space disperses spontaneously, which means that it has a space-like rather than a time-like world line. In perfect analogy a mass point can be viewed as a local distortion of space-time. In euclidean space it can be smoothed away without leaving any trace, but not on a curved manifold. Mass generation therefore resembles distortion of a euclidean cover when spread across a non-euclidean surface. A given degree of curvature then corresponds to creation of a constant quantity of matter, or a constant measure of misfit between cover and surface, that cannot be smoothed away. Associated with the misfit (mass) a strain field appears in the curved surface.

4.5 Gauge Fields

The concept of a gauge field and the notion of gauge invariance originated with a premature suggestion by Weyl [42] how to accommodate electromagnetic variables, in addition to the gravitational field, as geometric features of a differential manifold.

Vector transplantation in an arbitrary coordinate system, formulated in differential form, is described by (41)

$$d\xi^i = \Gamma^i_{mj} dx^m \xi^j$$

where the Γ are symmetric connections of the manifold, ξ^i are the components of the vector and dx^m the local displacement vector. In Riemannian space the length of the vector ξ, specified as

$$l^2 = |\xi|^2 = g_{\mu\nu}\xi^\mu\xi^\nu$$

remains invariant under transplantation. The suggestion made by Weyl was that a differential manifold which allows transplantation with non-constant

l could be considered instead. If the vector ξ is interpreted as the length of a measuring rod it means that this length (gauge) could change under dispacement. If it was assumed that the increment in length is proportional to the length itself and a linear homogeneous function of the displacement vector, then

$$dl = (\phi_m dx^m)l$$

where the covariant vector ϕ_m takes the role of the connection Γ.

For a time-independent gauge vector ϕ_0, like an electrostatic field, the gauge transformation after time x^4 would yield

$$l = l_0 \exp\left[\int_0^{x^4} \phi_0 dx^4\right] = l_0 \exp(\phi_0 x^4)$$

Since the gauge factor is in general dependent on the path and therefore not integrable, circulation vectors (compare 1.3.4) such as

$$f_{ik} = \frac{\partial \phi_i}{\partial x^k} - \frac{\partial \phi_k}{\partial x^i}$$

will be non-zero. The divergence

$$\frac{\partial f_{ik}}{\partial x^l} + \frac{f_{kl}}{\partial x^i} + \frac{\partial f_{li}}{\partial x^k} = 0 \quad , \quad i \neq k \neq l \quad , \quad i,k,l = 1,4.$$

The formal resemblance[15] with the electromagnetic field equations (5) and (3)

$$\begin{aligned} \text{curl}\,\boldsymbol{A} &= \boldsymbol{B} \\ \text{div}\,\boldsymbol{B} &= 0 \end{aligned}$$

prompted Weyl to identify the gauge factor with the potentials of the electromagnetic field.

Although Weyl's conjecture could not be substantiated in its original form it was pointed out soon afterwards by Schrödinger [43] and London [44] that the classical quantum conditions could be deduced from Weyl's world geometry by choosing complex components for the gauge factor, *i.e.*

$$l = l_0 \exp\left[\frac{2\pi i}{h}\int \phi_m dx^m\right]$$

[15] Compare 1.3.4 and 1.3.3

In view of this new insight Weyl [45] reinterpreted his geometric theory as a change of phase rather than scale. In the original proposal the scale was proposed to change from unity to

$$1 + S_\mu dx^\mu \tag{4.44}$$

in going from a point x^μ to a neighbouring point point $x^\mu + dx^\mu$ of space-time. In the new interpretation the factor S_μ is replaced by the complex quantity iS_μ. This has the far-reaching consequence of changing (44) into[16]

$$1 + i\alpha \to e^{i\alpha}$$

which is a phase change, not a scale change. The terms gauge invariance and gauge field were retained, have persisted to this day and should be understood to mean phase invariance and phase field respectively.

In the theory of special relativity the gauge factor α is a constant since there exists only one coordinate system for all space. The transformation[17]

$$\psi(x) \to \psi'(x) = e^{i\alpha}\psi(x) \tag{4.45}$$

is called a global gauge transformation and it has the special property that the derivative of the field

$$\partial_\mu \psi \to e^{i\alpha}\partial_\mu \psi$$

transforms like the field itself. It is assumed here that the complex field $\psi(x)$ describes an electron, following the suggestion of Schrödinger [43] (1922).

A charge (q) is therefore associated with the wave field $\psi(x)$ and the charge density is given by

$$\rho(x,t) = q|\psi_t|^2$$

which is clearly invariant under the phase transformation (45). This invariance implies by Noether's theorem, the global conservation of the total charge, and shows that the overall phase factor is not measurable. The phase can be chosen arbitrarily, but is a constant and, once chosen it remains fixed over the entire universe and for all time.

The situation is entirely different within the theory of general relativity, which is based on a curved manifold rather than flat space with a globally

---

[16]For small α $\cos\alpha \to 1$ and $\sin\alpha \to \alpha$, hence $e^{i\alpha} = \cos\alpha + i\sin\alpha \to 1 + i\alpha$.

[17]The family of phase transformations $U(\alpha) \equiv \exp i\alpha$, where a single parameter α may run continuously over real numbers, forms a unitary Abelian (commutative) group known as the $U(1)$ group. Compare 2.8.2.5

fixed coordinate system. Each point now has its own coordinate system and hence its own gauge factor. By doing away with the rigid coordinate system the gauge factor necessarily becomes an arbitrary function of position, $\alpha(x)$. Since the phase has no real physical significance, it may be redefined locally by an arbitrary rotation, at every space-time point of the manifold, without changing the physical situation. This stipulation may seem to rule out local charge conservation, unless there is some compensating field that restores the invariance under local phase transformation.

Suppose that the compensating field is represented by a set of functions $A_\mu(x)$ that depend on all space-time coordinates. Further suppose that the phases of the function that represents the particle at points x and $x + dx^\mu$ become parallel in the presence of the compensating field if the local values of the phase at these two points differ by an amount $qA_\mu dx^\mu$. The vector field A_μ that couples to the charge is called the gauge field.

Now consider two neighbouring points with local phases differing by the amount $qA_\mu dx^\mu$. Perform a gauge transformation by rotating the phase of the ψ function at x by an amount $q\alpha(x)$, *i.e.*

$$\psi(x) \rightarrow \psi'(x) = e^{iq\alpha(x)}\psi(x)$$

Since the gauge factor is no longer constant the field derivative no longer transforms like the field itself, *i.e.*

$$\begin{aligned}
\frac{\partial\psi'}{\partial x_\mu} &= \frac{\partial\psi}{\partial x_\mu}e^{iq\alpha(x)} + iq\psi e^{iq\alpha(x)} \\
&= e^{iq\alpha(x)}\left(\partial_\mu\psi + iq\psi\right) \equiv e^{iq\alpha(x)}D_\mu\psi
\end{aligned}$$

D_μ which does transform like the field is known as the covariant derivative.

At the same time the phase at neighbouring point $x + dx^\mu$ will change by an amount $q[\alpha(x) + \partial_\mu\alpha(x)dx^\mu]$, starting from $qA_\mu dx^\mu$. To ensure that the phases remain parallel the difference in phase after the transformation, *i.e.*

$$q[\alpha(x) + \partial_\mu\alpha(x)dx^\mu + A_\mu(x)dx^\mu - \alpha(x)]$$

should be equal to $qA'_\mu(x)dx^\mu$, where $A'_\mu(x)$ is the transformed value of $A_\mu(x)$. Hence

$$A_\mu(x) \rightarrow A'_\mu(x) = A_\mu(x) + \partial_\mu\alpha(x)$$

The identical transformation, equation (6), of the electromagnetic vector potentials was found before to leave the fields unaffected or gauge invariant. The fields A_μ are not gauge invariant, but the fields described by the tensor, equation (33)

$$F_{\mu\nu} = \partial_\nu A_\mu - \partial_\mu A_\nu$$

are. Gauge theory therefore leads directly to the electromagnetic field [46].

In summary, the principle of local invariance in a curved Riemannian manifold leads to the appearance of compensating fields. The electromagnetic field is the compensating field of local phase transformation and the gravitational field is the compensating field of local Lorentz transformations. Such compensating fields cannot be scalars but are vector or tensor fields whose associated rest mass vanishes. In modern physics not only electroweak but also strong interactions are understood in terms of theories which combine local gauge invariance with spontaneous symmetry breaking, so that the gauge vector bosons acquire a mass.

Gauge theories are commonly formulated in terms of group theory. Under a phase rotation

$$\psi \rightarrow \psi' = e^{iq\alpha}\psi$$

the factor $\exp iq\alpha$ is regarded as an element of the gauge group. However, these factors are not always single-valued and any two factors between which α differs by $2\pi n/q$, for integral n, give the same ψ' and are physically indistinguishable. The elements of the gauge group therefore do not require α to range from $-\infty$ to ∞, but only in the interval $[0, 2\pi/q]$, with the end points of the range identified. The gauge group is called the compact $U(1)$ group with the topology of a circle, and not the non-compact group with the topology of the real line[18]. The same is true for a set of particles with charges $q_r = n_r q$, where n_r is an integer and q is an elementary charge. All values of α differing by $2\pi n/q$ will give the same

$$\psi_r' = e^{iq_r\alpha}\psi_r$$

and the gauge group will be compact.

The electromagnetic field may now formally be interpreted as the gauge field which must be introduced to ensure invariance under local $U(1)$ gauge transformation. In the most general case the field variables are introduced in terms of the Lagrangian density of the field, which itself is gauge invariant. In the case of the electromagnetic field, as before,

$$L = -\frac{1}{4}F_{\mu\nu}F^{\mu\nu} + \mu_0 J^\mu A_\mu$$

the field tensor has six components, three electric and three magnetic. What has been done here, therefore, was to show that the field arises naturally

[18] U(1) is homomorhic with O(2) - see eqn. 2.25

by demanding invariance of the action under local gauge transformation. The gauge potential A_μ couples to the conserved current and the coupling strength can be interpreted as the charge, q of the field. The divergence (34), reformulated in terms of a covariant derivative and conserved charge becomes

$$\begin{aligned} \mu_0 J^\mu &= D_\mu F_{\mu\nu} \\ &= (\partial_\mu + iqA_\mu)F_{\mu\nu} \end{aligned}$$

If the electron is represented by a complex field[19], as assumed here, it is necessary to put

$$\begin{aligned} \mu_0 J^\mu &= (\partial_\mu\psi + iqA_\mu\psi)(\partial^\mu\psi^* - iqA^\mu\psi^*) \\ &= iq(\psi^* D_\mu\psi - \psi D_\mu\psi^*) \\ &= \partial_\mu F_{\mu\nu} \end{aligned}$$

The Lagrangian of the electromagnetic field reduces to

$$L = (\partial_\mu\psi + ieA_\mu\psi)(\partial^\mu\psi^* - ieA^\mu\psi^*) - \frac{1}{4}F_{\mu\nu}F^{\mu\nu}$$

or

$$L = (\partial_\mu\psi)(\partial^\mu\psi^*) - ie(\psi^*\partial^\mu\psi - \psi\partial^\mu\psi^*)A_\mu + e^2A_\mu A^\mu\psi^*\psi - \frac{1}{4}F_{\mu\nu}F^{\mu\nu}$$

The second term represents the conserved current and the third is a momentum term. Noting that $\partial L/\partial\dot{\boldsymbol{q}} = \boldsymbol{P}$, the field momentum bscomes

$$\frac{\partial L}{\partial\dot{\boldsymbol{q}}} = \frac{\partial}{\partial\dot{\boldsymbol{q}}}(\mu_0 J^\mu A_\mu) \equiv \frac{\partial}{\partial\boldsymbol{v}}(e\boldsymbol{v}A_\mu)$$

and hence the world momentum scalar $\boldsymbol{P_\mu}\cdot\boldsymbol{P_\mu} = e^2A_\mu^2$, represents a rest mass of zero. The interaction between the electron and the electromagnetic field is said to be mediated by a massless photon. For a massive particle the Lagrangian would contain a term like $M^2A_\mu A^\mu$. It is important to note that this term is not invariant under gauge transformation and so does not describe an observable quantity.

[19] ψ^* with covariant derivative $D_\mu\psi^* = (\partial_\mu - iqA_\mu)\psi^*$ describes a field with charge $-q$.

4.5.1 The Higgs Field

The existence of matter appertains to the structure of world space and all models designed to account for the appearance of matter have one assumption in common: the aether can exist in one of two forms, void or vacuum, depending on the absence or presence of matter; a phase transition between the forms amounts to creation. A detailed account of creation by such a mechanism has been formulated in terms of gauge fields, developed in the first instance to describe superconductivity [47] in the context of quantum field theory. With the superconductor the ionic lattice lowers the symmetry of an electron cloud to create bosonic excitations and screen long-range magnetic interactions. In the cosmological case the role of the lattice is assigned to an all-pervading (Higgs) field, postulated to become structured below a transition temperature where the aether transforms from void into vacuum.

Symmetry Breaking

The idea of spontaneous symmetry breaking refers to rearrangement of a system from a situation of high symmetry to a (degenerate) situation of lower symmetry, induced by some parameter that exceeds a critical value. In thermodynamics such a rearrangement is known as a phase transition and the phase of lowest symmetry is the ground state. The critical variable could be a parameter like temperature, pressure or magnetic field strength. The symmetry may refer to the orientation of some field vector such as molecular dipoles in a crystal or the magnetization of a ferromagnet.

In the most general case the Lagrangian density of a field suffers a reduction of symmetry at some critical value of an interaction parameter. Suppose that

$$\begin{aligned} L &= (\partial_\mu\phi)(\partial^\mu\phi^*) - m^2\phi^*\phi - \lambda(\phi^*\phi)^2 \\ &= (\partial_\mu\phi)(\partial^\mu\phi^*) - V(\phi,\phi^*) \end{aligned}$$

where m and λ are variable parameters, and L is invariant under $U(1)$ global gauge symmetry, $\phi \to \exp(i\alpha)\phi$. The ground state, which in field theory is regarded as the vacuum, is obtained by minimizing the potential V,

$$\frac{dV}{d\phi} = m^2\phi^* + 2\lambda\phi^*(\phi^*\phi)$$

so that, when $m^2 > 0$, the minimum occurs at $\phi^* = \phi = 0$. This is a scalar field with mass m. The potential function $V(\phi)$ has the shape shown in figure 3(a).

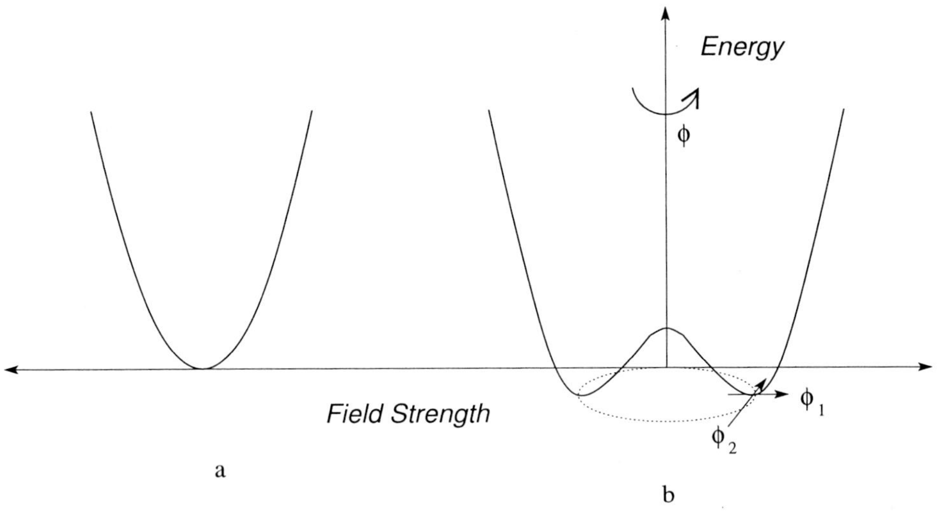

Figure 3: *Model potential to illustrate a transition between a symmetrical excited state and a structured ground state.*

If $m^2 < 0$ however, there is a local maximum at $|\phi| = 0$, and a minimum at $|\phi|^2 = \phi^*\phi = -m^2/2\lambda = a^2$, *i.e.* at $|\phi| = a$. This model potential has a shape like the curve in figure 3(b). The minima of V lie along the circle $|\phi| = a$ and the shape of the continuous potential function is generated by rotation of the curve about the energy axis by an angle ϕ. Any transition from the false minimum on the symmetry axis (Figure 3b) into the global minimum forces the selection of a special direction, for instance when the transition is modeled by a ball rolling into the annular depression. The selection of a special direction means that the field acquires a non-zero value in this direction.

The reflection symmetry of the Lagrangian has apparently been broken by choosing the ground state $|\phi| = a$ and the correct way to proceed is to explore the Lagrangian around the stable vacuum. This is achieved by defining the field

$$\phi(x) = a + \frac{1}{\sqrt{2}}\left[\phi_1(x) + i\phi_2(x)\right]$$

It is easy to see that, ignoring constant terms

$$L = \frac{1}{2}\left(\partial_\mu\phi_1\right)^2 + \frac{1}{2}\left(\partial_\mu\phi_2\right)^2 - 2\lambda a^2\phi_1^2 - \sqrt{2}\lambda\phi_1\left(\phi_1^2 + \phi_2^2\right) - \frac{\lambda}{4}\left(\phi_1^2 + \phi_2^2\right)^2$$

The third term has the form of a mass term $(-\frac{1}{2}m^2\phi^2)$ for the ϕ_1 field, given by $m^2 = 4\lambda a^2$. The second term represents the kinetic energy of the ϕ_2-field, but there is no corresponding mass term for ϕ_2. As a result of spontaneous symmetry breaking, what would otherwise be two massive fields (the real parts of ϕ), become one massive and one massless field. It clearly costs energy to displace ϕ_1 against the restoring forces of the potential, but there are no restoring forces corresponding to the displacement along the circular valley (see figure 3b) $|\phi| = a$, in view of the vacuum degeneracy.

The ϕ_2 particle is known as a Goldstone boson. The important point is that this phenomenon is general. The Goldstone theorem [48] which states that massless scalars occur whenever a continuous symmetry of a physical system is spontaneously broken (or, more accurately, is not apparent in the ground state) will be accepted without further proof, however, compare [49].

The Higgs Phenomenon

It was demonstrated by Higgs [50] that the appearance of massless bosons can be avoided by combining the spontaneous breakdown of symmetry under a compact Lie group with local gauge symmetry. The potential $V(\phi)$ which is invariant under the local transformation of the charged field

$$\phi \to e^{i\alpha(x)}\phi$$

is combined with the Hermitian vector gauge field $A_\mu(x)$ and symmetry breakdown is induced by translation.

The simplest example is $U(1)$ gauge symmetry. The gauge invariant Lagrangian is

$$L = (\partial_\mu + ieA_\mu)\phi(\partial_\mu - ieA^\mu\phi^* - m^2\phi^*\phi - \lambda(\phi^*\phi)^2 - \frac{1}{4}F_{\mu\nu}F^{\mu\nu} \qquad (4.46)$$

As before, m^2 is treated as a parameter, so that in the case $m^2 < 0$, and in the absence of the gauge field, the vacuum is at

$$|\phi| = a = \left(\frac{-m^2}{2\lambda}\right)^{\frac{1}{2}}$$

Then, as before, setting

$$\phi(x) = a + \frac{1}{\sqrt{2}}[\phi_1(x) + i\phi_2(x)]$$

gives for the Lagrangian, in terms of the physical fields ϕ_1 and ϕ_2

$$\begin{aligned} L &= -\frac{1}{4}F_{\mu\nu}F^{\mu\nu} + e^2a^2A_\mu A^\mu + \frac{1}{2}(\partial_\mu\phi_1)^2 + \frac{1}{2}(\partial_\mu\phi_2)^2 \\ & -2\lambda a^2\phi_1^2 + \sqrt{2}eaA^\mu\partial_\mu\phi^2 + \text{ higher terms.} \end{aligned}$$

The particle spectrum consists of a massless Goldstone boson ϕ_2, a massive scalar ϕ_1, and more crucially a massive vector A_μ. The Goldstone boson can be eliminated by gauge transformation. For infinitesimal gauge factor $\alpha(x)$,

$$\begin{aligned} \phi_1' &= \phi_1 - \alpha\phi_2 \\ \phi_2' &= \phi_2 + \alpha\phi_1 + \sqrt{2}\alpha a \end{aligned}$$

which shows that ϕ_2, like A_μ, has no physical meaning and α can be chosen to make $\phi_2 = 0$. The resulting Lagrangian

$$L = -\frac{1}{4}F_{\mu\nu}F^{\mu\nu} + e^2a^2A_\mu A^\mu + \frac{1}{2}(\partial_\mu\phi_1)^2 - 2\lambda a^2\phi_1^2 + \text{ coupling terms}$$

describes just two interacting massive particles, a vector gauge boson A_μ and a massive scalar, called the Higgs particle.

Superconductivity

Superconductivity provides an illustration of the Higgs mechanism. It is the property of materials that show no electrical resistance, usually at low temperatures. Such materials are capable to carry persistent currents. These currents effectively screen out magnetic flux, which is therefore zero in a superconductor (the Meisner effect). Another way of describing the Meisner effect is to say that the photons are effectively massive, as in the Higgs phenomenon. These conclusions can be shown to follow from the Lagrangian (46). In this instance it is sufficient to consider a static situation, *i.e.* $\partial_4\phi = 0$, *etc*, leading to the Lagrangian

$$L = -\frac{1}{2}(\nabla - ie\boldsymbol{A})\phi \cdot (\nabla + ie\boldsymbol{A})\phi^* - m^2|\phi|^2 - \lambda|\phi|^4 - \frac{1}{4}(\boldsymbol{\nabla} \times \boldsymbol{A})^2$$

or

$$-L = \frac{1}{4}(\boldsymbol{\nabla} \times \boldsymbol{A})^2 + \frac{1}{2}|(\nabla - ie\boldsymbol{A})\phi|^2 + m^2|\phi|^2 + \lambda|\phi|^4$$

Now $-L$ is the Landau-Ginzburg free energy, where $m^2 = a(T - T_c)$ near the critical temperature, ϕ is a macroscopic many-particle wave function, introduced by Bardeen-Cooper-Schrieffer, according to which an attractive force between electrons is mediated by bosonic electron pairs. At low temperature these fall into the same quantum state (Bose-Einstein condensation), and because of this, a many-particle wave function ϕ may be used to describe the macroscopic system. At $T > T_c$, $m^2 > 0$ and the minimum free energy is at $|\phi| = 0$. However, when $T < T_c$, $m^2 < 0$ and the minimum free energy is at

$$|\phi|^2 = -\frac{m^2}{2\lambda} > 0$$

which is, of course an example of spontaneous symmetry breaking. Now L is invariant under the usual local gauge transformation and the associated conserved current is

$$\boldsymbol{J} = -\frac{1}{2}(\phi^*\nabla\phi - \phi\nabla\phi^*) - e|\phi|^2\boldsymbol{A}$$

When $T < T_c$ and ϕ varies only slightly over the sample, the second term on the right dominates and

$$\boldsymbol{J} = \frac{em^2}{2\lambda}\boldsymbol{A} = -k^2\boldsymbol{A} \tag{4.47}$$

where k is a positive constant. The electric field $\boldsymbol{E} = -\partial\boldsymbol{A}/\partial t = 0$, and by Ohm's law $\boldsymbol{E} = R\boldsymbol{J}$ or $R = 0$, which defines superconductivity.

The Meisner effect follows from $\boldsymbol{\nabla} \times \boldsymbol{B} = \boldsymbol{J}$. Taking the curl, remembering that $\boldsymbol{\nabla} \cdot \boldsymbol{B} = \boldsymbol{0}$, and combining with (47) gives

$$\nabla^2\boldsymbol{B} = k^2\boldsymbol{B} \tag{4.48}$$

In one dimension $d^2\boldsymbol{B}/dx^2 = k^2\boldsymbol{B}$, $B_x = B_0\exp(-kx)$ and the magnetic field is seen to penetrate the sample only to a characteristic depth $1/k$.

Finally it is clear that the vector bosons have a mass k, which is a characteristic feature of the Higgs phenomenon.

Matter from the Void

When big-bang cosmologists ascribe the creation of matter to a Higgs mechanism on a cosmic scale it means that an actual matter field is postulated to be present in the vacuum. It is argued [51] that a cooling universe produced in a hot big bang must undergo a phase transition at a temperature of $10^{15}K$ reached 10^{-10} seconds after the bang. As an extra condition it is assumed that the energy released into the environment by the restructuring Higgs field had to be sufficient to compensate for the energy required for its own creation from the void. The aether with Higgs field hence contains less energy than aether without field and the lowest-energy ground-state of space-time is the void with Higgs field[20]. The formation of the Higgs field that occurs from nothing below the transition temperature, is a spontaneous process that yields sufficient energy to drive the pair production of all types

[20]The reasoning here is reminiscent of the negative mass that had to be postulated to support the phlogiston theory.

of particle from the radiation field of virtual particles that exists above the transition point. This physics of mass generation reduces to interference by the structured Higgs field with free motion of virtual particles that otherwise proceeds at an even uniform velocity c in the symmetrical excited state aether.

Interpreted, as it is, within the standard model, Higgs theory has little meaning in the real world, failing, as it does to relate the broken symmetry of the field to the chirality of space, time and matter. Only vindication of the conjecture is expected to be the heralded observation of the field bosons at stupendous temperatures in monstrous particle accelerators of the future. However, the mathematical model, without cosmological baggage, identifies important structural characteristics of any material universe. The most obvious stipulation is to confirm that inertial matter cannot survive in high-symmetry euclidean space.

4.6 The Arrow of Time

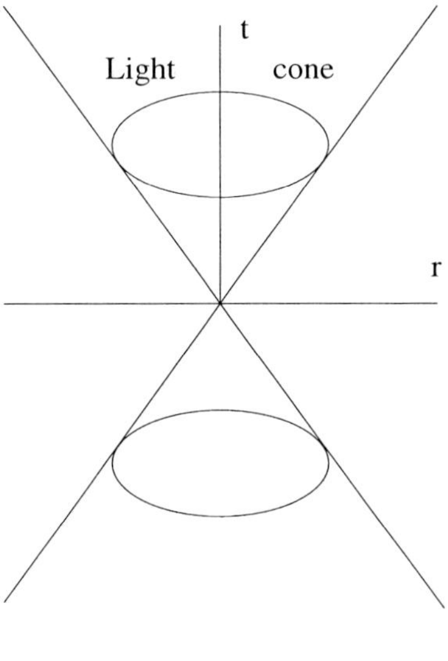

Figure 4: *Minkowski diagram of euclidean 4-space.*

There is no evidence that Minkowski space is flat on the large scale. The assumption of euclidean Minkowski space could therefore be, and probably is an illusion, like the flat earth. In fact, there is compelling evidence from observed spectroscopic red shifts that space is curved over galactic distances. These red shifts are proportional to distances from the source, precisely as required by a curved space-time[52]. An alternative explanation, in terms of an expanding-universe model that ascribes the red shifts to a Doppler effect, is still almost universally preferred. The more logical explanation will probably not be accepted until big-bang cosmology is finally disproved experimentally.

There also exists convincing internal evidence that real Minkowski space must be curved. Euclidean 4-space is commonly represented diagrammatically to distinguish between time and space axes as in figure 4.

Any stationary point is said to trace out a world line parallel to the time axis. A point in motion follows a worldline at an angle θ with respect to t, with a maximum of $\theta = \pi/4$ rad for a velocity of $v = c$. The space-like region outside the so-called light cone is usually ignored as physically meaningless since $v > c$. There is no explanation of why a stationary point has a world line and why it's always time-like and never space-like.

It is of interest to note how mass generation according to quantum field theory relates to special relativity: In the absence of a Higgs field only motion in the limiting surface that defines the light cone is allowed[21]. For any world line to occur within the light cone a Higgs field is required. This stipulation implies reduced symmetry (*i.e.* curvature) of space-time.

[21] Recall that the Higgs mechanism depends on local gauge symmetry, which is only defined within general relativity theory.

Chapter 5

Quantum Theory

5.1 Basic Concepts

5.1.1 Polarized Light: A familiar example of quantum behaviour

Experimental chemists are rarely concerned with quantum effects and it's not unusual to find them ignoring this fundamental theory altogether. Even when an effort is made to explore the topic more deeply traditional quantum phenomena like black-body radiation, Compton scattering and even the photoelectric effect may appear to be of somewhat limited importance. Experimentalists who rely on spectroscopic measurements get by with interpretations based on a few simple semi-classical rules, and without ever appreciating the deep significance of quantum theory. Maybe there is a problem with the rigorous mathematical formulation of quantum theory and too little emphasis on quantum effects routinely encountered in chemistry.

One quantum effect that chemists cannot ignore consistently is molecular chirality and the interaction of chiral molecules with polarized light. Although a detailed understanding of this issue will, in the final analysis, be beyond the scope of this preliminary discussion, it provides an easy introduction and a useful guide. While the discussion of molecular chirality only becomes possible at a much later stage, a phenomenological discussion of polarimetry is a common topic even for discussion at the elementary level.

A light ray is described equally well as either a beam of photons or a propagating electromagnetic disturbance, called a wave. In the wave description the disturbance consists of electric and magnetic fields, vibrating at right angles to each other and in a plane perpendicular to the direction of propagation. Any wave with an electric field that vibrates exclusively in a single direction is said to be polarized. In general however, natural light from a source does not show this behaviour. The total emission is the sum of contributions from many emitters (like atoms) and it follows that these contributions are uncorrelated in both phase and their directions of vibration.

Because of translational symmetry special directions occur in crystals and it often happens that the electric fields associated with light rays transmitted through a crystal are channeled to vibrate in a special direction that provides an easy passage. This means that the light becomes polarized.

When this happens a monochromatic light beam passing through the crystal is resolved into two linearly polarized components with electric displacement vectors vibrating at right angles to each other. The two vibration directions depend on the relative orientation between the crystal and the incident light beam and follow the ellipsoid of wave normals, or optical indicatrix of the crystal. The plane perpendicular to any direction of wave propagation intersects the ellipsoid in an ellipse and the principle axes of this ellipse correspond to the two possible polarizations.

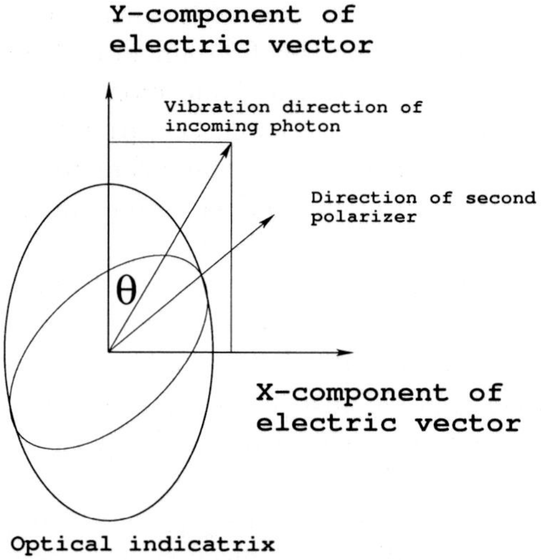

In the theory of optics this phenomenon is accounted for in terms of geometrical construction, but the physical picture is less convincing. Double refraction is a well-documented property of most crystals, at its most spectacular in Iceland spar. The double image of an object viewed through the crystal indicates the existence of two independent rays and not the components of a single ray. In mathematical terms the two rays are linearly independent and therefore orthogonal. Any intermediate situation represents a linear combination of the two orthogonal basis vectors and can be resolved into two components. What happens to an individual photon is however, not clear.

On entering the crystal it faces a choice of two vibration directions and a random chance of following one or the other.

This is a recurrent theme of quantum theory. Many quantum systems can be formulated exactly in terms of a wave equation and the behaviour of the system will be described exactly by the wavefunction, the solution to the wave equation. What is not always appreciated is that this is a mathematical description only, which does not ensure understanding of the event in terms of a comprehensible physical model. The problem lies therein that the description is only possible in terms of a wave formalism. Understanding of the physical behaviour however, requires reduction to a particle model. The wave description is no more than a statistically averaged picture of the behaviour of many particles, none of which follows the actual statistically predicted course. The wave description is non-classical, and the particle model is classical. Mechanistic understanding is possible only in terms of the classical approach, and a mathematically precise description only in terms of the wave formalism. The challenge of quantum theory is to reconcile the two points of view.

Phenomena like this, which cannot be analyzed in terms of common logic are appropriately known as non-classical or quantum phenomena. In the case of the unpolarized light beam passing through the crystal, it is resolved into two, and only two, independent polarized rays with displacement vectors vibrating at right angles to the direction of propagation and with each other. All intermediate situations are excluded, which in a sense justifies the description as a quantum event.

In some birefringent materials (e.g. quinine sulphate periodate, also known as herapathite) one of the rays is completely absorbed and therefore only a single ray of polarized light emerges. A device based on this principle is known as a polaroid filter. A surprising characteristic of polarization is observed on passing a light beam through successive polaroid filters in different orientation, such as the xz and yz polarizers described before (3.1.5). The polarized light passed by an xz filter and propagating in the z-direction is found to be blocked completely by a second, yz filter, since the x-polarized light produced by the first filter contains no y-component. However, when a third filter, inclined at $\pm 45°$ with respect to the xz and yz filters is inserted between these two, polarized light is observed to be transmitted by the yz filter.

The diagonally oriented filter (d) decomposes any incoming light into a component vibrating in the direction of d and another component in the perpendicular plane, pz. Both of these have x and y components, which are resolved by the third (yz) filter in the sequence. In mathematical terminology

d-polarized light is a linear combination

$$\boldsymbol{E} = E_0 \left[\frac{1}{\sqrt{2}} \boldsymbol{i} \cos(kz - \omega t) + \frac{1}{\sqrt{2}} \boldsymbol{j} \cos(kz - \omega t) \right]$$

of x and y polarized components[1].

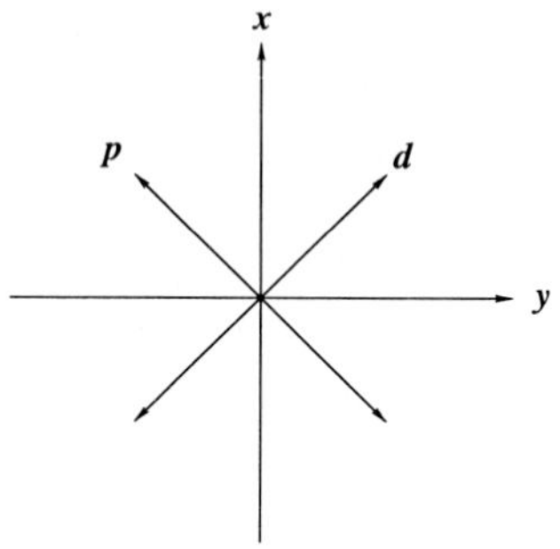

Once the diagonally oriented filter intervenes and selects a 45° polarized beam, it is immaterial whether the beam was previously xz. The diagonal filter creates a yz component that passes through the third filter. Any previous information on light polarization is destroyed by a polarizer, in this case the intermediate filter.

If an electric vector, **E** of a given wave is at an angle θ with respect to the direction of polarization it has a component $E_y = E \cos\theta$ in the polarization direction and another component $E \sin\theta$ vibrating perpendicular to the first. In the case where this second ray is absorbed the amplitude emerging from the

[1]This is not the same as circularly polarized light which is obtained by passing linearly polarized light through a quarter-wave plate [53]. The plate produces a superposition of x and y components which are 90° out of phase, *i.e.*

$$\boldsymbol{E} = \frac{1}{\sqrt{2}} E_0 \left[\boldsymbol{i} \cos(kz - \omega t) \pm \boldsymbol{j} \cos(kz - \omega t + \frac{\pi}{2}) \right] \tag{5.1}$$

Noting that $e^{i\pi/2} = \cos\left(\frac{\pi}{2}\right) + i \sin\left(\frac{\pi}{2}\right) = i$ it follows that

$$e^{i(kz-\omega t+\pi/2)} = e^{i(kz-\omega t)}.e^{i\pi/2} = i\cos(kz - wt) - \sin(kz - \omega t)$$

The real part of

$$\epsilon = \frac{1}{\sqrt{2}} \left[\boldsymbol{i} e^{i(kz-\omega t)} + i\boldsymbol{j} e^{i(kz-\omega t)} \right] \tag{5.2}$$

therefore corresponds to right-circular polarized light as defined in (1). The imaginary part represents left circular polarization.

polarizer will be reduced by the factor $\cos\theta$ and the intensity, $I = I_0 \cos^2\theta$. Mechanistically formulated the polarized wave is interpreted as consisting of a beam of polarized photons. Suppose it hits a second polarizer at an angle ϕ. To account for the measured intensity it is necessary to postulate that the probability of any polarized photon transmitted through the second polarizer is proportional to $\cos^2(\theta - \phi)$. Although all photons approach the polarizer with the same polarization, P_θ, the probability of any photon passing the second polarizer, with polarization P_ϕ, is $\cos^2(\theta - \phi)$.

The photon emerging from the first polarizer is said to be in the polarization state P_θ[2]. On emerging from the second polarizer each photon is in the polarization state P_ϕ, with the same energy $h\nu$ as before, but the intensity of the beam has now decreased by the factor $\cos^2(\theta - \phi)$.

Each polarization state, considered as a vector can be described in terms of its components, i.e.

$$P_\theta \leftarrow (\cos\theta, \sin\theta), P_\phi \leftarrow (\cos\phi, \sin\phi), \text{ etc.}$$

The direct (dot) product of these two state vectors

$$P_\theta.P_\phi = \cos\theta\cos\phi + \sin\theta\sin\phi = \cos(\theta - \phi)$$

and the square of the inner product, $\cos^2(\theta - \phi)$ is seen to be exactly the quantum-mechanical probability inferred before. This is no accident and all quantum events have such a scalar product as the only link between theory and experiment.

5.1.2 Electron Spin

A phenomenon closely related to polarization is observed in a Stern-Gerlach experiment [54] that passes a beam of silver atoms through an inhomogeneous magnetic field between the poles of a magnet. Silver atoms are made up of a nucleus and 47 electrons, 46 of which can be considered to make up a spherically symmetrical core with no net angular momentum. All observed magnetic effects (ignoring nuclear spin) can be linked to electron 47. If it is assumed that this electron has spin $\boldsymbol{S}$, it would impart a magnetic moment $\boldsymbol{\mu}$ to the atom, where

$$\boldsymbol{\mu} = \frac{e}{mc}\boldsymbol{S}.$$

[2]There is a polarization for each $\theta, 0 < \theta < \pi$ and $P_\theta = P_{\theta+\pi}$. Note that P_θ and $P_{\theta+\pi}$ are the same because the sense of the vector is not important.

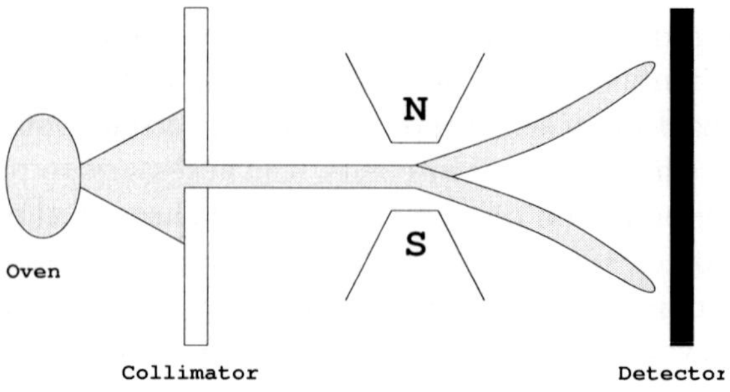

Interaction between the magnetic moment and the field generates a force on the atom, given by

$$F_z = \frac{\partial}{\partial z}(\boldsymbol{\mu} \cdot \boldsymbol{B}) \simeq \mu_z \frac{\partial B_z}{\partial z} \quad ,$$

if the components of $\boldsymbol{B}$ in directions other than z are ignored. If the electron were like a classical spinning object all values of μ_z between $|\boldsymbol{\mu}|$ and $-|\boldsymbol{\mu}|$ should be realized. A continuous beam of atoms, evenly spread around the direction of the incoming beam, is predicted to emerge from the SG magnet; some atoms displaced in the direction of the field and some in the opposite direction. What is observed instead is that the beam is split into two distinct components. This phenomenon used to be referred to as space quantization. The observation implies that the component S_z of the electron spin $\boldsymbol{S}$ has only two possible values, S_z^+ (up) and S_z^- (down). The two possible values of S_z are multiples of some fundamental unit of angular momentum; numerically $S_z = \pm\hbar/2$, where $\hbar = 1.054 \times 10^{-34}$ Js.

There is, of course, nothing special about the z-direction. The magnet could, for the same price, be turned through 90° to resolve the S_y components of a beam moving in the x-direction.

By performing sequential SG experiments it can be shown that electron spin has much in common with photon polarization. In a typical arrangement the beam from the oven is first passed through an SG$_z$ apparatus with the inhomogeneous magnetic field in the z-direction. Suppose the one (S_z^-) component from this experiment is blocked and the second (S_z^+), is allowed to pass into another SG$_z$ apparatus. The beam emerging from the second device will be found to have no S_z^- component, exactly as for two parallel polaroids.

This experiment is now repeated, but with the second device set up as SG$_y$. The S_z^+ beam that enters the second device is now split into two components, an S_y^+ and an S_y^- component, with equal intensities. This result is the analoque of two sequential polarizers mutually inclined at 45°.

In a third experiment a third SG_z device is added to the previous arrangement and the S_y^- component from the second (SG_y) device is blocked. Despite the fact that the S_z^- component emerging from the first device had been blocked, the beam emerging from the third device will now consist of S_z^+ and S_z^- components again. It is therefore wrong to infer that having passed through SG_z and SG_y and having the S_z^- and S_y^- components blocked out, the beam entering the third device consists of S_z^+ and S_y^+[3]. As in the case of polarization, here the SG_y device completely destroys any previous information about S_z.

As with polarization, the orientation of spin is represented by state vectors. Just like $\boldsymbol{i}x$ and $\boldsymbol{j}y$ which serve as base vectors to decompose polarization vectors such as the diagonal vector considered above, a spin state like S_y^+ can be represented by a vector $\boldsymbol{s}_y^+$, which is a linear combination of two base vectors $\boldsymbol{s}_z^+$ and $\boldsymbol{s}_z^-$. This decomposition could be formulated, by analogy as

$$\begin{aligned} \boldsymbol{s}_y^+ &= \frac{1}{\sqrt{2}}\left[\boldsymbol{s}_z^+ + \boldsymbol{s}_z^-\right] \\ \boldsymbol{s}_y^- &= -\frac{1}{\sqrt{2}}\left[\boldsymbol{s}_z^+ + \boldsymbol{s}_z^-\right] \end{aligned} \tag{5.3}$$

The unblocked component emergent from the second (SG_y) device in the three-sequence arrangement can thus be regarded as a superposition of S_z^+ and S_z^- in the sense of (3). It is for this reason that two components emerge from the third (SG_z) device.

An immediate question arising from the above concerns the definition of the equally feasible $S_x^\pm$ states. Symmetry suggests that an $S_z^\pm$ beam going in the y-direction through an SG_x apparatus should be equivalent to an $S_z^\pm$ beam going along x into SG_y. The state vectors of $S_x^\pm$ should then be regarded as a linear combination of $\boldsymbol{s}_z^\pm$, but from (3) it appears that all available options have been used up in forming $\boldsymbol{s}_y^\pm$. The problem is resolved by analogy with the difference between diagonally oriented plane polarized light and circular polarization. The difference implies that the S_x^+ and S_x^- spin states of silver atoms could be described as analogous to right and left circularly polarized light respectively.

Using equation (2) of footnote 1 as the basis for the analogy suggests that independent $\boldsymbol{s}_x^\pm$ state vectors could be defined by allowing complex

[3]This example is often used to demonstrate that in quantum mechanics S_z and S_y cannot be determined simultaneously.

coefficients, as in

$$s_x^{\pm} = \frac{1}{\sqrt{2}}\left(s_z^{+} \pm i s_z^{-}\right).$$

This linear combination is clearly different from (3). The implication is that the two-dimensional vector space needed to describe the spin states of silver atoms must be a complex vector space; an arbitrary vector in this space is written as a linear combination of the base vectors $s_z^{\pm}$ with, in general complex coefficients. This is the first example of the fundamental property of quantum-mechanical states to be represented only in an abstract complex vector space [55].

5.1.3 Quantum States

The polarized-light and spin examples have shown that, even though a quantum system may be in a definite state, as established by an exhaustive measurement, a subsequent observation does not necessarily yield a definite result. Knowing the result of an observation therefore does not reveal the state, the system was in at the time of the measurement, and neither does knowing the state of a system predict the exact outcome of any observation. Quantum theory only predicts the statistical outcome of many measurements of some property. To achieve this, a physical state is represented by a column vector or (equivalently) by the Hermitian conjugate row vector:

$$w = \{c_1, c_2, \cdots, c_n\} \text{ or } (c_1^*, c_2^*, \cdots, c_n^*), \text{ with } \sum | c_k |^2 = 1$$

If the system is subjected to an observation (e.g. by probing with a polarizer), $\mathbf{O}(s)$, where s is represented by the state vector v, then the probability of a positive observation (i.e. of finding the system in state v) is $| w^\dagger v |^2$.

The representation of a quantum system is never unique. For instance, the polarization state observation can also be described in terms of the representations:

$$P_\theta = \frac{1}{\sqrt{2}}\begin{pmatrix} e^{i\theta} \\ e^{-i\theta} \end{pmatrix} \text{ or } \frac{1}{\sqrt{2}}\left(e^{-i\theta}, e^{i\theta}\right)$$

When the photon in state P_θ falls on an analyzer at an angle ϕ

$$w^\dagger = \frac{1}{\sqrt{2}}\left(e^{-i\phi}, e^{i\phi}\right), v = \frac{1}{\sqrt{2}}\begin{pmatrix} e^{i\theta} \\ e^{-i\theta} \end{pmatrix}$$

and the probability of a signal from the detector is

$$\left| \frac{1}{\sqrt{2}}\left(e^{-i\phi}, e^{i\phi}\right) \cdot \frac{1}{\sqrt{2}}\begin{pmatrix} e^{i\theta} \\ e^{-i\theta} \end{pmatrix} \right|^2$$

$$= \frac{1}{4} \left| e^{i(\theta-\phi)} + e^{-i(\theta-\phi} \right|^2$$

$$= \frac{1}{4} \left| \cos(\theta - \phi) + i\sin(\theta - \phi) + \cos(\theta - \phi) - i\sin(\theta - \phi) \right|^2$$

$$= \cos^2(\theta - \phi),$$

as before. In general there is an infinite number of choices all leading to the same physical result. In practice one should choose only one and use it consistently.

It is instructive to compare the states represented by normalized vectors v and $ve^{i\theta}$ respectively. The probabilities of observing each of these vectors in the state w are $|w^\dagger v|^2$ and $|w^\dagger ve^{i\theta}|^2$ respectively. Whatever the choice of w, $|e^{i\theta}|^2 = e^{i\theta} \cdot e^{-i\theta} = 1$ and the two results are always the same. Any state vector therefore has associated with it an arbitrary phase factor.

5.1.4 Observables

Many observations which measure magnitudes produce a more complicated result, compared to the photon that either passes the polarizer, or does not. Measurements of this kind can however, still be related to elementary observations. The question: "What is the magnitude of X ?", is rephrased as a set of simple questions:

"Is χ_1 the magnitude of X ?"

"Is χ_2 the magnitude of X ?"

. . . .

"Is χ_n the magnitude of X ?"

where the χ_i are the possible outcomes of the measurements of X. A physical quantity that may be measured in this way is called an *observable*.

The possible outcomes χ_i can be chosen in a special way, such that the state corresponding to each of the χ_i may be represented by a unit vector, such as

$$\begin{aligned} \boldsymbol{u}_1 &= \{\ 1,0,0\ldots,0\} \\ \boldsymbol{u}_2 &= \{\ 0,1,0\ldots,0\} \\ &\quad \ldots\ldots\ldots\ldots \\ \boldsymbol{u}_n &= \{\ 0,0,0\ldots,1\} \end{aligned}$$

Together these unit vectors define an n-dimensional vector space and they are said to constitute an orthonormal set. State vectors of this type are called eigenvectors and the special χ_i are the eigenvalues. Taken together the unit eigenvectors define an $n \times n$ unit matrix

$$\boldsymbol{I} = \begin{pmatrix} 1 & 0 & 0 & \dots & 0 \\ 0 & 1 & 0 & \dots & 0 \\ . & . & . & \dots & . \\ 0 & 0 & 0 & \dots & n \end{pmatrix}$$

The matrix

$$\chi_i \boldsymbol{I} = \begin{pmatrix} \chi_1 & 0 & 0 & \dots & 0 \\ 0 & \chi_2 & 0 & \dots & 0 \\ . & . & . & \dots & . \\ 0 & 0 & 0 & \dots & \chi_n \end{pmatrix}$$

contains all the eigenvalues of the system.

Consider a quantity X for which each of the questions is an elementary observation $\mathbf{O}(u_j)$, and suppose that the system is in one of the eigenstates u_1. Then, the answer to the question "Is χ_1 the magnitude of X ?", is certainly yes, while the answer to "Is χ_k the magnitude of X ?", is certainly no, unless $k = 1$. Hence, the probability that measurement of X yields

$$\chi_k = \begin{cases} 1 & \text{if } k = 1 \\ 0 & \text{if } k \neq 1 \end{cases}, = \left| u_1^\dagger u_k \right|^2$$

Starting with each of the special states in turn, one finally has

$$\left| u_j^\dagger u_k \right|^2 = \begin{cases} 1 & \text{if } j = k \\ 0 & \text{if } j \neq k \end{cases}$$

for any pair of state vectors.

In a more general formulation the outcome of an elementary measurement is represented by an arbitrary vector with more than one non-zero component, different from unity, such as

$$\boldsymbol{v} = \{c_1 \boldsymbol{e}_1, c_2 \boldsymbol{e}_2, \dots, c_n \boldsymbol{e}_n\}$$

where the $\boldsymbol{e}_i$ are unit vectors. From the known properties of scalar products it follows that

$$\boldsymbol{v}.\boldsymbol{v} = \boldsymbol{v}^\dagger \boldsymbol{v} = \sum (c_i \boldsymbol{e}_i)^2 = \sum |c_i|^2 = 1.$$

These are the vectors assumed before (5.1.3) to represent polarization and spin states.

5.1.5 Mean Values

Consider a system in state w and that a measurement of the physical quantity X is made with the possible outcomes $\chi_1, \chi_2, \cdots$, and that this experiment is repeated a large number of times for the same state w. If the result χ_j occurs a proportion p_j of the times, then (classically) the mean (or average or expectation) value is defined as

$$\langle \chi \rangle_w = \sum_j p_j \chi_j$$

Quantum mechanically this probability is related to the squares of scalar products. If X can have any of the maximum number n, of possible values, then each value χ_i occurs with certainty for just one state v_i. The probability of finding χ_j for the system in state w is $|v_j^\dagger w|^2 = w^\dagger v_j v_j^\dagger w$. Thus $\langle X \rangle_w = \sum_j w^\dagger v_j \chi_j v_j^\dagger w$, written as $w^\dagger X w$, where X must be an $n \times n$ matrix,in order to multiply into both a row and a column vector

$$X_{ij}(i = 1, 2, \cdots, n; j = 1, 2, \cdots, n) = \begin{pmatrix} x_{11} & x_{12} & \cdots & x_{1n} \\ x_{21} & x_{22} & \cdots & x_{2n} \\ \cdots & \cdots & \cdots & \cdots \\ x_{n1} & x_{n2} & \cdots & x_{nn} \end{pmatrix}$$

The Hermitian conjugate

$$\mathbf{X}^\dagger = \begin{pmatrix} x_{11}^* & x_{21}^* & \cdots & x_{n1}^* \\ x_{12}^* & x_{22}^* & \cdots & x_{n2}^* \\ \cdots & \cdots & \cdots & \cdots \\ x_{1n}^* & x_{2n}^* & \cdots & x_{nn}^* \end{pmatrix}$$

This matrix is the appropriate representation of an observable such as X. A Hermitian matrix is its own hermitian conjugate. The diagonal elements of a Hermitian matrix are real and each element is symmetry related to its complex conjugate across the main diagonal.

Recognizing that any state v_i can show up in a measurement with the magnitude λ_i allows the total measurement problem to be represented by the equation $Av = \lambda v$, where A is the same Hermitian matrix, i.e.

$$\sum_{j=1}^{n} a_{ij} v_j = \lambda_i v_i (i = 1, \cdots, n)$$

which represents a system of n linear homogeneous equations. In general the only solution of these equations is $v_i = 0$ for all i. For certain *eigenvalues*, λ however, a non-zero solution v exists (as the column *eigenvector* corresponding to the eigenvalue). There is at least one eigenvalue and there may be as many as n. Since A is Hermitian all the eigenvalues are real.

5.1.6 Eigenvectors

A complete orthonormal set of n-component vectors is a set $v_1, v_2, \cdots, v_n$, satisfying the orthonormality relations

$$v_i^\dagger v_j = \delta_{ij} = \begin{cases} 1 & \text{if } i = j \\ 0 & \text{if } i \neq j \end{cases}, \quad \begin{matrix} i = 1, 2, \cdots, n \\ j = 1, 2, \cdots, n \end{matrix}$$

Any other n-component vector u, whatever may be expressed in exactly one way as a linear combination (superposition) of the vectors of a given complete orthonormal set

$$u = \sum_{i=1}^{n} c_i v_i$$

The coefficients are given[4] by $c_i = v_i^\dagger u$. A Hermitian $n \times n$ matrix always has as many as n linearly independent eigenvectors. Although it is always possible to select a complete orthonormal set of eigenvectors, this choice is never unique, since each v_i may be multiplied by an arbitrary phase factor without altering the orthonormality. If the corresponding eigenvalues $\lambda_1, \lambda_2, \cdots, \lambda_n$ are also distinct, there is an even greater arbitrariness in the choice, on the account of *degeneracy*. If v_1 and v_2 are two eigenvectors with eigenvalues λ_1 and λ_2, then $v_1^\dagger v_2 = 0$ whenever $\lambda_1 \neq \lambda_2$.

The eigenvalue-eigenvector decomposition of a Hermitian matrix with the complete orthonormal set of eigenvectors v_i and eigenvalues λ_i, is written as

$$\mathbf{A} = \sum_{i=1}^{n} v_i \lambda_i v_i^\dagger$$

commonly using the same symbol as for the matrix itself. If u is a column vector then the product

$$\mathbf{Au} = \left\{ \sum_{i=1}^{n} v_i \lambda_i v_i^\dagger \right\} u = \sum_{i=1}^{n} v_i (\lambda_i v_i^\dagger u)$$

This is a superposition of the vectors v_i with coefficients $\lambda_i v_i^\dagger u$, which are ordinary numbers. This is the way the matrix enters into the calculation of

[4]Substitute $c_i = v^\dagger u$ into $u = \sum_{i=1}^{n} c_i v_i$, *i.e.*

$$u = \sum_{i=1}^{n} \left(v_i^\dagger v_i \right) u = \sum_{i=1}^{n} u |c_i|^2 = u$$

mean values. It is custom to refer to both the physical quantity X and the matrix X which represents it as the observable X. The eigenvectors represent those eigenstates for which the corresponding eigenvalues are certain to be observed.

To summarize, suppose that all possible states for any given observable (spin, polarization, energy, momentum, etc.) are known and that each can be formulated in terms of a column vector $\mathbf{a} = \{a_1, a_2, \cdots a_n\}$. These vectors form an orthonormal set and are represented by an $n \times n$ matrix

$$\mathbf{A} = \begin{pmatrix} a_{11} & a_{12} & \cdots & a_{1n} \\ a_{21} & a_{22} & \cdots & a_{2n} \\ \cdots & \cdots & \cdots & \cdots \\ a_{n1} & a_{n2} & \cdots & a_{nn} \end{pmatrix}$$

Likewise, any system of interest can be in any one of a number of possible states and each state is represented by a column vector to produce a matrix X whose columns correspond to the possible states. The probable outcome of any measurement of the observable A on system X is described by the scalar products of all possible $\sum a^{\dagger}x$, representing individual probabilities.

All of these possible outcomes are contained in the product of the two matrices, noting that the matrix product $\mathbf{AX}$ has the elements

$$(ax)_{ij} = \left(\sum_{k=1}^{n} a_{ik} x_{kj} \right) ,$$

i.e. all of the required scalar products in the form of a matrix. If each of the states is an eigenstate $Ax_i = \lambda_i x_i$, the matrix product reduces to $\mathbf{AX} = \mathbf{\Lambda X}$, where $\mathbf{\Lambda}$ is the diagonal matrix with elements $\lambda_{ij} = \delta_{ij}$, i.e. $\lambda_i \mathbf{I}$. The solutions are obtained by solving the *secular* equation

$$(\mathbf{A} - \lambda_i \mathbf{I})\mathbf{X} = O$$

5.1.7 Incompatibility

The fact that quantum observables are represented by matrices immediately suggests problems of non-commutation. For instance, the observables can be measured at the same time only if they have a complete orthogonal set of eigenvectors in common. This happens only when they commute, i.e. $\mathbf{XY} = \mathbf{YX}$, or the commutator $[\mathbf{X}, \mathbf{Y}] = \mathbf{XY} - \mathbf{YX} = 0$. This is a central feature of the matrix formulation of quantum theory discovered by Heisenberg, Born and Jordan while trying to explain the observed spectral transitions of the hydrogen atom in a more fundamental way than the quantization prescription $\oint pdq = nh$, based on classical variables and Planck's constant. They ordered the possible values of momentum and position for a multiply periodic system in numbered square arrays with Fourier components of the type

$$q(nm)e^{2\pi i\nu(nm)t}$$

as elements. If this is differentiated and the time factor is dropped, one has

$$\dot{q} = 2\pi i\nu(nm)q(nm)$$

Now consider the diagonal array of energy values E_n, and let $\mathbf{Q}$ be any other square array. Then forming the function $[\mathbf{EQ} - \mathbf{QE}]$ it has as the (nm)th element the sum

$$\sum_k E_n\delta_{nk}q(km) - \sum_k q(nk)E_k\delta_{km}$$

The only surviving term is for $k = n$ in the first and for $k = m$ in the second sum, to give

$$(E_n - E_m)q_{nm} = (Eq - qE)(nm)$$

or, by the Bohr frequency condition

$$h\nu(nm)q(nm) = (Eq - qE)(nm)$$

From

$$q(nm) = \frac{1}{2\pi i\nu(nm)}\dot{q}$$

then follows the equation of motion

$$\frac{h}{2\pi i}\dot{q}(nm) = (Eq - qE)(nm)$$

The procedure followed here is clearly that of matrix multiplication and $\mathbf{Q}$ can be any matrix whatever, for the proof does not require it to be a coordinate. It follows almost immediately that if $\mathbf{F}$ is a matrix function of two other matrices $\mathbf{P}$ and $\mathbf{Q}$, then

$$\mathbf{FQ} - \mathbf{QF} = \frac{h}{2\pi i}\frac{\partial \mathbf{F}}{\partial \mathbf{P}}$$

$$\mathbf{PF} - \mathbf{FP} = \frac{h}{2\pi i}\frac{\partial \mathbf{F}}{\partial \mathbf{Q}}$$

provided that $\mathbf{P}$ and $\mathbf{Q}$ satisfy the relation

$$(\mathbf{PQ} - \mathbf{QP}) = \frac{h}{2\pi i}\mathbf{I} = \frac{h}{2\pi i}\delta_{pq}$$

To demonstrate this, set $\mathbf{F} = \mathbf{P}$ or $\mathbf{Q}$. The two basic postulates of matrix mechanics follow directly from this, in terms of the momentum-position commutator,

$$[p, x] = i\hbar \qquad (\hbar = h/2\pi)$$

and the equations of motion

$$\dot{p}_k = \frac{2\pi i}{h}(Hp_k - p_k H)$$

$$\dot{g}_k = \frac{2\pi i}{h}(Hq_k - g_k H)$$

In terms of a state vector $\mathbf{u}$ this reads

$$\mathbf{Hu} = i\hbar\dot{u} \ ,$$

where $\mathbf{H}$ is a Hermitian matrix

Photon Spin

To illustrate some of these principles the angular momentum of a photon will be examined [56]. Suppose a beam of circularly polarized light falls on a perfectly black absorbing surface, which not only heats up ($E = h\nu$) but also acquires a torque, on account of the angular momentum it absorbs. Circular polarization means that the probability of an elementary observation $\mathbf{O}(P_\theta) = \frac{1}{2}$. The ratio of energy/torque $= \omega(= 2\pi\nu)$, the angular frequency of each photon. Let the state of circular polarization be represented by $\begin{pmatrix} a \\ b \end{pmatrix}$, then for all θ,

$$\left| (\cos\theta, \sin\theta) \begin{pmatrix} a \\ b \end{pmatrix} \right|^2 = \frac{1}{2}$$

i.e. $|a\cos\theta + b\sin\theta|^2 = \frac{1}{2}$, for all θ.

This can be shown to hold only if $|a| = |b| = \frac{1}{\sqrt{2}}$ and $b = \pm ia$. The arbitrary phase factor can be chosen to make $a = \frac{1}{\sqrt{2}}$, to give the two vectors $\frac{1}{\sqrt{2}}\begin{pmatrix} 1 \\ i \end{pmatrix}$ and $\frac{1}{\sqrt{2}}\begin{pmatrix} 1 \\ -i \end{pmatrix}$, viz. $u_\pm$, pertaining to left and right-circular polarization respectively, compare (2). This means that the observable angular momentum of a photon can be formulated as $\langle M \rangle_\pm = (u^\dagger M u)_\pm$.

Classically, a circularly polarized light beam with angular frequency $\omega(= 2\pi\nu)$ transfers angular momentum at a rate of E/ω, where E is the rate of energy transfer. Considered as a beam of photons, $E = Nh\omega/2\pi$, so that the angular momentum of each photon is $h/2\pi = \hbar$.

To determine the form of M it is assumed to have a general Hermitian 2×2 matrix structure,

$$M = \begin{pmatrix} a & c \\ c^* & b \end{pmatrix},$$

where a and b are real. Then

$$\langle M \rangle_+ = \frac{1}{\sqrt{2}}(1, -i) \begin{pmatrix} a & c \\ c^* & b \end{pmatrix} \cdot \frac{1}{\sqrt{2}} \begin{pmatrix} 1 \\ i \end{pmatrix} = \hbar$$

$$= \frac{1}{2}(a - ic^*, c - ib) \begin{pmatrix} 1 \\ i \end{pmatrix}$$

$$= \frac{1}{2}(a + b + 2ic)$$

This requires $a = b = 0$, $c = -i\hbar$, so that the angular momentum, or spin, of a photon is represented by

$$M \leftarrow \hbar \begin{pmatrix} 0 & -i \\ i & 0 \end{pmatrix}$$

It can be tested against u_-, to give

$$\langle M \rangle_- = \frac{\hbar}{\sqrt{2}}(1, i) \begin{pmatrix} 0 & -i \\ i & 0 \end{pmatrix} \cdot \frac{1}{\sqrt{2}} \begin{pmatrix} 1 \\ -i \end{pmatrix} = -\hbar$$

5.1.8 Matrix Mechanics

Heisenberg's formulation of quantum theory consisted of associating with each observable a square Hermitian matrix, with the view of determining its measurable values. Applied to the quantity H, which, classically is a function of the cartesian coordinates q_i and the momenta p_i,

$$H = (q_1, q_2, \cdots, q_n; p_1, p_2, \cdots, p_n) \; ,$$

the prescription is to find a set of matrices $\mathbf{Q}_1, \mathbf{Q}_2, \cdots, \mathbf{Q}_n; \mathbf{P}_1, \mathbf{P}_2, \cdots, \mathbf{P}_n$, which satisfy the commutation rules

$$[\mathbf{Q}_m, \mathbf{Q}_n] = [\mathbf{P}_n, \mathbf{P}_m] = 0 \; ; \; [\mathbf{P}_m, \mathbf{Q}_n] = -i\hbar\delta_{nm}\mathbf{I} = \frac{h}{2\pi i} \cdot \delta_{nm}$$

and at the same time render the matrix $\mathbf{H}(\mathbf{Q}_1, \mathbf{Q}_2, \cdots, \mathbf{Q}_n; \mathbf{P}_1, \mathbf{P}_2, \cdots, \mathbf{P}_n)$ diagonal.

The Harmonic Oscillator

The classical potential energy of an harmonic oscillator, $V = \frac{1}{2}kx^2 = \frac{1}{2}\cdot(4\pi^2 m\nu^2 x^2)$, and

$$-\frac{\partial V}{\partial x} = -kx = F ,$$

where F is the restoring force and the force constant $k = m\omega^2$, where ω is the classical frequency, $\omega = 2\pi\nu$. The total energy, expressed in terms of the classical Hamiltonian

$$H = T + V = \frac{p^2}{2m} + 2\pi^2 m\nu^2 q^2$$

Hence, if **P** and **Q** are matrices

$$\mathbf{H}(\mathbf{Q}, \mathbf{P}) = \frac{1}{2m}(\mathbf{P}^2 + 4\pi^2 m^2\nu^2\mathbf{Q}^2)$$

Suppose that the matrices satisfy the Heisenberg conditions and diagonalize **H**, then

$$H_{kl} = \frac{1}{2m}(P^2 + 4\pi^2 m^2\nu^2 Q^2)_{kl} = E_k\delta_{kl}$$

where $H_{kk} = E_k$.

Construct the matrix

$$(\mathbf{P} + 2\pi im\nu\mathbf{Q})(\mathbf{P} - 2\pi im\nu\mathbf{Q}) = P^2 + 4\pi^2 m^2\nu^2 Q^2 - 2\pi im\nu(\mathbf{PQ} - \mathbf{QP})$$
$$= 2m\mathbf{H} - hm\nu\mathbf{I}$$

Similarly

$$(\mathbf{P} - 2\pi im\nu\mathbf{Q})(\mathbf{P} + 2\pi im\nu\mathbf{Q}) = 2m\mathbf{H} + hm\nu\mathbf{I}$$

Multiply the first construct from the left by $\mathbf{A} = \mathbf{P} - 2\pi im\nu\mathbf{Q}$, and the second by **A** from the right. Then

$$\mathbf{A}(2m\mathbf{H} - hm\nu\mathbf{I}) = (2m\mathbf{H} + hm\nu\mathbf{I})\mathbf{A}$$

This is written in component form by noting that on the left the kth row of **A** multiplies into the lth column of the second matrix while on the right the kth row of the first matrix multiplies into the lth column of **A**, so that

$$2m\sum_i A_{ki}(H - \frac{h\nu}{2})\delta_{il} = 2m\sum_i (H + \frac{h\nu}{2})\delta_{ki}A_{il}$$

Writing E_l , $E_k = H_{ii}$, this reduces to components

$$A_{kl}(E_l - \frac{h\nu}{2}) = (E_k + \frac{h\nu}{2})A_{kl}$$

Hence

$$(E_k - E_l + h\nu)A_{kl} = 0$$

The possible solutions are $A_{kl} = 0$ or $E_l - E_k = h\nu$. By definition, $A_{kl} = 0$ implies $(\mathbf{P} - 2\pi im\nu\mathbf{Q}) = 0$, and hence

$$\mathbf{H} = \frac{h\nu}{2}\mathbf{I}$$

whence $E_l = \frac{1}{2}h\nu$ or $E_l = E_k + h\nu$.

The lowest (zero-point) energy follows as $E_0 = \frac{1}{2}h\nu$ and the higher levels are $(n + \frac{1}{2})h\nu$.

5.1.9 Quantum Particles

When extending the quantum formalism from photons to more complicated systems one meets with the practical problem that an observable can no longer be represented by a finite matrix. Consider the position of a particle on a line segment. It can be specified as being within a certain interval, or box, on the line. To get a more precise location the size (number) of the boxes must decrease (increase), towards the infinitesimal (infinite). The general state G of the system is then represented by a normalized column vector with an infinite number of components

$$G \leftarrow \{\cdots c_{n-1}, c_n, c_{n+1}, \cdots\}, \sum_{n=-\infty}^{\infty} |c_n|^2 = 1$$

where $|c_n|^2$ is the probability of finding the particle in box n. The probability density at x is $\lim_{\delta\to 0}\delta^{-1}|c_{n(\delta)}|^2$, and this suggests that a state function specifying the state of the particle at point x, should be defined as

$$\Psi(x) = \lim_{\delta\to 0}\delta^{-\frac{1}{2}}c_{n(\delta)}$$

and used to represent $G \leftarrow \Psi$. When dealing with a continuous state function it is no longer practical to represent observables by a matrix. One expects the definition of of a mean value still to be of shape $\langle X\rangle = u^\dagger Xu$, but where X now takes the form of a continuous function or operator. For instance, the mean position of a particle

$$\langle x\rangle_\Psi = \int_{-\infty}^{\infty} x|\Psi(x)|^2dx = \int_{-\infty}^{\infty}\Psi^*(x)\cdot x\Psi(x)dx$$

The observable x is now the multiplication factor. The probability density in this terminology is defined as $\rho = \Psi^*\Psi$. The normalization condition, in three dimensions, is written

$$\int_{-\infty}^{\infty} \Psi^*\Psi d\tau = 1 \quad ,$$

and interpreted to mean that the total probability density, integrated over all space, must be unity.

From the Heisenberg formalism momentum should be represented by an operator that does not commute with x, i.e. $[x, p] = i\hbar$. The momentum operator can therefore not also be multiplicative, but can be a differential operator. The representation $p \leftarrow -i\hbar\frac{\partial}{\partial x}$ gives the correct form when operating on a continuous function $u(x)$.

$$-i\hbar\left[x\frac{\partial u}{\partial x} - \frac{\partial(ux)}{\partial x}\right] = -i\hbar\left[x\dot{u} - \dot{u}x - u\right] = i\hbar u$$

Introducing the same representation into the classical Hamiltonian $H = p^2/2m + V(x)$, one has

$$H \leftarrow -\frac{\hbar^2}{2m}\frac{\partial^2}{\partial x^2} + V$$

This gives the equation of motion (in one dimension), known as Schrödinger's equation:

$$i\hbar\frac{\partial\Psi}{\partial t} = -\frac{\hbar^2}{2m}\frac{\partial^2\Psi}{\partial x^2} + V\Psi$$

In three dimensions

$$i\hbar\frac{\partial\Psi}{\partial t} = -\frac{\hbar^2}{2m}\nabla^2\Psi + V\Psi \tag{5.4}$$

The energy, like the momentum is seen to be represented by an operator,

$$E_{op} = H \leftarrow i\hbar\frac{\partial}{\partial t}$$

5.1.10 Stationary States

The Hamiltonian will in general be a function of time, but the special case of a time-independent Hamiltonian is of central importance. Suppose that H has an eigenvalue of E and an eigenvector v at $t = 0$. Suppose that the state of the system at any other time t is $u(t) = v\exp(-iat)$, where the argument of the phase factor is to be established. The derivative

$$i\hbar\dot{u} = \hbar av\exp(-iat),$$

and therefore

$$= Hu = Eu = Ev\exp(-iat)$$

This can only hold for $a = E/\hbar$. Since $u(t)$ differs from v only by the phase factor $\exp(-iEt/\hbar)$ it is physically the same at all times and therefore represents a *stationary* state or *energy eigenstate.* The frequency of oscillation of the phase factor is $\nu = E/2\pi\hbar = E/h$, which confirms that E is an energy eigenvalue.

Since H is Hermitian, the eigenvectors v_j of H form a complete orthonormal set and the vector representing a general state at $t = 0$ may be expressed as a linear superposition of these eigenvectors, $u(0) = \sum_j c_j v_j$, $u(t) = \sum_j c_j v_j \exp(-iEt/\hbar)$.

The general solution is a superposition of stationary solutions with constant coefficients c_j. In this sense, a knowledge of the stationary states provides the complete solution to the problem. The probability that measuring the energy of the system would yield E_j, follows immediately as $|c_j|^2$.

A time-independent Schrödinger equation is obtained by substituting

$$\Psi = \psi e^{-iEt/\hbar}$$

$$\frac{\partial\Psi}{\partial t} = -(iE/\hbar)\psi e^{-iEt/\hbar}$$

Hence

$$i\hbar\frac{\partial\Psi}{\partial t} = E\psi e^{-iEt/\hbar} = \left(-\frac{\hbar^2}{2m}\frac{\partial^2}{\partial x^2} + V\right)\psi e^{-iEt/\hbar}$$

which is commonly rearranged into the form

$$\frac{\partial^2\psi}{\partial x^2} + \frac{8\pi^2 m}{h^2}(E - V)\psi = 0$$

and used to analyze most problems of chemical interest.

5.2 Wave Mechanics

Schrödinger's equation is widely known as a wave equation and the quantum formalism developed on the basis thereof is called wave mechanics. This terminology reflects historical developments in the theory of matter following various conjectures and experimental demonstration that matter and radiation alike, both exhibit wave-like and particle-like behaviour under appropriate conditions. The synthesis of quantum theory and a wave model was first achieved by De Broglie. By analogy with the dual character of light as revealed by the photoelectric effect and the incoherent Compton scattering of X-rays, it was proposed that all matter has associated with it momentum and wavelength related through the Planck quantum of action, $h = \lambda$p. The wavelength, being inversely proportional to mass (momentum) is not evident for massive objects and becomes measurable only for sub-atomic particles, such as electrons. An immediate success of De Broglie's model was the reasonable explanation to the *ad hoc* assumption of quantized angular momentum in the Bohr atomic model.

According to De Broglie an electron in a Bohr orbit is associated with a standing wave. To avoid self destruction by interference an integral number of wavelengts are required to span the orbit of radius r, which implies $n\lambda = 2\pi r$, or $nh/2\pi = \mathrm{p}r$, which is the Bohr condition. As a physical argument the wave conjecture is less plausible, but not indefensible. One possible interpretation considers the superposition of several waves rather than a single monochromatic wave to simulate the behaviour of a particle.

To represent observables in n-dimensional space it was concluded before that Hermitian matrices were required to ensure real eigenvalues, and orthogonal eigenvectors associated with distinct eigenvalues. The first condition is essential since only real quantities are physically measurable and the second to provide the convenience of working in a cartesian space. The same arguments dictate the use of Hermitian operators in the wave-mechanical space of infinite dimensions, which constitutes a Sturm-Liouville problem in the interval $[a, b]$, with differential operator $\mathcal{L}(x)$ and eigenvalues λ,

$$\mathcal{L}(x)y(x) = \lambda y(x) \tag{5.5}$$

If y_i and y_j are solutions to the problem, under boundary conditions $y(a) = y(b) = 0$, and $\mathcal{L}$ has the property

$$\int_a^b y_i^* \mathcal{L} y_j dx = \int_a^b (\mathcal{L} y_i)^* y_j dx$$

for all i and j, $\mathcal{L}$ is Hermitian [39]. To show how this condition leads to real eigenvalues and orthogonal eigenfunctions, g_i, use (5) to form the product integrals

$$\int_a^b y_i^* \mathcal{L} y_i dx = \lambda_i \int_a^b y_i^* y_i dx$$

$$\int_a^b y_i (\mathcal{L} y_i^*) dx = \lambda_i^* \int_a^b y_i y_i^* dx$$

Because of the Hermitian property it follows that

$$(\lambda_i - \lambda_i^*) \int_a^b y_i^* y_i dx = 0.$$

Unless the integral vanishes, it is implied that $\lambda_i - \lambda_i^* = 0$, hence that λ_i is real.

For different eigenvalues one has

$$\int_a^b y_j^* \mathcal{L} y_i dx = \lambda_i \int_a^b y_j^* y_i dx$$

$$\int_a^b y_i (\mathcal{L} y_j)^* dx = \lambda_j^* \int_a^b y_i y_j^* dx$$

Thus

$$(\lambda_i - \lambda_j^*) \int_a^b y_j^* y_i dx = 0.$$

Since λ_i is real and $\lambda_i \neq \lambda_j$, it follows that

$$(\lambda_i - \lambda_j^*) = (\lambda_i - \lambda_j) \neq 0,$$

hence

$$\int_a^b y_j^* y_i dx = 0,$$

which is the orthogonality condition.

Because of these properties of Hermitian functions it is accepted as a basic postulate of wave mechanics that operators which represent physical quantities or observables must be Hermitian. The normalized eigenfunctions of a Hermitian operator constitute an orthonormal set, *i.e.*

$$\int \psi^* \psi d\tau = \delta_{ij},$$

the Kronecker delta.

The most general solution to a Sturm-Liouville problem is a function

$$f(x) = \sum_i a_i y_i(x)$$

where the y_i are orthogonal solutions to the problem. The coefficients a_i are characterized by forming the product integral

$$\int_{-\infty}^{\infty} y_j^*(x) f(x) dx = \sum_i a_i \int_{-\infty}^{\infty} y_j^*(x) y_i(x) dx.$$

Because of orthogonality all terms vanish except for $i = j$, which leaves

$$a_j = \frac{\int_{-\infty}^{\infty} y_j^*(x) f(x) dx}{\int_{-\infty}^{\infty} y_j^*(x) y_j(x) dx}$$

The range of coefficients constitute a function $\phi(j)$, say, which is the Fourier transform of $f(x)$, *e.g.*

$$\phi(k) = \frac{1}{\sqrt{2\pi}} \int_{-\infty}^{\infty} \psi(x) e^{ikx} dx$$

According to Fourier theory the wave function, in turn, is the Fourier transform of $\phi(k)$,

$$\psi(x) = \frac{1}{\sqrt{2\pi}} \int_{-\infty}^{\infty} \phi(k) e^{-ikx} dk$$

The function $\phi(k)$ is known as the wave function in momentum space. The Fourier integral represents the superposition of many waves of different wave vectors. This construct defines a wave packet, once considered as the theoretically most acceptable description of a wave-mechanical particle[5].

Schrödinger's dynamical equation (4) for a free particle

$$-\frac{\hbar^2}{2m} \nabla^2 \Psi = \frac{\hbar}{i} \frac{\partial \Psi}{\partial t}$$

is the quantum-mechanical analogue of Newton's equations of motion in classical microphysics. The equation may also be written in complex conjugate form

$$-\frac{\hbar^2}{2m} \nabla^2 \Psi^* = \frac{\hbar}{i} \frac{\partial \Psi^*}{\partial t}.$$

Multiply the free-particle equation (4) on the left by Ψ^* and the conjugate equation by Ψ. Subtract the first result from the second, such that

$$\frac{\hbar}{i}\left(\Psi \frac{\partial \Psi}{\partial t} + \Psi^* \frac{\partial \Psi}{\partial t}\right)$$
$$= -\frac{\hbar^2}{2m}\left(\Psi \frac{\partial^2 \Psi^*}{\partial x^2} - \Psi^* \frac{\partial^2 \Psi}{\partial x^2} + \Psi \frac{\partial^2 \Psi^*}{\partial y^2} - \Psi^* \frac{\partial^2 \Psi}{\partial y^2} + \Psi \frac{\partial^2 \Psi^*}{\partial z^2} - \Psi^* \frac{\partial^2 \Psi}{\partial z^2}\right)$$

[5]Following a suggestion by Eddington [30] one could speculate that wave packets, identified as particles, imply the existence of all-pervading high-frequency vibrations in the sub-aether. To quote: "Individual ripples are beyond our ken; what we can appreciate is a combined effect - when by convergence and coalescence the waves conspire to create a disturbed area of extent large compared with individual ripples but small from our own Brobdingnagian point of view. Such a disturbed area is recognized as a material particle; in particular it can be an electron". "*The equations for the motion of a wave-group with given frequency and potential frequency are the same as the classical equations of motion of a particle with the corresponding energy and potential energy.*" "It is the group-velocity that is observed by us as the motion of the material particle".

or

$$\begin{aligned}\frac{\partial}{\partial t}(\Psi\Psi^*) &= -\frac{i\hbar}{2m}\left[\frac{\partial}{\partial x}\left(\Psi\frac{\partial\Psi^*}{\partial x}-\Psi^*\frac{\partial\Psi}{\partial x}\right)\right.\\ &+ \left.\frac{\partial}{\partial y}\left(\Psi\frac{\partial\Psi^*}{\partial y}-\Psi^*\frac{\partial\Psi}{\partial y}\right)+\frac{\partial}{\partial z}\left(\Psi\frac{\partial\Psi^*}{\partial z}-\Psi^*\frac{\partial\Psi}{\partial z}\right)\right] \end{aligned}\tag{5.6}$$

The product $\Psi\Psi^*$ is the probability density function ρ. The r.h.s. of (6) can be simplified by introducing the vector $\boldsymbol{j}$ with components

$$j_x = \frac{i\hbar}{2m}\left(\Psi\frac{\partial\Psi^*}{\partial x}-\Psi^*\frac{\partial\Psi}{\partial x}\right)$$

and similar expressions for j_y and j_z.

Equation (6) is thereby transformed into

$$-\frac{\partial\rho}{\partial t} = \frac{\partial j_x}{\partial x}+\frac{\partial j_y}{\partial y}+\frac{\partial j_z}{\partial z} = \boldsymbol{\nabla}\cdot\boldsymbol{j}$$

the divergence of the vector $\boldsymbol{j}$. Hence

$$\frac{\partial\rho}{\partial t}+\operatorname{div}\boldsymbol{j}=0. \tag{5.7}$$

Equation (7) is an exact analogue of the continuity equation (1.7) of hydrodynamics, and this allows definition of a probability current density

$$\boldsymbol{j} = \frac{\hbar}{2mi}\left(\Psi^*\nabla\Psi - \Psi\nabla\Psi^*\right).$$

Schrödinger's equation

$$\nabla^2\psi + \frac{2m}{\hbar^2}[E-V(x)]\psi = 0$$

is of the same form as the optical wave equation for a wave of angular frequency ω,

$$\nabla^2 A + \frac{\omega^2 n^2}{c^2}A = 0$$

The ψ equation hence resembles the equation for light in a medium of refractive index

$$n = \frac{c}{\hbar\omega}\sqrt{2m[E-V(x)]}$$

or, in other words, a medium in which n is a function of the position.

5.2.1 The Hydrogen Atom

The first application of quantum theory to a problem in chemistry was to account for the emission spectrum of hydrogen and at the same time explain the stability of the nuclear atom, which seemed to require accelerated electrons in orbital motion. This planetary model is rendered unstable by continuous radiation of energy. The Bohr postulate that electronic angular momentum should be quantized in order to stabilize unique orbits solved both problems in principle. The Bohr condition requires that

$$\boldsymbol{p}r = \frac{nh}{2\pi} \quad , \quad n = 1, 2, \cdots$$

for an electron in an allowed stationary orbit. Together with the energy balance between orbital acceleration and coulombic attraction to the nucleus,

$$\frac{e^2}{4\pi\epsilon_0 r^2} = \frac{\boldsymbol{p}^2}{mr}$$

the model predicts allowed orbits of radius

$$r = \frac{4\pi\epsilon_0(\boldsymbol{p}r)^2}{me^2} = \frac{\epsilon_0(nh)^2}{\pi me^2}.$$

The total energy $E = -e^2/8\pi\epsilon_0 r$ follows as

$$E = -\frac{e^2}{8\pi\epsilon_0} \cdot \frac{\pi me^2}{\epsilon_0(nh)^2} = -\frac{1}{n^2}\left(\frac{me^4}{8\epsilon_0^2 h^2}\right)$$

The Laplacian operator in spherical polar coordinates is

$$\nabla^2 = \frac{1}{r^2}\frac{\partial}{\partial r}\left(r^2\frac{\partial\psi}{\partial r}\right) + \frac{1}{r^2\sin\theta}\frac{\partial}{\partial\theta}\left(\sin\theta\frac{\partial\psi}{\partial\theta}\right) + \frac{1}{r^2\sin^2\theta}\frac{\partial^2\psi}{\partial\phi^2}$$

and the Schrödinger equation for an electron in the central coulombic field of a hydrogen atom is

$$\nabla^2\psi + \frac{8\pi^2\mu}{h^2}\left(E + \frac{e^2}{4\pi\epsilon_0 r}\right)\psi = 0$$

In order to separate the variables, the substitution $\psi(r,\theta,\phi) = R(r)\cdot Y(\theta,\phi)$ is made. After multiplication by r^2/YR the resulting equation rearranges into

$$\begin{aligned}\frac{1}{R}\frac{\partial}{\partial r}\left(r^2\frac{\partial R}{\partial r}\right) \quad &+ \quad \frac{8\pi^2\mu}{h^2}\left(E + \frac{e^2}{4\pi\epsilon_0 r}\right)r^2 \\ &= \quad -\frac{1}{Y\sin\theta}\frac{\partial}{\partial\theta}\left(\sin\theta\frac{\partial Y}{\partial\theta}\right) - \frac{1}{Y\sin^2\theta}\frac{\partial^2 Y}{\partial\phi^2}\end{aligned}$$

The l.h.s. of the expression now depends on r only and the r.h.s. is only angle dependent. The two sides are therefore separately equal to a constant λ. To ensure convergence of the Θ (Legendre) equation (1.17) λ is chosen equal to $l(l+1)$, for integer l.

The two separated equations are

$$\frac{1}{\sin\theta}\frac{\partial}{\partial\theta}\left(\sin\theta\frac{\partial Y}{\partial\theta}\right)+\frac{1}{\sin^2\theta}\frac{\partial^2 Y}{\partial\phi^2}+l(l+1)Y=0 \tag{5.8}$$

and

$$\frac{1}{R^2}\frac{d}{dr}\left(r^2\frac{dR}{dr}\right)+\left[\frac{8\pi^2\mu}{h^2}\left(E+\frac{e^2}{4\pi\epsilon_0 r}\right)-\frac{l(l+1)}{r^2}\right]R=0$$

or

$$\frac{d^2R}{dr^2}+\frac{2}{r}\frac{dR}{dr}+\left[\frac{8\pi^2\mu}{h^2}\left(E+\frac{e^2}{4\pi\epsilon_0 r}\right)-\frac{l(l+1)}{r^2}\right]R=0 \tag{5.9}$$

For a wave function with no angle dependence one therefore has a one-dimensional Schrödinger equation

$$\frac{d^2\psi}{dr^2}+\frac{2}{r}\frac{d\psi}{dr}+\frac{8\pi^2\mu}{h^2}\left(E-\frac{e^2}{4\pi\epsilon_0 r}\right)\psi=0 \tag{5.10}$$

For large r the equation becomes

$$\frac{d^2\psi}{dr^2}+\frac{8\pi^2\mu E\psi}{h^2}=0$$

This equation has a solution of the form e^{-ra} which, substituted into (10) gives:

$$\frac{d\psi}{dr}=-ae^{-ra}=-a\psi$$

and

$$\frac{d^2\psi}{dr^2}=-a\frac{d\psi}{dr}=a^2\psi$$

Hence

$$a^2-\frac{2a}{r}+\frac{8\pi^2\mu E}{h^2}+\frac{2\pi\mu e^2}{\epsilon_0 h^2 r}=0$$

Since this expression is valid for all r the sums that are respectively dependent and independent of r should be individually equal to zero. Separation of these terms gives the two equations:

$$a^2+\frac{8\pi^2\mu E}{h^2}=0$$

$$-2a + \frac{2\pi\mu e^2}{\epsilon_0 h^2} = 0$$

It is noticed that $a = \pi\mu e^2/h^2\epsilon_0 = 1/r_1$ is inversely proportional to the first Bohr radius. The corresponding energy

$$E = -\left(\frac{1}{r_1}\right)^2 \frac{h^2}{8\pi^2\mu} = \frac{\mu e^4}{8\epsilon_0^2 h^2}$$

In terms of the previous result convenient atomic units can be defined by

$$r = \frac{h^2\epsilon_0\rho}{\pi\mu e^2} \quad \text{and} \quad E = \frac{\mu e^4\varepsilon}{8\epsilon_0^2 h^2}$$

where the units of ρ and ε are *bohr* (radius) and *rydberg* respectively. It follows from

$$\rho = \frac{\pi\mu e^2 r}{\epsilon_0 h^2}$$

that

$$\frac{d\rho}{dr} = \frac{\pi\mu e^2}{\epsilon_0 h^2} \quad \text{and}$$

$$\frac{d}{dr} = \frac{d}{d\rho}\cdot\frac{d\rho}{dr} = \frac{\pi\mu e^2}{\epsilon_0 h^2}\cdot\frac{d}{d\rho}$$

$$\frac{d^2}{dr^2} = \frac{\pi^2\mu^2 e^4}{\epsilon_0^2 h^4}\cdot\frac{d^2}{d\rho^2}$$

In terms of the new variables the R-equation (9) becomes

$$\frac{d^2R}{d\rho^2} + \frac{2}{\rho}\frac{dR}{d\rho} + \left[\varepsilon + \frac{2}{\rho} - \frac{l(l+1)}{\rho^2}\right] R = 0$$

To simplify this equation, substitute $S = \rho R$.

$$R = \frac{S}{\rho} \quad ; \quad \frac{dR}{d\rho} = -\frac{S}{\rho^2} + \frac{1}{\rho}\frac{dS}{d\rho}$$

$$\text{Hence} \quad \rho^2\frac{dR}{d\rho} = -S + \rho\frac{dS}{d\rho}$$

$$\frac{d}{d\rho}\left(\rho^2\frac{dR}{d\rho}\right) = -\frac{dS}{d\rho} + \frac{dS}{d\rho} + \rho\frac{d^2S}{d\rho^2}$$

$$\text{and} \quad \frac{1}{\rho^2}\frac{d}{d\rho}\left(\rho^2\frac{dR}{d\rho}\right) = \frac{1}{\rho}\frac{d^2S}{d\rho^2}$$

The equation becomes

$$\frac{d^2S}{d\rho^2} + \left[\varepsilon + \frac{2}{\rho} - \frac{l(l+1)}{\rho^2}\right] S = 0 \tag{5.11}$$

For very small ρ the first two terms in the square brackets can be neglected to give

$$\frac{d^2S}{d\rho^2} - \frac{l(l+1)}{\rho^2} S = 0$$

$$\text{with solutions} \qquad S(\rho) = \rho^{l+1} \quad \text{or} \quad \rho^{-l}$$

Since S must be finite at $\rho = 0$ the second solution can be discarded and S must be of the form ρ^{l+1} for small ρ. For very large values of ρ the last two terms in the square brackets can be neglected to give

$$\frac{d^2S}{d\rho^2} + \varepsilon S = 0$$

with solutions

$$S(\rho) = e^{\pm\sqrt{\varepsilon}\rho}$$

The positive sign gives a function that goes infinite as $\rho \to \infty$ and is discarded. It is therefore reasonable to assume a solution of the form

$$S(\rho) = \rho^{l+1} e^{-\alpha\rho} \sum_k A_k \rho^k \quad , \text{ where} \quad \alpha^2 = -\varepsilon \quad ,$$

which has the correct behaviour for both small and large ρ. To establish the form of the infinite series the accepted solution is substituted into (11).

$$\frac{dS}{d\rho} = (l+1)\rho^l e^{-\alpha\rho} \sum A_k \rho^k - \alpha\rho^{l+1} e^{-\alpha\rho} \sum A_k \rho^k + \rho^{l+1} e^{-\alpha\rho} \sum k A_k \rho^{k-1}$$

$$\begin{aligned}
\frac{d^2S}{d\rho^2} &= l(l+1)\rho^{l-1} e^{-\alpha\rho} \sum A_k \rho^k - \alpha(l+1)\rho^l e^{-\alpha\rho} \sum A_k \rho^k \\
&+ (l+1)\rho^l e^{-\alpha\rho} \sum k A_k \rho^{k-1} - \alpha(l+1)\rho^l e^{-\alpha\rho} \sum A_k \rho^k \\
&+ \alpha^2 \rho^{l+1} e^{-\alpha\rho} \sum A_k \rho^k - \alpha\rho^{l+1} e^{-\alpha\rho} \sum k A_k \rho^{k-1} \\
&+ (l+1)\rho^l e^{-\alpha\rho} \sum k A_k \rho^{k-1} - \alpha\rho^{l+1} e^{-\alpha\rho} \sum k A_k \rho^{k-1} \\
&+ \rho^{l+1} e^{-\alpha\rho} \sum k(k-1) A_k \rho^{k-2}
\end{aligned}$$

After collecting similar powers of ρ, substitution into (11) yields

$$2[1-\alpha(l+1)]\sum A_k\rho^{k+l} - 2\alpha\sum kA_k\rho^{k+l} + 2(l+1)\sum kA_k\rho^{k+l-1} + \sum k(k-1)A_k\rho^{l+k-1} = 0$$

The power ρ^{k+l-1} occurs in the $(k-1)$ term of the first two sums and in the kth term of the last two sums. Since the coefficient of this term must be zero,

$$2[1-\alpha(l+1)A_{k-1}] - 2\alpha(k-1)A_{k-1} + 2(l+1)kA_k + k(k-1)A_k = 0$$

to give the recursion formula

$$A_k = -\left[\frac{1-(k+l)\alpha}{k(k+2l+1)}\right]A_{k-1}$$

The series goes infinite as $\rho\to\infty$ unless it terminates to give a polynomial. This only happens if

$$\alpha = \frac{1}{k+l} = \frac{1}{n} = \sqrt{-\varepsilon} \quad \text{for integral } n.$$

It follows that for S to be an acceptable wave function the energy must be

$$\varepsilon = -\frac{1}{n^2} \quad \text{rydberg.}$$

To solve for the wave function it is convenient to introduce a new variable $x = 2\rho/n = 2\alpha\rho$. Then

$$\frac{dx}{d\rho} = \frac{2}{n} \quad ; \quad \frac{dS}{d\rho} = \frac{dS}{dx}\frac{dx}{d\rho} = \frac{2}{n}\frac{dS}{dx}$$

$$\frac{d^2S}{d\rho^2} = \frac{d}{dx}\left[\frac{2}{n}\frac{dS}{dx}\right]\frac{2}{n} = \frac{4}{n^2}\frac{d^2S}{dx^2}$$

With these substitutions equation (11) reads

$$\frac{d^2S}{dx^2} + \left[-\frac{1}{4} + \frac{n}{x} - \frac{l(l+1)}{x^2}\right]S = 0 \tag{5.12}$$

This equation has the solution

$$S(x) = e^{-x/2}x^{l+1}L(x) \quad \text{where } L(x) \text{ is to be determined}$$

Differentiation gives

$$\frac{dS}{dx} = -\frac{1}{2}e^{-x/2}x^{l+1}L + (l+1)e^{-x/2}x^{l}L + e^{-x/2}x^{l+1}\frac{dL}{dx}$$

$$\begin{aligned}\frac{d^2S}{dx^2} &= \frac{1}{4}e^{-x/2}x^{l+1}L - \frac{1}{2}(l+1)e^{-x/2}x^lL - \frac{1}{2}e^{-x/2}x^{l+1}\frac{dL}{dx} \\ &- \frac{1}{2}(l+1)e^{-x/2}x^lL + l(l+1)e^{-x/2}x^{l-1}L + (l+1)e^{-x/2}x^l\frac{dL}{dx} \\ &- \frac{1}{2}e^{-x/2}e^{l+1}\frac{dL}{dx} + (l+1)e^{-x/2}x^l\frac{dL}{dx} + e^{-x/2}x^{l+1}\frac{d^2L}{dx^2}\end{aligned}$$

i.e.

$$\frac{d^2S}{dx^2} = \left[x\frac{d^2L}{dx^2} + 2(l+1)\frac{dL}{dx} - x\frac{dL}{dx} + \frac{1}{4}xL - (l+1)L + l(l+1)\frac{L}{x}\right]e^{-x/2}x^l$$

Substitution of these values of $\frac{d^2S}{dx^2}$ and S into (12) and division by $e^{-x/2}x^l$ give

$$x\frac{d^2L}{dx^2} + [2(l+1) - x]\frac{dL}{dx} + (n-l-1)L = 0 \tag{5.13}$$

Equation (13) is familiar from before (1.8). It is the associated Laguerre equation with the usual s replaced by $2l+1$ and n by $n+l$. It follows that the associated Laguerre polynomial $L_{n+l}^{2l+1}(x)$ is a solution of (12) and also that

$$S'(\rho) = x^{l+1}e^{-x/2}L_{n+l}^{2l+1}(x).$$

The n and l labels show that these solutions are characterized by two sets of quantum numbers. By definition $n+l \geq 2l+1$, hence $n-l-1 \geq 0$. n and l can have the values

$$n = 1, 2, \ldots\ldots$$

$$l = 0, 1, \ldots(n-1)$$

Finally the wave function must be normalized to unity, using (1.22)

$$\int_0^\infty e^{-x}x^{2l+2}\left[L_{n+l}^{2l+1}(x)\right]^2 dx = \frac{2n[(n+l)!]^3}{(n-l-1)!}$$

Since $S'(\rho)$ is expressed in terms of ρ and ($d\rho = \frac{n}{2}dx$) the normalization integral has the form

$$\frac{n}{2}\int_0^\infty e^{-x}x^{2l+2}\left[L_{n+l}^{2l+1}(x)\right]^2 dx = \frac{n^2[(n+l)!]^3}{(n-l-1)!}$$

After some rearrangement the following expression is obtained directly

$$R_{nl}(\rho) = \frac{S}{\rho} = -\sqrt{\left(\frac{2}{n}\right)^3\frac{(n-l-1)!}{2n[(n+l)!]^3}}e^{-x/2}x^lL_{n+l}^{2l+1}(x)$$

It is noticed that x represents a different variable for each energy state. The quantum number n appears as a scale factor in the dimensionless variable,

$$x = \frac{2}{n}\rho = \frac{2}{n}\frac{r}{a_0}$$

As a function of r the general expression for the radial factors of the wave function becomes

$$R_{nl}(r) = -\left(\frac{2}{na_0}\right)^{3/2}\left\{\frac{(n-l-1)!}{2n[(n+l)!]^3}\right\}^{\frac{1}{2}} e^{x/2}x^l L_{n+l}^{2l+1}(x) \tag{5.14}$$

Substituting from the table of associated Laguerre polynomials (1.17) the first few normalized radial wave functions are:

$$R_{10} = -2e^{-x/2}L_1^1 = 2\left(\frac{1}{a_0}\right)^{3/2} e^{-r/a_0}$$

$$R_{20} = -\frac{1}{2}e^{-x/2}L_2^1 = \frac{1}{\sqrt{2}}\left(\frac{1}{a_0}\right)^{3/2}\left(1-\frac{r}{2a_0}\right)e^{-r/2a_0}$$

$$R_{21} = -\frac{1}{2}\sqrt{\frac{1}{6}}e^{x/2}xL_3^3 = \frac{1}{2\sqrt{6}}\left(\frac{1}{a_0}\right)^{5/2} re^{-r/2a_0}$$

The angle-dependent equation can be separated into two equations by introducing the separation constant m^2,

$$\frac{d^2\Phi}{d\phi^2} + m^2\Phi = 0$$

$$\Phi = Ae^{\pm im\phi}$$

Normalization requires $\int_0^{2\pi} \Phi\Phi^* d\phi = 1$

$$\Phi\Phi^* = A^2 e^{im\phi(\pm 1 \mp 1)} = A^2$$

Therefore $$\int_0^{2\pi} \Phi\Phi^* = A^2\int_0^{2\pi} d\phi = A^2 2\pi \quad , \quad A = \frac{1}{\sqrt{2\pi}}$$

It follows that $\Phi = \frac{1}{\sqrt{2\pi}}e^{im\phi}$ where m can be negative.

The Θ equation is:

$$\frac{1}{\sin\theta}\frac{d}{d\theta}\left(\sin\theta\frac{d\Theta}{d\theta}\right) + \left[l(l+1) - \frac{m^2}{\sin^2\theta}\right]\Theta = 0$$

one form of Legendre's equation, with the associated Legendre functions as solutions; normalized they are:

$$\Theta(\theta) = \sqrt{\frac{(2l+1)}{2}\frac{(l-m)!}{(l+m)!}}P_l^m(\cos\theta)$$

where m and l are positive integers, $m \leq l$.

In summary, the wave mechanical treatment of the hydrogen atom assumes that the electronic motion is described by ψ, obtained as solutions of the equation (in atomic units)

$$\nabla^2\psi + \left[E + \frac{2}{r}\right]\psi = 0$$

The problem is solved by the separation of variables, *i.e.* in polar coordinates

$$\psi(r,\theta,\phi) = R(r)\cdot\Theta(\theta)\cdot\Phi(\phi)$$

The finite and single-valued solutions are

$$\left.\begin{array}{l} R(r) = f(n,l) \\ \Theta(\theta) = f(l,|m|) \\ \Phi(\phi) = f(m) \end{array}\right\} \text{where} \left\{\begin{array}{l} n = 1,2,\ldots\infty \\ l = 0,1,\ldots(n-1) \\ m = 0,\pm 1,\cdots \pm l \end{array}\right.$$

If each variable is separately normalized the total normalized wave function is

$$\psi_{nlm}(r\theta\phi) = \sqrt{\frac{(2l+1)(l-|m|)!(n-l-1)!}{4\pi(l+|m|)!\,2n[(n+l)!]^3}\left(\frac{2}{n}\right)^3}\, x^l L_{n+l}^{2l+1} P_l^{|m|} e^{im\phi}$$

with $x = 2r/na_0$.

Angular Momentum Eigenfunctions

Angular momentum is classically defined as

$$\boldsymbol{L} = \boldsymbol{r}\times\boldsymbol{p} = \begin{vmatrix} \boldsymbol{i} & \boldsymbol{j} & \boldsymbol{k} \\ x & y & z \\ p_x & p_y & p_z \end{vmatrix}$$

$$\begin{aligned} \boldsymbol{r} &= \boldsymbol{i}x + \boldsymbol{j}y + \boldsymbol{k}z \\ \boldsymbol{p} &= \boldsymbol{i}p_x + \boldsymbol{j}p_y + \boldsymbol{k}p_z \end{aligned}$$

The components of $\boldsymbol{L}$ are

$$\begin{aligned}
L_x &= yp_z - xp_y \\
L_y &= zp_x - xp_z \\
L_z &= xp_y - yp_x \\
\boldsymbol{L}\cdot\boldsymbol{L} &= L^2 = L_x^2 + L_y^2 + L_z^2
\end{aligned}$$

In terms of operators $p_x \rightarrow \frac{h}{2\pi i}\frac{\partial}{\partial x}$, *etc.*, the wave-mechanical components of angular momentum becomes

$$\begin{aligned}
L_x &= \frac{h}{2\pi i}\left(y\frac{\partial}{\partial z} - z\frac{\partial}{\partial y}\right) \\
L_y &= \frac{h}{2\pi i}\left(z\frac{\partial}{\partial x} - x\frac{\partial}{\partial z}\right) \\
L_z &= \frac{h}{2\pi i}\left(x\frac{\partial}{\partial y} - y\frac{\partial}{\partial x}\right)
\end{aligned}$$

Transformed to polar coordinates

$$\begin{aligned}
x &= r\sin\theta\cos\phi \\
y &= r\sin\theta\sin\phi \\
z &= r\cos\theta
\end{aligned}$$

Let

$$\begin{aligned}
q &= x, y, z \\
\frac{\partial}{\partial q} &= \frac{\partial r}{\partial q}\frac{\partial}{\partial r} + \frac{\partial\theta}{\partial q}\frac{\partial}{\partial\theta} + \frac{\partial\phi}{\partial q}\frac{\partial}{\partial\phi}
\end{aligned}$$

It is not too difficult to show that

$$\begin{aligned}
L_x &= \frac{ih}{2\pi}\left(\sin\phi\frac{\partial}{\partial\theta} + \cot\theta\cos\phi\frac{\partial}{\partial\phi}\right) \\
L_y &= \frac{ih}{2\pi}\left(-\cos\phi\frac{\partial}{\partial\theta} + \cot\theta\sin\phi\frac{\partial}{\partial\phi}\right) \\
L_z &= -\frac{ih}{2\pi}\frac{\partial}{\partial\phi}
\end{aligned}$$

as shown in the footnote[6] . The operator for total angular momentum

$$\hat{L}^2 = L_x^2 + L_y^2 + L_z^2 = -\frac{h^2}{4\pi^2}\left[\frac{1}{\sin\theta}\frac{\partial}{\partial\theta}\left(\sin\theta\frac{\partial}{\partial\theta}\right) + \frac{1}{\sin^2\theta}\frac{\partial^2}{\partial\phi^2}\right]$$

The angular-dependent part of the Schrödinger equation (8)

$$\frac{1}{\sin\theta}\frac{\partial}{\partial\theta}\left(\sin\theta\frac{\partial Y}{\partial\theta}\right) + \frac{1}{\sin^2\theta}\frac{\partial^2 Y}{\partial\phi^2} + l(l+1)Y = 0$$

has a similar form and the two expressions are identical if

$$\hat{L}^2 Y_{lm}(\theta,\phi) = l(l+1)\frac{h^2}{4\pi^2}Y_{lm}(\theta,\phi)$$

---

[6]All the required derivatives are obtained as shown for three examples only: Using $r^2 = x^2 + y^2 + z^2$

$$2r\frac{\partial r}{\partial x} = 2x; \quad \frac{\partial r}{\partial x} = \frac{x}{r} = \sin\theta\sin\phi$$

Using $\cos\theta = z/r$

$$\frac{\partial}{\partial z}(\cos\theta) = \frac{\partial}{\partial\theta}(\cos\theta)\frac{\partial\theta}{\partial z} = \frac{1}{r}\left(1 - \frac{z^2}{r^2}\right) = \frac{1}{r}\sin^2\theta$$

$$\frac{\partial\theta}{\partial z} = -\frac{1}{r}\sin\theta$$

Using $\tan\phi = y/x$

$$\frac{\partial}{\partial y}(\tan\phi) = \sec^2\phi\frac{\partial\phi}{\partial y} = \frac{1}{x}$$

$$\frac{\partial\phi}{\partial y} = \frac{\cos^2\phi}{r\sin\theta\cos\phi} = \frac{1}{r}\frac{\cos\phi}{\sin\theta}$$

From these derivatives

$$\frac{\partial}{\partial x} = \sin\theta\cos\phi\frac{\partial}{\partial r} + \frac{1}{r}\cos\theta\cos\phi\frac{\partial}{\partial\theta} - \frac{1}{r}\frac{\sin\phi}{\sin\theta}\frac{\partial}{\partial\phi}$$

with similar expressions for $\partial/\partial y$ and $\partial/\partial z$.

This means that $Y_{lm}(\theta, \phi)$ is an eigenfunction of $\hat{L}^2$, with eigenvalues $l(l+1)\frac{h^2}{4\pi^2}$. Also, since

$$\begin{aligned}
\Phi(\phi) &= \frac{1}{\sqrt{2\pi}} e^{im\phi} \\
\frac{\partial \Phi}{\partial \phi} &= im\frac{1}{\sqrt{2\pi}} e^{im\phi} \\
&= im\Phi \\
\text{But} \qquad L_z &= -\frac{ih}{2\pi}\frac{\partial}{\partial \phi} \\
\text{Therefore} \qquad L_z(\Phi) &= \frac{mh}{2\pi}\Phi
\end{aligned}$$

Thus $\Phi(\phi)$ is an eigenvalue of L_z with eigenvalue $\frac{mh}{2\pi}$. The angle-dependent part of the wave equation is seen to contain wave functions which are eigenfunctions of both the total angular momentum as well as the component of angular momentum along the polar axis.

There is nothing special about the z-axis, since all directions in space are equivalent. The only reason why it is easier to solve the L_z eigenvalue equation (compared to L_x and L_y) is because it has a simpe form in spherical polar coordinates, which involves only the angle of rotation ϕ about the z-axis.

The eigenfunctions of L^2 and L_z are spherical harmonics (or surface harmonics):

$$Y_l^m(\theta, \phi) = (-1)^m \sqrt{\frac{2l+1}{4\pi}\frac{(l-m)!}{(l+m)!}} P_l^m(\cos\theta) e^{im\phi}$$

The first few are

$$\begin{aligned}
Y_0^0 &= \frac{1}{\sqrt{4\pi}} \\
Y_1^0 &= \sqrt{\frac{3}{4\pi}}\cos\theta \\
Y_1^{\pm 1} &= \mp\sqrt{\frac{3}{8\pi}}\sin\theta e^{\pm i\phi}
\end{aligned}$$

Polar plots of the probability distributions

$$|Y_l^m|^2 = \frac{1}{\sqrt{2\pi}} |\Theta_l^m(\theta)|^2 \quad \text{are shown below.}$$

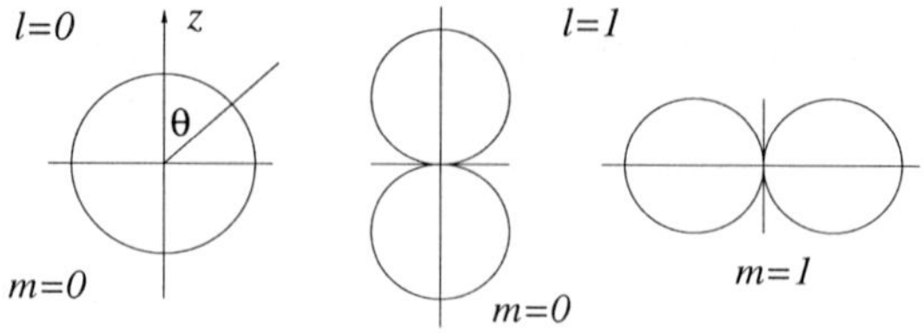

The plots are to be imagined as having rotational symmetry around the z-axis.

Since $l \geq |m|$, the magnitude $\sqrt{l(l+1)}\hbar$ of the orbital angular momentum $\boldsymbol{L}$ is greater than the magnitude $|m|\hbar$ of its z-component L_z, except for $l = 0$. If it were possible to have the angular-momentum magnitude equal to its z-component, this would mean that the x and y components were zero, and all three components of $\boldsymbol{L}$ would be specified. However, since the components of angular momentum do not commute, this cannot be done, except for $l = 0$. Still, all the eigenfunctions of L_z cannot also be eigenfunctions of L_x. Since L_x and L_y cannot be specified, the vector $\boldsymbol{L}$ can lie anywhere on the surface of a cone around the z-axis, with altitude $m\hbar$ and slant height

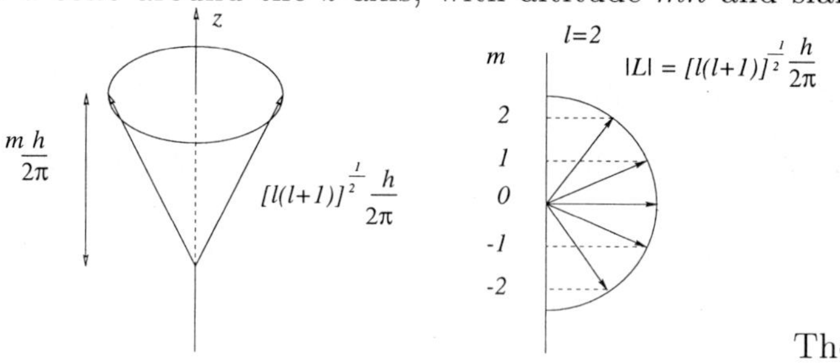

$\sqrt{l(l+1)}\hbar$. The orbital angular momentum vector $\boldsymbol{L}$, of length $\sqrt{l(l+1)}\hbar$ precesses about the z-axis, the $2l+1$ allowed projections of $\boldsymbol{L}$ on the z-axis being given by $m\hbar$, with $m = -l, -l+1, \ldots, +l$.

Orbital Energies

The energy eigenvalues of the hydrogen electronic bound states are inversely proportional to the square of the principal quantum number, in SI units,

$$E_n = -\frac{\mu e^4}{8\epsilon_0^2 h^2}\left(\frac{1}{n^2}\right)$$

This formula for hydrogenlike atoms with nuclear charge Z, is often formulated in terms of the dimensionless fine-structure constant $\alpha = 4\pi\epsilon_0\hbar c/e^2$ as

$$E_n = -\frac{1}{2}mc^2\left(\frac{Z\alpha}{n}\right)^2.$$

It is common practice to use the electronic mass, m rather than the reduced mass μ. This amounts to the assumption of an infinitely heavy nucleus, and numerically makes a difference of less than 0.0005. For heavier nuclei the difference is even less.

Since the energy is independent of the quantum numbers l and m_l the energy eigenstates are specified in the first instance only by the radial factors of the atomic wave function[7]. The energy levels are said to be degenerate with respect to l and m_l. For each value of n the orbital quantum number l may take the values $0, 1, \ldots, n-1$, and for each value of l the magnetic quantum number has $2l+1$ possible values, ranging from $-l$ to l. The total degeneracy is therefore given by

$$\sum_{l=0}^{n-1}(2l+1) = 2\frac{n(n-1)}{2} + n = n^2.$$

In this sense the electronic energy is a function of its radial distance from the nucleus.

The Radial Functions

The quantity $|R_{nl}(r)|^2$ represents the electron density as a function of r. The probability of finding the electron at a distance between r and $r + dr$, with no restrictions on θ and ϕ is obtained by integrating the probability density

$$|\psi|^2 d\tau = [R_{nl}(r)]^2 \left|Y_l^m(\theta, \phi)\right|^2 r^2 \sin\theta dr\, d\theta\, d\phi$$

over all angles, *i.e.*

$$[R_{nl}(r)]^2 r^2 dr \int_0^{2\pi} \int_0^{\pi} \left|Y_l^m(\theta, \phi)\right|^2 \sin\theta\, d\theta\, d\phi$$

$= [R_{nl}(r)]^2 r^2 dr$, since the spherical harmonics are normalized,

$$\int_0^{2\pi} \int_0^{\pi} |Y_l^m|^2 \sin\theta\, d\theta\, d\phi = 1.$$

The function $D_{nl}(r) = R^2 r^2$ is called the radial distribution function. The appearance of the factor r^2 can be understood because the volume enclosed between two spheres of radii r and $r + dr$ is proportional to that factor.

---

[7]In the case of heavier atoms the situation is more complicated.

The s-States Only for s-states[8] are the radial functions different from zero at $r = 0$. Since $Y_0^0 = \frac{1}{\sqrt{4\pi}}$, is independent of θ and ϕ, the density at $r = 0$ follows from (14) as

$$|\psi_{n0}(0)|^2 = \frac{1}{4\pi}|R_{n0}(0)|^2 = \frac{1}{\pi}\left(\frac{1}{na_0}\right)^3 .$$

An important property of s-states can be examined in one dimension. From $r = \sqrt{x^2 + y^2 + z^2}$ it follows that for points on the x-axis, where $y = z = 0$, $r = (x^2)^{1/2} = |x|$ and

$$\psi_{1s}(x,0,0) = \frac{1}{\sqrt{\pi}}\left(\frac{1}{a_0}\right)^{3/2} e^{-|x|/a_0}$$

The ψ_{1s} wave function is therefore symmetrical along x through the origin, where it has a cusp.

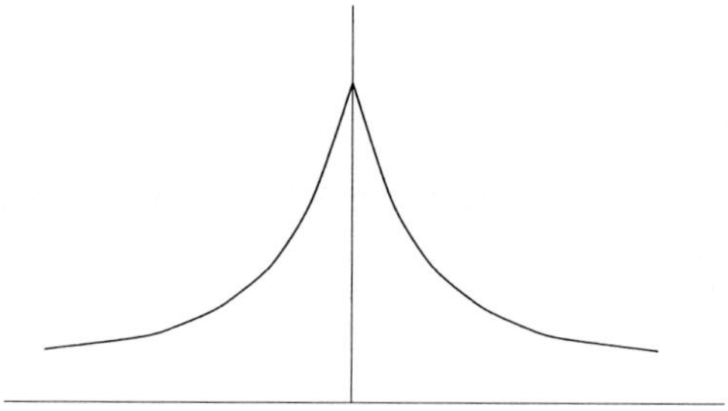

Although ψ is continuous at the origin its first derivative is discontinuous and $d\psi/dx \neq 0$. This peculiar behaviour is a result of the infinite potential energy at the origin.

Centrifugal Potential When $l \neq 0$, the equation for S (11)

$$\frac{d^2S}{d\rho^2} + \left[\epsilon + \frac{2}{\rho} - \frac{l(l+1)}{\rho^2}\right] S = 0$$

is the same as a one-dimensional Schrödinger equation in which the potential function is $V(\rho)+l(l+1)/\rho^2$. The system thus acts as if there were a repulsive

[8]The old spectroscopic notation of s,p,d,f, referring to states with quantum number $l = 0, 1, 2, 3$, is still in common use.

potential

$$\begin{aligned}\frac{l(l+1)}{\rho^2} &\equiv \left(\frac{\mu e^4}{8\epsilon_0^2 h^2}\cdot\frac{\epsilon_0^2 h^4}{\pi^2\mu^2 e^4 r^2}\right) l(l+1) \\ &= \frac{l(l+1)\hbar^2}{2\mu r^2},\end{aligned}$$

in addition to the usual potential $V(r)$. This repulsive potential may be thought of [57] as the term responsible for a centrifugal force that tends to keep particles of non-zero angular momentum away from the origin. For a Coulomb potential $V(r) = -e^2/4\pi\epsilon_0 r$, the effective potential for $l \neq 0$ has a shape like

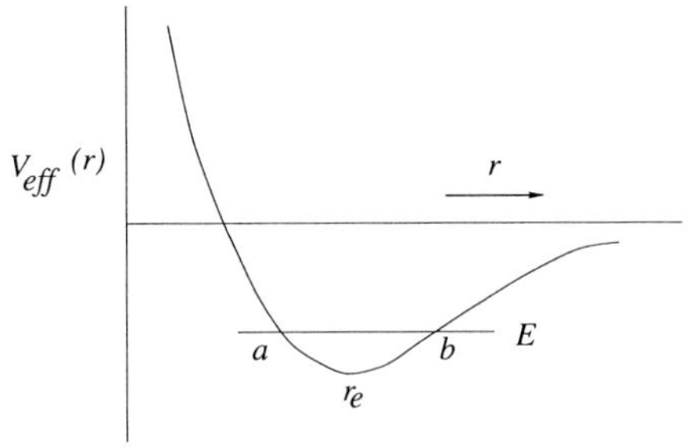

At large distances the Coulomb attraction predominates, but for small R it is overbalanced by the repulsive centrifugal term. The equilibrium occurs where the derivative of the effective potential is zero, or where

$$\frac{\partial V}{\partial r} - l(l+1)\frac{\hbar^2}{\mu r^3} = 0,$$

i.e. at

$$r_e = \frac{l(l+1)4\pi\epsilon_0\hbar^2}{\mu e^2} = l(l+1)a_0.$$

This is the classical radius of a circular orbit.

In general, if a particle is bound ($E < 0$) it will oscillate (classically) between some limits $r = a$, and $r = b$. For example, in an elliptic orbit of a hydrogen atom, the radius oscillates periodically between inner and outer limits. Only for a circular orbit is there no oscillation. Among the eigenvalues which have the same n, the one with lowest l has the largest amplitude in the vicinity of the nucleus.

Atomic Size The associated Laguerre polynomial $L_{n+l}^{2l+1}(x)$ is a polynomial of degree $n_r = n - l - 1$, which has n_r radial nodes (zeros). The radial distribution function therefore exhibits $n - l$ maxima. Whenever $n = l + 1$ and the orbital quantum number, l has its largest value, there is only one maximum. In this case $n_r = 0$ and from (14) follows

$$R_{n,n-1} \simeq r^{n-1}e^{-r/na_0}.$$

Hence $D_{n,n-1} = r^2 R_{n,n-1}^2$, has a maximum when

$$\frac{dD}{dr} = \left(2nr^{2n-1} - \frac{2}{na_0}r^{2n}\right)e^{2r/na_0} = 0$$

i.e. at

$$r = n^2 a_0 = \left(\frac{n^2 a_0}{Z}, \quad \text{for hydrogenic atoms with nuclear charge of } Z\right),$$

which is precisely the value appearing in the Bohr model. Although the notion of atomic size is less precise in wave mechanics than in Bohr theory, the most probable distance between electron and nucleus is proportional to n^2 and inversely proportional to Z. This observation probably gave rise to the idea of describing one-electron hydrogenic wave functions as *orbitals*.

Rydberg Atoms A highly excited atom (or ion) has an electron with a large principal quantum number n. The electron (or the atom) is said to be in a high Rydberg state and the highly excited atom is also referred to as a *Rydberg atom*. The most remarkable feature of rydberg atoms is their size, assumed proportional to $n^2 a_0$. A hydrogen atom with $n = 100$ has a Bohr radius $a_n = 5.3 \times 10^{-7}$ m, compared to $a_0 = 5.3 \times 10^{-11}$ m for $n = 1$; twelve orders of magnitude difference in volume. The binding energy however, is much lower at 1.36×10^{-3} eV, compared to 13.6 eV of the normal hydrogen atom. An important property of highly excited Rydberg atoms is their extremely long lifetimes of up to 1 s. Even thermal collisions can transfer enough energy to the atom to ionize it, but it is possible for a neutral object passing through the Rydberg atom to leave it undisturbed.

The existence of outsized hydrogen atoms was inferred early on from the observation that 33 terms of the Balmer series could be observed in stellar spectra, compared to only 12 in the laboratory [58]. More recently [59] Rydberg atoms have been produced by exciting an atomic beam with laser light. When the outer electron of an atom is excited into a very high energy level, it enters a spatially extended orbital which is far outside the orbitals of all other electrons. The excited electron interacts with the atomic core, consisting of the nucleus and all the inner electrons, which has a charge of $+e$, just the same as the charge of the hydrogen nucleus. If this electron has enough angular momentum, so that it does not approach the core too closely, it will essentially move in a Coulomb field corresponding to an effective charge of $+e$, and behaves as though it belonged to a hydrogen atom.

When a Rydberg atom reduces its principal quantum number by one unit, when emitting a photon, the light is in the microwave region of the electromagnetic spectrum. With this radiation isolated Rydberg atoms can be observed in interstellar space, where interatomic collisions are rare. Atoms with n up to 350 have been observed by radio astronomical methods.

Real Hydrogenlike Functions

The factor $e^{im\phi}$ makes the spherical harmonics complex, except when $m = 0$. In some applications it is convenient to use real hydrogenic wave functions, formed by taking linear combinations of the complex functions. Since the energy of the hydrogen electron does not depend on m the $2p_1$ and $2p_{-1}$ states belong to a degenerate energy eigenvalue. Any linear combination of them is therefore also an eigenfunction of the Hamiltonian with the same energy eigenvalue. One way to obtain a normalized real function is to form

$$\psi_{2p_x} \equiv \frac{1}{\sqrt{2}}\left(\psi_{2p_{-1}} + \psi_{2p_1}\right) = \frac{1}{4\sqrt{2\pi}}\left(\frac{Z}{a_0}\right)^{5/2} re^{-Zr/2a_0}\sin\theta\cos\phi$$

Transformation from polar to Cartesian coordinates clarifies the $2p_x$ designation, noting that the new wave function becomes

$$\psi_{2p_x} = \frac{1}{4\sqrt{2\pi}}\left(\frac{Z}{a_0}\right)^{5/2} xe^{-Zr/2a_0}$$

A second way of combining the functions produces

$$\psi_{2p_y} = \frac{1}{i\sqrt{2}}\left(\psi_{2p_1} - \psi_{2p_{-1}}\right) = \frac{1}{4\sqrt{2\pi}}\left(\frac{Z}{a_0}\right)^{5/2} r\sin\theta\sin\phi e^{-Zr/2a_0}$$

$$= \frac{1}{4\sqrt{2\pi}}\left(\frac{Z}{a_0}\right)^{5/2} ye^{-Zr/2a_0}$$

The function ψ_{2p_0} is real and often denoted by

$$\psi_{2p_z} = \frac{1}{\sqrt{\pi}}\left(\frac{Z}{2a_0}\right)^{5/2} ze^{-Zr/2a_0}.$$

The spherical harmonics in real form therefore exhibit a directional dependence and behave like simple functions of Cartesian coordinates. Orbitals using real spherical harmonics for their angular part are therefore particularly convenient to discuss properties such as the directed valencies of chemical bonds. The linear combinations still have the quantum numbers n and l, but they are no longer eigenfunctions for the z component of the angular momentum, so that this quantum number is lost.

The hydrogenlike wave functions are commonly known as orbitals. It is important to note that the degeneracy with respect to l is lifted (the energy levels become dependent on l) if the potential is not proportional to r^{-1}, although it may still be spherically symmetrical. Such is the case for many-electron atoms. Real orbitals therefore do not provide an accurate representation of the wave function of a many-electron atom. The l degeneracy is also lifted, even for hydrogen, if the problem is treated relativistically. The m degeneracy is only lifted by superposition of a non-spherically symmetrical potential on the central potential of the atom, *i.e.* an electrical or a magnetic field. This reduction of field symmetry also occurs in molecules.

The shapes of atomic orbitals are routinely confused with graphs of the angular factors in wave functions [60] and shown incorrectly. The graph of a p_y orbital, for example, gives tangent spheres lying on the y-axis.

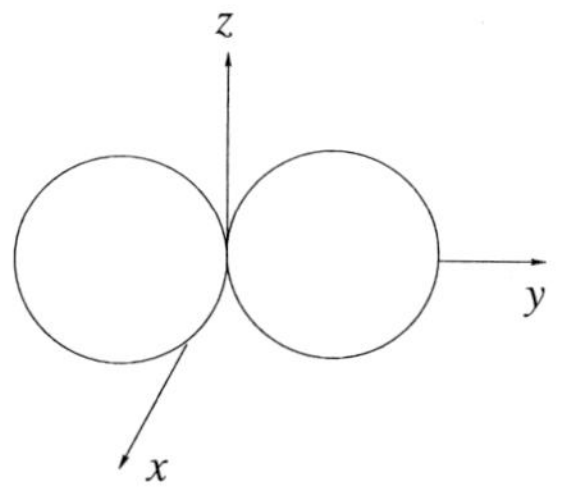

To get the orbital shape it is necessary to draw contour surfaces of constant probability density. For the p_y orbital the cross section in the yz plane is first obtained and rotated about the y-axis. In the yz plane, $\phi = \pi/2$, $\sin\phi = 1$, and hence

$$|\psi_{2p_y}| = k^{5/2}\frac{1}{\sqrt{\pi}}re^{-kr}|\sin\theta| = k^{5/2}\frac{1}{\sqrt{\pi}}ye^{-kr}$$

$k = Z/2a_0$. A contour of constant ψ has the shape

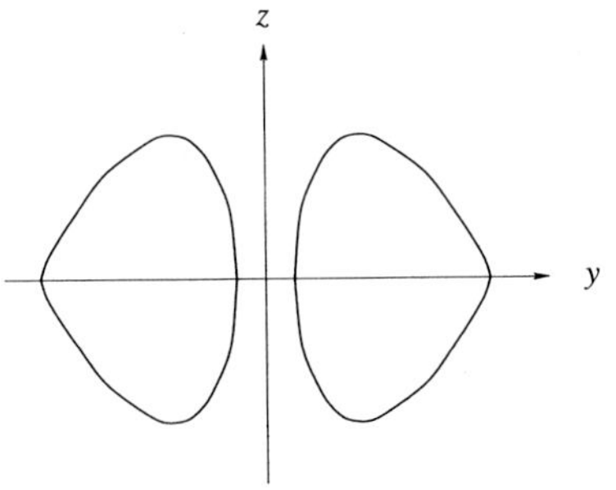

The three-dimensional probability density that represents the shape of the orbital consists of a pair of distorted ellipsoids, and not two tangent spheres.

The shape of the complex orbitals $\psi_{2p_{\pm 1}}$ derives from

$$|\psi_{2p_{\pm 1}}| = k^{5/2}\frac{1}{\sqrt{\pi}}e^{-kr}r|\sin\theta| = k^{5/2}\frac{1}{\sqrt{\pi}}e^{-kr}\sqrt{x^2+y^2}$$

which has the same cross section as ψ_{2p_y}, but the three-dimensional shape is now given by rotation about z, as a doughnut-shaped surface.

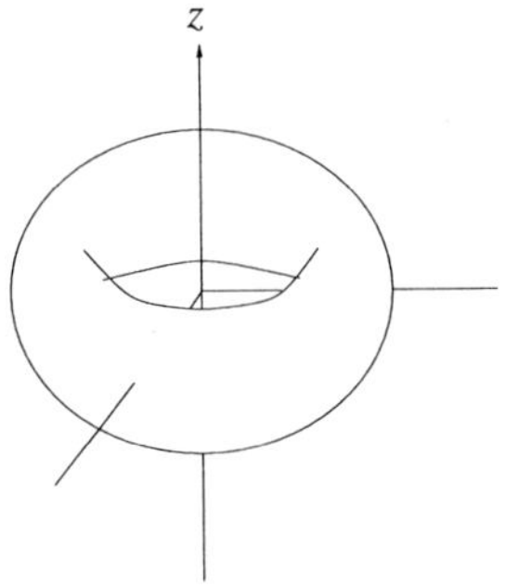

The use of spherical harmonics in real form is limited by the fact that, for $m \neq 0$ they are not eigenfunctions of L_z. They may be used to specify the angular distribution of electron density, but at the expense of any knowledge about angular momentum, and *vice versa*.

5.3 Relativistic Wave Equations

The relativistic expression for the energy of a free point particle with rest mass m and momentum $\boldsymbol{p}$ is

$$E = \sqrt{m^2c^4 + \boldsymbol{p}^2c^2}.$$

Any attempt to replace p_i by the differential operators $-i\hbar\partial/\partial x_i$ faces the problem that the square root of the linear operator is not uniquely defined. One way out is to square the energy expression into

$$E^2 = \boldsymbol{p}^2c^2 + m^2c^4$$

before using the correspondence principle

$$E \to i\hbar\frac{\partial}{\partial t} \quad , \quad \boldsymbol{p} \to -i\hbar\boldsymbol{\nabla}$$

that transforms the classical expression into the Schrödinger equation

$$i\frac{\partial}{\partial t}\Psi = -\frac{\hbar\nabla^2}{2m}\Psi.$$

This procedure leads to the Klein-Gordon equation

$$-\hbar^2\left(\frac{\partial^2}{\partial t^2} - c^2\nabla^2\right)\Psi = m^2c^4\Psi. \qquad (5.15)$$

Setting

$$\Box = \frac{\partial^2}{\partial x_1^2} + \frac{\partial^2}{\partial x_2^2} + \frac{\partial^2}{\partial x_3^2} - \frac{1}{c^2}\frac{\partial^2}{\partial t^2} = \nabla^2 - \frac{1}{c^2}\frac{\partial^2}{\partial t^2}$$

and $\kappa = mc/\hbar$, one has the common form of the KG equation

$$(\Box - \kappa^2)\Psi(\boldsymbol{x}, t) = 0.$$

Written in the form

$$\nabla^2 - \frac{1}{c^2}\frac{\partial^2\Psi}{\partial t^2} = \left(\frac{mc}{\hbar}\right)^2\Psi$$

it is recognized as a form of wave equation, with plane-wave solutions of the type $\Psi = \exp i(\boldsymbol{k}\cdot\boldsymbol{r} - \omega t)$.

Acceptable values of $\boldsymbol{k}$ and ω are established by substituting Ψ, $\partial^2\Psi/\partial t^2 = -\omega^2\Psi$, and $\nabla^2\Psi = -\boldsymbol{k}^2\Psi$ into (15), *i.e.*

$$\hbar^2\omega^2\Psi = \hbar^2c^2\boldsymbol{k}^2\Psi + m^2c^4\Psi.$$

Eigenvalues of the energy and momentum operators follow directly as

$$\begin{aligned} \hbar\omega &= \pm(\hbar^2c^2\boldsymbol{k}^2 + m^2c^4)^{\frac{1}{2}} \\ \hbar k &= \frac{1}{c}(\hbar^2\omega^2 - m^2c^4)^{\frac{1}{2}} \end{aligned}$$

where only positive eigenvalues of E are deemed physically reasonable.

The KG equation is Lorentz invariant, as required, but presents some other problems. Unlike Schrödinger's equation the KG equation is a second order differential equation with respect to time. This means that its solutions are specified only after an initial condition on both Ψ and $\partial\Psi/\partial t$ has been given. However, in contrast to Ψ, $\partial\Psi/\partial t$ has no direct physical interpretation [61]. Should the KG equation be used to define an equation of continuity, as was done with Schrödinger's equation (4), it is found to be satisfied by

$$\rho = \frac{i\hbar}{2mc^2}\left(\Psi^*\frac{\partial\Psi}{\partial t} - \frac{\partial\Psi^*}{\partial t}\Psi\right) .$$

Since, both Ψ and $\partial\Psi/\partial t$ can be prescribed arbitrary values at some time t_0, and since they are both functions of $\boldsymbol{x}$, the expression for the density ρ may assume both positive and negative values. Hence it cannot be interpreted as a probability density.

5.3.1 Dirac's Equation

In order to preserve the resemblance to Schrödinger's equation Dirac obtained another relativistic wave equation by starting from the form

$$i\hbar\frac{\partial\Psi}{\partial t} = H\Psi$$

which is linear in the time derivative. Since the relativistic spatial coordinates (x^1, x^2, x^3) must enter on the same footing as $x^4 = ct$, the Hamiltonian should likewise be linear in the space derivatives

$$\frac{\partial}{\partial x_k} \quad (k = 1, 2, 3)$$

and for the free particle may be written in the form [62]

$$H = c\boldsymbol{\alpha}\cdot\boldsymbol{p} + \beta mc^2$$

where $\boldsymbol{p}= i\hbar\boldsymbol{\nabla}$, as before. This is achieved by writing the wave function as a column vector $\Psi = \{\Psi_1, \Psi_2 \ldots \Psi_N\}$, together with the three components α^1, α^2 and α^3 of $\boldsymbol{\alpha}$ and β defined as non-commuting $N \times N$ matrices.

Using $E \to i\hbar\frac{\partial}{\partial t}$, these substitutions produce the Dirac wave equation

$$(E - c\boldsymbol{\alpha}\cdot\boldsymbol{p} - \beta mc^2)\Psi = 0 \tag{5.16}$$

or

$$i\hbar\frac{\partial\Psi}{\partial t} = -i\hbar c\boldsymbol{\alpha}\cdot\boldsymbol{\nabla}\Psi + \beta mc^2\Psi \quad ,$$

or more explicitly

$$i\hbar\frac{\partial\Psi_i}{\partial t} = -i\hbar c\sum_{j=1}^{N}\sum_{k=1}^{3}\alpha_{ij}^{k}\frac{\partial\Psi_j}{\partial x_k} + \sum_{j=1}^{N}\beta_{ij}mc^2\Psi_j. \tag{5.17}$$

The equations (17) are a set of N coupled equations for the N components Ψ_i of Ψ. Since the Hamiltonian is required to be Hermitian, the matrices must also be Hermitian, such that $\boldsymbol{\alpha} = \boldsymbol{\alpha}^\dagger$, $\beta = \beta^\dagger$.

Further conditions to be satisfied by $\boldsymbol{\alpha}$ and β follow from the requirement that each component of Ψ must separately satisfy the KG equation,

$$[E - \mathbf{p}^2c^2 - m^2c^4] = 0 \tag{5.18}$$

Multiplying (16) on the left by the operator $[E + c\boldsymbol{\alpha}\cdot\mathbf{p} + \beta mc^2]$ gives the second order equation

$$\left\{E^2 - c^2\left[\sum_{k=1}^{3}\left(\alpha^k\right)^2 p_k^2 + \sum_{k<l}\sum_{l}\left(\alpha^k\alpha^l + \alpha^l\alpha^k\right)p_kp_l\right]\right.$$
$$\left. -mc^3\left[\sum_{k=1}^{3}\left(\alpha^k\beta + \beta\alpha^k\right)p_k\right] - m^2c^4\beta^2\right\}\Psi = 0 \tag{5.19}$$

where $p_k(k = 1, 2, 3)$ denote the cartesian components of $\boldsymbol{p}$.

Comparing (17) to (16) shows that each component Ψ_i satisfies the KG equation providing

$$\begin{gathered}\left(\alpha^1\right)^2 = \left(\alpha^2\right)^2 = \left(\alpha^3\right)^2 = \beta^2 = 1 \\ \left[\alpha^1, \alpha^2\right]_+ = \left[\alpha^2, \alpha^3\right]_+ = \left[\alpha^3, \alpha^1\right]_+ = 0 \\ \left[\alpha^1, \beta\right]_+ = \left[\alpha^2, \beta\right]_+ = \left[\alpha^3, \beta\right]_+ = 0\end{gathered}$$

where $[A, B]_+$ denotes the anti-commutator

$$[A, B]_+ = AB + BA$$

It can be shown that the minimum dimensions for the various matrices required to satisfy all conditions are 4×4 and hence the wave function Ψ has at least four components. A representation of the matrices found useful is

$$\boldsymbol{\alpha} = \begin{pmatrix} 0 & \boldsymbol{\sigma} \\ \boldsymbol{\sigma} & 0 \end{pmatrix} \qquad \beta = \begin{pmatrix} \boldsymbol{I} & O \\ O & \boldsymbol{I} \end{pmatrix} \tag{5.20}$$

where $\boldsymbol{I}$ is the unit 2×2 matrix and

$$\sigma_1 = \begin{pmatrix} 0 & 1 \\ 1 & 0 \end{pmatrix}, \quad \sigma_2 = \begin{pmatrix} 0 & -i \\ i & 0 \end{pmatrix}, \quad \sigma_3 = \begin{pmatrix} 1 & 0 \\ 0 & -1 \end{pmatrix}$$

are the so-called Pauli matrices[9].

Free-particle Solutions

The solutions of the four coupled Dirac equations (17) must also be solutions of the KG equation and hence have the form of plane waves

$$\Psi_j(\boldsymbol{r},t) = u_j \exp i(\boldsymbol{k}\cdot\boldsymbol{r} - \omega t) \quad , \quad j = 1,2,3,4.$$

The u_{ij} are the eigenfunctions of the energy and momentum operators , with eigenvalues of $\hbar\omega$ and $\hbar\boldsymbol{k}$ respectively. Substitution of the plane-wave solutions and the 4×4 $\boldsymbol{\alpha}$ and β matrices into (17) give a set of secular equations [63]

$$\begin{bmatrix} u_1 \\ u_2 \\ u_3 \\ u_4 \end{bmatrix} \begin{bmatrix} E+mc^2 & 0 & cp_z & c(p_x+ip_y) \\ 0 & E+mc^2 & c(p_x-ip_y) & -cp_z \\ cp_z & c(p_x-ip_y) & E-mc^2 & 0 \\ c(p_x-ip_y) & -cp_z & 0 & E-mc^2 \end{bmatrix} = 0$$

Non-trivial solutions are obtained only if the determinant of coefficients vanishes. Reduced to a two-dimensional problem, *e.g.* in $\Psi(z,t)$, the secular equation becomes

$$\begin{bmatrix} u_1 \\ u_3 \end{bmatrix} \begin{bmatrix} E+mc^2 & cp_z \\ cp_z & E-mc^2 \end{bmatrix} = 0 \qquad (5.21)$$

[9]Dirac's equation is often formulated in a more symmetrical form by multiplying each component of (17) by $\beta/\hbar c$, and introducing the operators $\gamma_k = -i\beta\alpha_k$, $k = 1,2,3$, $\gamma_4 = \beta$, to give

$$\frac{\gamma_4}{ic}\frac{\partial\Psi}{\partial t} + \sum_{k=1}^{3}\gamma_k\frac{\partial\Psi}{\partial x_k} + \frac{mc}{\hbar}\Psi = 0.$$

Let $\quad \partial_k = \partial/\partial x_k, \quad \partial_4 = \partial/\partial x_4 = (1/ic)(\partial/\partial t), \quad \gamma_\mu\partial_\mu = \sum_{\mu=1}^{4}\gamma_\mu\partial_\mu, \quad \kappa = mc/\hbar,$

whereby Dirac's equation becomes

$$(\gamma_\mu\partial_\mu + \kappa)\Psi = 0.$$

The operators γ_μ satisfy the relations $\gamma_\mu^2 = 1$, $\gamma_\mu\gamma_\nu + \gamma_\nu\gamma_\mu = 0$, $(\mu \neq \nu)$, *i.e.*

$$\gamma_\mu\gamma_\nu + \gamma_\nu\gamma_\mu = 2g_{\mu\nu}$$

where g is the matrix equivalent of a δ function.

or

$$\begin{bmatrix} u_2 \\ u_4 \end{bmatrix} \begin{bmatrix} E+mc^2 & -cp_z \\ -cp_z & E-mc^2 \end{bmatrix} = 0 \tag{5.22}$$

In both cases, $E - m^2c^4 - (cp_z)^2 = 0$, *i.e.*

$$E = \pm c\sqrt{p_z^2 + m^2c^4}.$$

In general $E = \pm c\sqrt{\boldsymbol{p} + m^2c^4}$. Two sets of linearly independent solutions are obtained from (21) and (22), *i.e.*

$$u_3 = 1 \quad ; \quad u_1 = -\frac{cp_z}{E+mc^2} \qquad u_4 = 1 \quad ; \quad u_2 = \frac{cp_z}{E+mc^2}$$

$$u_1 = 1 \quad ; \quad u_3 = -\frac{cp_z}{E-mc^2} \qquad u_2 = 1 \quad ; \quad u_4 = \frac{cp_z}{E-mc^2}.$$

Dirac's equation therefore describes particles with the correct relativistic relation between energy and momentum eigenvalues, but as for the KG equation, both positive and negative energy eigenstates occur. The solutions u_1, u_2 correspond to positive energies and u_3, u_4 to negative energies. In the non-relativistic limit $E_+ = -E_-$ is close to mc^2 and large in comparison with $c|\boldsymbol{p}|$. Hence

$$u_1 \sim \frac{c(mv)}{2(\frac{1}{2}mv^2)} = \frac{c}{v} \sim \left(\frac{v}{c}\right) u_3 \quad ,$$

where v is the particle velocity, and $u_2 \sim \left(\frac{v}{c}\right) u_4$. The opposite is true for negative-energy solutions.

The difference between the two solutions with the same sign of energy can only be in their spin states. Since plane waves have no orbital angular momentum, the spin commutes with the Hamiltonian, and since $\sigma_3 u_1 = u_1$ but $\sigma_3 u_2 = -u_2$, the two solutions correspond to spin states with $S_z = \pm\hbar/2$ respectively.

Charged Particle in an Electromagnetic Field

The Hamiltonian for a charged particle in an electromagnetic field can be obtained from Hamilton's principle and Lagrange's equations of motion (Section 3.3):

$$\frac{d}{dt}\left(\frac{\partial L}{\partial \dot{q}_j}\right) - \frac{\partial L}{\partial q_j} = 0 \quad , \quad j = 1, 2, 3 \ldots$$

It is required to find L for a particle of mass m, charge e and velocity $\boldsymbol{v}$ in an electromagnetic field $\boldsymbol{E}(\boldsymbol{r},t)$ and magnetic field $\boldsymbol{B}(\boldsymbol{r},t)$ and subject to the Lorentz force. From eqn. 4.11

$$\boldsymbol{F} = e\left[-\nabla\phi - \frac{\partial \boldsymbol{A}}{\partial t} + \boldsymbol{v} \times (\boldsymbol{\nabla} \times \boldsymbol{A})\right].$$

The requirements are satisfied by defining

$$L = \frac{1}{2}mv^2 + e(-\phi + \boldsymbol{v}\cdot\boldsymbol{A})$$

Taking derivatives:

$$\begin{aligned}
\frac{\partial L}{\partial \boldsymbol{v}} &= m\boldsymbol{v} + e\left[\boldsymbol{A} + \boldsymbol{v}\frac{\partial \boldsymbol{A}}{\partial \boldsymbol{v}}\right] \\
\frac{d}{dt}\left(\frac{\partial L}{\partial \boldsymbol{v}}\right) &= m\ddot{\boldsymbol{r}} + e\left[\frac{\partial \boldsymbol{A}}{\partial t} + (\boldsymbol{v}\cdot\boldsymbol{\nabla})\boldsymbol{A}\right] \\
\text{grad } L &= e[-\phi + \boldsymbol{v}\times(\boldsymbol{\nabla}\times\boldsymbol{A}) + (\boldsymbol{v}\cdot\boldsymbol{\nabla})\boldsymbol{A}] \quad ,
\end{aligned}$$

using the vector identity (1.10)

$$\nabla(\boldsymbol{a}\cdot\boldsymbol{b}) = \boldsymbol{a}\times \text{ curl } \boldsymbol{b} + (\boldsymbol{a}\cdot\boldsymbol{\nabla})\boldsymbol{b} + \boldsymbol{b}\times \text{ curl } \boldsymbol{a} + (\boldsymbol{b}\cdot\boldsymbol{\nabla})\boldsymbol{a}$$

and noting that $\boldsymbol{v}$ is a constant vector. It follows immediately that the assumed form of L satisfies (4.11) for $\boldsymbol{F} = m\ddot{\boldsymbol{r}}$.

The generalized momenta are defined as $p_i = \partial L/\partial \dot{q}_i$, *i.e.* $\boldsymbol{p} = m\boldsymbol{v} + e\boldsymbol{A}$. The Hamiltonian,

$$\begin{aligned}
H &= \sum_{i=1}^{3} p_i\dot{q}_i - L = T + V \\
&= \frac{1}{2m}(\boldsymbol{p} - e\boldsymbol{A})^2 + e\phi
\end{aligned}$$

In the electromagnetic field the free particle operators therefore become

$$\boldsymbol{p} \to \boldsymbol{p} - e\boldsymbol{A} \quad , \quad E \to E - e\phi$$

and hence

$$(E - e\phi)^2 = m^2c^4 + c^2(\boldsymbol{p} - e\boldsymbol{A})^2,$$

or in Dirac notation

$$\left[(E - e\phi) - c\boldsymbol{\alpha}\cdot(\boldsymbol{p} - e\boldsymbol{A} - \beta mc^2\right]\Psi = 0 \quad ,$$

or

$$i\hbar\frac{\partial}{\partial t}\Psi = \left[-i\hbar c\boldsymbol{\alpha}\cdot\nabla - ce\boldsymbol{\alpha}\cdot\boldsymbol{A} + e\phi + \beta mc^2\right]\Psi.$$

The Dirac Hamiltonian in the presence of an external field is therefore given by

$$H = c\boldsymbol{\alpha}\cdot(\boldsymbol{p} - e\boldsymbol{A}) + e\phi + \beta mc^2.$$

To understand the physical implications of these equations it is customary to consider the non-relativistic limit of their stationary solutions, assuming

$$\Psi(r) = \eta(r)e^{-iEt/\hbar}.$$

Thus

$$E\eta(r) = [-i\hbar c\boldsymbol{\alpha}\cdot\boldsymbol{\nabla} - ce\boldsymbol{\alpha}\cdot\boldsymbol{A} + e\phi + \beta mc^2]\eta(r).$$

The 4-component spinor $\eta(r)$ is expressed in terms of two 2-component spinors

$$\eta(r) = \begin{pmatrix} \psi(r) \\ \chi(r) \end{pmatrix}.$$

Using the representations (20) for $\boldsymbol{\alpha}$ and β two coupled equations are obtained:

$$E\psi = c(i\hbar\nabla - e\boldsymbol{A})\cdot\boldsymbol{\sigma}\chi + (e\phi + mc^2)\psi$$
$$E\chi = c(i\hbar\nabla - e\boldsymbol{A})\cdot\boldsymbol{\sigma}\psi + (e\phi - mc^2)\chi$$

Substituting $E = E' + mc^2$,

$$E'\psi = c(i\hbar\nabla - e\boldsymbol{A})\cdot\boldsymbol{\sigma}\chi + e\phi\psi \tag{5.23}$$

$$(E' + 2mc^2)\chi = c(-i\hbar\nabla - e\boldsymbol{A})\cdot\boldsymbol{\sigma}\psi + e\phi\chi.$$

In the non-relativistic limit both E' and $e\phi$ are small compared to mc^2 and hence

$$\chi(r) = \frac{1}{2mc}(-i\hbar\nabla - e\boldsymbol{A})\cdot\boldsymbol{\sigma}\psi(r). \tag{5.24}$$

Now χ is smaller than ψ by a factor p/mc, *i.e.* v/c.

Substituting (24) into (23) yields

$$E'\psi = \frac{1}{2m}[(i\hbar\nabla - e\boldsymbol{A})\cdot\boldsymbol{\sigma}]^2\psi + e\phi\psi. \tag{5.25}$$

Using the identity

$$(\boldsymbol{\sigma}\cdot\boldsymbol{A})(\boldsymbol{\sigma}\cdot\boldsymbol{B}) = \boldsymbol{A}\cdot\boldsymbol{B} + i\boldsymbol{\sigma}\cdot(\boldsymbol{A}\times\boldsymbol{B})$$

for Pauli matrices, the first term on the right in (25) reduces to

$$\frac{1}{2m}[(-i\hbar\nabla - e\boldsymbol{A})\cdot\boldsymbol{\sigma}]^2\psi = \frac{1}{2m}(i\hbar\nabla - e\boldsymbol{A})^2\psi - \frac{e\hbar}{2m}\boldsymbol{\sigma}\cdot(\boldsymbol{\nabla}\cdot\boldsymbol{A})\psi.$$

Since $\boldsymbol{\nabla}\times\boldsymbol{A} = \boldsymbol{B}$, the magnetic field, (25) becomes

$$E'\psi(r) = \left[\frac{1}{2m}(-i\hbar\nabla - e\boldsymbol{A})^2 - \frac{e\hbar}{2m}(\boldsymbol{\sigma}\cdot\boldsymbol{B}) + e\phi\right]\psi(r) \tag{5.26}$$

known as the Pauli equation. The term $(e\hbar/2m)(\boldsymbol{\sigma} \cdot \boldsymbol{B})$ corresponds to an interaction $-\boldsymbol{M_s} \cdot \boldsymbol{B}$ between the magnetic field and an intrinsic magnetic moment $\boldsymbol{M_s}$ of the electron, due to its spin, with

$$\boldsymbol{M}_s = -g_s\mu_B \boldsymbol{S}/\hbar = g_s \frac{e}{2m}\boldsymbol{S}$$

and $\mu_B = e\hbar/2m$, the Bohr magneton. The gyromagnetic ratio $g_s = 2$.

Continuity Equation

Since the wave function may be interpreted as a column vector with four components, $\Psi^\dagger$ may be defined to be a row matrix with components Ψ_i^*, and which satisfies the adjoint equation

$$\begin{aligned} - \quad & i\hbar\frac{\partial \Psi^\dagger}{\partial t} = H\Psi^\dagger \\ = \quad & (i\hbar c\nabla - ce\boldsymbol{A})\Psi^\dagger \cdot \boldsymbol{\alpha} + e\phi\Psi^\dagger + mc^2\Psi^\dagger\beta \end{aligned}$$

The quantity

$$\rho(\boldsymbol{r}, t) = \Psi^\dagger\Psi = \sum_{i=1}^{4} |\Psi_i|^2$$

is clearly positive and can be interpreted as a position probability density, in the same way that $|\Psi|^2$ is the probability density for the non-relativistic Schrödinger equation.

Multiplying the Ψ equation on the left by $\Psi^\dagger$ and the adjoint equation on the right by Ψ, and taking the difference of the two results, it is found that

$$\frac{\partial \rho}{\partial t} + \nabla \cdot (\Psi^\dagger c\boldsymbol{\alpha}\Psi) = 0.$$

If the vector $\boldsymbol{j}(\boldsymbol{r}, t) = \Psi^\dagger c\boldsymbol{\alpha}\Psi$ is interpreted as a probability current density, one has the continuity equation

$$\frac{\partial \rho}{\partial t} + \boldsymbol{\nabla} \cdot \boldsymbol{j} = 0$$

and hence $c\boldsymbol{\alpha}$ can be interpreted as a velocity operator.

Hole Theory

The appearance of negative energy states was initially considered to be a fatal flaw in the Dirac theory, because it renders all positive energy states unstable. There is nothing to prevent an electron with positive energy from cascading down the negative levels *ad infinitum.* A solution to this dilemma was proposed by Dirac in terms of a many-particle theory. Observing that spin-$\frac{1}{2}$ particles, such as electrons, obey the Pauli exclusion principle, it was suggested that in the normal ground state (vacuum state) of Nature all possible negative energy levels are already occupied. This principle prevents any electron from falling into negative energy states and so ensures the stability of positive-energy physical states. The energy gap separating the negative and positive energy solutions is exactly $2mc^2$, and it will be possible to excite an electron from the negative energy sea into a positive energy state. It then leaves a hole in the sea. This hole in the negative energy, negatively charged states, behaves exactly as a particle with the same mass as that of an electron, but having a positive electric charge. It may therefore be identified with a positron. If the excitation is effected by absorption of a γ-ray photon the result will be transformation of radiation energy into the creation of an electron-positron pair. Positrons created in this way will disappear rather fast due to the fact that the existence of a positron is equivalent to a hole in the negative energy sea. An ordinary electron must soon fall down into the negative energy state causing both electron and positron to disappear and the equivalent energy to be regained in the form of electromagnetic radiation.

Awkward questions about the electromagnetic and gravitational fields of infinitely many particles in the vacuum remain unanswered. Also, the Dirac theory, amended by the hole proposition is certainly not a one-particle theory, and hence not a relativistic generalization of Schrödinger's equation.

Central Field Solutions

Although Dirac's equation does not directly admit of a completely self-consistent single-particle interpretation, such an interpretation is physically acceptable and of practical use, provided the potential varies little over distances of the order of the Compton wavelength ($\hbar/mc$) of the particle in question. It allows, for instance, first-order relativistic corrections to the spectrum of the hydrogen atom and to the core-level densities of many-electron atoms. The latter aspect is of special chemical importance. The required calculations are invariably numerical in nature and this eliminates the need to investigate central-field solutions in the same detail as for Schrödinger's equation. A brief outline suffices.

For a Dirac particle in a central field ϕ is spherically symmetrical and $\boldsymbol{A} = \boldsymbol{0}$. Setting the potential energy $V(r) = e\phi(r)$, the Dirac Hamiltonian becomes

$$H = c\boldsymbol{\sigma} \cdot \boldsymbol{p} + \beta mc^2 + V.$$

In non-relativistic Schrödinger theory every component of the orbital angular momentum $\boldsymbol{L} = \boldsymbol{r} \times \boldsymbol{p}$, as well as $\boldsymbol{L}^2$, commutes with the Hamiltonian $H = p^2/2m + V$ of a spinless particle in a central field. As a result, simultaneous eigenstates of the operators H, $\boldsymbol{L}^2$ and L_z exist in Schrödinger theory, with respective eigenvalues of E, $l(l+1)\hbar^2$ and $m\hbar$. In Dirac's theory, however, neither the components of $\boldsymbol{L}$, nor $\boldsymbol{L}^2$, commute with the Hamiltonian [10].

$$[\boldsymbol{L}, H] = i\hbar c\boldsymbol{\alpha} \times \boldsymbol{p} \tag{5.27}$$

Now consider the operator

$$\boldsymbol{S} = \frac{\hbar}{2}\boldsymbol{\Sigma} \tag{5.28}$$

where $\boldsymbol{\Sigma} \equiv (\Sigma^1, \Sigma^2, \Sigma^3)$, the Σ^k being the 4×4 matrices

$$\boldsymbol{\Sigma} = \begin{pmatrix} \boldsymbol{\sigma} & 0 \\ 0 & \boldsymbol{\sigma} \end{pmatrix}.$$

From the known properties of Pauli matrices it follows that

$$\boldsymbol{S}^2\Psi = s(s+1)\hbar^2\Psi, \quad s = \frac{1}{2}$$

and the two possible eigenvalues of $S_i (i = 1, 2, 3)$ are $\pm\hbar/2$.

Any component of $\boldsymbol{S}$ commutes with any component of $\boldsymbol{L}$ and, as for orbital angular momentum, it is readily shown that

$$[\boldsymbol{S}, H] = -i\hbar c\boldsymbol{\alpha} \cdot \boldsymbol{p} \tag{5.29}$$

---

[10]

$$\begin{aligned}
H &= c\boldsymbol{\alpha} \cdot \boldsymbol{p} + \beta mc^2 + V \\
\boldsymbol{L} &= \boldsymbol{r} \times \boldsymbol{p} \\
\boldsymbol{L}H &= (\boldsymbol{r} \times \boldsymbol{p})(c\boldsymbol{\alpha} \cdot \boldsymbol{p} + \beta mc^2 + V) \\
&= c(\boldsymbol{r} \times \boldsymbol{p})(\boldsymbol{\alpha} \cdot \boldsymbol{p}) + \beta mc^2(\boldsymbol{r} \times \boldsymbol{p}) + V(\boldsymbol{r} \times \boldsymbol{p}) \\
H\boldsymbol{L} &= (c\boldsymbol{\alpha} \cdot \boldsymbol{p} + \beta mc^2 + V)(\boldsymbol{r} \times \boldsymbol{p}) \\
&= c(\boldsymbol{\alpha} \cdot \boldsymbol{p})(\boldsymbol{r} \times \boldsymbol{p}) + \beta mc^2(\boldsymbol{r} \times \boldsymbol{p}) + V(\boldsymbol{r} \times \boldsymbol{p}) \\
[\boldsymbol{L}, H] &= \boldsymbol{L}H - H\boldsymbol{L} \\
&= c[(\boldsymbol{r} \times \boldsymbol{p})(\boldsymbol{\alpha} \cdot \boldsymbol{p}) - (\boldsymbol{\alpha} \cdot \boldsymbol{p})(\boldsymbol{r} \times \boldsymbol{p})] \\
&= c[(\boldsymbol{\alpha} \times \boldsymbol{p})(\boldsymbol{r} \cdot \boldsymbol{p}) - (\boldsymbol{p} \cdot \boldsymbol{r})(\boldsymbol{\alpha} \times \boldsymbol{p})], \quad \text{using (1.1)} \\
&= c[\boldsymbol{r}, \boldsymbol{p}](\boldsymbol{\alpha} \times \boldsymbol{p}) \\
&= i\hbar c(\boldsymbol{\alpha} \times \boldsymbol{p})
\end{aligned}$$

From (27) and (29) it follows that every component of the total angular momentum operator $\boldsymbol{J} = \boldsymbol{L} + \boldsymbol{S}$ and $\boldsymbol{J}^2$ commute with the Dirac Hamiltonian. The eigenvalues of $\boldsymbol{J}^2$ and J_z are $j(j+1)\hbar^2$ and $m_j\hbar$ respectively and they can be defined simultaneously with the energy eigenvalues E.

Despite the complication due to the interdependence of orbital and spin angular momenta, the Dirac equation for a central field can be separated in spherical polar coordinates [63]. The energy eigenvalues for the hydrogen atom ($V(r) = e^2/r$, in electrostatic units), are equivalent to the relativistic terms of the old quantum theory [64]

$$E = mc^2 \left[1 + \frac{\gamma^2}{(s+n')^2}\right]^{-\frac{1}{2}}$$

for quantum numbers s and n', and dimensionless fine-structure constant $\gamma = e^2/\hbar c \simeq 1/137$. Expanded in powers of γ^2

$$E = mc^2 \left[1 - \frac{\gamma^2}{2n^2} - \frac{\gamma^4}{2n^4}\left(\frac{n}{|k|} - \frac{3}{4}\right) - \ldots\right]$$

where $n = n' + |k|$ is the principal quantum number.

The second term is just the electronic binding energy of Schrödinger's equation.

Dirac Notation

In wave mechanics the wave function $\psi_\alpha(\boldsymbol{x})$ of a system in a state is specified by the label α. For example, in the case of a free particle α could be the momentum $\boldsymbol{p}$, and $\psi_\alpha(x)$ would be the de Broglie wave function,

$$\psi_\alpha(\boldsymbol{x}) = u_p(x) = e^{i\boldsymbol{p}\cdot\boldsymbol{x}/\hbar}.$$

The electronic state of a hydrogen atom is specified by the quantum numbers n, l and m, so in this case $\psi_\alpha(\boldsymbol{x}) = u_{nlm}(\boldsymbol{x}) = R_{nl}Y_{lm}(\theta,\phi)$. If the spin state of the electron is also to be specified this latter $\psi_\alpha(\boldsymbol{x})$ must be multiplied by a column vector χ_m, where

$$\chi_{1/2} = \begin{pmatrix} 1 \\ 0 \end{pmatrix} \quad \text{and} \quad \chi_{-1/2} = \begin{pmatrix} 0 \\ 1 \end{pmatrix}$$

represent spin-up and spin-down states respectively, and α comprises the four numbers $(nlmm_s)$.

Whereas the wave function is a mathematical construct, with interpretational rules for obtaining the answers to physical problems, the physical specification of the state is given by the labels α, which are usually the eigenvalues of commuting observables, as in the above examples. The last example shows that in order to incorporate electron spin an uneasy mixture of wave functions and matrices is required. In the Dirac formulation wave functions and spin vectors are put on an equal footing with emphasis on the state specification. In this formalism it becomes possible to make contact with both wave mechanics and matrix mechanics.

In terms of this notation the physical state of a system is represented in a dual complex vector space by *kets*, like $|\alpha\rangle$ and *bras*, $\langle\alpha|$. The state ket (or bra) carries complete information about the physical state. The sum of two kets is another ket.

$$|\alpha\rangle + |\beta\rangle = |\gamma\rangle.$$

The product $c|\alpha\rangle$ is another ket, and $c|\alpha\rangle = |\alpha\rangle c$. It is postulated that $|\alpha\rangle$ and $c|\alpha\rangle$, $(c \neq 0)$ represent the same physical state, and only the direction in vector space is of significance. This is more like the property of a *ray* than a vector.

An observable is represented by an operator that acts on a ket

$$A \cdot (|\alpha\rangle) = A|\alpha\rangle$$

to yield another ket.

Ket notation is conveniently used to specify eigenfunctions by listing their eigenvalues,

$$A|a_i\rangle = a_i|a_i\rangle.$$

The physical state corresponding to an eigenket is called an eigenstate, *e.g.*

$$S_z|S_z^+\rangle = \frac{\hbar}{2}|S_z^+\rangle \quad , \quad S_z|S_z^-\rangle = -\frac{\hbar}{2}|S_z^-\rangle$$

where $|S_z^{\pm}\rangle$ are eigenkets of the spin operator S_z with eigenvalues of $\pm\hbar/2$.

The inner product of a bra and a ket $\langle\beta|\alpha\rangle$ is a complex number, and

$$\langle\beta|\alpha\rangle = \langle\alpha|\beta\rangle^*.$$

Two kets are said to be orthogonal if $\langle\alpha|\beta\rangle = 0$. A normalized ket $|\tilde{\alpha}\rangle$ can be constructed from any given ket in the form

$$|\tilde{\alpha}\rangle = \left(\frac{1}{\sqrt{\langle\alpha|\alpha\rangle}}\right)|\alpha\rangle, \text{ where } \langle\tilde{\alpha}|\tilde{\alpha}\rangle = 1.$$

The quantity $\sqrt{\langle\alpha|\alpha\rangle}$ is known as the norm of $|\alpha\rangle$, analogous to the properties of vectors in Euclidean space.

The outer product of $|\beta\rangle$ and $\langle\alpha|$, $|\beta\rangle\langle\alpha|$ is regarded as an operator. This is seen by forming the product

$$(|\beta\rangle\langle\alpha|) \cdot |\gamma\rangle = |\beta\rangle \cdot (\langle\alpha|\gamma\rangle)$$

where $\langle\alpha|\gamma\rangle$ is just a number. The outer product acting on a ket produces another ket, just like an operator.

The product $(\langle\beta|) \cdot (X|\alpha\rangle) = (\langle\beta|X) \cdot (|\alpha\rangle)$

$$= \langle\beta|X|\alpha\rangle$$

called the matrix element of the operator X. For Hermitian X

$$\langle\beta|X|\alpha\rangle = \langle\alpha|X|\beta\rangle^*.$$

The expectation or average value of A can be written as

$$\langle A\rangle = \langle\psi|A|\psi\rangle.$$

5.4 Angular Momentum and Spin

5.4.1 Measurement of Angular Momentum

Orbital angular momentum of an electron in an atom can be measured by the same type of Stern-Gerlach experiment described before for the measurement of electron spin. In this case it will be assumed to use a metal such as magnesium whose atoms have a total electron spin of zero. The magnetic moment

$$\boldsymbol{\mu} = \frac{e\boldsymbol{L}}{2mc}$$

therefore arises from the orbital angular momentum only. As before, the force on an atom is given by

$$F_z = \mu_z \frac{\partial B_z}{\partial z}.$$

Thus, each atom experiences a force which is proportional to the z-component of its electronic angular momentum. By measuring the deflection of the beam F_z can be calculated.

Since there is no preferred direction for L_z before the atoms enter the magnetic field, atoms with all possible values of L_z will randomly appear with equal frequency. Each value of L_z will give rise to a single spot on the detecting screen. Since the total number of permissible components of L_z is $2l + 1$, one can measure l simply by counting the number of spots on the screen. This SG experiment yields direct proof of the quantization of angular momentum, since, classically, there should be a continuous range of angular momenta and hence a continuous spread of atoms across the screen.

The relationship between different components of orbital angular momentum such as L_z and L_x can be investigated by multiple SG experiments as discussed for electron spin and photon polarization before. The results are in fact no different. This is a consequence of the noncommutativity of the operators L_x and L_z. The two observables cannot be measured simultaneously. While total angular momentum is conserved, the components vary as the applied analyzing field changes. As in the case of spin or polarization, measurement of L_x, for instance, disturbs any previously known value of L_z. The structure of the wave function does not allow L_x to be made definite when L_z has an eigenvalue, and *vice versa.*

5.4.2 General Theory

A little reflection shows that the commutation relationships, recognized as one of the fundamental differences between classical and quantum systems, are common to all forms of angular momentum, including orbital, polarization and spin. It is of interest to note that the eigenvalues for all forms of angular momentum can be obtained directly from the commutation rules, without using special differential operators. To emphasize the commonality, angular momentum $\boldsymbol{M}$ of all forms will be represented here by three linear operators M_x, M_y and M_z, that obey the commutation rules:

$$[M_x, M_y] = i\hbar M_z$$

$$[M_y, M_z] = i\hbar M_x$$

$$[M_z, M_x] = i\hbar M_y$$

together with $M^2 = M_x^2 + M_y^2 + M_z^2$.

Since any component of $\boldsymbol{M}$ commutes with M^2 it can be shown that if M^2 has a definite eigenvalue associated with the eigenfunction ψ, this value is not changed if any component of $\boldsymbol{M}$ operates on ψ.

Assume that

$$M^2\psi = c\psi$$

and operate on ψ with any component (x, y, z), such that

$$M^2 M_z\psi = M_z M^2\psi = M_z c\psi = cM_z\psi$$

or

$$M^2(M_z\psi) = c(M_z\psi)$$

It follows that $(M_z\psi)$ is also an eigenfunction of M^2, belonging to the same eigenvalue c.

Now suppose that

$$M_z\psi_m = \hbar m\psi_m,$$

i.e. that ψ_m is simultaneously an eigenfunction of M^2 and M_z. Multiplying the above equation by the operator $(M_x + iM_y)$, yields

$$(M_x + iM_y)(M_z\psi_m) = \hbar m(M_x + iM_y)\psi_m \tag{5.30}$$

Now use the commutation rules to define

$$(M_x + iM_y)M_z - M_z(M_x + iM_y) = -\hbar(M_x + iM_y)$$

and rewrite (30) as

$$[m_z(M_x + iM_y) - \hbar(M_x + iM_y)]\psi_m = \hbar m(M_x + iM_y)\psi_m$$

i.e.

$$M_z(M_x + iM_y) = \hbar(m + 1)(M_x + iM_y)\psi_m$$

This shows that for ψ_m an eigenfunction, in which $M_z = m\hbar$, the function $(M_x + iM_y)\psi_m$ is an eigenfunction of M_z belonging to $M_z = (m+1)\hbar$, but the same value of M^2. In the same way one can show that

$$M_z(M_x - iM_y)\psi_m = \hbar(m - 1)(M_x - iM_y)\psi_m \tag{5.31}$$

The operator $M - iM_y$ therefore reduces the value of m by unity, but leaves L^2 unchanged[11].

Repeated application of the operator $(M_x + iM_y)$ will generate eigenfunctions of a fixed M^2 belonging to indefinitely large eigenvalues of M_z unless there is some value of m for which $(M_x + iM_y)\psi_m$ vanishes. Similarly, repeated application of $M_x - iM_y$ will lead to indefinitely large negative values unless $(M_x - iM_y)\psi_m$ vanishes for some ψ_m.

From the definition of

$$M^2 = M_x^2 + M_y^2 + M_z^2 = M_x^2 + M_y^2 + m\hbar^2$$

[11]The operators $(M_x + iM_y)$ and $(M_x - iM_y)$ are, for obvious reasons, called ladder operators.

it follows that there must be some $|m|$ for which the relationship $\hbar^2 m^2 \leq M^2$ does not hold, and hence, for which the mean value of M_x^2 and M_y^2 is not positive. This means that $|\hbar m|$ must not be allowed to exceed $\sqrt{M^2}$. Therefore, in a state in which $|m|$ has its maximum, consistent with a given M^2, either

$$(M_x + iM_y)\psi_{m_1} = 0 \quad \text{or} \quad (M_x - iM_y)\psi_{m_2} = 0$$

In order to find the relation between M^2 and m_1 and m_2 consider the expression

$$\begin{aligned} M^2\psi_{m_1} &= (M_x^2 + M_y^2 + M_z^2)\psi_{m_1} = [(M_x - iM_y)(M_x + iM_y) + M_z^2 + \hbar M_z]\psi_{m_1} \\ &= [(M_x - iM_y)(M_x + iM_y) + \hbar^2(m_1^2 + m_1)]\psi_{m_1} \qquad (5.32) \end{aligned}$$

Since $(M_x + iM_y)\psi_{m_1} = 0$, it follows that

$$M^2\psi_{m_1} = \hbar^2 m_1(m_1 + 1)\psi_{m_1} \qquad (5.33)$$

It is readily shown in a similar way that when $(M_x - iM_y)\psi_{m_2} = 0$, one finds

$$M^2\psi_{m_2} = \hbar^2 m_2(m_2 - 1)\psi_{m_2} \qquad (5.34)$$

If (33) and (34) are to hold simultaneously then either $m_2 = -m_1$, or $m_2 = m_1 + 1$. The second option is not admissible since m_1 is already defined as the allowed maximum. The condition $m_2 = -m_1$ ($= -l$, an assumed integer) leads to the well-known relationship

$$M^2 = \hbar^2 l(l + 1)$$

For any given value of l, the allowed value of m is any positive or negative integer between $\pm l$, including zero, as found for the eigenvalues of orbital angular momentum of the hydrogen atom. Since the progression from m_1 to m_2 proceeds by integral steps it is implied that $m_1 - m_2$ must be an integer. But, $m_2 = -m_1$, so that $m_1 - m_2 = 2m_1$, is an integer and $m_1 = l$ may be either an integer or a half-integer.

It is commonly argued that half-integer values of m are ruled out by the requirement of single-valued wave functions, *e.g.* the boundary condition

$$\Phi(\phi + 2\pi) = \Phi(\phi)$$

requires

$$Ae^{im\phi} \cdot e^{2\pi i m} = Ae^{im\phi}$$

i.e.

$$e^{2\pi im} = 1 \quad or \quad m = 0, \pm 1, \ldots$$

However, it is noted that with half-integral m the wave function only changes sign when advanced by 2π, which means that an observable quantity depending on $\phi^*\phi$ will be single-valued. When mixed integral and half-integral quantum numbers are included, not even the probabilities are single-valued.

It seems that a sensible theory for orbital angular momentum could be made if m was either integral or half-integral, but not both of these together. Experiment shows that only integral orbital angular momenta actually occur. The same criterion applies to intrinsic angular momentum of the electron, but in that case experiment rules out integral values.

The similarity between orbital and spin angular momenta is illustrated even better by reformulating the general theory in matrix representation. To achieve this the matrix elements of some operator A are written as

$$a_{nm} = \int \psi_n^*(x)\psi_m(x)dx.$$

Since M_x and M_y are Hermitian, $M_x + iM_y$ and $M_x - iM_y$ are Hermitian conjugates and equation (32) written in matrix notation becomes

$$[(M_x - iM_y)(M_x + iM_y)]_{mm'} + \hbar^2 m(m+1)\delta_{mm'} = \hbar^2 l(l+1)\delta_{mm'}$$

In terms of individual eigenvalues

$$[(M_x - iM_y)(M_x + iM_y)]_{mm'} = C_m^* C_{m'}\delta_{mm'}.$$

For $m = m'$ it follows that

$$\begin{aligned} C_m^* C_m &= \hbar^2[l(l+1) - m(m+1)] \\ &= \hbar^2(l-m)(l+m+1) \end{aligned}$$

which gives

$$[M_x + iM_y]_{mm'} = \hbar\sqrt{(l-m')(l+m'+1)}\delta_{m,m'+1} \tag{5.35}$$

$$[M_x - iM_y]_{mm'} = [M_x + iM_y]^\dagger_{mm'} = \hbar\sqrt{(l-m)(l+m+1)}\delta_{m',m+1}$$

This means that $M_x + iM_y$ and $M_x - iM_y$ are represented by matrices in which all elements are one space off the diagonal in m. For $l = \frac{1}{2}$, it follows from (35) that

$$(M_x + iM_y) = \hbar \begin{pmatrix} m'(\frac{1}{2}) & m'(-\frac{1}{2}) \\ m'(\frac{1}{2}) & m'(-\frac{1}{2}) \end{pmatrix} = \hbar \begin{pmatrix} 0 & 1 \\ 0 & 0 \end{pmatrix}$$

$$(M_x - iM_y) = \hbar \begin{pmatrix} 0 & 0 \\ 1 & 0 \end{pmatrix}$$

$$2M_x = \hbar \begin{pmatrix} 0 & 1 \\ 1 & 0 \end{pmatrix} \quad ; \quad M_x = \frac{\hbar}{2} \begin{pmatrix} 0 & 1 \\ 1 & 0 \end{pmatrix}$$

$$2iM_y = \hbar \begin{pmatrix} 0 & 1 \\ -1 & 0 \end{pmatrix} \quad ; \quad M_y = \frac{\hbar}{2} \begin{pmatrix} 0 & -i \\ i & 0 \end{pmatrix}$$

$$[M_x, M_y] = \frac{\hbar^2}{2} \begin{pmatrix} i & 0 \\ 0 & -i \end{pmatrix} = i\hbar M_z$$

$$M_z = \frac{\hbar}{2} \begin{pmatrix} 1 & 0 \\ 0 & 1 \end{pmatrix}$$

The angular momentum matrices for spin, $\hbar/2$, were first worked out by Pauli. These three matrices, called the Pauli matrices are written as

$$M_x = \frac{\hbar}{2}\sigma_x \qquad M_y = \frac{\hbar}{2}\sigma_y \qquad M_z = \frac{\hbar}{2}\sigma_z$$

where the σ define the matrices obtained above.

Because the σ matrices are proportional to angular momentum operators, they satisfy commutation rules of the type

$$\sigma_x\sigma_y - \sigma_y\sigma_x = 2i\sigma_z$$

It is readily shown by direct computation that $\sigma_x\sigma_y + \sigma_y\sigma_x = 0$, so that

$$\sigma_x\sigma_y = i\sigma_z \quad \text{or more generally} \quad \boldsymbol{\sigma} \times \boldsymbol{\sigma} = 2i\boldsymbol{\sigma}.$$

Also, $\sigma_x^2 = \sigma_y^2 = \sigma_z^2 = 1$, so that $\sigma^2 = 3$, and

$$M^2 = \frac{\hbar^2}{4}\left(\sigma_x^2 + \sigma_y^2 + \sigma_z^2\right) = \frac{3}{4}\hbar^2,$$

the correct value for $l = \frac{1}{2}$.

5.4.3 Schrödinger's Equation and Spin

Schrödinger's equation has solutions characterized by three quantum numbers only, whereas electron spin appears naturally as a solution of Dirac's relativistic equation. As a consequence it is often stated that spin is a relativistic effect. However, the fact that half-integral angular momentum states, predicted by the ladder-operator method, are compatible with non-relativistic systems, refutes this conclusion. The non-appearance of electron spin as a solution of Schrödinger's equation shows that in some sense the equation is incomplete, and it is important to identify this defect.

The Schrödinger equation for a free electron consists of a time and a space part

$$i\hbar\frac{\partial\Phi}{\partial t} = -\frac{\hbar^2}{2m}\nabla^2\Phi$$

or rearranged into

$$\left(2mi\hbar\frac{\partial}{\partial t} + \hbar^2\nabla^2\right)\Phi = 0$$

it appears as a combination of one linear and one squared operator, $\hat{E}$ and $\hat{p}^2$, where

$$\hat{E} \leftarrow i\hbar\frac{\partial}{\partial t} \quad \text{and} \quad \hat{p} \leftarrow -i\hbar\nabla$$

For wave functions like $\Phi = \Phi' \exp[if(x,t)]$, the squared operator would mask the phase information, since $\Phi^*\Phi = |\Phi'|^2$, and to avoid this, a linear Schrödinger operator would be preferred. This has the immediate advantage of a wave equation which is linear in both space and time derivatives. The most general equation with the required form is

$$S = A\hat{E} + \mathbf{B}\cdot\hat{\mathbf{p}} + C = 0$$

such that $S^2 = 2m\hat{E} - \hat{p}^2$, *i.e.*

$$(A\hat{E} + \mathbf{B}\cdot\hat{\mathbf{p}} + C)^2 = A^2(\hat{E}^2) + 2AB(\hat{E}\hat{p}) + 2AC(\hat{E}) + 2BC(\hat{p}) + B^2(\hat{p}^2) + C^2$$

To match the Schrödinger formulation, $(2m\hat{E} - \hat{p}^2)$, this requires that

$$A^2 = AB = BC = C^2 = 0$$

$$2AC = 2m$$

$$B^2 = \pm 1$$

These conditions are clearly impossible if A, B and C are complex numbers, but could be satisfied by square matrices.

To establish the nature of the coefficient matrices, the square operator is formed with $\mathbf{B}\cdot\mathbf{p}$ in component form, as

$$S^2 = (A\hat{E} + B_ip_i + C)(A\hat{E} + B_jp_j + C)$$

$$= A^2\hat{E}^2 + (AB_i + B_iA)p_i + (AC + CA) + \frac{1}{2}(B_iB_j + B_jB_i)p_ip_j$$

The restrictions on the matrices can now be recognized more precisely from the conditions:

$$\begin{array}{ll} A^2 = O & AB_i + B_iA = O \\ AC + CA = 2m & B_iB_j + B_jB_i = 2\delta_{ij} \\ C^2 = O & B_iC + CB_i = O \end{array}$$

If A, C and B_i are square matrices, the zeros represent $n \times n$ null matrices and the wave function on which S operates must be an n-component column vector, $\Phi = (\varphi_1, \varphi_2, \cdots \varphi_n)$.

These conditions define a matrix algebra which requires at least four anti-commuting, traceless (hence even-dimensioned) matrices. The smallest even dimension, $n = 2$, can only accommodate three anti-commuting matrices, the Pauli matrices:

$$\sigma_1 = \begin{pmatrix} 0 & 1 \\ 1 & 0 \end{pmatrix} \quad \sigma_2 = \begin{pmatrix} 0 & -i \\ i & 0 \end{pmatrix} \quad \sigma_3 = \begin{pmatrix} 1 & 0 \\ 0 & -1 \end{pmatrix}$$

Each pair has an anti-commutator, $\{\sigma_j, \sigma_k\} = \sigma_j\sigma_k + \sigma_k\sigma_j = 2\delta_{jk}$. Based on this the required representation of the B_i matrices follows immediately as

$$B_i = \begin{pmatrix} \sigma_i & O \\ O & \sigma_i \end{pmatrix}$$

where O is the 2×2 null matrix. Then, $B_iB_j + B_jB_i = 2\delta_{ij}I_4$, as required. For A and C, with A^2 and C^2 null, a representation of the form

$$A = \begin{pmatrix} O & O \\ I & O \end{pmatrix} \quad C = \begin{pmatrix} O & 2mI \\ O & O \end{pmatrix}$$

where I is the 2×2 unit matrix, satisfies the requirement, i.e.

$$AC + CA = 2mI_4$$

It is readily demonstrated that the anti-commutators

$$\{C, B_i\} \text{ and } \{A, B_i\} = O$$

In line with the definition of the B_i, the four-component wave function is written in the form of two two-dimensional objects

$$\Phi = \begin{pmatrix} \psi \\ \chi \end{pmatrix} = \{\phi_1, \phi_2, \phi_3, \phi_4\}$$

to produce the Schrödinger equation

$$E\begin{pmatrix} O & O \\ I & O \end{pmatrix}\begin{pmatrix} \psi \\ \chi \end{pmatrix} + \begin{pmatrix} \sigma_i & O \\ O & \sigma_i \end{pmatrix} p_i \begin{pmatrix} \psi \\ \chi \end{pmatrix} + \begin{pmatrix} O & 2mI \\ O & O \end{pmatrix}\begin{pmatrix} \psi \\ \chi \end{pmatrix} = 0$$

i.e.

$$E\begin{pmatrix} O \\ \psi \end{pmatrix} + (\sigma \cdot p)\begin{pmatrix} \psi \\ \chi \end{pmatrix} + 2m\begin{pmatrix} \chi \\ O \end{pmatrix} = 0$$

which is

$$(\sigma \cdot p)\psi + 2m\chi = 0$$

$$E\psi + (\sigma \cdot p)\chi = 0$$

This shows that ψ and χ are not linearly independent,

$$\chi = -\frac{(\sigma \cdot p)\psi}{2m}$$

whereby

$$E\psi = \frac{(\sigma \cdot p)^2\psi}{2m}$$

Since $\sigma^2 = 1$, this is simply the Schrödinger equation

$$(2mE - p^2)\psi = 0$$

$$\implies \left(2mi\hbar\frac{\partial}{\partial t} + \hbar^2\nabla^2\right)\psi = 0$$

This may well appear not to produce anything new until the electron is examined in an external electromagnetic field, represented by a scalar potential V, and a vector potential $\mathbf{A}$. The appropriate operators then become

$$E \leftarrow (\hbar i\frac{\partial}{\partial t} - V)$$

$$p \leftarrow (-\hbar i\nabla - e\mathbf{A}/c$$

These are substituted into $[2mE - (\sigma \cdot p)^2]\psi = 0$. *i.e.*

$$2m(\hbar i\frac{\partial}{\partial t} - V)\psi = \left[\sigma\left(-\hbar i\nabla - \frac{e\mathbf{A}}{c}\right)\right]\left[\sigma\left(-\hbar i\nabla - \frac{e\mathbf{A}}{c}\right)\right]\psi$$

$$= \left(-\sigma hi\nabla - \frac{e(\sigma \cdot \mathbf{A})}{c}\right)\left(-\sigma hi\nabla - \frac{e(\sigma \cdot \mathbf{A})}{c}\right)\psi$$

$$= -\hbar^2\nabla^2\psi + \frac{\sigma e\hbar i\psi}{c}\nabla(\sigma \cdot \mathbf{A}) + \frac{2e\mathbf{A}\hbar i}{c}\nabla\psi + \frac{e^2\mathbf{A}^2}{c^2}\psi$$

$$= -\left[\left(\hbar\nabla - \frac{ie\mathbf{A}}{c}\right)^2 \psi - \frac{\hbar i\sigma e\psi}{c}\nabla(\sigma\cdot\mathbf{A})\right]$$

It follows from vector identities that

$$\nabla(i\sigma\cdot\mathbf{A}) = \nabla\times\mathbf{A} = \mathrm{curl}\mathbf{A} = \mathbf{B} \ ,$$

the magnetic field strength. The equation therefore reduces to

$$i\hbar\frac{\partial\psi}{\partial t} = -\frac{\hbar^2}{2m}\left(\nabla - \frac{ie\mathbf{A}}{\hbar c}\right)^2\psi + \frac{\hbar e}{2mc}(\sigma\cdot\mathbf{B})\psi + V\psi$$

which is the Pauli equation, first obtained empirically by the addition of an extra operator to the Schrödinger Hamiltonian to account for electron spin.

The importance of this equation lies in the term that reflects an intrinsic magnetic moment

$$\mu = \frac{\hbar e}{2mc}\sigma$$

which is exactly as required to explain the fine structure of electronic spectra in terms of a spin angular momentum of $\sigma/2$ and a Landé factor of 2.

The matrix representation of the spin operator requires the spin state of a particle to be represented by row vectors, commonly interpreted as spin 'up' or 'down'. An arbitrary state function Ψ must be represented as a superposition of spin up and spin down states

$$|\uparrow\rangle = \begin{pmatrix}1\\0\end{pmatrix} \quad , \quad |\downarrow\rangle = \begin{pmatrix}0\\1\end{pmatrix}$$

$$\Psi = \psi_+ \ |\uparrow\rangle + \psi_- \ |\downarrow\rangle = \begin{pmatrix}\psi_+\\ \psi_-\end{pmatrix}$$

This two-component wave function is known as a spinor.

5.4.4 Addition of Angular Momenta

The total angular momentum of a quantum particle consists of orbital and spin components. Only orbital angular momentum

$$\boldsymbol{L} = \boldsymbol{r}\times\boldsymbol{p}$$

is a function of ordinary space coordinates, whereas spin represents intrinsic angular momentum, with commutation relations formally identical with those obeyed by the orbital components. All components of spin therefore commute with those of $\boldsymbol{r}$ and $\boldsymbol{p}$, and hence with all those of $\boldsymbol{L}$. The total angular mometum

$$\boldsymbol{J} = \boldsymbol{L} + \boldsymbol{S}$$

satisfies commutation relations

$$[J_x, J_y] = i\hbar J_z \quad , \quad [J_y, J_z] = i\hbar J_x \quad , \quad [J_z, J_x] = i\hbar J_y$$

with J_x, J_y, J_z defined as linear self-adjoint operators. The eigenvalues of $\boldsymbol{J}$ and J_z can be obtained through ladder operators and shown to be of the form $j(j+1)\hbar^2$ ad $m\hbar$. Since $\boldsymbol{J}$ is the sum of orbital and spin angular momenta, both integral and half-integral quantum numbers are allowed, ($j = 0, 1/2, 1, 3/2, \ldots$) and $m_j = -j, \ldots, j$. For a given value of l and s the possible values of j are given by $|l - s|$, $|l - s| + 1$, $\ldots, l + s$. For a given j the quantum number m_j can take on $2j + 1$ values between $-j$ and j.

For systems of two or more particles orbital and spin angular momenta are added separately by the rules of vector addition. Integral projections of the shorter vector on the direction of the longer vector, are added to the long vector, to give the possible eigenvalues of total angular momentum.

In the case of many-electron atoms the total electronic orbital angular momentum is defined as the vector sum over n electrons

$$\boldsymbol{L} = \sum_{i=1}^{n} \boldsymbol{L}_i$$

and the spin

$$\boldsymbol{S} = \sum_{i=1}^{n} \boldsymbol{S}_i$$

The orbital angular momentum state is characterized by the quantum number L where $L(L+1)\hbar^2$ is the square of the magnitude of the total electronic orbital angular momentum

In an atomic *term symbol* the value of $L = 1, 2, \ldots,$*etc.* is coded according to spectroscopic notation as $S, P, D, F, G, \ldots$ *etc.* (alphabetically for $L \geq 3$). Likewise, each atomic state can be characterized by a total electronic spin quantum number S,

$$\hat{S}^2\psi = S(S+1)\hbar^2\psi$$

The numerical value of the quantity $2S + 1$ (total multiplicity) is used as a left superscript with the corresponding term code to define the total atomic term symbol, *e.g.* for $L = 2$ and $S = 1$, the term symbol is ^3D. Closed subshells make no contribution to L and S and may be ignored.

To derive which of the terms arising from a given electron configuration is lowest in energy *Hund's empirical rule* states that the term with the largest value of S lies lowest; if there is more than one term with the largest S, then, from these, the term with largest L lies lowest.

Hund's rule works well for the ground-state configuration, but occasionally fails for an excited configuration. For partly filled subshells Hund's rule dictates that electrons are assigned to orbitals so as to give the highest number of parallel spins. Based on this, the traditional explanation of Hund's rule is that electrons with the same spin tend to keep apart and so minimize the Coulombic repulsion between them. The term with the greatest number of parallel spins (highest S) will therefore be lowest in energy. This explanation has never been substantiated by detailed calculation and the rule remains an empirical one.

5.4.5 Exclusion Principle

To understand how electrons with spin interact, it is useful to examine a system consisting of two electrons, such as the helium atom. Let this two-electron system be described by the wave function, in space coordinates, $\Phi(r)$, If the electrons are interchanged the wave function will in general be different, (Φ'), but since the electrons are identical (ignoring spin) the energy of the system will not be affected. The wave functions therefore belong to degenerate levels.

Suppose that the effect of interchanging of electrons is correctly described by some operator, π, *i.e.* $\pi\Phi = \Phi'$. By repeating the operation the situation is reversed and the system returns to its initial state, *i.e.*

$$\pi\Phi' = \pi^2\Phi = \Phi$$

which shows that $\pi = \pm 1$. The wave function is said to be either symmetrical $(\pi = 1)$ or anti-symmetrical with respect to the interchange of electron coordinates.

Now let $\Phi_a(r_1)$ and $\Phi_b(r_2)$ be one-electron wave functions for electrons a and b at coordinates r_1 and r_2 respectively. The combined function can then be formulated in terms of the product functions

$$\psi_1 = \Phi_a(r_1)\Phi_b(r_2) \quad \text{and} \quad \phi_2 = \Phi_a(r_2)\Phi_b(r_1)$$

or a linear combination of these. It is noted that the functions

$$\psi_\pm = \frac{1}{\sqrt{2}}(\psi_1 \pm \psi_2)$$

correctly describe wave functions that are either symmetrical or anti-symmetrical with respect to electron interchange. The factor $1/\sqrt{2}$ ensures normalization.

The physical meaning of the linear combination is that individual electrons are not necessarily in either state Φ_a or Φ_b and that interference between ψ_1 and ψ_2 has an important influence on the physical properties of the system.

For two identical particles ($\Phi_a = \Phi_b$) it is noted that ψ_+ approaches a maximum at $r_1 = r_2$, whereas ψ_- approaches zero. This means that two electrons in the same state would tend to stay together if the wave function is symmetrical and to avoid each other when the wave function is anti-symmetrical under exchange. It also follows that the function $\Phi_a\Phi_b$ is automatically symmetrical and that the anti-symmetric function vanishes identically.

To arrive at the correct formulation of the ground state of the helium atom it is necessary to also take into account the effect of spin, represented by the functions α and β. There are four possibilities, according to the electrons having the same spin, either up or down:

$$\alpha(1)\alpha(2) \quad \text{or} \quad \beta(1)\beta(2)$$

and states with the spins opposed:

$$-\frac{1}{\sqrt{2}}\{\alpha(1)\beta(2) \pm \beta(1)\alpha(2)\}$$

The linear combination is used instead of the unsymmetrical states $\beta(1)\alpha(2)$ and $\beta(2)\alpha(1)$. It is reasonable to expect that each of these spin states could occur in combination with the ground-state function $\psi_\pm(r)$ to yield four different levels at the ground state. However, for the helium atom only one ground-state function can be identified experimentally and it is significant to note that only one of the spin functions is anti-symmetrical, *i.e.*

$$\psi_-(s) = \frac{1}{\sqrt{2}}\{\alpha(1)\beta(2) - \beta(1)\alpha(2)\}$$

Since $\psi_+(r)$ is symmetrical, this suggests that the total wave function $\psi = \psi_+(r)\psi_-(s)$, and anti-symmetrical.

There is no theoretical ground for this conclusion, which is a purely empirical result based on a variety of experimental measurements. However, it seems to apply everywhere and to represent a law of Nature, stating that systems consisting of more than one particle of half-integral spin are always represented by anti-symmetric wave functions. It is noted that if the space function is symmetrical, the spin function must be anti-symmetrical to give an anti-symmetrical product. When each of the three symmetrical states is combined with the anti-symmetrical space function this produces what is known as a triplet state, in which the two spins are parallel. Opposite spins produce the singlet state described above.

For an N-electron system anti-symmetrical wave functions are conveniently represented by determinants. If $w_i(x_i)$ represents an electronic wave function with space and spin components, a typical wave function for an N-electron system can be formulated as

$$\psi = w_1(x_1)w_2(x_2)\dots w_N(x_N).$$

This wave function is degenerate in the sense that the same energy is obtained whenever two particles are interchanged. A total of $N!$ different wave functions can be obtained in this way. Since the total wave function must be anti-symmetrical in the exchange of any two particles, it is correctly represented by the determinant

$$\psi = \begin{vmatrix} w_1(x_1) & w_2(x_1) & \cdots & w_N(x_1) \\ w_1(x_2) & w_2(x_2) & \cdots & w_N(x_2) \\ \dots & \dots & \dots & \dots \\ w_1(x_N) & w_2(x_N) & \cdots & w_N(x_N) \end{vmatrix}$$

which is a linear combination of the degenerate functions.

Any determinant changes sign when any two columns are interchanged. Moreover, no two of the product functions (columns) can be the same since that would cause the determinant to vanish. Thus, in all nonvanishing completely anti-symmetric wave functions, each electron must be in a different quantum state. This result is known as Pauli's exclusion principle, which states that no two electrons in a many-electron system can have all quantum numbers the same. In the case of atoms it is noted that since there are only two quantum states of the spin, no more than two electrons can have the same set of orbital quantum numbers.

5.5 Quantum Mechanics of the Photon

5.5.1 Introduction

This topic is considered in detail by Akhiezer and Berestetskii [65] according to the Schrödinger picture and their treatment serves as a guide in the following short summary. An essentially equivalent treatment according to the Heisenberg picture is described by Gottfried[66].

Quantum theory started as an explanation of electromagnetic effects (black-body radiation) in terms of corpuscular properties of light, expressed in terms of the energy quantum

$$w = h\nu = \hbar\omega$$

and the frequency of the electromagnetic field. Subsequent development was mainly by reference to atomic systems, in terms of non-relativistic theory. The photon has zero rest mass and therefore no non-relativistic energy region. The theory of the photon must therefore be relativistic from the beginning. To describe a massive particle, quantum mechanics introduces a field of one or more wave functions which determine the probability distribution and the expectation values of the various physical quantities related to the particle. To turn this description into a relativistic theory it is necessary to ensure that the field equations are invariant under Lorentz transformation. In the case of the electromagnetic field however, the field equations are available as the Maxwell relations, directly in relativistic form.

5.5.2 The Quantized Field

The electromagnetic field in free space is described by the electric field vector **E** and the magnetic field vector **H**, which in the absence of charges satisfy Maxwell's equations

$$\begin{aligned}\nabla \times \mathbf{E} &= -\mu_0 \frac{\partial \mathbf{H}}{\partial t} \\ \nabla \cdot \mathbf{H} &= 0 \\ \nabla \times \mathbf{H} &= \epsilon_0 \frac{\partial \mathbf{E}}{\partial t}\end{aligned}$$

$$\nabla \cdot \mathbf{E} = 0 \tag{5.36}$$

It is now assumed that the vectors **E** and **H** describe the quantum mechanical state of a photon, or quantum of light, as the counterpart of the particle in atomic systems. This assumption is conveniently formulated in terms the Fourier transforms of the field equations

$$\mathbf{E} = \int \mathbf{E}_{\mathbf{k}} e^{i\mathbf{k}\cdot\mathbf{r}} d\mathbf{k}$$

$$\mathbf{H} = \int \mathbf{H}_{\mathbf{k}} e^{i\mathbf{k}\cdot\mathbf{r}} d\mathbf{k} \tag{5.37}$$

in **k** space , for comparison with Schrödinger's equation of ordinary quantum mechanics.

To transform the Maxwell equations into **k** space the field is considered as a function of a space coordinate **r** measured along a line whose direction is defined by the unit vector $\mathbf{k}$. The rectangular components k_x, k_y, k_z are clearly the direction cosines of this coordinate. The gradient operator in this notation becomes

$$\nabla = \left(\boldsymbol{i}\frac{\partial}{\partial x} + \boldsymbol{j}\frac{\partial}{\partial y} + \boldsymbol{n}\frac{\partial}{\partial z} \right) = d\mathbf{k}$$

The Maxwell divergence equations in $\mathbf{k}$ space therefore integrate to

$$\begin{aligned} \mathbf{k} \cdot \mathbf{E_k} &= 0 \\ \mathbf{k} \cdot \mathbf{H_k} &= 0 \end{aligned}$$

These relationships show that the Fourier components are perpendicular to the propagating vector $\mathbf{k}$, and consequently the field components can be combined in the complex plane as $\mathbf{E_k} + i\mathbf{H_k}$, and together define the so-called tranversality condition. Equivalent arguments applied to the curl operator produce the remaining Maxwell equations in $\mathbf{k}$ space:

$$\mu_0 \dot{\mathbf{H}}_\mathbf{k} = -\mathbf{k} \times \mathbf{E_k} \tag{5.38}$$

$$\epsilon_0 \dot{\mathbf{E}}_\mathbf{k} = i(\mathbf{k} \times \mathbf{H_k}) \tag{5.39}$$

Although not explicitly shown, the functions $\mathbf{E_k} \equiv \mathbf{E}(\mathbf{k}, t)$ and $\mathbf{H_k} \equiv \mathbf{H}(\mathbf{k}, t)$ are temporal dependent. The description is completed by adding the condition that the fields should be real,

$$\mathbf{E_{-k}} = \mathbf{E_k^\star} \quad , \quad \mathbf{H_{-k}} = \mathbf{H_k^\star} \tag{5.40}$$

Instead of condition (40) a transformation that automatically implies (40) is applied in the form[12]

$$\mathbf{E_k} = N(k)(\mathbf{f_k} + \mathbf{f^\star_{-k}}) \tag{5.41}$$

$$\dot{\mathbf{E}}_k = -ikN(k)(\mathbf{f_k} - \mathbf{f^\star_{-k}})$$

where $N(k)$ is some normalizing factor, for later convenience chosen as

$$N(k) = \sqrt{\frac{k}{2(2\pi)^3}} \tag{5.42}$$

[12]The functions $\boldsymbol{f_k}$ and $\boldsymbol{f_k^\star}$ are the counterparts of the so-called destruction (annihilation) and creation operators in the Heisenberg-Dirac picture. It is noted in anticipation that these operators occur as the solutions $\boldsymbol{a_k}(t) = \boldsymbol{a_k}e^{-i\omega_k t}$ of the Hamiltonian equation

$$i\hbar\frac{d}{dt}\boldsymbol{a_k}(t) = [\boldsymbol{a_k}(t), H] = \hbar\omega_k\boldsymbol{a_k}(t)$$

Instead of using two field vectors one can be eliminated as follows. From (39) follows

$$\begin{aligned}\epsilon_0 \mathbf{k} \times \dot{\mathbf{E}}_{\mathbf{k}} &= i(\mathbf{k} \times \mathbf{k} \times \mathbf{H}_{\mathbf{k}}) \\ &= ik^2 \mathbf{H}_{\mathbf{k}}\end{aligned}$$

i.e.

$$\mathbf{H}_{\mathbf{k}} = i\epsilon_0 \left[\frac{\mathbf{k}}{k^2} \times \dot{\mathbf{E}}_{\mathbf{k}}\right] \tag{5.43}$$

Therefore

$$\dot{\mathbf{H}}_{\mathbf{k}} = \frac{\partial \mathbf{H}_{\mathbf{k}}}{\partial t} = i\epsilon_0 \left(\frac{\mathbf{k}}{k^2} \times \frac{\partial^2 \mathbf{E}_{\mathbf{k}}}{\partial t^2}\right)$$

But

$$\dot{\mathbf{H}}_{\mathbf{k}} = -\frac{i}{\mu_0} (\mathbf{k} \times \mathbf{E}_{\mathbf{k}})$$

Therefore

$$\epsilon_0 \left(\frac{\mathbf{k}}{k^2} \times \frac{\partial^2 \mathbf{E}_{\mathbf{k}}}{\partial t^2}\right) + \frac{1}{\mu_0} (\mathbf{k} \times \mathbf{E}_{\mathbf{k}}) = 0$$

i.e.

$$\left(\frac{\partial^2}{\partial t^2} + \frac{k^2}{\epsilon_0 \mu_0}\right) \boldsymbol{E}_{\boldsymbol{k}} = 0$$

The product of three constants can conveniently be replaced by a single constant k^2, which amounts to setting $c = 1$. The final expression

$$\left(\frac{\partial^2}{\partial t} + k^2\right) \mathbf{E}_{\mathbf{k}} = 0$$

may be written in the form

$$\left(\frac{\partial}{\partial t} + ik\right)\left(\frac{\partial}{\partial t} - ik\right) \mathbf{E}_k = 0$$

This second-order equation may be transformed into a first-order equation for $\mathbf{f}_{\mathbf{k}}$. By utilizing the relation

$$\left(\frac{\partial}{\partial t} - ik\right) \mathbf{E}_{\mathbf{k}} = -2iN(k)k\mathbf{f}_{\mathbf{k}}$$

which follows from (41), one gets

$$i\frac{\partial \mathbf{f}_{\mathbf{k}}}{\partial t} = k\mathbf{f}_{\mathbf{k}} \tag{5.44}$$

In terms of the new variables the transversality condition becomes

$$\mathbf{k} \cdot \mathbf{f}_{\mathbf{k}} = 0 \tag{5.45}$$

The vector $\mathbf{k}$ is in the direction of propagation and it follows that longitudinal modes do not occur. To combine the two expressions they are labeled separately, as α and β, say. Multiply (45) by the unit vector $\mathbf{k}/k$ and add to (44), *i.e.*

$$i\frac{\partial \mathbf{f}_k}{\partial t} = (k\mathbf{f}_k)_\alpha - \left(\frac{\mathbf{k}}{k} \cdot \mathbf{k} \cdot \mathbf{f}_k\right)_\beta$$

The r.h.s. rearranges to either

$$k\left[\mathbf{f}_k^\alpha - \frac{\mathbf{k}^2}{k_\alpha k_\beta}\mathbf{f}_k^\beta\right]$$

or

$$k\left[\frac{k_\alpha k_\beta}{k^2}\mathbf{f}_k^\alpha - \frac{\mathbf{k}^2}{k^2}\mathbf{f}_k^\beta\right]$$

which is

$$k\left(\boldsymbol{\delta}_{\alpha\beta} - \frac{k_\alpha k_\beta}{k^2}\right)\mathbf{f}_k = w_{\alpha\beta}\mathbf{f}_k^\beta = (w\mathbf{f}_k)_\alpha$$

The equation

$$i\frac{\partial \mathbf{f}_k}{\partial t} = w\mathbf{f}_k \tag{5.46}$$

together with the initial condition (44) is equivalent to the system of Maxwell's equations, and has the same form as Schrödinger's equation in which w is the Hamiltonian operator defined in units[13] for which $\hbar = 1$. The eigenvalues of this operator are equal to k. This result defines the quantum relationship between the energy and the frequency of the photon, provided the Hamiltonian operator introduced here can be identified with the physical operator for the photon energy.

5.5.3 Energy

The energy of the electromagnetic field is given by

$$\bar{w} = \frac{1}{2}\int \left(\epsilon_0\mathbf{E}^2 + \mu_0\mathbf{H}^2\right) d^3\mathbf{r} \tag{5.47}$$

---

[13]With a different choice of the normalization constant the equivalent expression in common use is $\hbar i\dot{\mathbf{f}}_k = w\mathbf{f}_k$, with eigenvalues $w = \hbar k$.

Space integrals of expressions quadratic in the wave function are interpreted as expectation values of the corresponding physical quantities. This interpretation suggests that (47) should be interpreted as the expectation value of the photon energy, which would mean that

$$\bar{w} = \int \mathbf{f}_k^\star \, w \, \mathbf{f}_k \, d\mathbf{k} \tag{5.48}$$

To show that this is the case, consider the expressions (37) substituted into (47):

$$\bar{w} = \frac{1}{2} \int \{\epsilon 0 \mathbf{E}_\mathbf{k} \cdot \mathbf{E}_{\mathbf{k}'} + \mu_0 \mathbf{H}_\mathbf{k} \cdot \mathbf{H}_{\mathbf{k}'}\} \, e^{i(\mathbf{k}+\mathbf{k}')\mathbf{r}} d\mathbf{k} d\mathbf{k}' d\mathbf{r}$$

This integration can be done by using the relation

$$\int e^{i(\mathbf{k}+\mathbf{k}')\mathbf{r}} d\mathbf{r} = (2\pi)^3 \delta(\mathbf{k}+\mathbf{k}')$$

where $\delta(\mathbf{k}+\mathbf{k}')$ is a three-dimensional delta function. Then, using (43) and again setting $\epsilon_0 \mu_0 = 1$

$$\bar{w} = 4\pi^3 \int \left(\mathbf{E}_\mathbf{k} \cdot \mathbf{E}_{-\mathbf{k}} + \frac{1}{k^2} \dot{\mathbf{E}}_\mathbf{k} \cdot \dot{\mathbf{E}}_{-\mathbf{k}} \right) d\mathbf{k}$$

Now express $\mathbf{E}$ and $\dot{\mathbf{E}}$ in terms of $\mathbf{f}_\mathbf{k}$ according to (41), to yield

$$\bar{w} = 16\pi^3 \int N^2 \mathbf{f}_\mathbf{k} \cdot {\mathbf{f}_\mathbf{k}}^\star d\mathbf{k}$$

Now, choosing the value for $N(k)$ according to (42), equation (48) is obtained. Next consider a monochromatic solution of (46)

$$\mathbf{f}_\mathbf{k} \equiv \mathbf{f}(\mathbf{k}, t) = \mathbf{f}_o(\mathbf{k}) e^{-i\omega t} \tag{5.49}$$

where ω is the eigenvalue of the photon energy operator w. It is known from Fourier theory that $\mathbf{f}_\mathbf{o}(\mathbf{k})$ will always be zero, unless $k = \omega$, when the expression for the energy becomes

$$\bar{w} = \omega \int \mathbf{f}_\mathbf{k} \cdot {\mathbf{f}_{-\mathbf{k}}}^\star d\mathbf{k}$$

By stipulating that this expression be identical with the quantum relationship [14] $\bar{w} = \omega$ the normalization condition of the photon wave function is obtained as

$$\int \mathbf{f}_\mathbf{k} \cdot {\mathbf{f}_\mathbf{k}}^\star d\mathbf{k} = 1 \tag{5.50}$$

[14]The quantum condition $\hbar\omega = \bar{w}$ reduces to this expression by setting $\hbar = 1$.

Thus, if the normalization condition is satisfied, the energy of the electromagnetic field becomes identical with the photon energy.

5.5.4 Momentum

The momentum of the electromagnetic field is given by

$$\bar{\mathbf{p}} = \epsilon_0\mu_0 \int (\mathbf{E}\times\mathbf{H})d^3\mathbf{r} \tag{5.51}$$

It can be shown that when the normalization condition for the photon wave function (50) is satisfied, the vector $\bar{\mathbf{p}}$ can be interpreted as the expectation value of the photon momentum. To do this it is necessary to express $\bar{\mathbf{p}}$ in terms of $\mathbf{f_k}$. Substitution of (37) into (51) yields

$$\bar{\mathbf{p}} = \epsilon_0\mu_0 \int (\mathbf{E_k}\times\mathbf{H_k})\, e^{i(\mathbf{k}+\mathbf{k}')\mathbf{r}} d\mathbf{k}d\mathbf{k}'d\mathbf{r}$$

$$= (2\pi)^3\epsilon_0\mu_0 \int (\mathbf{E_k}\times\mathbf{H_{-k}})\, d\mathbf{k} = -(2\pi)^3 i \int \frac{\mathbf{k}}{k^2}\left(\mathbf{E_k}\times\dot{\mathbf{E}}_{-\mathbf{k}}\right) d\mathbf{k}$$

In the final step the constant $\epsilon_0\mu_0$ has been absorbed into k^2 as before. By expressing $\mathbf{E_k}$ and $\dot{\mathbf{E}}_{-\mathbf{k}}$ in terms of $\mathbf{f_k}$, according to (41)

$$\bar{\mathbf{p}} = (2\pi)^3 \int N^2 \left(\mathbf{f}^\star_\mathbf{k}\cdot\mathbf{f_k} - \mathbf{f}^\star_{-\mathbf{k}}\cdot\mathbf{f}_{-\mathbf{k}} + \mathbf{f}^\star_{-\mathbf{k}}\cdot\mathbf{f}^\star_\mathbf{k} - \mathbf{f_k}\cdot\mathbf{f}_{-\mathbf{k}}\right)\frac{\mathbf{k}}{k}d\mathbf{k}$$

On replacing $\mathbf{k}$ by $-\mathbf{k}$ in the integrand, it is noted that the integral of the last two terms vanishes, while the integrals of the first two terms are identical. By using the definition of $N(k)$ according to (42) one finally gets

$$\bar{\mathbf{p}} = \int f^\star_{k\alpha}\mathbf{k} f_{k\alpha} d\mathbf{k} \tag{5.52}$$

where the subscript $\alpha = 1,2,3$ denotes the components of the vector $\mathbf{f_k}$ and summation over α is implied. It follows that the operator which defines multiplication by the propagating vector $\mathbf{p} = \mathbf{k}$, identifies with the photon momentum operator, and $\mathbf{k}$-space with momentum space. It might appear that by subjecting the function $\mathbf{f_k}$ to the inverse Fourier transformation

$$\frac{1}{(2\pi)^3}\int \mathbf{f_k} e^{i\mathbf{k}\cdot\mathbf{r}} d\mathbf{k} = \mathbf{F}$$

it should be possible to obtain the photon wave function $\mathbf{F}$ in the coordinate representation, normalized in the usual manner

$$\int \mathbf{F}^\star\cdot\mathbf{F}d\mathbf{r} = 1.$$

However, the quantity $\mathbf{F}^{\star} \cdot \mathbf{F}$ will not have the meaning of the probability density of finding the photon at a given point of space. Indeed, the presence of a photon can be established only as a result of its interaction with charges. This interaction is determined by the values of the electromagnetic field vectors $\mathbf{E}$ and $\mathbf{H}$ at the given point, but the latter are not determined by the value of the wave function $\mathbf{F}$ at that point, but by its value over all space. The localization of a photon in a region smaller than a wavelength has no meaning, and the concept of probability density for the localization of a photon does not exist.

5.5.5 Polarization

The time-dependence of the photon wave equation in momentum space

$$\mathbf{f}(\mathbf{k},t) = \mathbf{f}_o(\mathbf{k})e^{ikt}$$

defines a time-independent function $\mathbf{f}_o(\mathbf{k})$ which is restricted by the tranversality condition (46). By treating $\mathbf{k}$-space as discrete, the photon state $\mathbf{f}_o(\mathbf{k})$ differs from zero only at point $\mathbf{p}$ which is a state of definite momentum and the corresponding wave function will be an eigenfunction of the momentum operator $\mathbf{p}$. The eigenfunction has the form

$$\mathbf{f}(\mathbf{k},t) = i\mathbf{e}N_f\delta_{\mathbf{k},\mathbf{p}}e^{i\omega t}$$

where $\omega = |\mathbf{p}|$, N_f is a normalizing function and $\mathbf{e}$ is the photon polarization vector which has absolute value of unity and is perpendicular to $\mathbf{p}$: $|\mathbf{e}|^2 = 1, \quad \mathbf{e}\cdot\mathbf{p} = 0$. [The phase factor i is introduced for convenience].

For a given $\mathbf{p}$ two linearly independent vectors $\mathbf{e}$ are possible. If the z-axis is taken to be directed along $\mathbf{p}$, these two vectors can be defined in terms of the unit vectors $\boldsymbol{\chi}_1$ and $\boldsymbol{\chi}_2$, along the x and y-axes respectively.

$$\chi_{1x} = 1, \quad \chi_{1y} = \chi_{1z} = 0$$
$$\chi_{2y} = 1, \quad \chi_{2x} = \chi_{2z} = 0 \tag{5.53}$$
$$\boldsymbol{\chi}_1 \cdot \boldsymbol{\chi}_2 = 0$$

If $\mathbf{e} = \boldsymbol{\chi}_\mu$, $(\mu = 1, 2)$ the photon is said to be linearly polarized. One can also choose the following two mutually perpendicular unit vectors $\boldsymbol{\chi}'_1$ and $\boldsymbol{\chi}'_2$:

$$\chi'_{1x} = \frac{1}{\sqrt{2}}, \quad \chi'_{1y} = \frac{i}{\sqrt{2}}, \quad \chi'_{1z} = 0$$
$$\chi'_{2x} = \frac{1}{\sqrt{2}}, \quad \chi'_{2y} = -\frac{i}{\sqrt{2}}, \quad \chi'_{2z} = 0 \tag{5.54}$$

$$\chi_1'^{*} \cdot \chi_2' = 0$$

In this case, if $\mathbf{e} = \chi'_\mu$, the photons will be circularly polarized[15]. It follows that photon states of definite momentum are two-fold degenerate. In order to specify a state uniquely it is necessary to specify its type of polarization. The function $\mathbf{f}(\mathbf{k}, t)$ should therefore be provided with two quantum numbers $\mathbf{p}$ and $\mu(\mu = 1, 2)$. The quantities p_x, p_y, p_z and μ form a complete set of quantum numbers for the photon. (of course, the energy is thereby also determined, since $\omega = p$).

It is also possible to represent the electric and magnetic fields corresponding to a photon state of definite momentum $\mathbf{p}$ and polarization μ. The expressions are[16]

$$\mathcal{E}_{\mathbf{p}\mu} = i\sqrt{\frac{\omega}{2\Omega}}\,\chi_\mu e^{i(\mathbf{p}\cdot\mathbf{r}-\omega t)}$$

$$\mathcal{H}_{\mathbf{p}\mu} = i\sqrt{\frac{\omega}{2\Omega}}\left[\frac{\mathbf{p}}{p}\cdot\chi_\mu\right] e^{i(\mathbf{p}\cdot\mathbf{r}-\omega t)} \tag{5.55}$$

The photon state for a given momentum is characterized by the polarization unit vector $\mathbf{e}$. On expanding $\mathbf{e}$ in terms of the unit vectors (53) and (54)

$$\mathbf{e} = e_1\chi_1 + e_2\chi_2 \tag{5.56}$$

The state of polarization is determined by the pair of complex numbers e_1 and e_2; the quantities $|e_1|^2$ and $|e_2|^2$ represent probability densities of a definite (linear or circular) polarization of the photon as determined by the unit vectors χ_1 and χ_2. Since e_1 and e_2 are related by the normalization condition

$$|e_1|^2 + |e_2|^2 = 1$$

and since, morover, the common phase factor of $\mathbf{e}$ is arbitrary, the polarization is determined by two real parameters and the expansion can be written as

$$\mathbf{e} = \chi_1 \cos\alpha + (\chi_2 \sin\alpha)e^{i\beta} \tag{5.57}$$

With χ_1 and χ_2 as defined before, $\beta = 0$ denotes linear polarization at an angle α to the x-axis; $\beta = \pi/2$ and $\alpha = \pi/4$ denote circular polarization, and arbitrary α and β correspond to elliptical polarization.

---

[15] χ_1' refers to right circular and χ_2' to left circular polarization.
[16] $\mathbf{k}$-space is here assumed to be divided into cells of volume $(2\pi)^3/\Omega$.

Partial Polarization

There are situations in which a definite wave function cannot be ascribed to a photon and hence cannot quantum-mechanically be described completely. One example is a photon that has previously been scattered by an electron. A wave function exists only for the combined electron-photon system whose expansion in terms of the free photon wave functions contains the electron wave functions. The simplest case is where the photon has a definite momentum, *i.e.* there exists a wave function, but the polarization state cannot be specified definitely, since the coefficients e_μ depend on parameters characterizing the other system. Such a photon state is referred to as a state of partial polarization. It can be described in terms of a density matrix

$$\rho = \frac{1}{2}\begin{pmatrix} 1+\xi_3 & \xi_1 - i\xi_2 \\ \xi_1 + i\xi_2 & 1-\xi_3 \end{pmatrix}$$

or

$$\rho = \frac{1}{2}\left(\mathbf{I} + \sum_{j=1}^{3} \xi_j \tau_j\right)$$

where $\mathbf{I}$ is the unit matrix and τ_j are the two-dimensional Pauli matrices. The quantities τ, known as the Stokes parameters can be determined experimentally by measuring polarization at various angles and the degree of polarization determined as a linear combination of a completely unpolarized and a completely polarized state.

Angular Momentum and Spin

To express the angular momentum $\bar{\mathbf{M}}$ of the electromagnetic field corresponding to one photon in terms of the wave function for the photon, $\bar{\mathbf{M}}$ is identified with the expectation value of the angular momentum in the state $\mathbf{f}_\mathbf{k}$, *i.e.*

$$\bar{\mathbf{M}} = \int [\mathbf{r} \times (\mathbf{E} \times \mathbf{H})] d\mathbf{r}$$

By substituting the Fourier expansions of the field into this expression

$$\bar{\mathbf{M}} = \int \left[\mathbf{r} \times (\mathbf{E}_\mathbf{k} \times \mathbf{H}_{\mathbf{k}'}) e^{i(\mathbf{k}+\mathbf{k}')\cdot\mathbf{r}}\right] d\mathbf{k} d\mathbf{k}' d\mathbf{r}$$

The long but straightforward integration[65] produces the expression for the angular momentum in terms of a new vector operator, $\mathbf{s}$, in the form

$$\bar{\mathbf{M}} = \int f_\alpha^\star \{-i(\mathbf{k} \times \nabla) + \mathbf{s}\} f_\alpha d\mathbf{k}$$

This expression has the structure of quantum-mechanical expectation values, defining the operator for photon angular momentum as

$$\hat{M} = -i(\mathbf{k} \times \nabla) + \mathbf{s} \tag{5.58}$$

This operator can now be shown to be identical with the operator for an infinitesimal rotation of the vector field multiplied by i, *i.e.* $\mathbf{J} = -i\mathbf{M}$. The components of the angular momentum operator satisfy the commutation relations

$$M_x M_y - M_y M_x = iM_z$$

$$M_\alpha \mathbf{M}^2 - \mathbf{M}^2 M_\alpha = 0$$

which hold for all quantum systems. The eigenvalues of $\mathbf{M}^2$ are equal to

$$\mathbf{M}^2 = j(j+1)$$

where $2j+1$ is a positive integer. For a photon it can be shown that j is also an integer. The operator $\mathbf{M}$ commutes with the energy and hence photon states with definite values of w, $\mathbf{M}^2$ and M_z are possible (quantum numbers ω, j, M).

Formula (58) shows that the angular momentum operator for the photon consists of two terms. The first term is identical with the usual quantum-mechanical operator $\hat{L}$ for the orbital angular momentum in the momentum presentation

$$\hat{L} = -i(\mathbf{k} \times \nabla_{\mathbf{k}}).$$

The second term $\mathbf{s}$ may be called the operator for spin angular momentum of the photon. However, the separation of the angular momentum of the photon into an orbital and a spin part has restricted physical meaning. Firstly, the usual definition of spin as the angular momentum of a particle at rest is inapplicable to the photon since its rest mass is zero. More importantly, it will be seen that states with definite values of orbital and spin angular momenta do not satisfy the condition of transversality.

The eigenfunctions of the $\hat{L}$ operators will be denoted by the spherical harmonics $Y_{\ell m}$.

$$\mathbf{L}^2 Y_{\ell m} = \ell(\ell+1) Y_{\ell m}$$

$$\mathbf{L} Y_{\ell m} = m Y_{\ell m}$$

The eigenfunction of the $\mathbf{s}$ operator is $\boldsymbol{\chi}_{s\mu}$ with eigenvalues of $\mathbf{s}^2$ equal to $s(s+1)$, and of s_z equal to μ. The quantity s is found to assume only the value $s = 1$. The magnitude μ of the spin component may assume the three values $\mu = 0, \quad \pm 1$.

The functions $\boldsymbol{\chi}_\mu$ satisfy the following equations:

$$\mathbf{s}^2\chi_\mu = 2\chi_\mu$$

$$s_z\chi_\mu = \mu\chi_\mu$$

with solutions

$$\chi_o = \begin{pmatrix} 0 \\ 0 \\ 1 \end{pmatrix} \quad ; \quad \chi_1 = \frac{1}{\sqrt{2}}\begin{pmatrix} 1 \\ i \\ 0 \end{pmatrix} \quad , \quad \chi_{-1} = \frac{1}{\sqrt{2}}\begin{pmatrix} 1 \\ -i \\ 0 \end{pmatrix}$$

The unit vectors $\boldsymbol{\chi}_\mu$ provide three basic unit vectors along which any arbitrary vector $\mathbf{f}$ may be decomposed:

$$\mathbf{f} = \sum_{\mu=-1}^{1} f^\mu \boldsymbol{\chi}_\mu$$

The f^μ are referred to as the contravariant components of $\mathbf{f}$ along these basic unit vectors. It is found that

$$f^o = f_z$$

$$f^{\pm 1} = \mp\frac{1}{\sqrt{2}}(f_x \mp i f_y)$$

Since the spin operator commutes with the momentum operator, it is possible to speak of states of definite momentum $\mathbf{p}$ and spin component μ. The components of the polarization vector may be chosen in such a way that $\mathbf{e} = \boldsymbol{\chi}\mu$. The two possible polarizations correspond to only two values of the component of spin angular momentum μ. The third value is excluded by the condition of tranversality. If the z-axis is directed along $\mathbf{p}$, then $\boldsymbol{\chi}_o$ is excluded. The two vectors $\boldsymbol{\chi}'_1$ and $\boldsymbol{\chi}'_2$, corresponding to circular polarization are equivalent, respectively to $\boldsymbol{\chi}_1$ and $\boldsymbol{\chi}_{-1}$. Thus, the value[17] of the spin component $\mu = 1$ corresponds to right circular polarization, while $\mu = -1$ corresponds to left circular polarization.

The eigenfunctions of the M operators are the physically more meaningful quantities. They are denoted by$\mathbf{Y}_{jM}$ and called vector spherical harmonics. In terms of the spin variable $\alpha(= x, y, z)$, and written as functions of the unit vector $\mathbf{n} = \mathbf{k}/k$,

$$[\mathbf{Y}_{jM}(\mathbf{n})]_\alpha = Y_{jM}(\mathbf{n}, \alpha).$$

[17]In macroscopic units the spin of a photon is restricted to values of $\pm\hbar$.

These functions satisfy the equations

$$\mathbf{M}^2 Y_{jM}(\mathbf{n},\alpha) = j(j+1)Y_{jM}(\mathbf{n},\alpha)$$

$$M_z Y_{jM}(\mathbf{n},\alpha) = \mathbf{M}Y_{jM}(\mathbf{n},\alpha)$$

Instead of solving these equations they can be analyzed according to the known rules for addition of angular momenta, applied to the orbital and spin degrees of freedom with the variables $\mathbf{k}$ and α. In general there are three different wave functions Y_{jM} corresponding to three orbital states, $Y_{j\ell M}$.

The vector spherical harmonics $\mathbf{Y}_{j\ell M}$ form an orthogonal system. The state of the photon with definite values of j and M is described by a wave function which in general is a linear combination of three vector spherical harmonics

$$\mathbf{f}_{jM} = \sum_{\ell=j-1}^{j+1} a_\ell \mathbf{Y}_{j\ell M}$$

The coefficients of this linear combination are not independent since the photon wave function must satisfy the tranversality condition

$$\mathbf{f}_{jM} \cdot \mathbf{n} = 0$$

Therefore there are two, not three, different photon states with given quantum numbers j, M. Call these wave functions $\mathbf{f}^\lambda_{jM}$, where $\lambda = 0, 1$. The photon wave function can now be written as a linear combination

$$\mathbf{f}_{jM} = a_1 \mathbf{Y}^1_{jM} + a_o \mathbf{Y}^0_{jM}$$

with arbitrary coefficients. This equation means that the states $\mathbf{f}_{jM}$ are two-fold degenerate. Alternatively each state can be assigned a definite parity; *i.e.* it should be an eigenstate of the inversion operator I, defined in the vector field by $I\mathbf{f}(\mathbf{k}) = -\mathbf{f}(-\mathbf{k})$. (The minus sign occurs since under inversion any direction goes over into the opposite one, $I\mathbf{n} = -\mathbf{n}$.)

Since $I^2 = 1$, the operator has two eigenvalues, ± 1. In summary, the photon state may be uniquely specified by giving four quantum numbers to quantify the energy ω, the angular momentum j, the component of angular momentum M and the parity λ. The normalized wave function is of the form

$$\mathbf{f}(\mathbf{k},t) = \frac{i}{\omega\sqrt{\delta}} \mathbf{Y}^\lambda_{jM} e^{-i\omega t} \delta_{k\omega} \tag{5.59}$$

It is not possible to ascribe a definite value of the orbital angular momentum to a photon state since the vector spherical harmonic $\mathbf{Y}^\lambda_{jM}$ may be a function of different values of ℓ. This provides the evidence that, strictly speaking, it is not possible to separate the total angular momentum of the photon into orbital and spin parts.

If $j = 0$ there is only one vector spherical harmonic which is identical with the longitudinal harmonic $\mathbf{Y}_{00}^{-1} = \mathbf{n}Y_{00}$. From this observation it follows that there are no transverse spherical harmonics for $j = 0$. It also means that the state with angular momentum zero represents a spherically symmetrical state, but a spherically symmetrical vector field can only be longitudinal. Thus, a photon cannot exist in a state of angular momemtum zero.

5.5.6 Spherical Waves

A major conceptual problem associated with the idea of a photon is how to reconcile its corpuscular nature with a wave expanding in three dimensions. One way to address the problem starts from the wave function of a photon in an arbitrary state, expanded in terms of the general wave function (59) as

$$\mathbf{f} = \sum_{\omega,j,M,\lambda} a_{\omega jM\lambda}\mathbf{f}_{\omega jM\lambda}$$

Since the system (59) is orthonormal the expansion coefficients will be determined in the following manner:

$$a_{\omega jM\lambda} = \int \mathbf{f}\cdot\mathbf{f}^{\star}_{\omega jM\lambda}d\mathbf{k} = \sum_{k} k^2\delta \int \mathbf{f}\cdot\mathbf{f}^{\star}_{\omega jM\lambda}d\mathbf{o}$$

where $\mathbf{o}$ denotes a solid angle variable. This formulation can be used to expand states of definite momentum $\mathbf{p}$ and polarization μ in terms of states of definite angular momentum. In this case

$$a^{(\mathbf{p}\mu)}_{\omega jM\lambda} = e_{\mu}\mathbf{Y}^{(\lambda)\star}_{jM}\left(\frac{\mathbf{p}}{p}\right)\sqrt{\delta_o}\delta_{\omega\mathbf{p}}$$

where δ_o is an element of solid angle. From this formula is calculated the probability that the direction of motion lies within a certain solid angle. When summed over the polarizations it is found to depend only on the angular momentum and not the parity.

Expressions for the electric and magnetic fields can likewise be obtained. These plane-wave solutions are then expanded in terms of spherical harmonics

$$e^{i\mathbf{k}\cdot\mathbf{r}} = \sum_{\ell m} g_{\ell}(kr)Y^{\star}_{\ell m}\left(\frac{\mathbf{k}}{k}\right)Y_{\ell m}\left(\frac{\mathbf{r}}{r}\right) ,$$

where

$$g_\ell(kr) = (2\pi)^{3/2} i^\ell \frac{J_{\ell+\frac{1}{2}}}{\sqrt{kr}}$$

and $J_{\ell+\frac{1}{2}}$ are Bessel functions.

Each of the fields is found to correspond to a sum of waves coming in to a central point and others going out from the central point. Using the electric field as an example

$$\mathcal{E} = \mathcal{E}^{in} + \mathcal{E}^{out}$$

$$\mathcal{E}^{in} = A\mathbf{Y}^{\lambda}_{jm}(\mathbf{n})\frac{e^{ik(r+t)}}{r}$$

$$\mathcal{E}^{out} = B\mathbf{Y}^{\lambda}_{jm}(\mathbf{n})\frac{e^{ik(r-t)}}{r}$$

Each of the fields separately satisfies Maxwell's equations.

Chapter 6

Quantum Chemistry

6.1 Introduction

The quantum theory of the previous chapter may well appear to be of limited relevance to chemistry. As a matter of fact, nothing that pertains to either chemical reactivity or interaction has emerged. Only background material has been developed and the quantum behaviour of real chemical systems remains to be explored. If quantum theory is to elucidate chemical effects it should go beyond an analysis of atomic hydrogen. It should deal with all types of atom, molecules and ions, explain their interaction with each other and predict the course of chemical reactions as a function of environmental factors. It is not the same as providing the classical models of chemistry with a quantum-mechanical gloss; a theme not without some common-sense appeal, but destined to obscure the non-classical features of molecular systems.

The difficulty of finding a link between classical and quantum variables in chemistry is well documented. Eloquent arguments have been advanced, particularly by Primas [67], to redefine the theory of chemistry as comprising of classical and non-classical aspects. In this scheme the Hilbert space of quantum theory becomes embedded, together with a classical kernel in a W^* algebra of observables. This bold attempt, apart from identifying serious errors of abstraction that may be caused by separation of holistic systems, has not resulted in a useful synthesis of classical and quantum variables as yet. The problem is, first of all, to distinguish between purely classical, partially holistic and totally non-classical concepts. The issue is clouded by the role of statistical mechanics, which is supposed to bridge the gap between non-classical microsystems and classical macrosystems. The confusion caused by semi-empirical approximations to mathematically intractable quantum-mechanical problems is even worse. Although the immediate error may appear to be trivial, an holistic non-classical structure, once destroyed, can never be restored. Many of the elegant semi-classical models, apparently so useful in chemistry, in fact, stifles theoretical progress. Some of the atomic and molecular models to be discussed in the next two chapters belong to this class.

6.2 Quantum Aspects of General Chemistry

The failure of classical mechanics in the analysis of physical phenomena, such as black-body radiation, is routinely discussed in elementary texts to emphasize the need of a quantum theory. The failure of classical mechanics to deal correctly with simple chemical systems, although rarely stated, is equally dramatic.

One of the main objectives of mechanics applied to chemical systems, is to predict how an amount of energy, transferred to a closed system, would be distributed amongst individual molecules. This problem is akin to the prediction of how a large number of particles are distributed in a potential field. A typical example is a vertical column of ideal gas in the gravitational field of the earth. The force exerted by a column of length l and cross section A is $A\rho lg$, where ρ is the density and g the gravitational acceleration. The force per unit area, *i.e.* the pressure, is $p = \rho lg$. For a differential height increase of dh in the field, the pressure drops by

$$dp = -\rho g\,dh = -\frac{M}{V}g\,dh$$

For one mole of an ideal gas ($pV = RT$)

$$dp = -\frac{Mp}{RT}g\,dh$$

Integrating between sea level ($p = p_0$) and h

$$\int_{p_0}^{p(h)} \frac{dp}{p} = -\frac{Mg}{RT}\int_0^h dh$$

$$\ln\left(\frac{p}{p_0}\right) = -\frac{Mgh}{RT} = -\frac{\Delta E}{RT}$$

The barometric formula,

$$p = p_0 e^{-\frac{\Delta E}{RT}}$$

is one example of the Boltzmann distribution that relates the populations at different energy levels to the difference in energy between those levels, *i.e.*

$$N_i = N_j \exp[-(\epsilon_i - \epsilon_j)/kT]$$

where ϵ_i are the energies of individual molecules, $k = R/L$ is the Boltzmann constant and L is Avogadro's number. The Boltzmann distribution applies, not only to the total energy of a molecule, but also to the individual contributions such as electronic, vibrational, rotational and translational energies. Any molecule can increase its energy by absorbing energy from an external source. According to classical mechanics this absorption is a continuous process and should lead to a monotonic increase of all energies, as postulated by the principle of equipartition of energies.

6.2.1 The Equipartition Principle

The mechanical modes whereby molecules may absorb and store energy are described by *quadratic* terms. For translational kinetic energy it involves the square of the linear momentum ($E = p^2/2m$), for rotational motion it is the square of angular momentum ($E = L^2/2I$) and for vibrating bodies there are both kinetic and potential energy ($kx^2/2$) terms. The equipartition principle states that the total energy of a molecule is evenly distributed over all available quadratic modes.

According to the Boltzmann distribution the number of particles at an energy level ε_j above ground state, $\varepsilon_0 = 0$, is given by

$$N_j = N_0 e^{-\varepsilon_j/kT}$$

The total number of particles

$$N = \sum_{j=0}^{\infty} N_j = N_0 \sum_{j=0}^{\infty} e^{-\varepsilon/kT}$$

The total energy, summed over all energy levels is

$$E = \sum_{j=0}^{\infty} N_j \varepsilon_j$$

For a continuous energy spectrum the sums may be replaced by the integrals

$$N = \int_0^{\infty} dN_z = N_0 \int_0^{\infty} e^{-\varepsilon_z/kT} dz$$

$$U_z = N_0 \int_0^\infty \varepsilon_z e^{\varepsilon_z/kT} dz$$

The mean particle energy is given by

$$\bar{\varepsilon}_z = \frac{U_z}{N}$$

If the energy is a quadratic function of z, then $\varepsilon_z = \alpha z^2$, and

$$\bar{\varepsilon}_z = \frac{\alpha \int_0^\infty z^2 e^{-\left(\frac{\alpha}{kT}\right)z^2} dz}{\int_0^\infty e^{-\left(\frac{\alpha}{kT}\right)z^2} dz}$$

Using the standard integrals

$$\int_0^\infty x^{2n} e^{-ax^2} dx = \frac{1}{4a}\sqrt{\frac{\pi}{a}}$$

$$\int_0^\infty e^{-a^2x^2} dx = \frac{\sqrt{\pi}}{2a}$$

the mean energy follows as

$$\bar{\varepsilon}_z = \frac{\frac{\alpha}{4}\left(\frac{kT}{\alpha}\right)\sqrt{\frac{\pi}{a}}}{\frac{1}{2}\sqrt{\frac{\pi}{a}}} = \frac{1}{2}kT$$

For a molecule of N atoms, $3N$ coordinates are needed to describe its position. It is said to have $3N$ degrees of freedom. Considered as a rigid unit however, only 3 coordinates are needed to describe its centre of mass. Translational motion is therefore defined to have 3 degrees of freedom. The energy for each degree of freedom is described by a quadratic term, *e.g.* $E_x = p_x^2/2m$. It follows that the kinetic energy of an ideal gas molecule is $\frac{1}{2}kT$ per degree of freedom. The remaining $3N - 3$ coordinates represent the internal degrees of freedom of the molecule.

The internal degrees of freedom are associated with the rotation and vibration of the molecule. A linear molecule has 2 degrees of rotational motion and a non-linear molecule has three. The remaining $(3N - 5)$ or $(3N - 6)$ degrees of freedom describe the motion of the nuclei with respect to each other. For example, the linear CO_2 has $(3N - 5) = 4$ vibrational degrees of freedom and the non-linear SO_2 has three. The mode

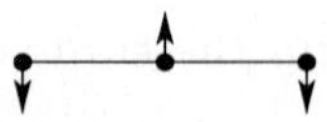

of the linear CO_2 molecule becomes a rotation

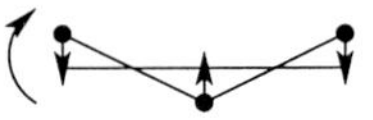

of SO_2.

By the equipartition principle it now follows that each rotational degree of freedom can absorb energy of $\frac{1}{2}kT$ while each vibrational mode can absorb kT. By the same principle the heat capacity of an ideal gas

$$C_V = \left(\frac{\partial U}{\partial T}\right)_V$$

is predicted to be $C_V = R/2$ per degree of freedom. For a monatomic gas with three degrees of translational freedom only, $C_V = 3R/2 = 12.47$ JK^{-1}mol^{-1}. This result is in good agreement with experiment. The quantity C_V/R is dimensionless and provides an easy measure for comparison. The comparison between theory and experiment for a few simple molecules is shown in the table below.

Gas	$\frac{C_V}{R}$	Observed, T(K)		
		300	1000	1500
He	1.5	1.5	1.5	1.5
H_2	3.5	2.5	2.6	2.9
O_2	3.5	2.5	3.2	3.4
H_2O	6.0	3.0	4.0	4.7
CO_2	6.5	3.5	5.5	6.0

For ethane gas with a non-linear molecule C_2H_6, $N = 8$, $3N - 6 = 18$, the average energy $\bar{\epsilon} = 3(\frac{1}{2}kT) + 3(\frac{1}{2}kT) + 18(kT) = 21(kT)$ and $C_V/R = 21$. The measured (C_V/R) heat capacity at 1000 K$= 3.35$.

For monatomic He the agreement is good at all temperatures. For other small molecules it is reasonable, especially at high temperatures and for ethane it is completely wrong. It seems that the agreement improves at high temperatures and deteriorates for more degrees of vibrational freedom. The conclusion is that the prediction only holds for translational degrees of freedom, that the prediction for rotational modes is correct at high temperatures and that vibrational freedom does not absorb energy as predicted by the equipartition principle. The theory, in short, is inadequate to account for observed heat capacities and its temperature dependence.

The situation becomes even worse when the Boltzmann formula is used to interpret the absorption of radiant energy by molecules. Electromagnetic radiation considered as a fluctuating electric field interacts with electrons in molecules, which absorb energy from the radiation. The energy and wavelength of the absorbed light can be measured spectroscopically. The surprising result is that any pure compound only absorbs light at special wavelengths. Assuming an inverse relationship between wavelength and energy this observation means that not any amount of energy can be absorbed by an electron in a molecule. The energy spectrum of electrons in atoms and molecule is therefore not continuous as suggested by the Boltzmann distribution. Only some fixed energy levels are allowed. This is demonstrated well by measuring the wavelength of light emitted by hot gases. The observed line spectrum arises from electrons (e) jumping between discrete energy levels:

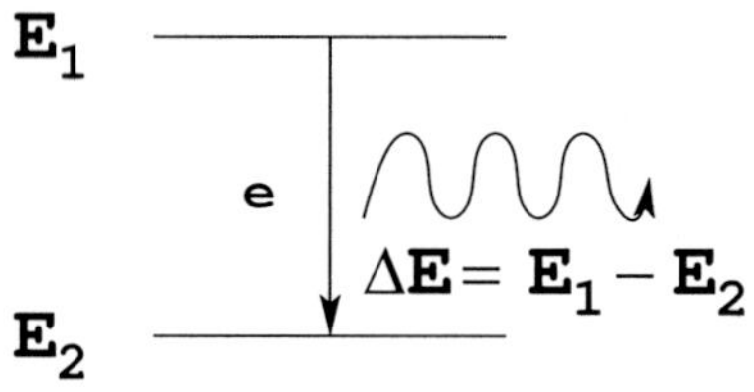

By using light at different wavelenghts it can be demonstrated that vibrational and rotational energies of molecules also have discrete spectra. Detailed analysis shows that the spacing between energy levels decreases in the order: electronic > vibrational > rotational > translational. Whereas translational levels are so closely spaced as to be almost continuous, electronic levels are separated by the order of 1 kJ mol^{-1}. This difference explains why the equipartition principle is not valid: translational energy can increase continuously but rotational and other forms of energy can only increase in bigger bursts or quanta. It often requires high temperatures for any but the lowest vibrational levels to be populated. A separate Boltzmann distribution operates for each mode and the transitions are triggered by energy quanta of different magnitude.

In order to understand these observations it is necessary to resort to quantum mechanics, based on Planck's postulate that energy is quantized in units of $E = h\nu$ and the Bohr frequency condition that requires an exact match between level spacings and the frequency of emitted radiation, $h\nu = E_{upper} - E_{lower}$. The mathematical models are comparatively simple and in all cases appropriate energy levels can be obtained from one-dimensional wave equations.

Translational Motion

The simplest macroscopic chemical system is a chemically unreactive ideal gas that corresponds to an ensemble of freely moving structureless particles.

Between collisions individual particles are described by the one-dimensional Schrödinger equation

$$-\frac{\hbar^2}{2m}\frac{d^2\psi}{dx^2} = E\psi$$

with solutions

$$\psi = Ae^{-ikx} + Be^{ikx}$$

or equivalently

$$\psi = a\sin kx + b\cos kx,$$

$$\text{where} \qquad k^2 = \frac{2mE}{\hbar^2}.$$

The two solutions represent particles of positive and negative momentum respectively, hence particles moving in positive and negative directions along x.

Since a sine function is continuous, single-valued and finite for all real values of its argument, the only restriction on $E_x = p_x^2/2m$ is that it be positive. The energy spectrum is therefore continuous as for a classical particle and the constant $k = p/\hbar$, or in terms of the de Broglie wavelength, $k = 2\pi/\lambda$ is also known as the wave vector of the particle.

Despite its classical energy spectrum it would be wrong to simply treat the free particle as a classical entity. As soon as its motion is restricted, for instance, when confined to a fixed line segment $0 \leq x \leq L$ something remarkable happens to the spectrum.

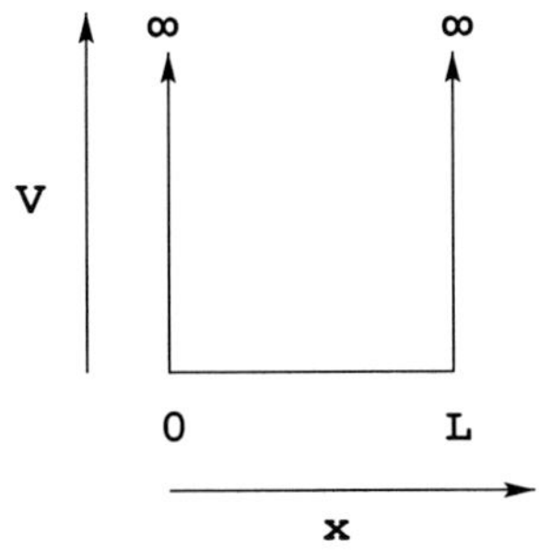

The restriction clearly implies the boundary conditions on the wave function, $\psi(0) = \psi(L) = 0$. Since $\cos 0 = 1 \neq 0$, the first boundary condition requires that $b = 0$ and hence that

$$\psi(x) = a\sin kx \quad , \quad a \neq 0$$

The condition $\psi(L) = 0$ next requires that $\sin kL = 0$, *i.e.* $kL = n\pi$, for n a positive integer. The corresponding energy spectrum is no longer continuous but quantized,

$$E = \frac{n^2h^2}{8mL^2}$$

and the normalized wave function

$$\psi_n = \sqrt{2/L}\sin\left(\frac{n\pi x}{L}\right) \quad , \quad n = 1, 2\ldots$$

Note that the condition $n = 0$ is not allowed since that would imply $\psi = 0$, everywhere, which is not square integrable as required by the Born condition. The ground-state, or zero-point energy

$$E_1 = \frac{h^2}{8mL^2} \neq 0$$

is a purely quantum-mechanical effect. Completely free particles exist nowhere, however tenuous their confinement may be and the only classical feature of an ideal gas therefore is the energy spectrum, and only because the energy quanta are too small for detection.

It is easy to show by separating variables that the energy of a particle confined in a rectangular box is given by

$$\epsilon_k = \frac{h^2}{8m}\left(\frac{n_1^2}{a^2} + \frac{n_2^2}{b^2} + \frac{n_3^2}{c^2}\right)$$

For a cubical box, $a = b = c$, $a^3 = V$,

$$\epsilon_k = \frac{h^2}{8mV^{2/3}}\left(n_1^2 + n_2^2 + n_3^2\right)$$

If two or more of the three quantum numbers are equal different levels will have the same energy and define a degenerate set. In the case $n_1 = n_2 = n_3$ the energy levels $n_1n_1n_3$, $n_1n_3n_1$, $n_3n_1n_1$ define a three-fold degenerate set. The degeneracy of translational energy levels in general may be estimated with reference to the diagram of section 6.3.2, and defining an effective quantum number $n^2 = n_x^2 + n_y^2 + n_z^2$. For large n the number of states between n and $n + 1$ defines the degeneracy g_n as

$$\frac{1}{8}\left[\frac{4}{3}\pi(n+1)^3\right] - \frac{1}{8}\left(\frac{4}{3}\pi n^3\right) \simeq \frac{1}{2}\pi n^2.$$

As an example, the ground state kinetic energy of a hydrogen molecule in a volume of 1m^3 follows as

$$\epsilon(111) = \frac{(6.6 \times 10^{-34})^2 \times 3}{8 \times 3.347 \times 10^{-27}} = 4.88 \times 10^{-41}\text{J}$$

From kinetic theory the average energy of a hydrogen molecule at 300K, calculated as

$$\frac{3}{2}kT = 6 \times 10^{-21}\mathrm{J} >>> \epsilon(111),$$

is enormous compared to the translational quantum. For all practical purposes the energy spectrum therefore is continuous, but the perpetual motion of a gas molecule is a quantum effect.

Rotational Motion

Each atom of a molecule that rotates about an axis through its centre of mass, describes a circular orbit. The total rotational energy must therefore be a function of the molecular moment of inertia about the rotation axis and the angular momentum. The energy calculation for a complex molecule is of the same type as the calculation for a single particle moving at constant (zero) potential on a ring.

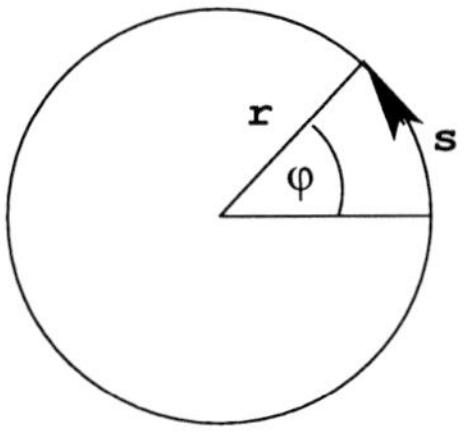

A rigid diatomic molecule, as the simplest example will be used to demonstrate the calculation.

A rotating diatomic molecule consists of masses m_1 and m_2 circling the centre of mass at distances r_1 and r_2 respectively. The moment of inertia is $I = m_1r_1^2 + m_2r_2^2$. By definition, the centre of mass is located such that $m_1r_1 = m_2r_2$, and hence

$$(m_1r_1 - m_2r_2)^2 = 0$$

i.e.

$$m_1^2r_1^2 + m_2^2r_2^2 = 2m_1m_2r_1r_2$$

Completing a square on the right by adding the term $m_1m_2\left(r_1^2 + r_2^2\right)$ to both sides, the equation rearranges to

$$m_1\left(m_1r_1^2 + m_2r_2^2\right) + m_2\left(m_2r_2^2 + m_1r_1^2\right) = m_1m_2\left(r_1 + r_2\right)^2$$

Thus

$$I = m_1r_1^2 + m_2r_2^2 = \frac{m_1m_2}{m_1 + m_2}\left(r_1 + r_2\right)^2 = \mu\left(r_1 + r_2\right)^2$$

A single particle of (reduced) mass μ in an orbit of radius $r = r_1 + r_2$ (= interatomic distance) therefore has the same moment of inertia as the diatomic molecule. The classical energy for such a particle is $E = p^2/2m$ and the angular momentum $L = pr$. In terms of the moment of inertia $I = mr^2$, it follows that $L^2 = 2mEr^2 = 2EI$. The length of arc that corresponds to particle motion is $s = r\varphi$, where φ is the angle of rotation. The Schrödinger equation is[1]

$$-\frac{\hbar^2}{2m}\frac{d^2\psi}{ds^2} = E\psi \quad \text{or} \quad -\frac{\hbar^2}{2mr^2}\frac{d^2\psi}{d\varphi^2} = -\frac{\hbar^2}{2I}\frac{d^2\psi}{d\varphi^2} = E\psi$$

This equation is of the form

$$\frac{d^2\psi}{d\varphi^2} + k^2\psi = 0 \quad , \text{ where } \quad k^2 = \frac{8\pi^2 mr^2 E}{h^2}$$

with solutions $\psi(\varphi) = Ae^{\pm ik\varphi}$.

For ψ to be single valued the boundary condition is that $\psi(\varphi) = \psi(\varphi + 2\pi)$ *i.e.* $\exp(ik\varphi) = \exp[ik(\varphi + 2\pi)]$.

This condition requires that $\exp(2\pi ki) = 1$ *i.e.* $\cos(2\pi k) + i\sin(2\pi k) = 1$ which means that k is an integer. Hence

$$E = \frac{h^2k^2}{8\pi^2 mr^2} = \frac{\hbar^2 k^2}{2I} \quad , \quad k = 0, \pm 1, \pm 2, \ldots$$

The fact that there is no zero-point energy may seem to conflict with the uncertainty principle. However, when the rotor stops, it could happen anywhere on the circle and that accounts for the uncertainty. The expression $L^2 = 2EI = \hbar^2 k^2$ shows that the angular momentum, L is restricted to integral multiples of $\hbar$.

In three dimensions the rotating diatomic molecule is equivalent to a particle moving on the surface of a sphere. Since $V = 0$ the Schrödinger equation is

$$-\frac{\hbar^2}{2\mu}\nabla^2\psi = E\psi$$

[1] The change of variable implies:

$$\frac{d\psi}{ds} = \frac{d\psi}{d\varphi}\cdot\frac{d\varphi}{ds} = \frac{1}{r}\frac{d\psi}{d\varphi}$$

$$\frac{d^2\psi}{ds^2} = \frac{d}{ds}\left(\frac{1}{r}\frac{d\psi}{d\varphi}\right) = \frac{1}{r}\frac{d}{d\varphi}\cdot\frac{d\phi}{d\varphi}\cdot\frac{d\varphi}{ds} = \frac{1}{r^2}\frac{d^2\psi}{d\varphi^2}$$

Since r is constant the first term of the Laplacian

$$\nabla^2 = \frac{1}{r^2}\frac{\partial}{\partial r}\left(r^2\frac{\partial}{\partial r}\right) + \frac{1}{r^2\sin\theta}\frac{\partial}{\partial\theta}\left(\sin\theta\frac{\partial}{\partial\theta}\right) + \frac{1}{r^2\sin^2\theta}\frac{\partial^2}{\partial\phi^2}$$

vanishes and since $I = \mu r^2$ the wave equation reduces to

$$\left[\frac{1}{\sin\theta}\frac{\partial}{\partial\theta}\left(\sin\theta\frac{\partial}{\partial\theta}\right) + \frac{1}{\sin^2\theta}\frac{\partial^2}{\partial\phi^2}\right]\psi(\theta,\phi) = -\frac{2IE}{\hbar^2}\psi(\theta\phi)$$

This equation is the same as the angular momentum equation (5.8) of the hydrogen atom, with $2IE/\hbar^2$ instead of $l(l+1)$. For molecular rotation it is conventional to symbolize the Θ quantum number as J, rather than l. The energy eigenvalues are therefore determined by setting

$$J(J+1) = 2IE/\hbar^2 \quad \text{or} \quad E = \frac{J(J+1)\hbar^2}{2I} \qquad J = 0, 1, 2\ldots$$

Using CO as an example

$m_C = 12.000 \times 10^{-3}/L = 1.993 \times 10^{-26}\,\text{kg}$

$m_O = 15.995 \times 10^{-3}/L = 2.656 \times 10^{-26}\,\text{kg}$

$\mu = 1.139 \times 10^{-26}\,\text{kg}$ For $J = 0$, $\epsilon_r = 0$, for $J = 1$

$r_e(\text{C-O}) = 112.8\,\text{pm}$

$I = \mu r_e^2 = 1.449 \times 10^{-46}\,\text{kg}\,\text{m}^2$

$$\epsilon_r = \frac{(6.63 \times 10^{-34})^2 \times 2}{8\pi^2(1.449 \times 10^{-46})} = 7.685 \times 10^{-23}\,\text{J}$$

Rotational quanta are seen to be larger than the translational by many orders of magnitude, but they are still small relative to average classical energies ($kT = 4 \times 10^{-21}\,\text{J}$ at $300\,\text{K}$). The quanta are large enough to be observed, but even at room temperature rotational energies approach classical predictions. At low temperatures however, classical predictions can be seriously in error.

Vibrational Motion

For small displacements molecular vibrations obey Hooke's law for simple harmonic motion of a system that vibrates about an equilibrium configuration. In this case the restoring force on a particle of mass m is proportional to the displacement x of the particle from its equilibrium position, and acts in the opposite direction. In terms of Newton's second law:

$$m\frac{d^2x}{dt^2} = -kx$$

The differential equation

$$\frac{d^2x}{dt^2} + \frac{k}{m}x = 0$$

has the general solution $x = D\cos\left(\sqrt{\frac{k}{m}} \cdot t\right) + B\sin\left(\sqrt{\frac{k}{m}} \cdot t\right)$.

Using $\sin(\theta + \phi) = \sin\theta\cos\phi + \cos\theta\sin\phi$, and putting $D = A\sin\varphi$, $B = A\cos\varphi$, the solution takes the form

$$x = A\sin\left(\sqrt{\frac{k}{m}} \cdot t + \varphi\right)$$

where A and φ are arbitrary constants of *amplitude* and *phase* respectively. The function x is periodic with time and repeats with period $\tau = 2\pi\sqrt{m/k}$, or with an oscillation frequency of $\nu = 1/\tau = \sqrt{k/m}/2\pi$. The *force constant* takes the form $k = 4\pi^2 m\nu^2$ and $x = A\sin(2\pi\nu t + \varphi)$.

In the one-dimensional problem the force

$$F = -\frac{dV}{dx} = -kx$$

and the potential energy

$$V = \int_0^x kx\,dx = \frac{1}{2}kx^2 \quad , \quad \text{or} \quad V = 2\pi^2\nu^2 mx^2.$$

The graph of $V(x)$ is a parabola:

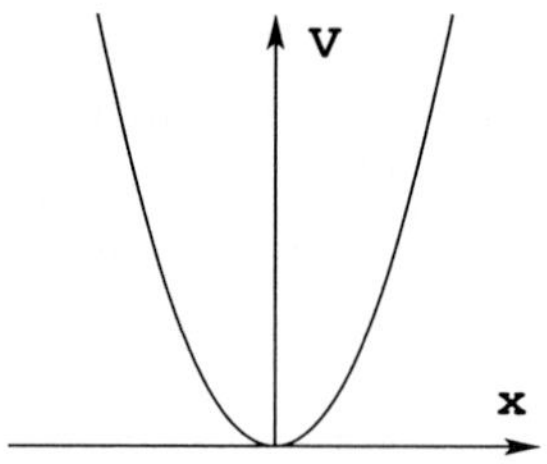

Schrödinger's equation for the one-dimensional harmonic oscillator is

$$\frac{d^2\psi}{dx^2} + \frac{2m}{\hbar^2}\left(E - \frac{1}{2}kx^2\right)\psi = 0$$

It is convenient to simplify this equation by introducing the dimensionless quantities

$$\alpha = \frac{2E}{\hbar}\sqrt{\frac{m}{k}} = \frac{2E}{h\nu} \quad \text{and} \quad y = \sqrt{\frac{2\pi m\nu}{\hbar}}x$$

to yield

$$\frac{d^2\psi}{dy^2} + \left(\alpha - y^2\right)\psi = 0 \tag{6.1}$$

In eigenvalue form

$$\left(-\frac{d^2}{dy^2} + y^2\right)\psi = \alpha\psi$$

To solve for α Schrödinger's operator method will be used by way of illustration. The Hamiltonian operator is factored into two operators containing first derivatives only, using the relationship[2]

$$\left(\frac{d^2}{dy^2} - y^2\right)\psi = \left[\left(\frac{d}{dy} - y\right)\left(\frac{d}{dy} + y\right) - 1\right]\psi$$

whereby (1) may be written in the form

$$\left(\frac{d}{dy} - y\right)\left(\frac{d}{dy} + y\right)\psi_\alpha = -(\alpha - 1)\psi_\alpha \tag{6.2}$$

where the eigenfunction ψ_α corresponds to the eigenvalue α. Equation (2) is now operated on from the left with $\left(\frac{d}{dy} + y\right)$ to give

$$\left(\frac{d^2}{dy^2} - y^2 - 1\right)\left(\frac{d}{dy} + y\right)\psi_\alpha = -(\alpha - 1)\left(\frac{d}{dy} + y\right)\psi_\alpha$$

Writing $\left(\frac{d}{dy} + y\right)\psi_\alpha = \varphi_\alpha$ (a new function)

$$\left(\frac{d^2}{dy^2} - y^2\right)\varphi_\alpha = -(\alpha - 2)\varphi_\alpha$$

or

$$\left[\frac{d^2}{dy^2} + \left(\alpha - y^2 - 2\right)\right]\varphi_\alpha = 0 \tag{6.3}$$

[2]Noting,

$$\begin{aligned}\left(\frac{d}{dy} - y\right)\left(\frac{d\psi}{dy} + y\psi\right) &= \frac{d^2\psi}{dy^2} + \psi\frac{dy}{dy} + y\frac{d\psi}{dy} - y\frac{d\psi}{dy} - y^2\psi \\ &= \left(\frac{d^2}{dy^2} + 1 - y^2\right)\psi\end{aligned}$$

Comparing (1) and (3) shows that when ψ_α is an eigenfunction corresponding to the eigenvalue α, then φ_α is an eigenfunction of the same equation corresponding to the an eigenvalue of $\alpha - 2$. Thus, given any solution, one can always derive another. Furthermore if α is an eigenvalue, so is $\alpha - 2$. By repeating this process indefinitely it follows that any $\alpha - 2n$, where n is an integer must be an eigenvalue.

If n is made large enough the eigenvalue, and hence the energy, must become negative. Since it is not physically acceptable for an harmonic oscillator to have negative energy it follows that the lowest possible value of α must be such that $\varphi_\alpha = (d/dy + y)\psi_\alpha = 0$. In this case no solutions of negative α can be obtained. Multiplication of this equation by $(d/dy - y)$ yields

$$\left(\frac{d}{dy} - y\right)\left(\frac{d}{dy} + y\right)\psi_\alpha = \frac{d^2\psi_\alpha}{dy^2} - y^2\psi_\alpha + \psi_\alpha = 0 \tag{6.4}$$

Comparison of (4) and (1) shows that the required lowest allowed eigenvalue is $\alpha = 1$, which occurs for n=0. The only allowed values of α must then be such that $\alpha - 2n = 1$, where n is an integer.

This result determines the energy spectrum of the harmonic oscillator through

$$\alpha = \frac{2E}{h\nu} = 2n + 1 \quad \text{as} \quad E = \left(n + \frac{1}{2}\right)h\nu.$$

The vibrational levels corresponding to $n = 0, 1, 2\ldots$ are evenly spaced. Like the particle confined to a line segment, the harmonic oscillator also has zero-point energy $E_0 = \frac{1}{2}h\nu$.

The wave functions of the harmonic oscillator can be obtained by the same method, starting from the lowest state, described by

$$\frac{d\psi_0}{dy} + y\psi_0 = 0 \quad \text{or} \quad \frac{d\psi_0}{\psi_0} + y\,dy = 0$$

which can be integrated directly, giving

$$\psi_0 = Ae^{-y^2/2} \quad , \quad A = \text{constant}$$

The normalized function $\psi_0 = \frac{1}{\sqrt{\pi}}e^{-y^2/2}$, is a Gaussian function.

Without repeating the factorization, it can be shown [57] that a function $\varphi = (d/dy - y)\psi_\alpha$ also satisfies the Schrödinger equation, but corresponds to an eigenvalue $\alpha + 2$. Hence, any eigenfunction ψ_α can be used to generate the next higher eigenfunction by operating on it with the operator $(d/dy - y)$.

The first few wave functions generated by this method are:

$$\begin{aligned}\psi_0 &\sim e^{-y^2/2}\\ \psi_1 &\sim 2ye^{-y^2/2}\\ \psi_2 &\sim (2y^2-1)e^{-y^2/2}\end{aligned}$$

Each eigenfunction consists of a (*Hermite*) polynomial $H_n(y)$, the Gaussian factor and a normalization factor, *i.e.*

$$\psi_n = \left(\frac{2m\nu}{\hbar}\right)^{\frac{1}{4}} (2^n n!)^{-\frac{1}{2}} H_n(y)e^{-y^2/2}$$

The vibration of a diatomic molecule, or any vibrational mode in a polyatomic molecule, may be approximated by two atoms of mass m_1 and m_2 joined by a Hooke's law bond that allows vibration relative to the centre of mass. The frequency of such a two-body oscillator is given by

$$\nu_0 = \frac{1}{2\pi}\sqrt{\frac{k}{\mu}} \quad \text{for reduced mass} \quad \mu = \frac{m_1 m_2}{m_1 + m_2}$$

This result is the same as for the harmonic oscillator and the allowed vibrational energy levels follow immediately as

$$E_v = \left(v + \frac{1}{2}\right) h\nu_0 = \left(v + \frac{1}{2}\right) \hbar\sqrt{\frac{k}{\mu}}$$

A solution of CO in tetrachloromethane absorbs infrared radiation of frequency 6.42×10^{13} Hz. If this is interpreted as a vibration quantum both the force constant of the C-O bond and the spacing of vibrational levels can be calculated directly. The reduced mass $\mu_{CO} = 1.14 \times 10^{-26}$ kg. The force constant $k = 4\pi^2\nu_0^2\mu = (4\pi^2)(6.42 \times 10^{13})(1.14 \times 10^{-26}) = 1.86 \times 10^3$ Nm^{-1}. The separation between the vibrational levels of CO is

$$\Delta E = h\nu_0 = 6.63 \times 10^{-34} \times 6.42 \times 10^{13} = 4.26 \times 10^{-20} \text{ J}$$

This interval is considerably more than the spacing between rotational levels, and since $\Delta E_V >> kT$ at room temperature it is safe to conclude that most molecules in such a sample exist in the lowest vibrational state with only their zero-point energies. Here is the real reason for the breakdown of the classical equipartition principle.

Boltzmann Distribution

The Boltzmann distribution for quantum systems is totally different in concept from the principle of classical equipartition of energy. The main difference comes from the way in which the energies of the various modes vary with quantum number:

$$E_{el} \propto \frac{1}{n^2} \quad , \quad E_{vib} \propto n + \frac{1}{2} \quad , \quad E_{rot} \propto n(n+1) \quad , \quad E_{tr} \propto n^2$$

Electronic levels are spaced more closely together at higher quantum numbers as the ionization limit is approached, vibrational levels are evenly spaced, while rotational and translational levels are spaced further apart at high energies. The classical principle assumes continuous variation of all energies.

The quantum Boltzmann distribution only applies between allowed energy levels of the same family and each type of energy has its own characteristic *partition function*, that can be established by statistical methods and describes the response of a system to thermal excitation. If the total number of particles N, is known, the Boltzmann distribution may be used to calculate the number, or fraction, of molecules in each of the allowed quantum states. For any state i

$$N_i = N_0 e^{-(\epsilon_i - \epsilon_0)/(kT)}$$

and hence,

$$N = \sum N_i = N_0 \sum \left[e^{-(\epsilon_i - \epsilon_0)/(kT)} \right]$$

written as $N = N_0 q$, where q is the partition function or *sum over states*

$$q = 1 + e^{-(\epsilon_1 - \epsilon_0)/(kT)} + e^{-(\epsilon_2 - \epsilon_0)/(kT)} + \ldots$$

The partition function, defined in terms of (degenerate) energy levels, rather than states, is

$$q = \sum g_i e^{-(\epsilon_i - \epsilon_0)/(kT)}$$

The partition function is the extension of the Boltzmann distribution to a complete set of allowed energies. Because of their different quantum-mechanical structures, different energy modes have different partition functions.

For one-dimensional translational motion of gas molecules

$$\epsilon_n = \frac{n^2 h^2}{8ma^2} \quad , \quad n = 1, 2, 3 \ldots$$

and hence

$$\epsilon_n - \epsilon_0 = \frac{(n^2 - 1^2)\, h^2}{8ma^2} \simeq \frac{n^2 h^2}{8ma^2} \quad \text{for large } n.$$

The partition function for one-dimensional translational energies follows as

$$q_x = \sum_{n=0}^{\infty} e^{-n^2[h^2/(8ma^2)]/(kT)}$$

For three-dimensional motion an effective quantum number may be defined as

$$n^2 = n_x^2 + n_y^2 + n_z^2 \quad \text{and} \quad a^3 = V$$

The degeneracy of the energy levels, as shown before, is then given by $q_n = \frac{1}{2}\pi n^2$, and the partition function becomes

$$q = \sum_{n=0}^{\infty} \left(\frac{1}{2}\pi n^2\right) e^{-n^2[h^2/(8ma^2)]/(kT)}$$

For closely spaced levels the summation may be replaced by integration. Using a standard integral[3], the expression reduces to

$$\begin{aligned} q &= \left(\frac{\pi}{8}\right)^{3/2} \left[\frac{kT(8ma^2)}{h^2}\right]^{3/2} \\ &= \left(\frac{2\pi mkT}{h^2}\right)^{3/2} V \end{aligned}$$

The rotational energies for a linear molecule are given by

$$E_{rot} = \frac{J(J+1)h^2}{8\pi^2 I} \quad , \quad J = 0, 1, 2 \ldots$$

The degeneracies of the rotational energy levels are $g_J = 2J + 1$. In terms of these quantities the rotational partition function becomes

$$q_{rot} = \sum (2J+1) e^{-J(J+1)h^2/(8\pi^2 IkT)}$$

At moderate temperature such a large number of terms contribute to the summation that it may be replaced by integration

$$q_{rot} = \int (2J+1) e^{-J(J+1)h^2/(8\pi^2 IkT)} dJ$$

[3] $\int_0^\infty x^2 e^{-ax^2} dx = \frac{1}{2a}\sqrt{\frac{\pi}{a}}$

In terms of the new variable $x = J(J+1)h^2/(8\pi^2 IkT)$,

$$\frac{dx}{dJ} = \frac{(2J+1)h^2}{8\pi^2 IkT} \quad \text{or} \quad (2J+1)dJ = \frac{8\pi^2 IkT}{h^2}dx$$

such that

$$\begin{aligned} q_{rot} &= \frac{8\pi^2 IkT}{h^2}\int_0^\infty e^{-x}dx = \frac{8\pi^2 IkT}{h^2} - e^{-x}\Big|_0^\infty \\ &= \frac{8\pi^2 IkT}{h^2} \end{aligned}$$

For each vibrational mode

$$\epsilon_{vib} = \left(v + \frac{1}{2}\right)h\nu \quad , \quad v = 0, 1, 2 \ldots$$

and the vibrational partition function may be written

$$q_{vib} = 1 + e^{-x} + e^{-2x} + e^{-3x} + \ldots \tag{6.5}$$

where $x = \Delta\epsilon_{vib}/kT$. Rewriting equation (5) as

$$q_{vib}\, e^{-x} = e^{-x} + e^{-2x} + e^{-3x} + \ldots$$

and subtracting the two expressions,

$$q_{vib} - q_{vib}\, e^{-x} = q_{vib}\left(1 - e^{-x}\right) = 1$$

The vibrational partition function follows directly as

$$q_{vib} = \frac{1}{1 - e^{-\Delta\epsilon_{vib}/(kT)}}$$

From the known partition function the thermal energy for each mode can be calculated as the sum over all energy levels. Each term consists of the number of molecules in a quantum state N_i multiplied by the number of states g_i at that level and the energy of the level above the ground state, $\epsilon_i - \epsilon_0$. If the total energy is represented by the symbol U, the thermal energy is

$$\begin{aligned} U - U_0 &= g_0 N_0(0) + g_1 N_1(\epsilon_1 - \epsilon_0) + g_2 N_2(\epsilon_2 - \epsilon_0) + \ldots \\ &= \sum_i g_i N_i(\epsilon_i - \epsilon_0) \end{aligned}$$

With N_i expressed as a Boltzmann function

$$U - U_0 = N_0 \sum_i \left[g_i(\epsilon_i - \epsilon_0)e^{(\epsilon_i - \epsilon_0)/(kT)}\right]$$

In terms of partition function

$$U - U_0 = \frac{N}{q}\sum_i \left[g_i(\epsilon_i - \epsilon_0)e^{-(\epsilon_i-\epsilon_0)/(kT)}\right]$$

Since

$$\frac{dq}{dT} = \sum_i \left[g_i\frac{(\epsilon_i - \epsilon_0)}{kT^2}e^{-(\epsilon_i-\epsilon_0)/(kT)}\right]$$

$$kT^2\frac{dq}{dT} = \sum_i \left[g_i(\epsilon_i - \epsilon_0)e^{-(\epsilon_i-\epsilon_0)/(kT)}\right]$$

and hence

$$U - U_0 = \frac{N}{q}kT^2\frac{dq}{dT}$$

For one mole of sample

$$U - U_0 = \frac{RT^2}{q}\frac{dq}{dT} \tag{6.6}$$

This expression reflects the quantum-mechanical version of equipartition. Substituting the partition functions for different modes into (6) the classical equipartition principle can readily be recovered.

For translational motion

$$q_{tr} = \left(\frac{2\pi mk}{h^2}\right)^{3/2} VT^{3/2}$$

$$\frac{dq_{tr}}{dT} = \left(\frac{2\pi mk}{h^2}\right)^{3/2} V\left(\frac{3}{2}T^{1/2}\right)$$

Substitution into (6) gives

$$(U - U_0) = \frac{3}{2}RT,$$

the same as the classical result.

Likewise, substitution of $q_{rot} = 8\pi^2 IkT/h^2$ into (6) gives the result for a linear molecule:

$$(U - U_0)_{rot} = RT.$$

Finally

$$\frac{dq_{vib}}{dT} = \frac{\left[\Delta\epsilon_{vib}/(kT)^2\right]e^{-\Delta\epsilon_{vib}/(kT)}}{1 - e^{-\Delta\epsilon_{vib}/(kT)^2}} = \frac{(x/T)e^{-x}}{(1 - e^{-x})^2}$$

Hence, from (6)

$$(U - U_0)_{vib} = \frac{RTx}{e^x - 1} \quad , \quad x = \frac{\Delta\epsilon_{vib}}{kT}$$

For small x, $e^x = 1 + x + x^2/2! + \cdots \simeq 1 + x$. Therefore, in the limit as $\Delta\epsilon_{vib}$ approaches zero,

$$(U - U_0)_{vib} = RT \quad \text{for} \quad \Delta\epsilon_{vib} \to 0$$

Quantum theory therefore correctly predicts the equipartition principle in the classical limit.

6.3 Molecular Spectroscopy

One branch of chemistry where the use of quantum mechanics is an absolute necessity is molecular spectroscopy. The topic is interaction between electromagnetic waves and molecular matter. The major assumption is that nuclear and electronic motion can effectively be separated according to the Born-Oppenheimer approximation, to be studied in more detail later on. The type of interaction depends on the wavelength, or frequency of the radiation which is commonly used to identify characteristic regions in the total spectrum, ranging from radio waves to γ-rays.

In order to enable the interaction there must be some electromagnetic effect produced by molecular change, that couples with the field associated with the radiation. It is sufficient to consider electromagnetic radiation as consisting of two mutually perpendicular oscillating fields, one electric and the other magnetic, both perpendicular to the direction of propagation of the light. The most evident way in which radiation can interact with matter is by virtue of these radiation fields inducing oscillating dipoles in atoms or molecules. When this oscillation matches a natural frequency of the atom or molecule, resonance occurs and energy is transferred between the light wave and the atom or molecule. To create an electric dipole it is necessary to separate charges of opposite sign along an axis, whereas a magnetic dipole corresponds to a rotation of charge about an axis, as in a solenoid carrying a current.

A molecular spectrum is generated when radiation of smoothly varying wavelength is passed through a homogeneous molecular medium, and consists of all new features of the transmitted beam compared to the incoming beam. These spectral features could arise from transitions between nuclear, electronic, vibrational or rotational energy levels of the molecule. The various transitions happen in different regions of the electromagnetic spectrum and it is customary to differentiate between types of spectrum on this basis.

Rotational spectra are due to charges that occur on molecules with permanent electric dipole moments. As the molecule rotates the dipole moment in a given direction changes periodically with time and the rotation can therefore be driven by the fluctuating magnetic field of the radiation.

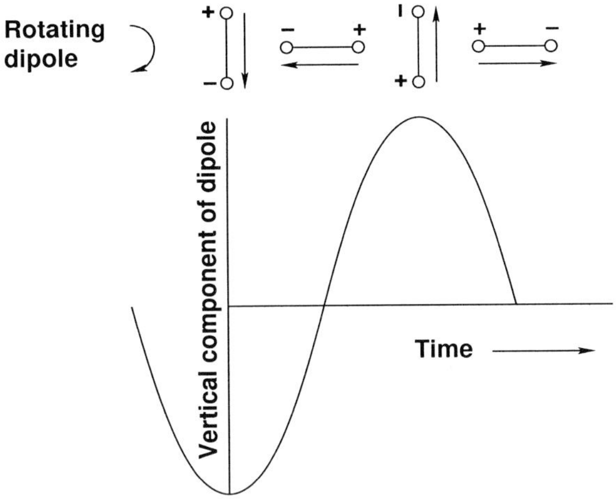

In the process the molecule absorbs sufficient energy from the radiation to excite it to a higher rotational energy level. This absorption is possible if the frequency of the radiation matches a rotational quantum energy gap of the molecule, *i.e.* $\nu = \Delta E_{rot}/h$. Separations between the rotational levels of molecules are of the order of hundreds of joules per mole and the corresponding frequencies occur in the microwave region of the spectrum ($3 \times 10^{10} - 3 \times 10^{12}$ Hz). Only molecules with permanent dipoles are ″microwave active″. A molecule that suffers a change in rotational energy experiences a parallel change in its angular momentum. It was noted that the quantum-mechanical angular momentum of a rigid rotor can only change in multiples of $J\hbar$, where J is an integer. It is significant to note that a photon also carries angular momentum of precisely $\hbar$. It is therefore found that pure rotational transitions are always characterised by the selection rule $\Delta J = \pm 1$, on absorption or emission of a photon. Microwave spectroscopy provides information about molecular moments of inertia.

Any periodic distortion that causes polarization of a molecule can also cause interaction with the electric field component of radiation. An example is the asymmetric stretching vibration of the CO_2 molecule, that creates a fluctuating dipole moment as shown below.

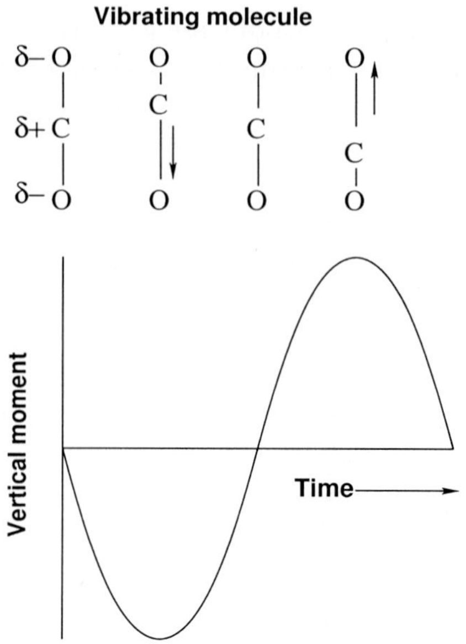

According to this mechanism the symmetric stretch will be "infrared inactive". In general, any vibration that gives rise to a change in dipole moment can absorb a photon in the infrared region ($3 \times 10^{12} - 3 \times 10^{14}$ Hz). This is the frequency region that matches vibrational energies. Because of the angular momentum of a photon any vibrational transition must always be accompanied by a change in rotational state, $\Delta J = \pm 1$, as well.

Vibrational levels are often analyzed in terms of an harmonic model, but in practice this is not a good approximation for real chemical bonds. An empirical expression that provides a more realistic description of a stretched bond, is the Morse function

$$E = D[1 - \exp\{a(r_e - r)\}]^2$$

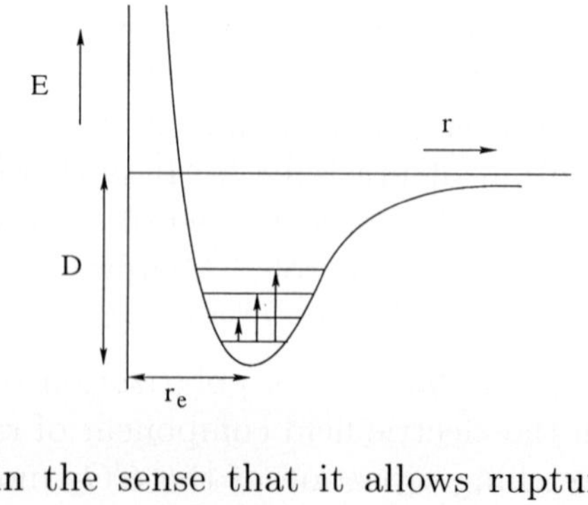

It is more realistic in the sense that it allows rupture of the bond when stretched to the limit. The quantity D represents the dissociation energy of the bond, r_e is the equilibrium bond length and a is a force constant. For small displacements the vibrational energy levels are comparable to the predictions of the harmonic model, but somewhat more complicated. As shown before, the Boltzmann distribution predicts low probabilities for vibrational levels higher than $v = 0$. Of the three transitions shown on the diagram only the first (fundamental) has appreciable intensity compared to the overtones, $v = 2, 3$.

The dipole and polarization selection rules of microwave and infrared spectroscopy place a restriction on the utility of these techniques in the study of molecular structure. However, there are complementary techniques that can be used to obtain rotational and vibrational spectrum for many other molecules as well. The most useful is Raman spectroscopy.

6.3.1 The Raman effect

Interaction between radiation and matter amounts to either absorption, that satisfies the Bohr frequency condition, or scattering. Scattering refers to the interaction of a freely moving particle (photon) with an obstacle that turns it into another free particle. During elastic (Raleigh) scattering the secondary and primary photons have the same energy. It should however, not be confused with a process of absorption and re-emission that requires bound states to match the energy of the photon. Discrete energy levels cannot exist with positive energy. The transient association between photon and target that causes the scattering, is therefore more like excitation to virtual, rather than actual energy levels.

Inelastic scattering occurs with a change of energy, as if the scattered particle encounters a finite potential barrier. During Compton scattering it is the recoil of a weakly bound electron that absorbs part of energy of the X-ray photon. The optical equivalent is known as Raman scattering. In this case inelastic scattering of visible or ultaviolet light occurs because of internal deformation of a target molecule. The inelastic energy component corresponds to either rotational or vibrational quanta of the molecule.

The exciting radiation, usually laser light, may be of any wavelength. Interaction is not by absorption and only requires the presence of polarizable molecules. The oscillating electric field of the incident radiation $E = E_0 \cos 2\pi\nu t$ induces a dipole moment

$$\mu_{ind} = \alpha E = \alpha E_0 \cos 2\pi\nu t$$

in the polarizable molecule. The polarizability α, formulated here as a scalar quantity, actually is a tensor. It is the oscillating dipole that emits secondary radiation and causes the molecule to be a scattering centre. The polarizability, except in the case of highly symmetrical molecules, is anisotropic and therefore changes when observed in an arbitrary fixed direction, as the molecule rotates or vibrates.

The induced dipole moment in a given direction fluctuates at double the rotational frequency of the molecule as shown schematically below. The upper diagram shows the electric field in phase with a rotating polar molecule.

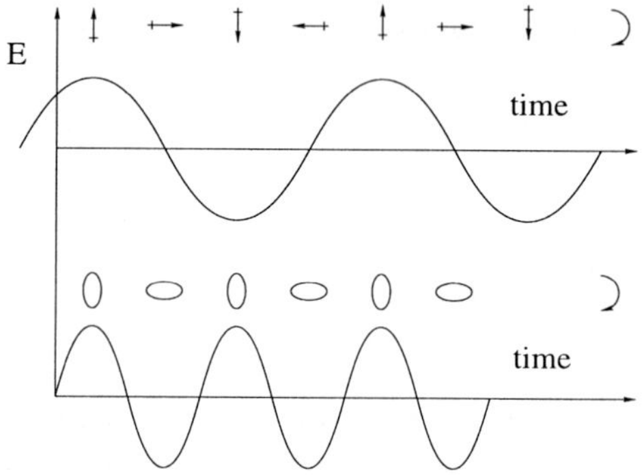

The lower part shows a fluctuating field in phase with a rotating induced dipole, which is always directed with the field.

The polarizability along the field direction fluctuates as

$$\alpha = \alpha_0 + \Delta\alpha \cos 2(2\pi\nu_R)t$$

where ν_R is the rotation frequency of the molecule. The induced dipole varies with time according to

$$\mu_{ind} = [\alpha_0 + \Delta\alpha \cos 2(2\pi\nu_R)t](E_0 \cos 2\pi\nu t)$$

The trigonometric relation $2\cos\theta\cos\phi = \cos(\theta+\phi) + \cos(\theta-\phi)$ shows that the previous expression is equivalent to

$$\mu_{ind} = \alpha_0 E_0 \cos 2\pi\nu t + (\Delta\alpha)E_0[\cos 2\pi(\nu + 2\nu_R) + \cos 2\pi(\nu - 2\nu_R)t]$$

The first term on the right corresponds to Raleigh scattering. The second term suggests an energy-exchange interaction that depends on the anisotropy of the polarizability and involves frequencies shifted from that of the incident radiation by an amount which depends on twice the rotational or vibrational frequency of the molecules in the sample.

The induced dipole is seen to oscillate at the frequencies ν, $\nu - 2\nu_R$ and $\nu + 2\nu_R$. As a consequence, the molecules scatter radiation at these frequencies, which in Raman spectroscopy are measured at right angles to the incident beam. The sample is exposed to radiation of a frequency not absorbed by the molecules and the spectrum consists of weak scattered lines on both sides of the intense Raleigh peak. The lines that appear on the low-frequency side are called *Stokes lines* and the other are anti-Stokes lines.

The selection rule for rotational Raman transitions are $\Delta J = \pm 2$. This result relates to the involvement of two photons, each with angular momentum $\hbar$, in the scattering process. Also allowed is $\Delta J = 0$, but since such a transition implies zero change in energy it represents Raleigh scattering only.

6.3.2 Electronic spectra

Under the Born-Oppenheimer scheme it is reasonable to assume that the total molecular energy is partitioned as a sum of contributions

$$E_{total} = E_{electronic} + E_{vibration} + E_{rotation}$$

and this assumption is supported to first approximation by the appearance of electronic molecular spectra as observed in the visible and ultraviolet spectral regions. If the energy changes underlying an observed transition also add up like

$$\Delta E_{tot} = \Delta E_{el} + \Delta E_{vib} + \Delta E_{rot}$$

and the approximate orders of magnitude involved are related like

$$\Delta E_{el} \approx \Delta E_{vib} \times 10^3 \approx \Delta E_{rot} \times 10^6$$

one expects, and finds, that vibrational changes produce a coarse structure and rotational changes a fine structure on the spectra of electronic transitions.

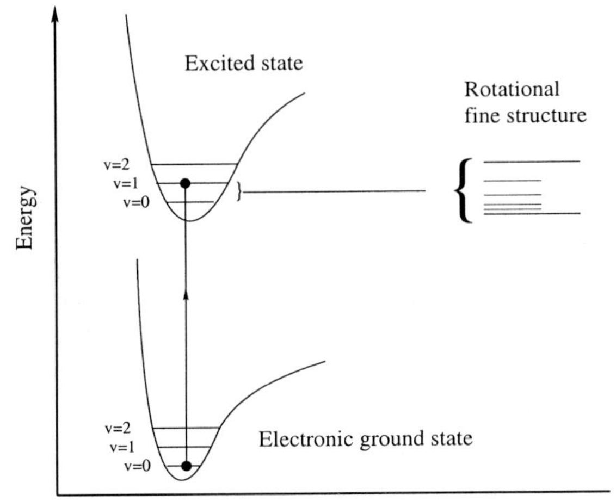

Electronic spectra usually appear to consist of broad bands, but when spectrographs of larger resolving power are used, most of the bands are found to exhibit a *fine structure* in the sense that they consist of many closely spaced individual lines.

Pure rotational spectra only appear for molecules with permanent dipole moments and vibrational spectra require a change of dipole during the motion. However, electronic spectra are observed for all molecules, and changes in the electron distribution in a molecule are always accompanied by dipole changes. As a result even homonuclear molecules (H_2 or N_2) which have no rotation or vibration spectra, do give electronic spectra with vibrational and rotational structure from which rotational constants and bond vibration frequencies may be derived.

The interpretation of molecular spectra in the visible and ultraviolet regions is based on a large number of empirical and semi-empirical rules and has been correlated extensively with many models of bond formation and molecular structure. The complexity of the problem however, is such that despite an impressive body of self-consistency detailed interpretation of most spectra is simply not feasible.

6.3.3 Spin Resonance Spectroscopy

An electron with spin $\frac{1}{2}$ has angular momentum of

$$\left\{\frac{1}{2}\left(\frac{1}{2}+1\right)\right\}^{1/2}\frac{h}{2\pi}=\frac{\sqrt{3}}{2}\quad\text{units of }\hbar$$

Like electrons, some nuclei also have spin. Protons and neutrons also have spins of $\frac{1}{2}$. Thus, if a nucleus consists of p protons and n neutrons its total spin will be a vector sum of $p+n$ spins of $\frac{1}{2}$. Each isotope will have its own spin value, but the laws governing the vector addition of nuclear spins are not known, and at the moment, nuclear spins are known only in terms of some empirical rules.

The spin of ^1H is $\frac{1}{2}$ as expected. Deuterium (^2H) could have spin 0 or 1, depending on the relative alignment of the proton and neutron spins. It is observed to be 1. ^4He has zero spin. From these observations a set of empirical rules specifying the spins of all important nuclei has been derived.

The angular momentum of a nucleus with spin $\boldsymbol{I}$ is $\{I(I+1)\}^{1/2}\hbar = \sqrt{I(I+1)}$ units.

The components of spin in a given direction are $I_z = I, I-1, \ldots, -I$, giving $2I+1$ components. The corresponding levels are usually degenerate, but in an external field the degeracy is lifted and $2I+1$ levels appear.

For a particle of charge q and mass m, the magnetic moment is given by

$$\boldsymbol{\mu} = G\left(\frac{q}{2m}\boldsymbol{I}\right)$$

where G is a proportionality constant, called the gyromagnetic ratio. For electrons $G = g = 2$, the Landé factor. Thus, for (a negatively charged) electron

$$\begin{aligned}\boldsymbol{\mu} &= g\frac{e}{2m}\sqrt{I(I+1)}\frac{h}{2\pi} \quad \text{in units of Am}^{-2} = \text{JT}^{-1}\ \textit{i.e.}\ \text{Joules/Tesla.} \\ &= -g\beta\sqrt{I(I+1)}\end{aligned}$$

where

$$\beta = \frac{(-e)h}{4\pi m} = 9.273 \times 10^{-24}\,\text{JT}^{-1}$$

is known as the *Bohr magneton.* Nuclear moments are more conveniently expressed in terms of a *nuclear magneton* β_N, defined in terms of the proton as

$$\beta_N = \frac{eh}{4m_p\pi} = 5.050 \times 10^{-27}\,\text{JT}^{-1}$$

For a nucleus of p protons, mass M and charge pe

$$\begin{aligned}\boldsymbol{\mu} &= \frac{Gpe}{2M}\sqrt{I(I+1)}\frac{h}{2\pi} = \frac{Gm_p p}{M}\beta_N\sqrt{I(I+1)} \\ &= g_n\beta_N\sqrt{I(I+1)}\,\text{JT}^{-1}\end{aligned}$$

where g_N is characteristic of each nucleus.

The components in a reference direction are

$$\mu_z = -g\beta I_z \quad \text{(for electrons)} \quad \mu_z = g_N\beta_N I_z \quad \text{(for nuclei).}$$

For a nucleus with spin $I = 1$, the angular momentum is

$$\boldsymbol{I} = \sqrt{1 \times 2} = \sqrt{2}\,\text{units, shown graphically below}$$

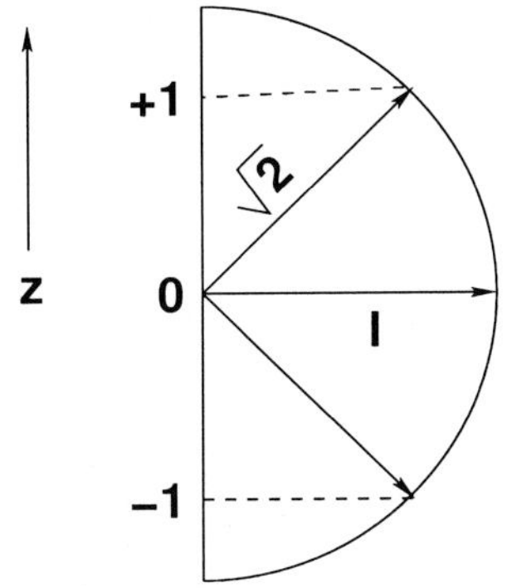

There are three possible orientations of the spin in the field and hence three energy levels. The energy of interaction between the magnetic dipole and the field (B_z) is $\mu_z B_z$. The separation between neighbouring energy levels is

$$\Delta E = |g_N \beta_N I_z B_z - g_N \beta_N (I_z - 1) B_z|$$

or, expressed as a frequency

$$\frac{\Delta E}{h} = \frac{g_N \beta_N B_z}{h} \text{ Hz}$$

This expression provides the basis of a spectroscopic method: The transition of electron or nuclear spins between energy levels ("change of spin") may be associated with the emission or absorption of energy in the form of radiation with the appropriate frequency. Since the frequency is proportional to the applied field, spin spectra can in principle be studied in any region of the electromagnetic spectrum, merely by choosing an appropriate field strength. For practical reasons the fields are normally of the order of 1.5 tesla for nuclei and 0.3 tesla for electrons.

For nuclei, $\beta_N = 5.05 \times 10^{-27}$ JT^{-1}, the g-factor for hydrogen is 5.858, and by choosing $B_z = 1.4092$ T

$$\frac{\Delta E}{h} = \frac{5.585 \times 5.05 \times 10^{-27} \times 1.4092}{6.63 \times 10^{-34}} = 60 \times 10^6 \text{ Hz} = 60 \text{ MHz}$$

This frequency is in the short-wave radio region. Many simpler spectrometers operate at this frequency.

For electrons, $\beta = 9.273 \times 10^{-24}$ JT^{-1}, $g = 2$, and assume $B_z = 0.33$ T. Then

$$\frac{\Delta E}{h} = \frac{2 \times 9.273 \times 10^{-24} \times 0.33}{6.63 \times 10^{-34}} \simeq 9000 \times 10^6 \text{ Hz} = 9 \text{ GHz}$$

This frequency is on the long wavelength side of the microwave region.

It was hinted before that the spin does not line up exactly in the field direction. If the moment associated with the spin is

$$\boldsymbol{\mu} = g_N \beta_N \sqrt{I(I+1)}$$

its component in the field direction must be either integral or half-integral, but the quantity $\sqrt{I(I+1)}$ can never assume these values.

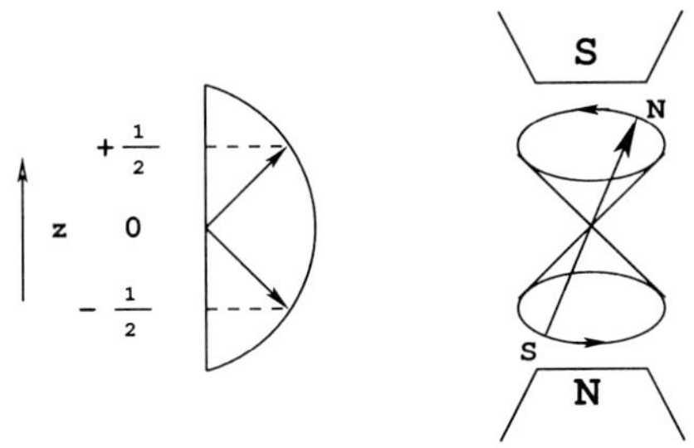

It lies at an angle to the field and is under the influence of a couple that works to align it along z. The net effect is that the vector precesses in the field at the *Larmor* frequency

$$\begin{aligned}\omega &= \frac{\text{magnetic moment}}{\text{angular momentum}} \times B_z \\ &= \frac{\boldsymbol{\mu} B_z}{2\pi \boldsymbol{I}} = \frac{g_N \beta_N B_z}{h} \text{ Hz}\end{aligned}$$

This expression is the same as that for the splitting between energy levels. It follows that this precessing angular momentum vector can interact with radiation of the same frequency while energy is exchanged. This match is known as *resonance*. For nuclei it is referred to as nuclear magnetic resonance (nmr) and for electrons as electron paramagnetic resonance (epr).

Experimentally there are two alternatives. One is to fix the magnetic field that determines the Larmor frequency and to sweep the radiation frequency through this value. Alternatively radiation at a fixed frequency is used while the field sweeps through the resonance value. When the resonance condition is reached absorption of radiation occurs.

The resonance frequency is characteristic of each nucleus for a given field strength. Nmr may hence be used to identify different nuclei in a sample. Since the effective field at a nucleus is modified by other nuclei and electrons in its vicinity, frequency shifts depending on the environment are observed. This is known as a *chemical shift*, and on the basis of this it is possible to map out the molecular environment of each spin in a system and to reconctruct details of the molecular structure. It is in this area that nmr finds its major application in chemistry. Epr is used to study radicals, *i.e.* molecules with unpaired electron spins.

6.3.4 Optical Rotation Spectroscopy

Optical activity refers to the ability of molecules to interact with linearly polarized light by rotating the plane of polarization. It has been demonstrated already (4.2.4) how to interpret linear polarization as the sum of two equal vectors rotating in opposite sense at the same rate.

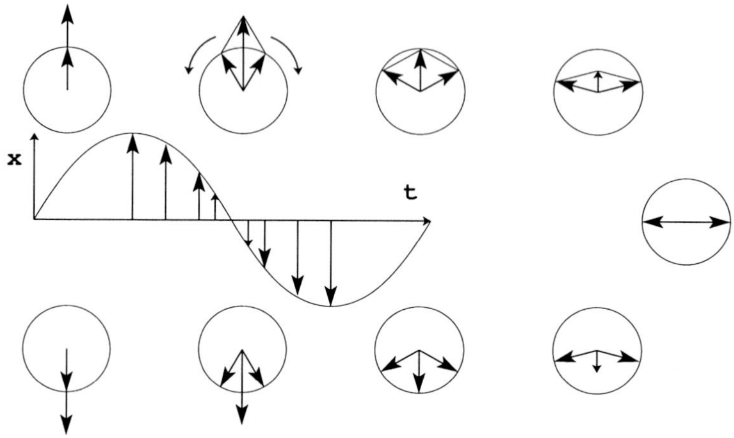

The graphic demonstration thereof is self-explanatory. The linear(resultant) vector defines a fluctuating electromagnetic field. Each of the rotating vectors describes a helix around the time axis.

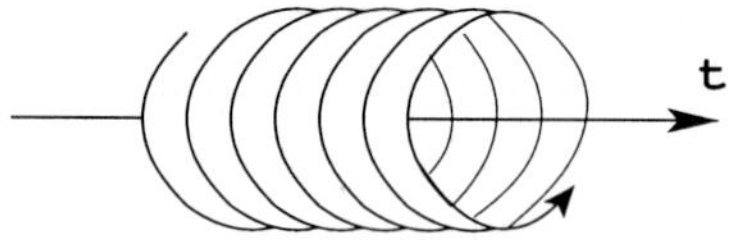

The rotating electric vectors generate equal, but opposite magnetic fields that cancel exactly, without any effect on most molecules in their path. However, molecules with non-zero orbital angular momentum interact preferentially with one of the two magnetization vectors of the radiation field. Only chiral molecules have this ability [69]. When such interaction occurs the two rotating vectors no longer propagate at the same rate through the sample and emerge out of phase. Their resultant is still a linear vector, but now rotated through an angle α with respect to the polarization plane of the incoming beam, as shown in figure 1.

The different velocities of propagation are commonly ascribed *e.g.* [70] to the medium exhibiting two different indices of refraction with respect to right and left circularly polarized light. Since the index of refraction is known experimentally to vary with the wavelength of incident light, the optical rotation must also be wavelength dependent. The spectroscopic technique known as *optical rotatory dispersion* (ORD) is based on this phenomenon. It is a generalization of of the effect demonstrated on the left of figure 1.

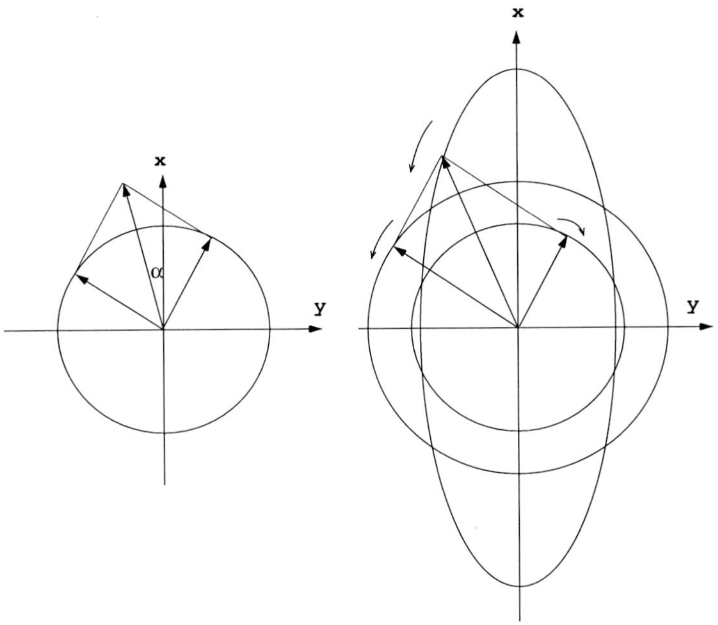

Figure 1 *The two diagrams show rotation of vectors equal in magnitude, on the left, and unequal vectors differentiated by absorption, on the right. The two situations lead to ORD and CD respectively.*

Apart from rotation, optical activity may also lead to differences in the absorptivity (ε) of right and left polarized components. The medium is said to exhibit *circular dichroism* (CD). Since the rotating vectors are no longer of the same magnitude, the resultant no longer oscillates along a single line but traces out an ellipse, as shown in figure 1 on the right.

An absorption band that occurs within a wavelength region of optical activity may be associated with either a positive or a negative CD response, since either the right or the left rotating vector may suffer preferential absorption. The difference ($\varepsilon_L - \varepsilon_R$) may be either positive or negative. Likewise, the ORD response may be either positive or negative on the long wavelength side of the absorption band. This anomalous feature is called the *Cotton effect*, which may be either positive or negative. Typical Cotton effects are shown schematically in figure 2.

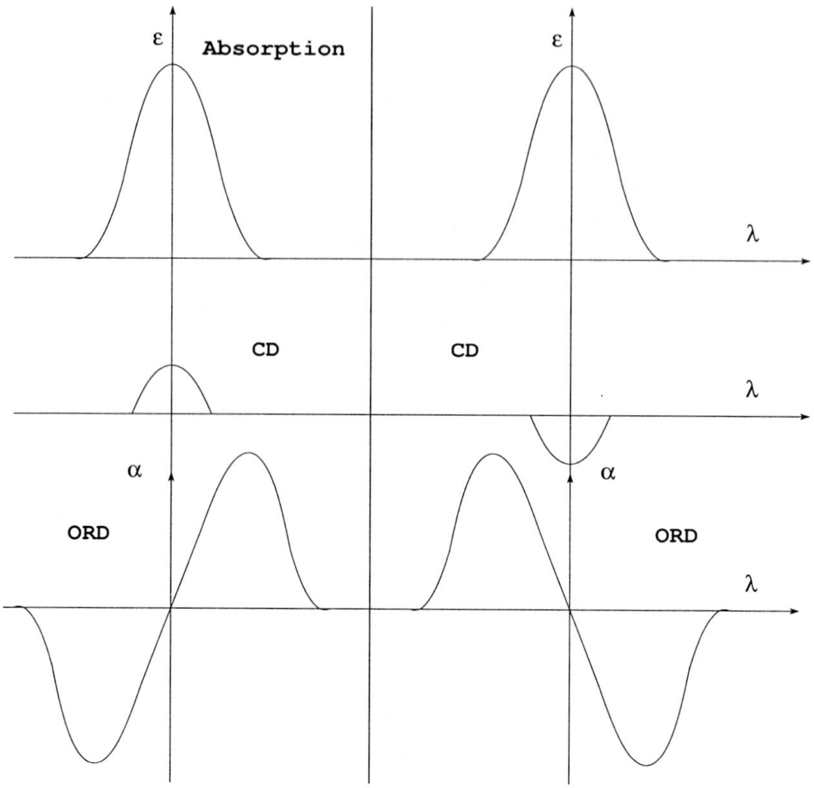

Figure 2 *Comparison of positive and negative Cotton effects.*

Both ORD and CD are important tools in the study of molecular conformation in solution. These techniques are of special importance in the study of biomolecules, their helical content, and helix-coil transitions.

6.3.5 Mössbauer Spectroscopy

Radioactive processes commonly leave atomic nuclei in an excited state from which they may relax to their ground state by the emission of γ-radiation. A typical example is the decay of cobalt through β-capture:

$$^{57}\text{Co} + \beta \rightarrow ^{57}\text{Fe}^* \rightarrow ^{57}\text{Fe} + \gamma$$

According to the Bohr frequency condition the emitted radiation should have the exact frequency to excite a second Fe atom from its ground state, in the reverse process

$$\text{Fe} + \gamma \rightarrow \text{Fe}^*$$

Non-occurrence of the inverse event was explained by Mössbauer in terms of the energy loss because of atomic recoil, during emission of the γ-ray photon. A simple calculation shows that a photon of frequency 10^{18} Hz has sufficient momentum to cause an Fe atom to recoil at a velocity of 10^2 ms^{-1}. Alternatively, the photon is Doppler shifted because of the recoil by an amount

$$\Delta\nu = \frac{\nu v}{c} = \frac{3.3 \times 10^{18} \times 10^2}{3 \times 10^8} \approx 10^{12}\,\text{Hz} \tag{6.7}$$

Although this shift is a minute fraction of the frequency, it exceeds the line width of the emission by orders of magnitude. Using $\Delta E \cdot \Delta t \approx h/2\pi \approx 10^{-34}$ Js, the uncertainty in energy that depends on the lifetime of the excited state, which for Fe* is about 1.5×10^{-7} s, implies a linewidth of

$$\begin{aligned}\Delta\nu &= \frac{\Delta E}{h} \approx \frac{h}{2\pi h \Delta t} \approx \frac{1}{2\pi \Delta t} \\ &\approx 10^6\,\text{Hz}\end{aligned}$$

The recoil therefore destroys the Bohr condition and prevents the re-absorption of the photon by a second Fe atom.

In order to dissipate the recoil energy Mössbauer was the first to use atoms in solid crystal lattices as emitters and also to cool both emitter and absorber. In this way it could be shown that the γ-ray emission from radioactive cobalt metal was absorbed by metallic iron. However, it was also found that if the iron sample were in any other chemical state, the different chemical surroundings of the iron nucleus produce a sufficient effect on the nuclear energy levels for absorption no longer to occur. To enable a search for the precisely required absorption frequency, a scan based on the Doppler effect was developed. It was noted that a velocity of 10^2 ms^{-1} produced an enormous Doppler shift and using the same equation (7) it follows that a readily attainable displacement of the source at a velocity of 1 cm s^{-1} produces a shift of 10^8 Hz. This shift corresponds to about 100 line-widths and provides a reasonable scan width.

In practice the source is oscillated back and forth at a velocity that varies between ± 1 cm s^{-1}. Samples with Fe in different chemical environments then absorb at different frequencies and the observed chemical shifts are correlated with electronic configurations in different environments.

6.3.6 Symmetry Aspects

The interpretation of molecular spectra consists of finding a correlation between observed spectral features and the symmetry of molecular species.

The procedure, described in more detail elsewhere [68], is demonstrated well by noting the relationship between molecular shape and the representation of symmetry groups.

Representations of Groups

Each symmetry operation of a group has the effect of turning the symmetrical object (molecule) into itself. An alternative statement of this observation is that each symmetry operation on the molecule is equivalent to multiplication by unity, as shown below for the operators of point group C_{2v}, using H_2O as example.

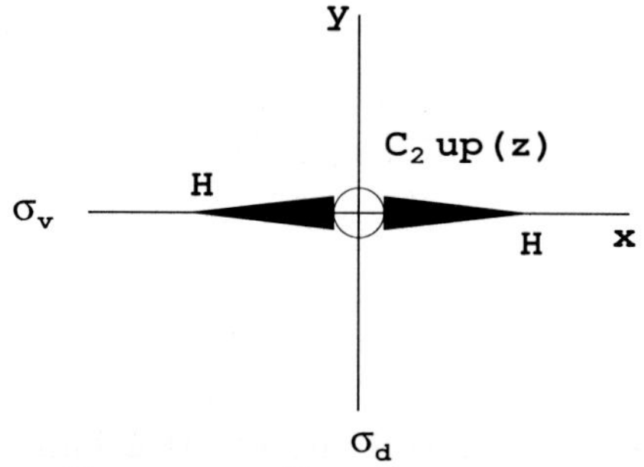

Operator	E	C_2	σ_v	σ_d
Effect	1	1	1	1

This exercise may appear pointless until one asks about the effect of the symmetry operations on other properties associated with the point group, such as cartesian coordinates and rotations.

Consider the effect of each operation on the cartesian coordinates x, y, z.

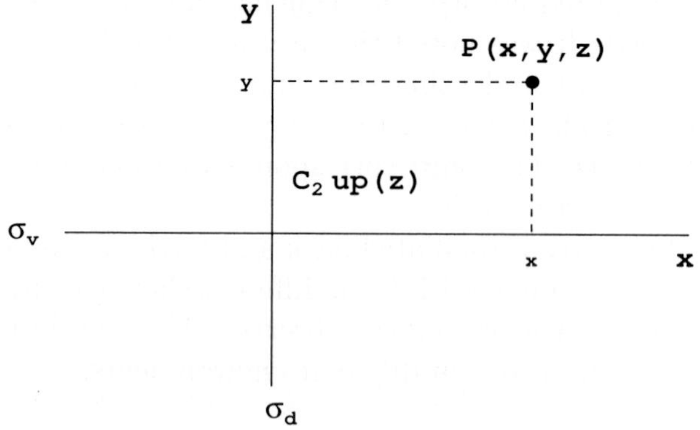

E(x)	$\rightarrow$	x	1	E(y)	$\rightarrow$	y	1	E(z)	$\rightarrow$	z	1
C_2(x)	$\rightarrow$	-x	-1	C_2(y)	$\rightarrow$	-y	-1	C_2(z)	$\rightarrow$	z	1
σ_v(x)	$\rightarrow$	x	1	σ_v(y)	$\rightarrow$	-y	-1	σ_v(z)	$\rightarrow$	z	1
σ_d(x)	$\rightarrow$	-x	-1	σ_d(y)	$\rightarrow$	y	1	σ_d(z)	$\rightarrow$	z	1

Next consider the effect of the symmetry operators on the clockwise rotations, defining $R_{cw} = 1$, $R_{acw} = -1$.

$E(R_x)$	$\rightarrow$	R_x	1	$E(R_y)$	$\rightarrow$	R_y	1	$E(R_z)$	$\rightarrow$	R_z	1
$C_2(R_x)$	$\rightarrow$	$-R_x$	-1	$C_2(R_y)$	$\rightarrow$	$-R_y$	-1	$C_2(R_z)$	$\rightarrow$	R_z	1
$\sigma_v(R_x)$	$\rightarrow$	$-R_x$	-1	$\sigma_v(R_y)$	$\rightarrow$	R_y	1	$\sigma_v(R_z)$	$\rightarrow$	$-R_z$	-1
$\sigma_d(R_x)$	$\rightarrow$	R_x	1	$\sigma_d(R_y)$	$\rightarrow$	$-R_y$	-1	$\sigma_d(R_z)$	$\rightarrow$	$-R_z$	-1

Each set of four numbers (± 1) constitutes an *irreducible representation* (i.r.) of the symmetry group, on the *basis* of either a coordinate axis or an axial rotation. According to a well-known theorem of group theory [2.7.4(v)], the number of i.r.'s is equal to the number of classes of that group. The four different i.r.'s obtained above therefore cover all possibilities for C_{2v}. The theorem thus implies that any representation of the symmetry operators of the group, on whatever basis, can be reduced to one of these four. In summary, the i.r.'s of C_{2v} are given by Table 1.

Table 1

C_{2v}	E	C_2	σ_v	σ_v'	
A_1	1	1	1	1	z
A_2	1	1	-1	-1	R_z
B_1	1	-1	1	-1	x,R_y
B_2	1	-1	-1	1	y,R_x

The labels A, B will be explained later on.

Whereas the number of i.r.'s is fixed, reducible representations are unlimited in number and generally made up of matrices. As an example, the symmetry operations of C_{2v} may be shown to correspond to the transformations described by the following 3rd order matrices:

$$E = \begin{bmatrix} 1 & 0 & 0 \\ 0 & 1 & 0 \\ 0 & 0 & 1 \end{bmatrix} \quad C_2 = \begin{bmatrix} -1 & 0 & 0 \\ 0 & -1 & 0 \\ 0 & 0 & 1 \end{bmatrix} \quad \sigma_v = \begin{bmatrix} 1 & 0 & 0 \\ 0 & -1 & 0 \\ 0 & 0 & 1 \end{bmatrix} \quad \sigma_v' = \begin{bmatrix} -1 & 0 & 0 \\ 0 & 1 & 0 \\ 0 & 0 & 1 \end{bmatrix}$$

$$\chi(E) = 3 \qquad \chi(C_2) = -1 \qquad \chi(\sigma_v) = 1 \qquad \chi(\sigma_v') = 1$$

Using the rules of matrix multiplication it is further shown that these matrices reproduce the C_{2v} multiplication table shown in Table 2 and readily obtained from the diagram below. As an example, from the table it follows that $\sigma_v C_2 = \sigma_v'$ as shown by the matrix multiplication.

Table 2

	E	C_2	σ_v	σ_v'
E	E	C_2	σ_v	σ_v'
C_2	C_2	E	σ_v'	σ_v
σ_v	σ_v	σ_v'	E	C_2
σ_v'	σ_v'	σ_v	C_2	E

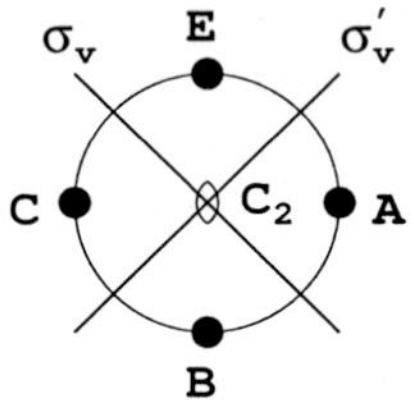

$$\begin{bmatrix} 1 & 0 & 0 \\ 0 & -1 & 0 \\ 0 & 0 & 1 \end{bmatrix} \begin{bmatrix} -1 & 0 & 0 \\ 0 & -1 & 0 \\ 0 & 0 & 1 \end{bmatrix} = \begin{bmatrix} -1 & 0 & 0 \\ 0 & 1 & 0 \\ 0 & 0 & 1 \end{bmatrix}$$

Again, each element of C_{2v} is its own inverse, and the same is readily demonstrated for the matrices, *e.g.*

$$\begin{bmatrix} 1 & 0 & 0 \\ 0 & -1 & 0 \\ 0 & 0 & 1 \end{bmatrix} \begin{bmatrix} 1 & 0 & 0 \\ 0 & -1 & 0 \\ 0 & 0 & 1 \end{bmatrix} = \begin{bmatrix} 1 & 0 & 0 \\ 0 & 1 & 0 \\ 0 & 0 & 1 \end{bmatrix}$$

The dimension of a representation is the same as the order of the matrix. To reduce a representation it is necessary to reduce its order. It is noted that the dimension of a matrix representation corresponds to the character of the identity (E) matrix.

The matrix representation of the C_{2v} group has characters

$$E : 3\, ;\; C_2 : -1\, ;\; \sigma_v : 1\, ;\; \sigma_v' : 1$$

The representation specified by these characters, *i.e.*

	E	C_2	σ_v	σ_v'
Γ	3	-1	1	1

is reducible to $\Gamma = A_1 + B_1 + B_2$, showing that the i.r.'s list the characters of their reduced matrices (one-dimensional in this case).

Character Tables

Choose any row of Table 1, the second say, implying the following association between symmetry operations and numbers:

$$\begin{array}{cccc} E & C_2 & \sigma_v & \sigma_v' \\ 1 & 1 & -1 & -1 \end{array}$$

Use this association to replace the symmetry elements in the body of Table 2 by the associated numbers to give:

Table 3

	1	1	−1	−1
1	1	1	−1	−1
1	1	1	−1	−1
−1	−1	−1	1	1
−1	−1	−1	1	1

Looked upon purely as an arithmetic multiplication table, the products are all correct. This condition remains valid whatever row is selected initially[4]. Because of the close association between tables 1 and 2 each row in Table 1 may indeed be regarded as a representation of the symmetry elements.

The A and B labels in Table 1 follow the convention that A's have characters of +1 for the rotation axis of highest order (C_2 in the present case) while B's have character -1. A_1, by convention, is the totally symmetric i.r., since all operations of the group turn something of A_1 symmetry into itself. Every group has a totally symmetric i.r. I.r.'s with suffix 1 are symmetric (character +1) under σ_v, whereas those with suffix 2 are antisymmetric (character -1). Table 1 is an example of a character table. Two-dimensional representations are denoted by symbols E.

6.3.7 Spectroscopic selection rules

In molecular spectroscopy the probability that a transition between energy states generates a signal, is formulated in terms of an integral such as

$$\int \psi_2 P \psi_1 \mathrm{d}\tau$$

[4]A search for alternative sets that give a correct multiplication table in the same sense, is abortive.

performed over all space. The wave functions ψ_1 and ψ_2 define the two states and P is an operator associated with the observable property of interest. Such integrals are difficult to evaluate, but many cases in which the integral vanishes identically because of symmetry factors can easily be recognized from point-group character tables.

In most forms of spectroscopy a beam of electromagnetic radiation is allowed to interact with a molecular sample. The radiation may be regarded as composed of two oscillating fields - an electric field and a magnetic field. Either, or both of these fields may interact with a molecule by coupling to an electric and/or magnetic fluctuation (electric dipole or magnetic moment) caused by distortion of the molecule. This happens when the field oscillation matches a natural oscillation (including rotation) frequency of the molecule.

Vibrations may be decomposed into three orthogonal components T_α ($\alpha = x, y, z$) in three directions. These displacements have the same symmetry properties as cartesian coordinates. Likewise, any rotation may be decomposed into components R_α. The i.r. spanned by translations and rotations must clearly follow the appropriate symmetry type of the point-group character table. In quantum formalism, a transition will be allowed only if the symmetry product of the initial and final-state wave functions contains the symmetry species of the operator appropriate to the transition process. Definition of the symmetry product will be explained in terms of a simple example.

Consider the product $\phi_1\phi_2$ and suppose that ϕ_1 has symmetry A_2 of the point group C_{2v}, written as $\phi_1(A_2)$ and that ϕ_2 has symmetry B_2, *i.e.* $\phi_2(B_2)$. It is required to find the symmetry of $\psi = \phi_1(A_2) \cdot \phi_2(B_2)$. The wave function ψ is subjected to all operations of C_{2v}, starting with C_2, noting from the character table that under this operation

$$\begin{aligned} \psi &= \phi_1(A_2) \cdot \phi_2(B_2) \rightarrow \phi_1(A_2) \cdot [-\phi_2(B_2)] \\ &= -\psi \end{aligned}$$

Under σ_v

$$\begin{aligned} \psi &= \phi_1(A_2) \cdot \phi_2(B_2) \rightarrow [-\phi_1(A_2)] \cdot [-\phi_2(B_2)] \\ &= \psi \end{aligned}$$

and under σ_v', by the same procedure, $\psi \rightarrow -\psi$.

The characters that transform the product function ψ follow from the foregoing as

E	C_2	σ_v	σ_v'
1	−1	1	−1

which correspond to the i.r. B_1.

This result can be expressed in terms of the *direct product* of the individual characters under A_2 and B_2, written $B_1 = A_2 \otimes B_2$. This procedure to establish the symmetry species of a product function is generally valid. The symmetry product in the previously stated selection rule is thereby defined as the direct product symmetry species of the initial and final state wave functions.

An electric dipole operator, of importance in electronic (visible and uv) and in vibrational spectroscopy (infrared) has the same symmetry properties as T_α. Magnetic dipoles, of importance in rotational (microwave), nmr (radio frequency) and epr (microwave) spectroscopies, have an operator with symmetry properties of R_α. Raman (visible) spectra relate to polarizability and the operator has the same symmetry properties as terms such as x^2, xy, *etc.* In the study of optically active species, that cause helical movement of charge density, the important symmetry property of a helix to note, is that it corresponds to simultaneous translation and rotation. Optically active molecules must therefore have a symmetry such that T_α and R_α $(\alpha = x, y, z)$ transform as the same i.r. It only occurs for molecules with an *alternating* or *improper* rotation axis, S_n.

When deriving selection rules from character tables it is noted that vibrations are usually excited from the ground state which is totally symmetric. The excited state has the symmetry of the vibration being excited. Hence: A vibration will be spectroscopically active if the vibration has the same symmetry species as the relevant operator.

6.4 Free-Particle Models

Only those problems that can be reduced to one-dimensional one-particle problems can be solved in closed form by the methods of wave mechanics, which excludes all systems of chemical interest. As shown before, several chemical systems can be approximated by one-dimensional model systems, such as a rotating diatomic molecule modelled in terms of a rotating particle in a fixed orbit. The trick is to find a one-dimensional potential function, V that provides an approximate model of the interaction of interest, in the Schrödinger formulation

$$\frac{d^2\psi}{dx^2} + k^2\psi \quad ; \quad k^2 = \frac{2m}{\hbar^2}(E - V)$$

In many cases it is convenient to start from the free-particle formulation, $V = 0$, and introduce suitable potential barriers to simulate specific interactions.

All models of this type have become known colloquially by the misnomer free-particle model. Diverse objects with formal resemblance to chemical systems are included here, such as an electron in an impenetrable sphere to model activated atoms; particle on a line segment to model delocalized systems; particle interacting with finite barriers to simulate tunnel effects; particle interacting with periodic potentials to simulate electrons in solids, and combinations of these.

6.4.1 Particle in a Sphere

Closely related to the problem of a particle on a line is that of a particle confined to a hollow sphere. Such a particle is described by the same Hamiltonian as a free particle ($V = 0$), *i.e.*

$$-\frac{\hbar^2}{2m}\nabla^2\psi = E\psi$$

$$\nabla^2\psi + k^2\psi = 0 \quad , \text{ where} \quad k = \sqrt{\frac{2mE}{\hbar^2}},$$

with plane-wave solutions

$$\psi = Ae^{i\boldsymbol{k}\cdot\boldsymbol{r}}$$

interpreted as the wavefronts of De Broglie waves normal to $\boldsymbol{k}$, $\lambda = 2\pi/k$. Described in spherical polar coordinates

$$\psi(r, \theta, \phi) = R(r) \cdot Y(\theta, \phi)$$

the radial part of Schrödinger's equation becomes

$$\frac{d^2R}{dr^2} + \frac{2}{r}\frac{dR}{dr} + \left[\frac{2mE}{\hbar^2} - \frac{l(l+1)}{r^2}\right] R = 0$$

$$\frac{d^2}{dr^2}(rR) + \left[k^2 - \frac{l(l+1)}{r^2}\right] rR = 0$$

By substituting $rR = y\sqrt{r}$, $x = kr$ and noting that

$$\frac{d^2}{dr^2}(y) = k^2\frac{d^2}{dx^2}(y), \quad etc.$$

the equation becomes

$$\frac{d^2y}{dx^2} + \frac{1}{x}\frac{dy}{dx} - \frac{y}{4x^2} + \left[1 - \frac{l(l+1)}{x^2}\right] y = 0$$

i.e.

$$\frac{d^2y}{dx^2}+\frac{1}{x}\frac{dy}{dx}+\left[1-\frac{(l+\frac{1}{2})^2}{x^2}\right]y=0$$

which is Bessel's equation for $n = l + \frac{1}{2}$. This equation is known to have solutions

$$y = J_l(x) = \sqrt{\frac{2x}{\pi}}j_l(x)$$

$$\text{which is}\quad y\sqrt{x} = \sqrt{\frac{2}{\pi}}xj_l(x)$$

$$\begin{aligned} i.e.\quad rR &= \sqrt{\frac{2}{\pi}}\cdot kr\, j_l(kr) \\ R &= \sqrt{\frac{2}{\pi}}\cdot k\, j_l(kr) \end{aligned}$$

The $j_l(kr)$ are spherical Bessel functions, *e.g.* $j_0(x) = (\sin x)/x$. Since k^2 can take any positive real value the energy spectrum

$$E = \frac{\hbar^2k^2}{2m}$$

is continuous, $0 < E < \infty$.

For every positive E there exists an eigenfunction

$$\psi_{E,l,m}(r,\theta,\phi) = c\, j_l(kr)\, Y_{lm}(\theta,\phi)$$

This ensemble of spherical waves forms a complete set. The plane-wave solution of a particle of momentum $\hbar k$ and energy E is therefore represented by

$$e^{i\boldsymbol{k}\cdot\boldsymbol{r}} = \sum_{l=0}^{\infty}\sum_{m=-l}^{l} c_{lm}(k)\, j_l(kr)\, Y(\theta,\phi)$$

For a particle enclosed in an impenetrable hollow sphere of radius a, $ka = \alpha$, the requirement that the wave function must vanish at a, means that α should coincide with some zero of j_l. In the ground state, at the first zero

$$j_0 = \frac{\sin\alpha}{\alpha}\quad,\ i.e.\quad \alpha = \pi$$

$$E_0 = \frac{\hbar^2\pi^2}{2ma^2} = \frac{h^2}{8ma^2}$$

or in general

$$E_n = \frac{n^2h^2}{8ma^2}$$

The wave function is normalized by putting

$$\begin{aligned} N^2 \int_0^a 4\pi r^2 R^2 dr &= 1 \\ \text{For} \quad R &= \sqrt{\frac{2}{\pi}} k\, j_0(kr) \\ I &= 8N^2 \int_0^a \sin^2(kr)dr \\ &= 8N^2 \left[\frac{r}{2} - \frac{\sin 2kr}{4k}\right]_0^a \\ &= 8N^2 \left(\frac{a}{2}\right) \end{aligned}$$

$$i.e. \quad N = \frac{1}{2\sqrt{a}}$$

It may be shown [5] that the set of functions $\psi(r, \theta, \phi)$ which vanishes at $r = a$ and is piecewise continuous can be expanded in the form

$$\sum_l Y_l \sum_n c_{l,k_n} r^{-\frac{1}{2}} J_{l+\frac{1}{2}}(k_{l,n}r)$$

For the special case $l = 0$ this expansion reduces to a Fourier series. The ground-state series is the Fourier transform of $\sin ka/ka$, which is the box function

$$f(r) = \begin{cases} \alpha & \text{if} \quad |r| < a \\ 0 & \text{if} \quad |r| > a \end{cases}$$

This result becomes important as the wave function of the valence electron of compressed atoms.

It is easy to show (3.6.3) that the reverse transform

$$\begin{aligned} g(k) &= \frac{a}{\sqrt{2\pi}} \int_{-a}^{a} \alpha e^{ikx} dx \\ &= \frac{\sin ka}{ka} \\ \text{for} \quad \alpha &= \frac{\sqrt{2\pi}}{2a} \end{aligned}$$

This result is consistent with describing the particle as a plane-wave in all directions, which implies that the energy is infinitely degenerate.

6.4.2 Electron Gas

Solids (crystals) are distinguished from liquids and gases by their regular periodic structures. Solids can be classified as metals with resistivities of about $10^{-8}\Omega$m, semiconductors (10^{-4} - $10^{-6}\Omega$m) and insulators ($> 10^{8}\Omega$m). These differences are thought to reflect different electronic distributions and therefore amenable to quantum-mechanical analysis. Since a crystal however, consists of an enormous number of particles, the solution of the complete Schrödinger equation over all nuclei and electrons is quite impossible. A simple model that has been used successfully to describe electrons in a metal is the Sommerfeld free-electron model.

Sommerfeld suggested that the potential in a metal crystal could be assumed constant. This assumption implies that the forces acting on an electron cancel to zero and that the electrons in a metal can be described like a non-interacting gas of electrons, confined to a box that represents the metal. The only restriction on electronic motion would be the Pauli principle. The electronic energy in a three-dimensional rectangular box is known as

$$E_k = \frac{h^2}{8\pi^2 m}\left(k_x^2 + k_y^2 + k_z^2\right)$$

with eigenfunctions $\psi(k) = A\sin k_x x.\sin k_y y.\sin k_z z$, where $k_x = l\pi/a$, $k_y = m\pi/b$, $k_z = n\pi/c$. For a cubic box with equal sides ($a = b = c$), the energy becomes

$$E_k = \frac{h^2}{8ma^2}\left(l^2 + m^2 + n^2\right)$$

Since a is of the order 10^{-2}m the energy levels are closely spaced ($\sim 10^{-15}$eV apart) and for most purposes can be taken as zero.

The number of electronic states (neglecting spin) with kinetic energies less than some energy E_k, is given by the number of integral points within the positive octant of a sphere of radius r, such that

$$E_k = \frac{(hr)^2}{8ma^2}, \quad r = \left(\frac{8ma^2 E_k}{h^2}\right)^{\frac{1}{2}}$$

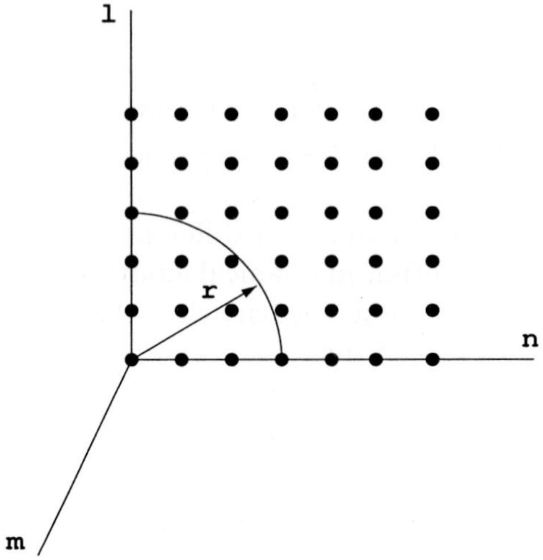

Figure 3: *Diagram to show the estimate of quantum states at energies less than E_k in terms of the lattice points in a limiting sphere.*

For large r, the number of points is the octant volume, $v_8 = \frac{1}{8}\left(\frac{4}{3}\pi r^3\right)$.

The number of electronic states with $E < E_k$ therefore is

$$\begin{aligned} M(E_k) &= \frac{1}{8}\left(\frac{4\pi}{3}\right)\left(\frac{8ma^2E_k}{h^2}\right)^{\frac{3}{2}} \\ &= \left(\frac{\pi a^3}{6}\right)\left(\frac{8mE_k}{h^2}\right)^{\frac{3}{2}} \\ &= \frac{\pi V}{6}\left(\frac{8mE_k}{h^2}\right)^{\frac{3}{2}} \end{aligned}$$

where V is the volume of the box. The number of states per unit energy range, called the density of states $N(E)$ is related to $M(E)$ by

$$\int_0^{E_k} N(E)dE = M(E_k)$$

Hence

$$N(E_k) = \left(\frac{dM}{dE}\right)_{E_k}$$

It follows that

$$N(E_k) = \frac{V\pi}{4}\left(\frac{8m}{h^2}\right)^{\frac{3}{2}} E_k^{\frac{1}{2}}$$

as shown graphically below.

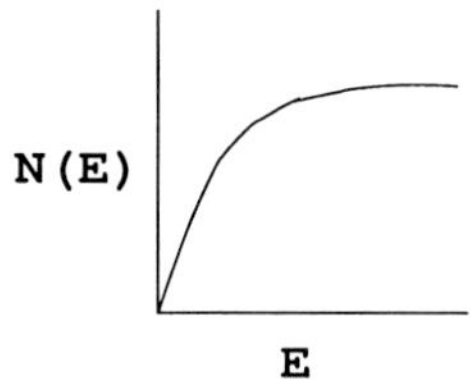

When spin is included these values of $N(E)$ and $M(E)$ are doubled.

The electrons that occupy the levels of a Fermi gas have energies $< \epsilon_F$ and may be considered as confined to a *(Fermi) sphere* of radius k_F in k-space. For large volumes the free-electron quantum numbers may be treated as continuous variables and the number of states in a range $d\boldsymbol{k} = dk_x dk_y dk_z$, is

$$dn_x dn_y dn_z = \frac{a}{2\pi} \cdot \frac{b}{2\pi} \cdot \frac{c}{2\pi} d\boldsymbol{k} = \frac{V}{2\pi^3} d\boldsymbol{k}$$

Unit volume in k-space therefore accommodates $V/4\pi^3$ electrons (two per level). The Fermi sphere accommodates

$$\frac{V}{4\pi^3} \cdot \frac{4}{3} \pi k_F^3 = \frac{V}{3\pi^2} k_F^3 = N$$

electrons, so that

$$k_F = (3\pi^2 \rho)^{\frac{1}{3}} \quad , \quad \text{where } \rho = \frac{N}{V} \tag{6.8}$$

is the number of electrons per unit volume. At the surface of the Fermi sphere, known as the *Fermi surface*, the energy is the Fermi energy

$$\epsilon_F = \frac{h^2}{8m} \left(\frac{3N}{\pi V} \right)^{2/3} = \frac{h^2}{8m} \left(\frac{3\rho}{\pi} \right)^{2/3} = \frac{h^2}{8m} \left(3\pi^2 \rho \right)^{2/3} \cdot \frac{1}{\pi^2} = \frac{\hbar^2}{2m} k_F^2$$

Fermi-Dirac Statistics

Electrons with their half-integral spins are known as Fermi particles or *fermions* and no more than two electrons can occupy a quantum state. At absolute zero the electrons occupy energy levels from zero to a maximum value of ϵ_F, defined by

$$\frac{N}{2} = \frac{\pi V}{6} \left(\frac{8m}{h^2} \right)^{\frac{3}{2}} \epsilon_F^{\frac{3}{2}}$$

i.e.

$$\epsilon_F = \frac{h^2}{8m} \left(\frac{3N}{\pi V} \right)^{\frac{2}{3}}$$

The value of the *Fermi energy* ϵ_F corresponding to the valence electron density in metals is of the order of a few eV.

The average energy of an electron at 0K in the Fermi sea is

$$\bar{\epsilon} = \frac{1}{N}\int_0^{\epsilon_F} E\,dN$$

$$\text{From} \quad N = \frac{\pi V}{3}\left(\frac{8mE}{h^2}\right)^{\frac{3}{2}} \quad \text{follows that}$$

$$\begin{aligned} \frac{dN}{dE} &= \frac{\pi V}{3}\left(\frac{8m}{h^2}\right)^{\frac{3}{2}} \cdot \frac{3}{2}E^{\frac{1}{2}} \\ \bar{\epsilon} &= \frac{\pi V}{2N}\left(\frac{8m}{h^2}\right)^{\frac{3}{2}} \int_0^{\epsilon_F} E^{\frac{3}{2}}dE \\ &= \frac{3}{5}\epsilon_F \end{aligned}$$

As the temperature increases some electrons are excited to higher levels. The probability of a system being in an allowed state with energy E is proportional to $\exp(-E/kT)$. This result is useful in quantum systems only if the many-electron eigenstates are known.

The Fermi-Dirac distribution law[5] gives the probability that a single-

---

[5] The energy distribution of electrons is described by the Fermi-Dirac distribution function. It embodies the facts that electrons are indistinguishable and obey an exclusion principle that allows only one electron to occupy each state. Consider N electrons with fixed total energy U. Imagine a set of levels with energies ϵ_i and degeneracies g_i. Let n_i be the number of electrons at each level i. The n_i are arbitrary except for the restrictions $\sum_i n_i = N$, $\sum_i n_i\epsilon_i = U$. The first particle can be placed in any of the g_i cells, the second in any of the remaining $g_i - 1$, and so on. The total number of microstates is given by the product of binomial coefficients $\Omega = \prod_i\{g_i!/[n_i!(g_i - n_i)!]\}$. Using Stirling's approximation $[\ln N! = N\ln N - N]$ one has

$$\begin{aligned} \ln\Omega &= \sum_i [g_i \ln g_i - n_i \ln n_i - (g_i - n_i)\ln(g_i - n_i)] \\ \frac{\partial \ln\Omega}{\partial n_i} &= \ln(g_i - n_i) - \ln n_i \end{aligned}$$

Since entropy is defined in terms of Ω as $S = k\ln\Omega$, the total differential $d\ln\Omega$ is exact, and for $\Omega = \Omega(N, U, V)$, can be written as

$$d\ln\Omega = \left(\frac{\partial\ln\Omega}{\partial N}\right)_{U,V} dN + \left(\frac{\partial\ln\Omega}{\partial U}\right)_{N,V} dU = \alpha dN + \beta dU = k^{-1}S \quad \text{(by definition)}$$

electron state is occupied at temperatute T as

$$f(\epsilon_i) = \left[\exp\left(\frac{\epsilon_i - \epsilon_F}{kT}\right) + 1\right]^{-1}$$

Noting that

$$\lim_{T\to 0} \exp\left(\frac{x}{T}\right) = \infty \quad \text{and} \quad \lim_{T\to 0} \exp\left(-\frac{x}{T}\right) = 0$$

it follows that for $\epsilon_i < \epsilon_F$, $f = 1$ at $T = 0$ and for $\epsilon_i > \epsilon_F$, $f = 0$.

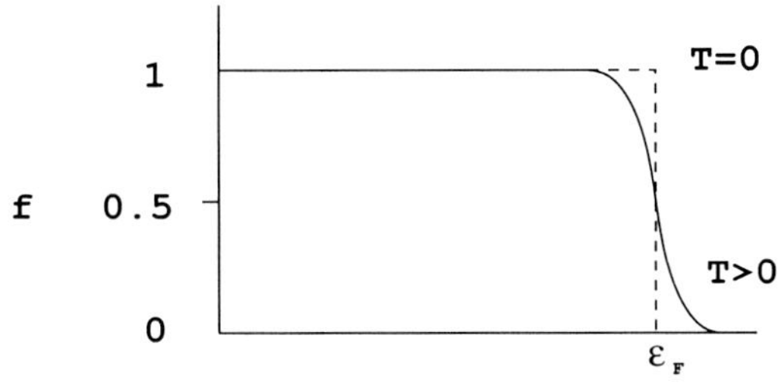

The only region where $f(E)$ is significantly different from 1 is in the neighbourhood of ϵ_F, over a range of energies of the order of kT.

The functions α and β are intensive because N, U and $d\ln\Omega$ are extensive. If the size of the system is increased m-fold, while keeping the intensive variables fixed, integration gives

$$\int_{\ln\Omega}^{m\ln\Omega} d\ln\Omega = \alpha\int_N^{mN} dN + \beta\int_U^{mU} dU$$

i.e. $\ln\Omega = \alpha N + \beta U$. To find the maximum in this equilibrium, put

$$\frac{\partial}{\partial n_i}(\ln\Omega - \alpha N - \beta U) = 0$$

$$\text{But} \quad N = \sum_i n_i \quad \text{and} \quad U = \sum_i n_i\epsilon_i, \quad \text{hence} \quad \frac{\partial \ln\Omega}{\partial n_i} - z\alpha - \beta\epsilon_i = 0$$

Therefore $\ln[(g_i - n_i/n_i] = \alpha + \beta\epsilon_i$, *i.e.* $(g_i - n_i)/n_i = \exp(\alpha + \beta\epsilon_i)$, which rearranges to the distribution $(n_i/g_i) = [1 + \exp(\alpha + \beta\epsilon_i)]^{-1}$, usually presented in the form $f(\epsilon) = [1 + \exp\{(\epsilon - \epsilon_F)/kT\}]^{-1}$. To evaluate α and β one uses the well-known results from thermodynamics: $dU = TdS - PdV + \mu\, dN$ *i.e.* $dS = (dU/T) + (P/T)dV - (\mu/T)dN = d(k\ln\Omega) = k\alpha\, dN = k\beta\, dU$. Thus,

$$\left(\frac{\partial S}{\partial N}\right)_{V,N} = \frac{1}{T} = k\beta \;\; ; \;\; \left(\frac{\partial S}{\partial N}\right)_{U,V} = -\frac{\mu}{T} = k\alpha$$

Hence

$$\alpha = -\mu/kT\,,\; \beta = 1/kT \quad \text{and} \quad n_i/g_i = [1 + \exp(\epsilon_i - \mu)/kT]^{-1}$$

which shows that the Fermi level ϵ_F is the same as the chemical potential, μ.

The Fermi Level

Since the energy of electrons in a material is specified by the Fermi level, ϵ_F, the flow of electrons across an interface must likewise depend on the relative Fermi levels of the materials in contact. Redox properties are therefore predicted to be a function of the Fermi energy and one anticipates a simple relationship between the Fermi level and redox potential. In fact, the Fermi level is the same as the chemical potential of an electron. Clearly when dealing with charged particles, the local energy levels ϵ_i are increased by qV, where q is the charge on the particle and V is the local electrostatic potential. The ϵ_i should therefore be replaced by $\epsilon_i + qV$ and so

$$\mu = \epsilon_F - qV \quad \text{or} \quad \epsilon_F = \mu + qV = \bar{\mu}$$

which is the electrochemical potential of the electron.

To show that the Fermi level can be interpreted as the redox potential, consider a redox equilibrium such as

$$\mathrm{Fe}^{3+} + e^- \leftrightharpoons \mathrm{Fe}^{2+}$$

which requires a balance of electrochemical potentials

$$\bar{\mu}(\mathrm{Fe}^{3+}) + \bar{\mu} = \bar{\mu}(\mathrm{Fe}^{2+})$$

In terms of concentrations: $\bar{\mu}(x) = \bar{\mu}(x)^{\ominus} + kT \ln c(x)$ where $x^{\ominus}$ refers to the standard state, it follows that

$$\bar{\mu}(\mathrm{Fe}^{+3})^{\ominus} + kT \ln c(\mathrm{Fe}^{3+}) + \bar{\mu} = \bar{\mu}(\mathrm{Fe}^{2+})^{\ominus} + kT \ln c(\mathrm{Fe}^{2+})$$

i.e.

$$\bar{\mu} = [\bar{\mu}(\mathrm{Fe}^{2+})^{\ominus} - \bar{\mu}(\mathrm{Fe}^{3+})^{\ominus}] + kT \ln[c(\mathrm{Fe}^{2+})/c(\mathrm{Fe}^{3+})]$$

This equation has the same form as the conventional expression for the redox potential of the $\mathrm{Fe}^{2+}/\mathrm{Fe}^{3+}$ couple:

$$V_r(\mathrm{Fe}^{2+}/\mathrm{Fe}^{3+}) = V_r^{\ominus}(\mathrm{Fe}^{2+}/\mathrm{Fe}^{3+}) + \frac{kT}{e} \ln[c(\mathrm{Fe}^{2+})/c(\mathrm{Fe}^{3+})]$$

where e is the electronic charge. The quantity $\bar{\mu}/e$ therefore represents the redox potential, except for the choice of reference level.

6.4.3 Potential Barriers

An important difference between classical and quantum particles is the way they interact with potential barriers. It is a principle of classical mechanics that the only way to overcome a potential barrier is with sufficient energy. Quantum-mechanically this is not always the case. The effect is illustrated by a beam of particles (*e.g.* electrons) approaching a potential barrier.

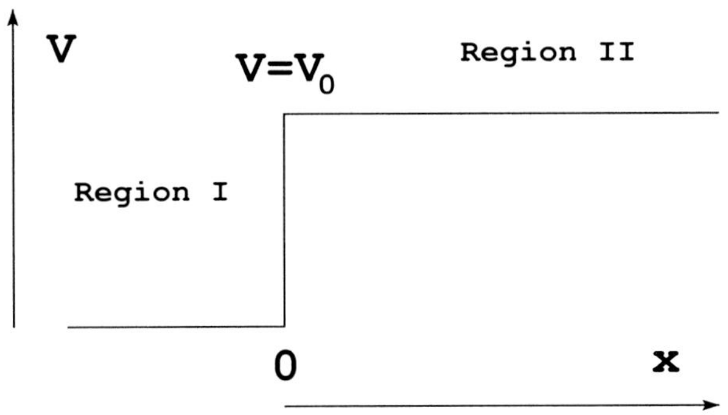

The potential energy of the particle in region I, $x < 0$, is $V(x) = 0$. In region II, $x > 0$, $V(x) = V_0$.

It is required to find the eigenfunction solutions of the equation

$$\frac{d^2\psi}{dx^2} + \frac{8\pi^2 m}{h^2}\left[E - V(x)\right]\psi = 0$$

The familiar solution in region I is

$$\psi_I(x) = Ae^{i\alpha x} + Be^{-i\alpha x} \quad , \quad x < 0$$

$$\text{with} \quad \alpha = \frac{2\pi}{h}(2mE)^{1/2} \quad \text{and} \quad E = \frac{h^2\alpha^2}{8\pi^2 m}$$

A and B are constants.

The first term represents particles (waves) moving towards the barrier and the second describes particles reflected from the barrier. To the right of the barrier the eigenfunctions are

$$\psi_{II}(x) = Ce^{i\beta x} + De^{-i\beta x}\,,\; x > 0\,,\; \beta = \frac{2\pi}{h}[2m(E - V_0]^{1/2}.$$

The particle density is given by $\psi^*\psi$. There are two important cases to be distinguished:

(i) $E > V_0$

Treated as a classical problem all particles would be transmitted across the barrier. Quantum-mechanically this is not necessarily true: The intensity of the incident beam is given by

$$\psi_i\psi_i^* = Ae^{i\alpha x} \cdot Ae^{-i\alpha x} = A^2,$$

and of the reflected beam proportional to B^2. The ratio of reflected to transmitted number of particles is B^2/A^2, which is non-zero, and there is a finite probability of finding reflected particles in region I.

The second term in region II represents particles moving to the left, and since this is zero by definition, $D = 0$. The wave function must be continuous at $x = 0$ and this requires that $\psi_I(0) = \psi_{II}(0)$, and also that the first derivatives

$$\frac{d\psi_I(0)}{dx} = \frac{d\psi_{II}(0)}{dx}.$$

It follows that $A + B = C$, $(e^0 = 1)$, and that

$$\alpha(A - B) = \beta C\,, \left[\frac{d}{dx}(e^{ax}) = ae^{ax}\right]$$

After elimination of C:

$$\begin{aligned}
\alpha(A - B) &= \beta(A + B) \\
\frac{B}{A} &= \frac{\alpha - \beta}{\alpha + \beta} = \frac{\sqrt{E} - \sqrt{E - V_0}}{\sqrt{E} + \sqrt{E - V_0}} \\
\frac{C}{A} &= \frac{2\alpha}{\alpha + \beta} = \frac{2\sqrt{E}}{\sqrt{E} + \sqrt{E - V_0}}
\end{aligned}$$

In the limit $E \to \infty$, $B = 0$, $C = A$. Quantum mechanics is then in agreement with classical mechanics and all particles are transmitted.

(ii) $E < V_0$

Classically no particles would be transmitted in this case. Quantum mechanically however,

$$\beta^2 = \frac{8\pi^2 m}{h}(E - V_0)$$

$$\beta = \frac{2\pi i}{h}\sqrt{2m(V_0 - E)}$$

and therefore

$$\psi_{II} = Ce^{-\beta x} \neq 0.$$

It is noted that the solutions are real rather than complex exponentials. Since the constant $D = 0$ the probability density remains finite as $x \to \infty$. In the region $x < 0$, as before,

$$A - B = \frac{\beta}{\alpha}C = iC\sqrt{\frac{V_0 - E}{E}}$$

$$\frac{B}{C} = \frac{\alpha - \beta}{2\alpha} = \frac{1}{2}\left(1 - \frac{\beta}{\alpha}\right) = \frac{1}{2}\left(1 - i\sqrt{\frac{V_0 - E}{E}}\right)$$

$$A = \frac{C}{2}\left(1 + i\sqrt{\frac{V_0 - E}{E}}\right)$$

Hence the intensity ratio of reflected to incident waves is

$$R = \frac{|B|^2}{|C|^2} = 1$$

This ratio implies that the entire wave is reflected, and that no particles are transmitted. However, there is an exponentially decaying part of the wave function in the region $x > 0$. The total wave function looks somewhat like

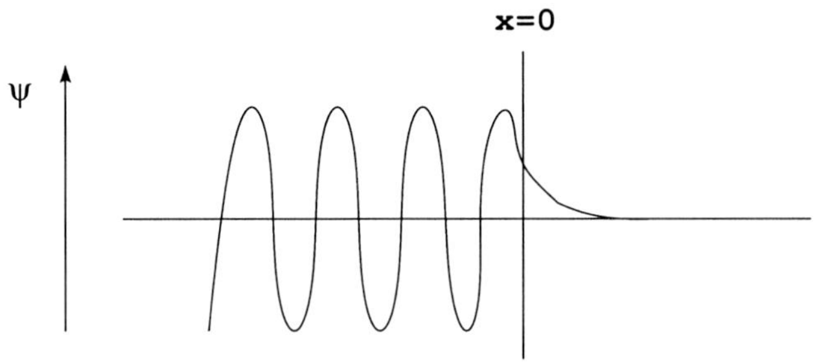

The exponentially decaying part of the function should therefore not be interpreted as describing a wave that moves to the left, but as an exponentially decaying absorption. The surprising non-classical result is the non-zero probability that the wave function in region II penetrates into the barrier. Penetration into the potential barrier makes sense only in terms of a wave model and not in terms of a particle model of (*e.g.*) an electron, at all. Recall that an electron has definite energy only in a stationary state when its wave function is an eigenfunction of the Hamiltonian operator, and therefore only when the electron is spread over a broad region of space. An electronic particle in the region $V > E$ would have negative kinetic energy and is undefined. It can be localized by increasing the electronic energy until $E > V$, *i.e.* the same condition as for a classical particle.

The region where $V > E$ corresponds to an imaginary index of refraction

$$n = \frac{c}{\hbar\omega}\sqrt{2m(E - V)}$$

and penetration into this region is very similar to the total internal reflection of light waves, that occurs for exactly the same reason.

6.4.4 The Tunnel Effect

When the potential barrier in case (ii) above is of finite extent and separates two regions at lower potential the possibility arises that a particle may tunnel through the barrier.

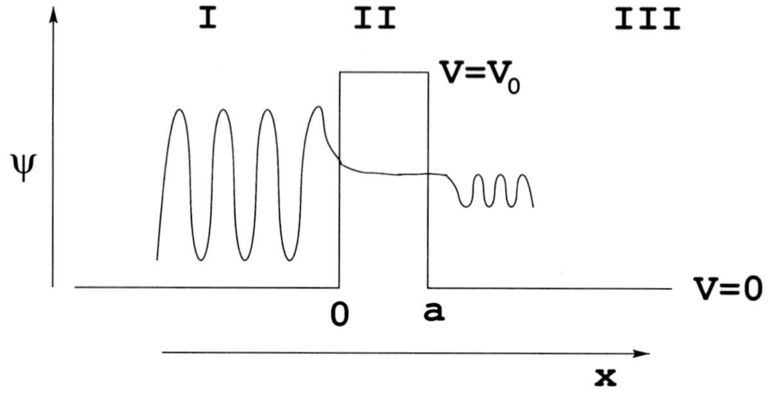

Tunnelling solutions may occur when the energy of particles, incident from the left, is less than the height of the potential barrier. Classically none of these would be transmitted, but this is not necessarily so in quantum mechanics. The Schrödinger equations to be solved are:

$$\frac{d^2\psi}{dx^2} + \frac{8\pi^2 mE}{h^2}\psi = 0 \quad ; \quad x < 0, x > a.$$

$$\frac{d^2\psi}{dx^2} + \frac{8\pi^2 m(E - V_0)}{h^2}\psi = 0 \quad ; \quad 0 < x < a.$$

When $E < V_0$ the solutions are:

$$\begin{aligned}\psi_I &= Ae^{i\alpha x} + Be^{-i\alpha x}\\ \psi_{II} &= Ce^{\beta x} + De^{-\beta x}\\ \psi_{III} &= Fe^{i\alpha x}\end{aligned}$$

$$\text{where} \quad \alpha = \frac{2\pi}{h}\sqrt{2mE}, \quad \beta = \frac{2\pi}{h}\sqrt{2m(V_0 - E)}, \quad \beta\,\text{real.}$$

The boundary conditions for continuity are that the wave functions and first derivatives should match at $x = 0$, a. These conditions determine the value of the constants relative to A. At $x = 0$:

$$A + B = C + D$$

$$i\alpha(A - B) = \beta(C - D)$$

At $x = a$:

$$Ce^{\beta a} + De^{-\beta a} = Fe^{i\alpha a}$$

$$\beta(Ce^{\beta a} - De^{-\beta a}) = i\alpha Fe^{-i\alpha a}$$

After elimination of C and D

$$\frac{F}{A} = \frac{4i\alpha\beta}{e^{i\alpha a}[4i\alpha\beta\cosh\beta\alpha - (\beta^2 - \alpha^2)\sinh\beta a]}$$

The number of particles transmitted through the barrier is proportional to $(F/A)^2$, *i.e.*

$$\left(\frac{F}{A}\right)^2 = \frac{4}{\left[4\cosh\beta a + \left(\frac{\beta}{\alpha} - \frac{\alpha}{\beta}\right)^2 \sinh^2\beta a\right]}$$

For large βa, $\cosh\beta a = \sinh\beta a \to (1/2)\exp\beta a$, and the expression reduces to

$$\left(\frac{F}{A}\right)^2 = \frac{16\alpha^2\beta^2}{(\alpha^2 + \beta^2)^2} \cdot e^{-2\beta a}$$

The number of particles transmitted falls off exponentially with barrier height as $e^{-2\beta a}$, *i.e.*

$$n \simeq \exp\left\{-\frac{4\pi}{h}\left[2m(V_0 - E)\right]^{\frac{1}{2}} a\right\}$$

In the classical limit $(m/h \to \infty)$, there is no transmission.

The way in which tunnelling affects the energy levels of a system is illustrated well by the behaviour of a particle in a one-dimensional box with a central potential barrier.

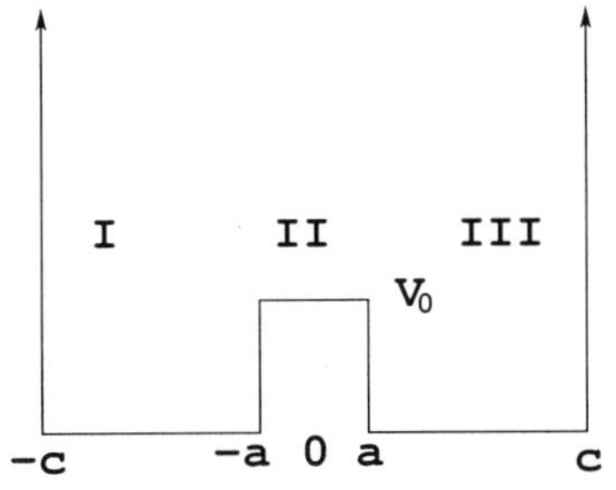

The potential function is defined as follows, for $c = a + b$

$$\begin{aligned} V(x) &= V_0 \quad , \quad -a < x < a \\ V(x) &= 0 \quad , \quad -(a+b), x, -a \quad , \quad a < x < a+b \\ V(x) &= \infty \quad , \quad x < -(a+b) \quad , \quad x > a+b \end{aligned}$$

The particle is described by the Schrödinger equation

$$\left(-\frac{\hbar^2}{2m}\frac{d^2}{dx^2}+V\right)\psi=E\psi$$

For $E<V_0$ the wave functions in the three distinct regions, described in terms of the constants

$$\alpha=\sqrt{\frac{2mE}{\hbar^2}}\quad\text{and}\quad\beta=\sqrt{\frac{-2m(E-V_0)}{\hbar^2}}\qquad\text{are}$$

$$\begin{aligned}\psi_I &= Ae^{i\alpha x}+Be^{-i\alpha x}\\ \psi_{II} &= Ce^{\beta x}+De^{-\beta x}\\ \psi_{III} &= Fe^{i\alpha x}+Ge^{-i\alpha x}\end{aligned}$$

Three of the arbitrary constants can be eliminated on grounds of the symmetry that requires $F=B$, $G=A$, $D=\pm C$. Because of the boundary condition $\psi_{III}(a+b)=0$, it further follows that $B=A\exp\{-2i\alpha(a+b)\}$. To ensure continuity of the wave function at a (and $-a$) it is required that

$$-Ae^{-2i\alpha(a+b)}\cdot e^{i\alpha a}+Ae^{-i\alpha a}=C\left(e^{\beta a}\pm e^{-\beta a}\right)$$

$$-i\alpha Ae^{-2i\alpha(a+b)}\cdot e^{i\alpha a}-i\alpha Ae^{-i\alpha a}=\beta C\left(e^{\beta a}\mp e^{-\beta a}\right)$$

It follows directly from the final two expressions that

$$\frac{\beta\left[e^{i\alpha a}-e^{-i\alpha(a+2b)}\right]}{i\alpha\left[e^{i\alpha(a+2b)}+e^{-i\alpha a}\right]}=\frac{\beta\left(1-e^{-2i\alpha b}\right)}{i\alpha\left(1+e^{-2i\alpha b}\right)}=-\frac{e^{\beta a}\pm e^{-\beta a}}{e^{\beta a}\mp e^{-\beta a}}$$

This expression rearranges to

$$\beta[1-\cos(2\alpha b)+i\sin(2\alpha b)]+i\alpha[1+\cos(2\alpha b)-i\sin(2\alpha b)]\left\{\frac{e^{\beta a}\pm e^{-\beta a}}{e^{\beta a}\mp e^{-\beta a}}\right\}=0$$

Separating real and imaginary terms, and choosing matching signs for the term in curly brackets, give rise to two equations:

$$\coth(\beta a)=-\frac{\beta[1-\cos(2\alpha b)]}{\alpha\sin(2\alpha b)}$$

$$\tanh(\beta a)=-\frac{\beta[\sin(2\alpha b)]}{\alpha[1+\cos(2\alpha b)]}$$

that simplify to:

$$\coth(\beta a)=-\frac{\beta}{\alpha}\tan(\alpha b)$$

$$\tanh(\beta a) = -\frac{\beta}{\alpha}\tan(\alpha b)$$

These two transcendental equations define a pair of closely spaced energy levels, respectively associated with symmetric and antisymmetric wave functions as defined by the arbitrary choice of $D = \pm C$.

For levels lying low with respect to the barrier height it may be assumed that $\beta >> 1$ and $\alpha/\beta << 1$. The even eigenstate may then be approximated[6] by

$$\tan(\alpha\beta) \simeq -\frac{\alpha}{\beta}[1 + 2\exp(-\beta a)]$$

and in zero-order approximation this equation reduces to

$$\tan(\alpha b) = 0 \quad \text{or} \quad \alpha b = n\pi$$

The solution for an isolated potential box emerges immediately as

$$E_n^0 = \frac{\alpha^2\hbar^2}{2m} = \frac{n^2h^2}{8mb^2}$$

It is readily shown that the same zeroth order eigenvalues appear for the odd eigenstate. Because of tunnelling through the barrier the degeneracy of the energy levels for the simple potential is therefore seen to be disturbed and split into a doublet. Writing $\tan\alpha = \beta$ for small values of β, it is implied that $\alpha \simeq n\pi + \beta$ and an approximate value for the level splitting can be calculated [71] as

$$\Delta E = \frac{8E^0}{\beta^0 b}e^{-\beta^0 a}$$

A number of chemical phenomena cannot be explained by any mechanism other than quantum-mechanical tunnelling. The more obvious of these include electrochemical processes that depend on the transfer of electrons across electrode surfaces, and solid-state rearrangements that involve the rotation of bulky moeities in sterically restricted space. Neither of these phenomena has been studied in quantitative detail.

---

[6]For $y >> 1$:

$$\tanh(y) = \frac{e^y - e^{-y}}{e^y + e^{-y}} = \frac{1 - e^{-2y}}{1 + e^{-2y}} \simeq \left(1 - e^{-2y}\right)^2 \simeq 1 - 2e^{-2y}$$

and

$$\coth(y) \simeq 1 + 2e^{-2y}$$

Electrode processes

According to the Sommerfeld model electrons in a metal electrode are free to move through the bulk of the metal at a constant potential, but not to escape at the edge. Within the metal electrons have to penetrate the potential barriers that exist between atoms, as shown schematically below.

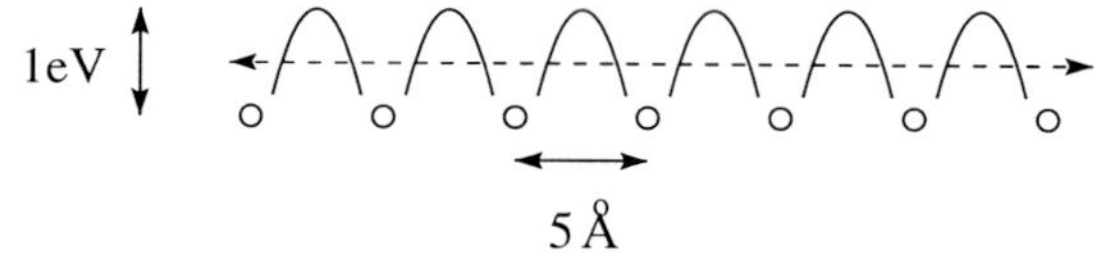

The barriers are about 1eV high and 5Åwide. The number of electrons expected to penetrate the barrier is calculated as

$$\begin{aligned} n &= \exp\left\{-\frac{4\pi}{h}\,[2m(V_0 - E)]^{\frac{1}{2}}\,a\right\} \\ &= \exp\left\{-\frac{4\pi}{6.6\times 10^{-34}}\left[2\times 9\times 10^{-31}(V_0 - E)1.6\times 10^{-19}\right]^{\frac{1}{2}}\times 5\times 10^{-10}\right\} \\ &\approx 0.02 \quad (\text{for a } \tfrac{1}{2} \text{ eV electron}) \end{aligned}$$

The electron therefore has about 2 percent probability of penetrating the barrier. However, if it is considered that the electron oscillates with a wavelength of about 2 bohr radii ($a_0 = 5.3 \times 10^{-11}$m), the electron velocity calculated from its momentum $p = h/\lambda = mv$, corresponds to $h/2ma_0$. It therefore collides $h/2ma_0^2$ times per second (*i.e.* 1.3×10^{17}) with the barrier and penetrates after 10^{-15} seconds, and moves freely through the crystal.

As soon as an electron should leave the surface an image charge is induced within the metal and the electron is pulled back across the edge. A characteristic amount of energy, known as the *work function* W, is required to remove an electron from a metal. Suppose now that the metal is placed in a strong electric field which is directed to pull electrons from the metal. The electric potential $-eEx$, where E is the electric field and x the distance from the edge, now modifies the potential at the surface to resemble the situation shown in figure 4.

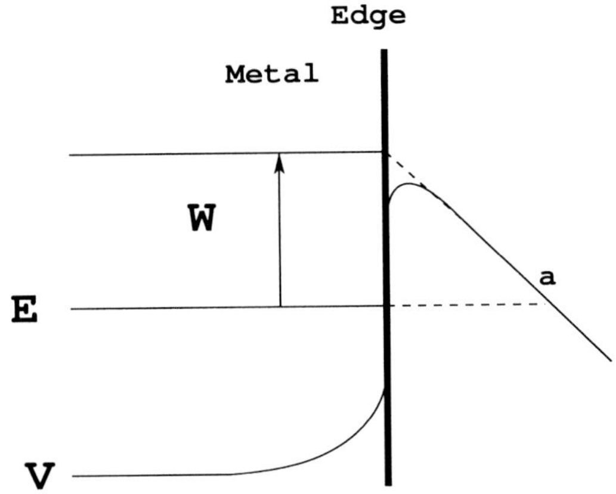

Figure 4: *Potential energy change at a metal surface.*

There will always be a position a, where the electron has positive kinetic energy even though it is outside the metal and there will be a finite probability that it leaks through the barrier and leaves the metal permanently, in a process of *cold emission.*

In electrochemical media the metal electrode is not in contact with the vacuum, as shown above, but with an electrolyte that could drastically modify the potential field near the interface.

Since the probability of tunnelling depends inversely on the square root of the mass, tunnelling effects are common for electrons and less so for protons, although many reaction mechanisms depend on proton transfer against potential barriers.

The Inversion of Ammonia

The most familiar argument that postulates quantum-mechanical tunelling to occur in an, otherwise classical, chemical system, is the common explanation of the ease whereby an ammonia molecule manages to invert itself.

The NH_3 molecule is generally agreed to have a trigonal pyramidal structure, characterized by three internal parameters: $d = 101$ pm, $z_0 = 38$ pm, $\alpha = 68°$, and with nitrogen at the apex, over a trigonal array of hydrogen atoms.

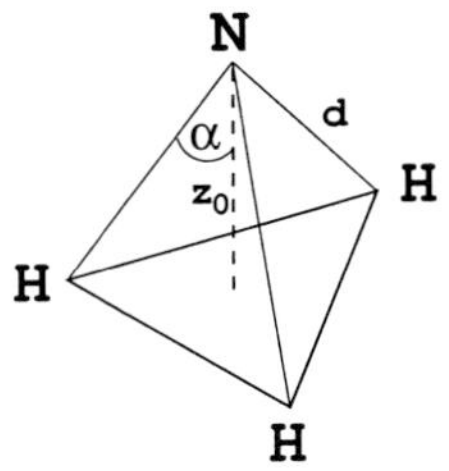

Although there is no direct experimental evidence to support this structure assignment, it is well in accord with all theories of chemical bonding and many experimental observations. However, it fails to account for all details of the observed vibrational spectrum of ammonia. In particular, to account for a closely spaced doublet observed around wave number $\bar{\nu} = 905\ \mathrm{cm}^{-1}$, it is necessary to consider a molecular mode, analogous to the movement of an umbrella that opens and closes, on a cycle that matches the oscillation of angle α about its equilibrium value. Neglecting all other degrees of freedom, the potential energy of this vibration is a function $V(z)$ of the distance between N and the plane at $z = 0$ containing the H atoms. Because the system is symmetric with respect to $z = 0$, the potential $V(z)$ must be an even function of z. A curve representing such a function is sketched in figure 5.

The potential $V(z)$ has a barrier around $z = 0$. This barrier represents steric interaction of the approaching N atom with the hydrogen atoms in the plane. If the barrier were infintely high, the nitrogen atom would never be able to penetrate the plane of the hydrogen atoms and be found on the other side of the plane.

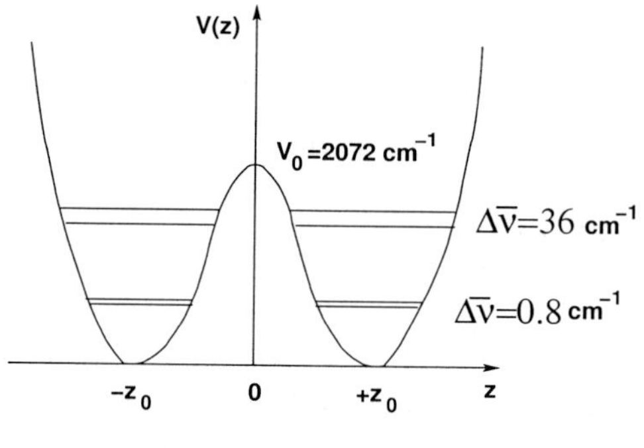

Figure 5: *Double-well potential function constructed to fit the vibrational spectrum of ammonia.*

As shown below there is reason to think that the barrier has a finite height of $V_0 = 2072\ \mathrm{cm}^{-1}$, so that there is a certain probability that the molecule will invert during the course of its vibrations. It is important to note that in both the ground state ($v = 0$) and the first excited state ($v = 1$) of the vibrational mode considered here, the energy of the molecule is lower than the potential barrier. Inversion of ammonia in its lowest vibrational states is therefore classically forbidden. Since inversion as a (hindered) vibrational mode is spectroscopically observed therefore means that it is due to a quantum-mechanical tunnelling effect.

The tunnelling motion can be understood in terms of the one-dimensional

Schrödinger equation for relative displacement along the z-axis:

$$-\frac{\hbar^2}{2m}\frac{d^2\psi^2(z)}{dz^2} + V(z)\psi(z) = E\psi(z)$$

A reduced mass of $m = 3M_H M_N/(3M_H + M_N)$ may be assumed, provided the hydrogen triangle is not distorted during the motion. If the potential barrier between the two wells were infinitely high the two wells would be totally independent and the energy spectrum would consist of the same set of energy eigenvalues in each well. Each energy eigenvalue of the system would be doubly degenerate, and the eigenfunctions corresponding to a given energy would be a linear combination of the wave functions describing nitrogen displacement along $+z$ and $-z$ respectively. These wave functions are shown on the left of the following diagram.

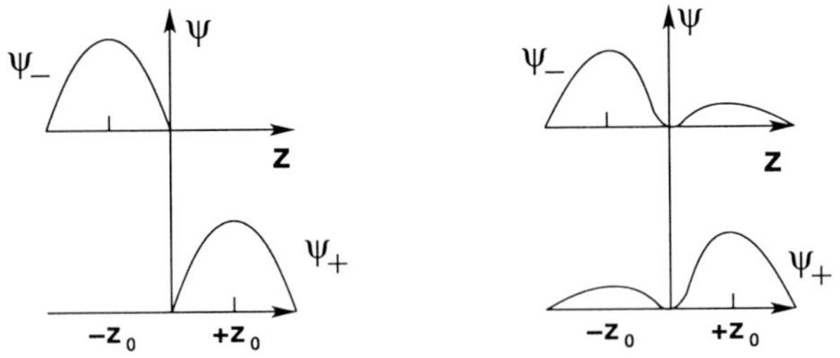

In the actual situation, with a finite barrier, the wave functions are not confined to a single side of the barrier and inversion can happen. The two wave functions now have the shapes shown schematically on the diagram at the right. Since the wells are now linked the functions $\psi_+(z)$ and $\psi_-(z)$ are not eigenfunctions and not orthogonal to each other. The true energy eigenfunctions are the linear combinations:

$$\psi_1 = -\frac{1}{\sqrt{2}}[\psi_+(z) + \psi_-(z)]$$

$$\psi_2 = -\frac{1}{\sqrt{2}}[\psi_+(z) - \psi_-(z)]$$

The eigenvalues of the doublet are such that $E_1 < E_2$. It is the coupling between the two wells that lifts the degeneracy, as shown before for the double square-well potential.

Writing $E_2 = E_1 + \Delta E$, the general time-dependent wave function of the two-level system may be formulated as

$$\Psi(z,t) = c_1\psi_1(z)e^{-(i/\hbar)E_1 t} + c_2\psi_2(z)e^{-(i/\hbar)E_2 t}$$

Suppose it is known that at time $t = 0$ the wave function that describes the system is ψ_+, so that the nitrogen atom is most probably above the plane at that time. Thus

$$\begin{aligned}\Psi(z,0) &= c_1\psi_1(z) + c_2\psi_2(z) = \psi_+(z) \\ &= \frac{1}{\sqrt{2}}\left[\psi_1(z) + \psi_2(z)\right]\end{aligned}$$

so that $c_1 = c_2 = 1/\sqrt{2}$, and

$$\Psi(z,t) = \frac{1}{\sqrt{2}}\left[\psi_1(z)e^{-(i/\hbar)E_1t} + \psi_2(z)e^{-(i/\hbar)E_1t}\cdot e^{-(i/\hbar)\Delta Et}\right]$$

The energy difference between the two levels may be written as $\Delta E = h\nu$, whereby

$$\Psi(z,t) = \frac{1}{\sqrt{2}}\left[\psi_1(z) + \psi_2 e^{-2\pi i\nu t}\right]e^{-(i/\hbar)E_1t}$$

At the time $t = 1/2\nu$ the wave function will be given by[7]

$$\begin{aligned}\Psi(z, t = 1/2\nu) &= \frac{1}{\sqrt{2}}\left[\psi_1(z) - \psi_2(z)\right]e^{-(i/\hbar)E_1t} \\ &= \psi_- e^{-(i/\hbar)E_1t}\end{aligned}$$

so that

$$\left|\Psi\left(z, \frac{1}{2\nu}\right)\right|^2 = |\psi_-(z)|^2$$

and the nitrogen atom will most probably be found below the plane. Since the energy difference corresponds to an inversion frequency of $\nu \simeq 23800$ MHz, the time required for inversion of the ammonia molecule is $t = 1/2\nu \simeq 2.1 \times 10^{-11}$ s.

The proper ground state and first excited state of the ammonia molecule

$$\Psi_1 = \frac{1}{\sqrt{2}}(\Psi_+ + \Psi_-)$$

$$\Psi_2 = \frac{1}{\sqrt{2}}(\Psi_+ - \Psi_-)$$

respectively, are *stationary* states of the Hamiltonian [72]. With respect to Ψ_+ and Ψ_- however, a single ammonia molecule does not have a nuclear

[7]Using the relationship $e^{-\pi i} = \cos\pi - i\sin\pi = -1$

frame anymore. Neither nitrogen nor hydrogen nuclei occupy fixed positions and only probability distributions of their positions can be given. The states Ψ_+ and Ψ_- are not stationary states and describe a tunelling process that interconverts these states. The molecular shape is fuzzy as shown in the diagram below.

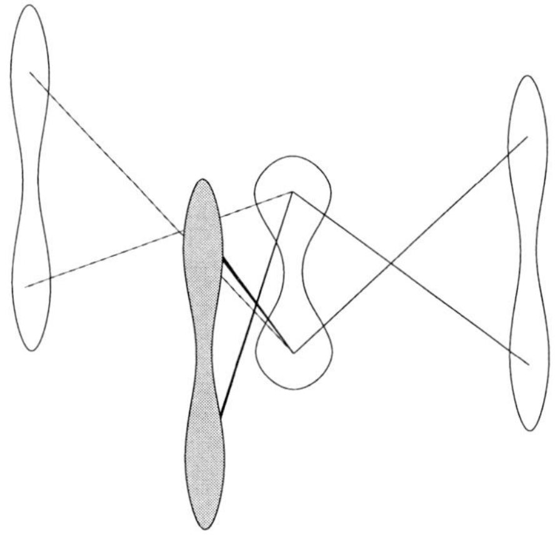

The ammonia molecule is not an exception and it will be shown later on that similar situations can be constructed quantum-mechanically for all molecules.

6.4.5 The Nearly-free electron model

Although the free electron model leads to a simple understanding of electrochemical phenomena, even in solution, it offers no explanation of the different conduction properties of different types of solid. In order to understand the conduction of solids it is necessary to extend the free electron model to take account of the periodic lattice of a solid.

For a one-dimensional lattice with translational period a, it follows that

$$\psi(x+a) = \psi(x) = C\psi(x)$$

$$\psi(x+Na) = \psi(x) = C^N\psi(x)$$

since $\psi(x)$ must be single-valued. It follows that C must be one of the N roots of unity[8]. Therefore $C = \exp(2\pi is/N)$; $s = 0, 1, 2, \ldots$. These boundary

[8]Note that $\exp(i\theta) = \cos\theta + i\sin\theta$. Hence $\exp(2\pi si = \cos 2s\pi + i\sin 2s\pi = 1$ for $s = 0, 1, 2 \ldots$. Thus, if $x^N = 1$, then $x = 1^{(1/N)} = \exp(2\pi is/N)$.

values define the wave function $\psi(x) = u_k \exp(ik.x)$, where $k = 2\pi s/Na$ and $u_k(x)$ has the period of the lattice. This is *Bloch's theorem*.

This result was used by Kronig and Penney to define a one-dimensional square well potential for solids.

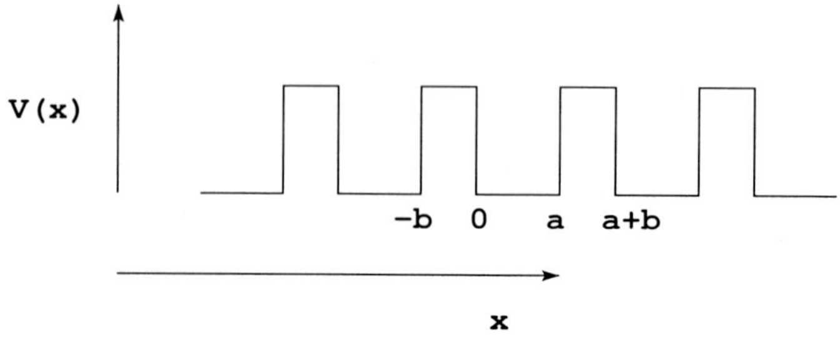

The wave equation

$$\frac{d^2\psi}{dx^2} + \frac{8\pi^2 m}{h^2}[E - V(x)]\psi - 0$$

has different solutions between the barriers and over the barriers. In the region $0 < x < a$ the solution is

$$\psi_I = Ae^{i\alpha x} + Be^{-i\alpha x} \quad , \quad \alpha^2 = 8\pi^2 mE/h^2$$

and in the region $a < x < (a + b)$

$$\psi_{II} = Ce^{\beta x} + De^{-\beta x} \quad , \quad \beta^2 = 8\pi^2 m(|V - E|)/h^2$$

Continuity at $x = 0$ requires $\psi_I(0) = \psi_{II}(0)$ and $\psi_I'(0) = \psi_{II}'(0)$, *i.e.*

$$A + B = C + D \qquad (6.9)$$

$$i\alpha(A - B) = \beta(C - D) \qquad (6.10)$$

To ensure continuity at a it is required that

$$Ae^{i\alpha a} + Be^{-i\alpha a} = Ce^{\beta a} + De^{-\beta a}$$

$$i\alpha\left(Ae^{1\alpha a} - Be^{-i\alpha a}\right) = \beta\left(Ce^{\beta a} - De^{-\beta a}\right)$$

These conditions do not take into account the periodicity of the potential that repeats at intervals of $(a + b)$. To ensure that the wave function has the same periodicity it is necessary to specify that

$$\psi(x) = u(x) \cdot e^{ikx} = \psi(x + a + b) = u(x) \cdot e^{ik(x+a+b)} = u(x) \cdot e^{ikx} \cdot e^{ik(a+b)}$$

The same condition relates $\psi(a)$ and $\psi(-b)$ and it can therefore be combined directly with the previous conditions on $\psi(a)$,

$$Ae^{i\alpha a}\cdot e^{-ik(a+b)}+Be^{-i\alpha a}\cdot e^{-ik(a+b)} = Ce^{-\beta b}+De^{\beta b} \quad (6.11)$$
$$i\alpha\left[Ae^{i\alpha a}\cdot e^{-ik(a+b)}-Be^{-i\alpha a}\cdot e^{-ik(a+b)}\right] = \beta\left(Ce^{-\beta b}-De^{\beta b}\right) \quad (6.12)$$

The four arbitrary constants can be eliminated from the numbered equations to yield a single expression. The required procedure is equivalent to setting the determinant of coefficients equal to zero. The final expression:

$$\frac{1}{4i}\left\{\frac{\beta^2-\alpha^2}{2\alpha\beta}\right\}\left(e^{\beta b}-e^{-\beta b}\right)\left(e^{i\alpha a}-e^{-i\alpha a}\right)+\frac{1}{4}\left(e^{\beta b}+e^{-\beta b}\right)\left(e^{i\alpha a}+e^{-i\alpha a}\right) = \frac{1}{2}\left\{e^{ik(a+b)}+e^{-ik(a+b)}\right\}$$

reduces to

$$\left\{\frac{\beta^2-\alpha^2}{2\alpha\beta}\right\}\sinh\beta b\cdot\sin\alpha a+\cosh\beta b\cdot\cos\alpha a=\cos k(a+b)$$

This equation can be solved graphically for given values of V, a and b. The lhs of the final expression is plotted against E.

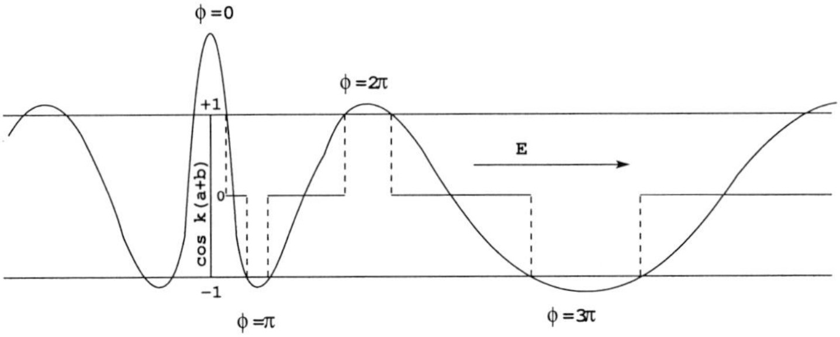

Note that the rhs is a cosine function that cannot exceed unity. Those energies for which the curve of the lhs falls outside the limits ±1 therefore represent forbidden regions or gaps in the energy function. These gaps occur whenever the function $\phi=\cos^{-1}k(a+b)=\pm1$. The graph of energy as a function of $k(a+b)$ in units of π, illustrates the appearance of the gaps more clearly.

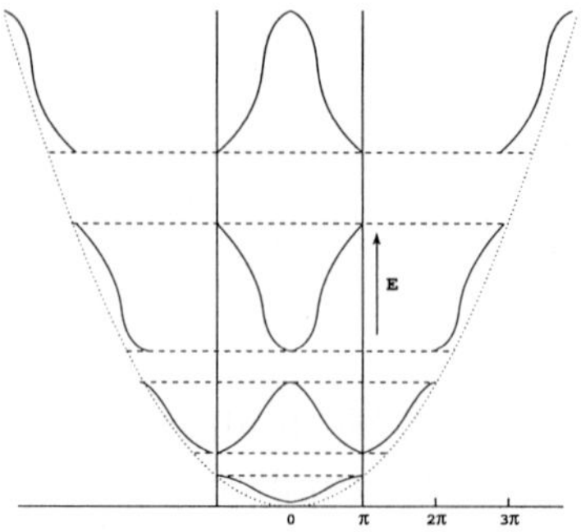

The outer profile is the curve for the corresponding free-electron model[9], *i.e.* of $k = (2\pi/h)(2mE)^{\frac{1}{2}}$. The band gaps are therefore shown to be a direct consequence of the periodic lattice. They can be interpreted to mean that electrons of certain specific wavelengths cannot move freely through the lattice. The gaps occur for all $k(a+b) = n\pi$, for integer n. Writing $a+b = d$ the condition $kd = n\pi$ turns out the same as Bragg's diffraction condition, $2d = n\lambda$ for $k = 2\pi/\lambda$.

In a three-dimensional lattice the Bragg condition, $n\lambda = 2d\sin\theta$ for spacing d between the atomic layers in a crystal and hence running electron waves in a solid must be reflected at appropriate lattice planes. Bands of forbidden and permitted energies therefore also define zones from which electron waves cannot escape. When the band structure is moved by multiples of 2π a plot that lies entirely within the first zone is generated, as in the second graph above. This convention has a peculiar consequence, since k is proportional to momentum. An electron can be accelerated until its wave vector reaches the zone boundary at $+\pi$. It then reappears from the other side of the diagram, with the opposite momentum corresponding to $k = -\pi/d$. This is the diffraction effect mentioned before.

The number of states in an allowed zone is determined by the allowed values of the wavelength. Consider N atoms in linear array over a distance L and a regular separation of a. The allowed values of the wave vector $k = \pm 2\pi n/L$. The series terminates at $N\pi/L = \pi/a$, the zone boundary and hence there are exactly $N = L/a$ states - one per atom. In three dimensions the zone is a geometrical body known as a *Brillouin* zone, rather than a line segment in k-space, and described by three rather than a single integer. The first Brillouin zone again encloses one state per atom.

[9]Let $ka = n\pi$ for the one-dimensional box with energy $E = (nh)^2/8ma^2 = (kh)^2/8\pi^2 m$.

Electrons in Solids

The band structure that appears as a consequence of the periodic potential provides a logical explanation of the different conductivities of electrons in solids. It is a simple case of how the energy bands are structured and arranged with respect to the Fermi level. In general, for any solid there is a set of energy bands, each separated from the next by an energy gap. The top of this set of bands (the valence band) intersects the Fermi level and will be either full of electrons, partially filled, or empty.

Conductivity means that an electron moves under the influence of an applied field, which implies that field energy transferred to the electron promotes it to a higher level. Should the valence level be completely filled there are no extra higher-energy levels available in that band. Promotion to a higher level would then require sufficient energy to jump across the gap into a conduction level in the next band. The width of the band gap determines whether the solid is a conductor, a semi-conductor or an insulator. It is emphasized that in three-dimensional solids the band structure can be much more complicated than for the illustrative one-dimensional model considered above and could be further complicated by impurity levels.

The situation described here is based on a simple one-electron model which can hardly be expected to predict the behaviour of complex many-electron systems in quantitative detail. There can be no doubt however, that the qualitative picture is convincing and probably that the broad principles of electronic behaviour in solids have been identified. The most significant feature of the model is the band structure that makes no sense except in terms of the electron as a wave. Important, but largely unexplored aspects of solid-state reactions and heterogeneous catalysis must also relate to the nearly-free models of electrons in solids.

6.4.6 Delocalized Chemical Systems

The Kronig-Penney model, although rather crude, has been used extensively to generate a substantial amount of useful solid-state theory [73]. Simple free-electron models have likewise been used to provide logical descriptions of a variety of molecular systems, by a method known in modified form as the Hückel Molecular Orbital (HMO) procedure [74].

Although a square-well potential, with energy levels proportional to n^2 is totally inappropriate to model the electronic levels of an atom, ($E \propto 1/n^2$), it provides a compelling qualitative rationalization of a whole class of chemical interactions. The simplest possible description of bond formation is in terms of two coalescing potential boxes [75].

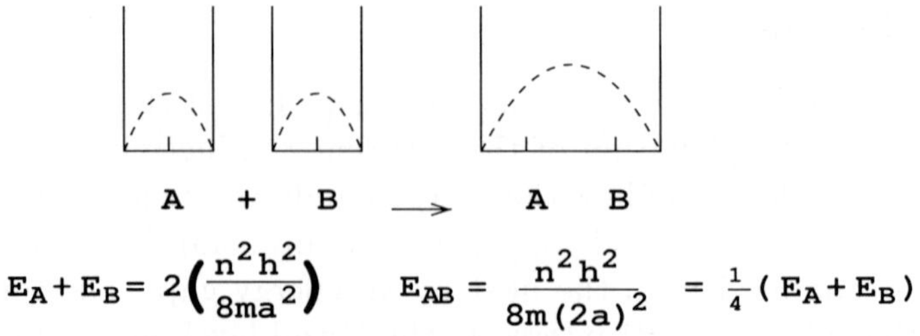

The total energy of two electrons confined to boxes A and B, representing identical atoms, decreases dramatically when potential barriers between the atoms disappear, since each electron gains freedom to move over twice the space available before. The ground-state $|\psi(x)|^2$ curves show an accumulation of electron charge between the nuclei when the bond is formed. Conservation of energy dictates that bond formation be accompanied by emission of energy. This dispersal of energy which is an increase in entropy drives the chemical reaction.

A more realistic use of free-electron simulation occurs with conjugated systems, assumed to be characterized by a number of electrons delocalized over the entire molecule. The simplest example is the ethylene molecule, C_2H_4. From the known planar structure

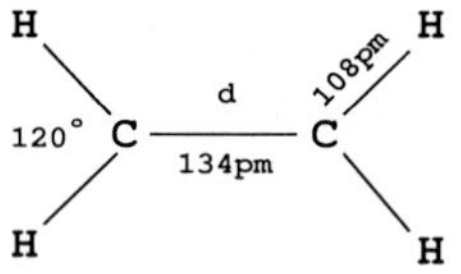

it is inferred that only ten of the available valence electrons are tied up in electron-pair bonds, leaving one pair of electrons free to roam.

Substituted ethylenes, with hydrogen atoms replaced by various alkyl groups, have a common feature in their electronic spectra, *i.e.* an absorption band at *ca.* 164 – 180 nm. This band is interpreted to show that the delocalization of the electron pair is largely confined to the vicinity of the unsaturated centre, commonly referred to as the C=C double bond. If the delocalization is assumed not to exceed a linear distance of one bond length on either side of the double bond, the electron pair remains in a linear potential box of width $3d$, with allowed energy levels of

$$E_n = \frac{n^2h^2}{8m(3d)^2} = \frac{n^2h^2}{72md^2}$$

By the Pauli principle each level can accommodate two electrons and therefore both electrons can be assumed to be in the ground state, $n = 1$. Excitation to the first excited state requires energy of

$$\Delta E = h\nu = E_2 - E_1 = \frac{h^2}{72md^2}(2^2 - 1) = \frac{hc}{\lambda}$$

Absorption of radiation for this transition should hence occur at the wavelength

$$\lambda = \frac{72md^2c}{3h} = 176\,\text{nm}$$

This free-electron model is readily extended to polyenes, such as hexatriene.

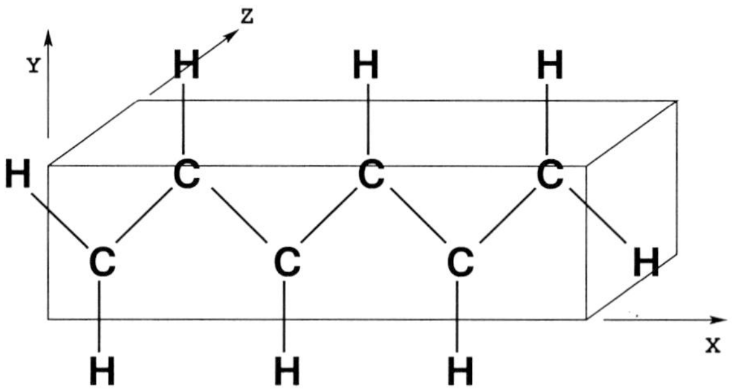

The delocalization may be assumed to occur in a box of dimensions $L_x \times L_y \times L_z$, containing the carbon skeleton, freely rotating about its x-axis. The dimensions $L_y = L_z$ are assumed equal to l, the diameter of the rotationally disordered skeleton. The length of the box, for a C_N chain is assumed to be $L_x = Nl$. The delocalized electrons are confined by the Pauli principle to $N/2$ energy levels with wave functions

$$\psi_n(x) = \sqrt{\frac{2}{L_x}}\sin\left(\frac{n\pi x}{L_x}\right)$$

The y and z components make constant contributions to the total energy. The total electron density $Q(x)$ is the sum of individual densities

$$Q(x) = -e\sum_n b_n\psi_n^2(x)$$

where $b_n = 0, 1$ or 2 is the occupation number of the E_n level, as dictated by the Pauli principle, and the total energy $E_T = \sum_n b_n E_n$. An actual calculation for hexatriene, assuming $l = 140$pm [74] shows a charge distribution of the form

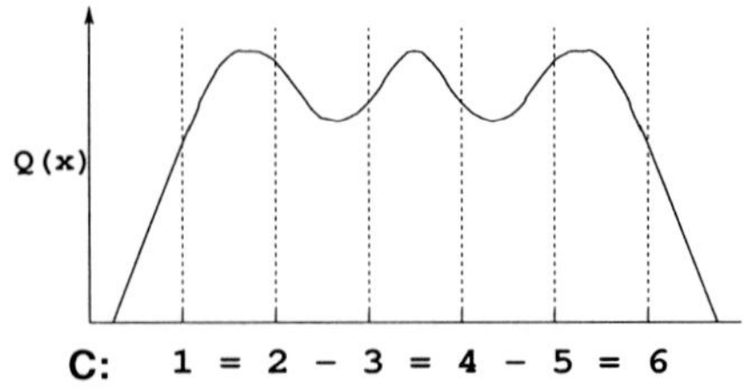

The density reaches maxima at $x/L = \frac{1}{5}, \frac{4}{5}$ and $\frac{1}{2}$, *i.e.* roughly between C atoms 1 and 2, 3 and 4, and 5 and 6, a result which suggests three essentially localized double bonds. However, there is considerable density in the region of the formal single bonds, suggesting a fairly delocalized system of so-called π-electrons.

The working model for polyenes is essentially one-dimensional, with kinetic energies

$$E_n = \frac{h^2}{8mL^2} \cdot n^2 = \frac{p^2}{2m} = \frac{1}{2m}\left(\frac{h}{\lambda}\right)^2$$

$$i.e. \quad \lambda^2 = \frac{4L^2}{n^2} \quad \text{or} \quad L = \frac{n\lambda}{2}$$

The electron behaves as a standing wave with an integral number of half wavelengths fitting into the one-dimensional box, with boundary conditions $\psi_n(0) = \psi_n(L) = 0$.

If the ends of such a one-dimensional box are joined to form a ring the wave functions ψ_n for n odd are not continuous at $x = 0, L$.

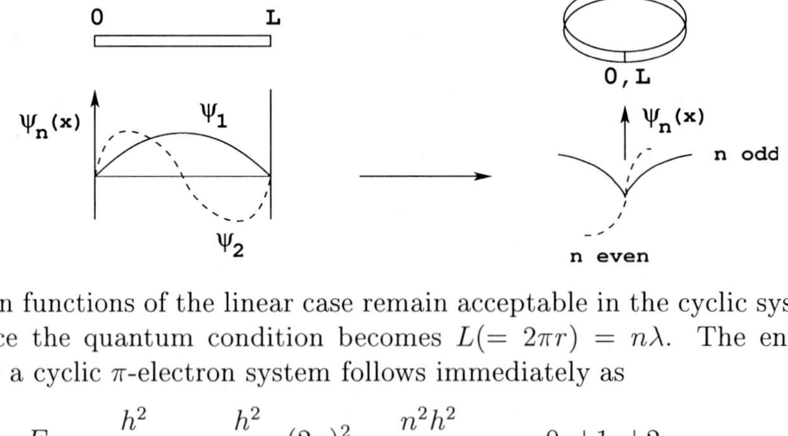

Only even functions of the linear case remain acceptable in the cyclic system and hence the quantum condition becomes $L(= 2\pi r) = n\lambda$. The energy levels for a cyclic π-electron system follows immediately as

$$E_n = \frac{h^2}{2m\lambda^2} = \frac{h^2}{8mL^2}(2n)^2 = \frac{n^2h^2}{2mL^2}, \; n = 0, \pm 1, \pm 2 \ldots$$

The different signs for quantum numbers $n \neq 0$ indicate that the electron can rotate either clockwise or counterclockwise, with either positive or negative angular momentum.

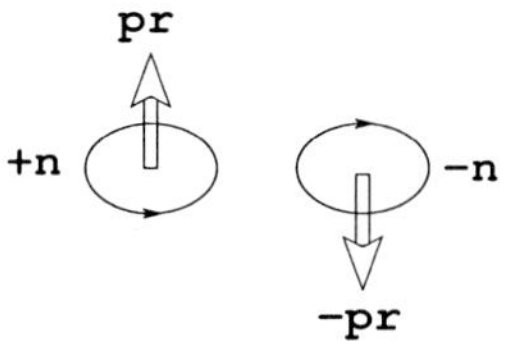

$$\pm pr = \pm\frac{hr}{\lambda} \quad \text{or} \quad \frac{nhr}{2\pi r} = n\left(\frac{h}{2\pi}\right), \ n = 0, \pm 1, \pm 2 \ldots$$

Each energy level ($n \neq 0$) is doubly degenerate since the energy depends on n^2 and therefore is independent of the sense of rotation. The quantum number $n = 0$ is no longer forbidden as in the linear case since the boundary conditions $\psi(0) = \psi(L) = 0$ no longer apply. In the cyclic case $n = 0$ implies infinite λ, *i.e.* $\psi_0 =$ constant, and $E_0 = 0$.

The linear and cyclic hexatrienes are compared in the table below.

	Linear	*Cyclic*
Quantum conditions	$n.\frac{\lambda}{2} = L$	$n.\lambda = L$
Quantum numbers	$n=1,2,3...$	$n=0, \pm 1 \pm 2...$
Eigenvalues	$E_n = \frac{h^2}{8mL^2}n^2$	$E_n = \frac{h^2}{8mL^2}(2n)^2$

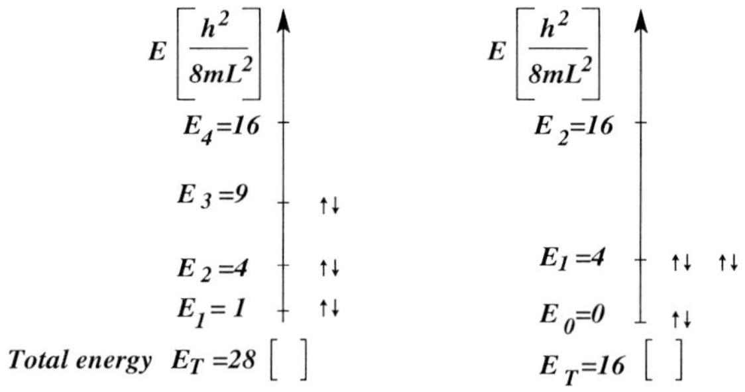

In general, linear π-electron systems with $Z_\pi = 2N$ electrons at the lowest energy levels have closed-shell singlet states while cyclic systems reach closed shell structures only when $Z_\pi = 4N + 2$. Cyclic π-electron systems with $Z_\pi \neq 4N + 2$ will therefore exhibit multiplet ground states according to Hund's rules, and should be chemically reactive because of the unpaired electrons. Hückel's rule that predicts pronounced stability for so-called *aromatic* ring systems with $4N + 2$ π-electrons is based on this shell structure. The comparison with cyclic systems further predicts that ring closure of linear π-electron systems should be exothermic by an amount

$$\Delta E = E_T^{linear} - E_T^{cyclic} = 12\left[\frac{h^2}{8mL^2}\right]$$

which roughly corresponds to the quantity often referred to as *resonance energy.*

Electronic Spectra

The spectra of linear polyenes are modelled well as one-dimensional free-electron systems. The cyanine dyes are a classical example. They constitute a class of long chain conjugated systems with an even number n of π-electrons distributed over an odd number $N = n - 1$ of chain atoms. The cyanine absorption of longest wavelength corresponds to promotion of an electron from the highest occupied energy level, $E_{n/2}$ to the lowest unoccupied level, such that in terms of a free-electron model

$$\begin{aligned}\Delta E = hc\tilde{\nu} &= E_{(n/2)+1} - E_{n/2} \\ &= \frac{h^2}{8mL^2}\left[\left(\frac{n}{2}+1\right)^2 - \left(\frac{n}{2}\right)^2\right] = \frac{h^2}{8mL^2}(n+1)\end{aligned}$$

Substituting $L = (N+1)l = nl$ gives

$$\Delta E = \frac{h^2}{8ml^2}\left(\frac{n+1}{n^2}\right) \qquad \tilde{\nu} = 1.55 \times 10^5 \left(\frac{n+1}{n^2}\right)\ \text{cm}^{-1}$$

The experimental values for $N = 9 \rightarrow 15$, compare extremely well with calculated spectra [74]. The method is sufficiently accurate to elucidate the mechanism of colour vision in terms of the calculated spectrum of the related chromophore, retinal [76].

In the case of planar conjugated aromatic systems delocalized electrons may be viewed as moving in a two-dimensional box. For cata-condensed ring systems, of general formula $C_{4n+2}H_{4n+2}$, no C atom belongs to more than two rings and each carbon atom is on the periphery of the conjugated system. Free electrons effectively travel in a one-dimensional loop at constant potential around the perimeter, assumed circular and of radius R. A suitable free-electron box has $L = \pi R$ since circular motion reduces to oscillation in one dimension. An alternative definition of R is to assume the circle of motion to have the same area as the aromatic hydrocarbon. This area assumption allows peri-condensed systems such as pyrene and perylene

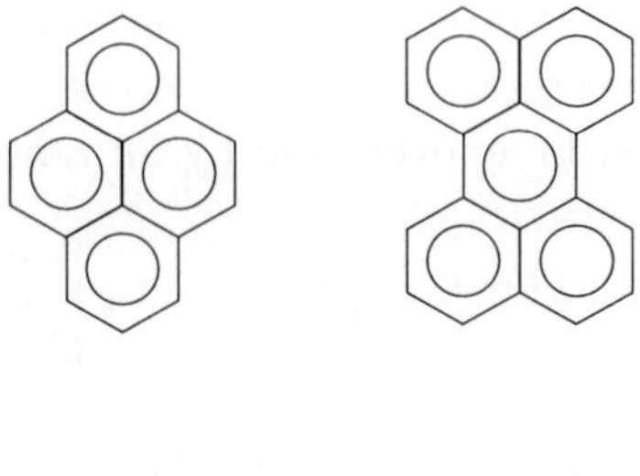

to be treated by the same method. The wave numbers for an $n \to n+1$ transition is given by

$$hc\tilde{\nu} = \frac{h^2}{8\pi^2 mR^2}(2n+1)$$

Although the correspondence with experimental spectra is less convincing than for cyanine dyes, qualitative trends are reproduced well.

As a final example the spectra of molecular charge-transfer complexes are considered next. Electron acceptors such as pyromellitic dianhydride, chloranil and tetracyanobenzene

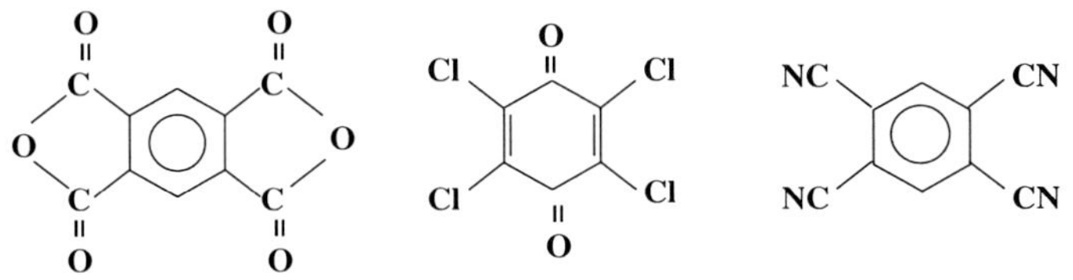

form highly coloured complexes with aromatic hydrocarbons, the electron donors. The complexes are of the sandwich type and in the solid state occur as mixed stacks with donor and acceptor molecules alternating in parallel sequence. The intermolecular distance of about 340 pm indicates that there is no chemical bond in the ordinary sense of the word between D and A molecules. The colour is thought to arise from charge transfer from the highest occupied energy level of the donor molecule to the lowest empty acceptor level.

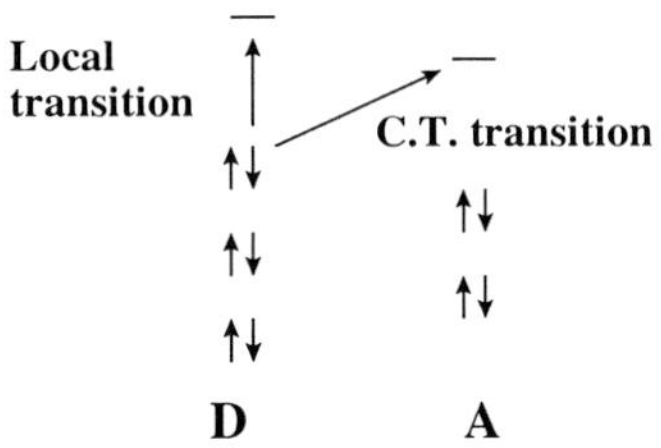

The electronic spectrum of the complex consists of a combination of the spectra of the parent compounds plus one or more higher wavelength transitions, responsible for the colour. Charge transfer is promoted by a low ionization energy of the donor and high electron affinity of the acceptor. A potential barrier to charge transfer of $V_0 = I_D - E_A$ is predicted. The width of the barrier is related to the intermolecular distance. Since the same colour develops in the crystal and in solution a single donor-acceptor pair should be adequate to model the interaction. A simple potential box with the shape

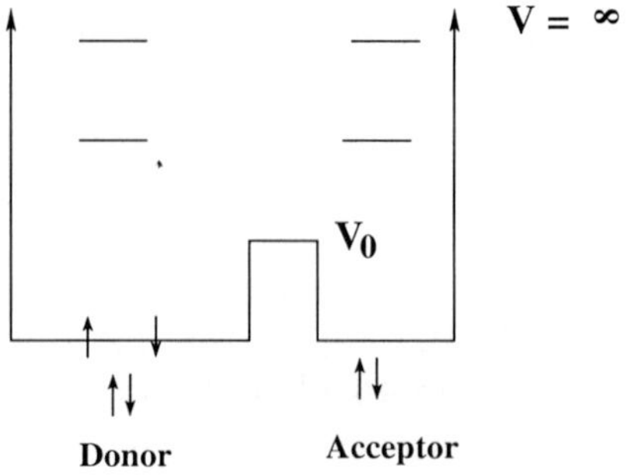

and choosing $V = 0$ at the highest occupied donor level, was found [77] to predict, not only the correct charge-transfer transitions, but also the first donor bands.

Once more, free-electron models correctly predict many qualitative trends and demonstrate the appropriateness of the general concept of electron delocalization in molecules. Free electron models are strictly one-electron simulations. The energy levels that are used to predict the distribution of several delocalized electrons are likewise one-electron levels. Interelectronic effects are therefore completely ignored and modelling the behaviour of many-electron systems in the same crude potential field is not feasible. Whatever level of sophistication may be aimed for when performing more realistic calculations, the basic fact of delocalized electronic waves in molecular systems remains of central importance

Chapter 7

Atoms and Molecules

Despite spectacular successes with the modelling of electron delocalization in solids and simple molecules, one-particle models can never describe more than qualitative trends in quantum systems. The dilemma is that many-particle problems are mathematically notoriously difficult to handle. When dealing with atoms and molecules approximation and simplifying assumptions are therefore inevitable. The immediate errors introduced in this way may appear to be insignificant, but because of the special structure of quantum theory the consequences are always more serious than anticipated.

7.1 Many-particle Systems

A system of N non-relativistic spinless particles is described, using standard terminology, by the classical Hamiltonian

$$H(\boldsymbol{r}_1, \dots \boldsymbol{r}_N, \boldsymbol{p}_1, \dots \boldsymbol{p}_n, t) = \sum_{i=1}^{N} \frac{p_i^2}{2m_i} + V(\boldsymbol{r}_1 \dots \boldsymbol{r}_N, t)$$

The total classical energy $E = H$. The Schrödinger equation for the wave function $\psi(\boldsymbol{r}_1 \dots \boldsymbol{r}_N, t)$ which describes the dynamical state of the system is obtained by defining E and $\boldsymbol{p}_i$ as the differential operators

$$\hat{E} = \hat{H} = i\hbar\frac{\partial}{\partial t} \quad , \quad \hat{\boldsymbol{p}}_i = -i\hbar\boldsymbol{\nabla}_r$$

such that

$$i\hbar\frac{\partial}{\partial t}\Psi = \hat{H}\Psi = \left[\sum_{i=1}^{N}\left(-\frac{\hbar^2}{2m_i}\nabla_{r_i}^2\right) + V(\boldsymbol{r}_1 \dots \boldsymbol{r}_N, t)\right]\Psi$$

The normalization condition takes the form

$$\int |\psi(\boldsymbol{r}_1 \ldots \boldsymbol{r}_N, t)|^2 \, d\boldsymbol{r}_1 \ldots d\boldsymbol{r}_N = 1$$

The total orbital angular momentum L of the system is the sum of the individual orbital angular momenta $\boldsymbol{L}_i = \boldsymbol{r}_i \times \boldsymbol{p}_i$,

$$\boldsymbol{L} = \sum_{i=1}^{N} \boldsymbol{L}_i$$

All the components of the position $\boldsymbol{r}_i$ and momentum $\boldsymbol{p}_i$ vectors of the particle i commute with all those pertaining to particle j provided that $i \neq j$, so that the fundamental commutation relations are

$$[x_i, p_{xj}] = i\hbar\delta_{ij} \quad , \quad [y_i, p_{xj}] = 0\,, \ etc.$$

7.1.1 Two-body Systems

Of particular interest is a system of two particles interacting by way of a time-independent potential $V(\boldsymbol{r}_1 - \boldsymbol{r}_2)$ that only depends on the relative coordinate $\boldsymbol{r}= (\boldsymbol{r}_1 - \boldsymbol{r}_2)$. The classical Hamiltonian of the system is given by

$$H = \frac{\boldsymbol{p}_1^2}{2m_1} + \frac{\boldsymbol{p}_2^2}{2m_2} + V(\boldsymbol{r}_1 - \boldsymbol{r}_2)$$

and the corresponding Schrödinger equation reads

$$i\hbar\frac{\partial}{\partial t}\Psi(\boldsymbol{r}_1, \boldsymbol{r}_2, t) = \left[-\frac{\hbar^2}{2m_1}\nabla_{r_1}^2 - \frac{\hbar^2}{2m_2}\nabla_{r_2}^2 + V(\boldsymbol{r}_1 - \boldsymbol{r}_2)\right]\Psi(\boldsymbol{r}_1, \boldsymbol{r}_2, t)$$

In terms of the relative coordinate $\boldsymbol{r}$, the parameter $M = m_1 + m_2$, the reduced mass

$$\mu = \frac{m_1 m_2}{m_1 + m_2},$$

and the vector

$$\boldsymbol{R} = \frac{m_1\boldsymbol{r}_1 + m_2\boldsymbol{r}_2}{m_1 + m_2}$$

that determines the position of the centre of mass (com) of the system, the Schrödinger equation becomes

$$i\hbar\frac{\partial}{\partial t}\Psi(\boldsymbol{R}, \boldsymbol{r}, t) = \left[-\frac{\hbar^2}{2M}\nabla_R^2 - \frac{\hbar^2}{2\mu}\nabla_r^2 + V(\boldsymbol{r})\right]\Psi(\boldsymbol{R}, \boldsymbol{r}, t)$$

Since the potential is independent of time, temporal dependence of the wave function can be factored out, writing

$$\Psi(\boldsymbol{R},\boldsymbol{r},t)=\psi(\boldsymbol{R},\boldsymbol{r},t)e^{iEt/\hbar}$$

At this stage the assumption that the wave function ψ can be factorized into com and relative-motion (rm) components, by defining $E = E_{com} + E_{rm}$, is commonly made. In terms of $\psi = \phi(\boldsymbol{R})\psi(\boldsymbol{r})$ the two-body problem is decoupled into two one-body problems:

$$-\frac{\hbar^2}{2M}\nabla_R^2\phi(\boldsymbol{R})=E_{com}\phi(\boldsymbol{R})$$

and

$$\left[-\frac{\hbar^2}{2\mu}\nabla_r^2+V(\boldsymbol{r})\right]\psi(\boldsymbol{r})=E_{rm}\psi(\boldsymbol{r})$$

The consequences of this assumption will be examined later on.

7.1.2 Particles with spin

The total angular momentum of a particle with spin, is

$$\boldsymbol{J}_i=\boldsymbol{L}_i+\boldsymbol{S}_i$$

For a system of N particles

$$\boldsymbol{S}=\sum_{i=1}^{N}\boldsymbol{S}_i \quad \text{and} \quad \boldsymbol{J}=\sum_{i=1}^{N}\boldsymbol{J}_i$$

where summation follows the rules of addition discussed before.

When the system is made up of identical particles (*e.g.* electrons in a molecule) the Hamiltonian must be symmetrical with respect to any interchange of the space and spin coordinates of the particles. Thus an interchange operator P_{ij} that permutes the variables q_i and q_j (denoting space and spin coordinates) of particles i and j commutes with the Hamiltonian, $[P_{ij},H]=0$. Since two successive interchanges of q_i and q_j return the particles to the initial configuration, it follows that $P_{ij}^2=I$, and the eigenvalues of P_{ij} are $\epsilon=\pm1$. The wave functions corresponding to $\epsilon=1$ are such that

$$\begin{aligned}P_{ij}\psi(q_1,\ldots q_i,\ldots q_j,\ldots q_N) &= \psi(q_1,\ldots q_j,\ldots q_i,\ldots q_N)\\ &= \psi(q_1,\ldots q_i,\ldots q_j,\ldots q_N)\end{aligned}$$

and are said to be *symmetric* under the interchange P_{ij}. On the other hand, wave functions which correspond to the eigenvalue $\epsilon = -1$ are such that

$$\begin{aligned} P_{ij}\psi(q_1, \ldots q_i, \ldots q_j, \ldots q_N) &= \psi(q_1, \ldots q_j, \ldots q_i, \ldots q_N) \\ &= -\psi(q_1, \ldots q_i, \ldots q_j, \ldots q_N) \end{aligned}$$

and are said to be *antisymmetric* under the interchange P_{ij}.

More generally there are $N!$ different permutations of the variables $q_1 \ldots q_N$. Defining P as the permutation that replaces q_1 by q_{P1}, q_2 by q_{P2}, ... q_N by q_{PN} and noting that P can be obtained as a succession of interchanges, it follows that $[P, H] = 0$.

A permutation is said to be *even* or *odd* depending on whether the number of interchanges leading to it is even or odd. The operator P acts on a wave function such that

$$P\psi(q_1, \ldots q_N) = \psi(q_{P1}, \ldots q_{PN})$$

It is important to note that except for $N = 2$ the $N!$ permutations P do not commute among themselves. The reason is that the interchange operators P_{ij} and P_{ik} $(k \neq j)$ do not commute. The eigenfunctions $\psi(q_1, \ldots q_N)$ are therefore not in general eigenfunctions of all the $N!$ permutation operators P. However, there are two exceptional states which are eigenstates of H and the $N!$ permutation operators: The totally symmetric state $\psi_S(q_1, \ldots q_N)$ satisfies, for all P, the operations

$$\begin{aligned} P\psi_S(q_1, \ldots q_N) &= \psi_S(q_{P1}, \ldots q_{PN}) \\ &= \psi_S(q_1, \ldots q_N) \end{aligned}$$

and the totally antisymmetric state which, for any interchange, satisfies

$$\begin{aligned} P\psi_A(q_1, \ldots q_N) &= \psi_A(q_{P1}, \ldots q_{PN}) \\ &= \begin{cases} \psi_A(q_1, \ldots q_N) & \text{for an even permutation} \\ -\psi_A(q_1, \ldots q_N) & \text{for an odd permutation} \end{cases} \end{aligned}$$

Since $[P, H] = 0$, P is a constant of the motion, which means that a system of particles represented by either ψ_S or ψ_A will keep that symmetry for all time. The particles of Nature that fall into the two classes with either symmetrical or antisymmetrical states, are known as *bosons* and *fermions* respectively.

7.2 Approximation Methods

The most common approximation introduced to deal with complicated quantum systems, known as perturbation theory, is not used that often to deal with atomic and molecular systems. The method is based on the introduction of a small disturbance to a system and examining the effects thereof on the wave functions and energy levels of the system.

It is assumed that the Hamiltonian of the system consists of two parts,

$$H = H_0 + \lambda H'$$

The unperturbed Hamiltonian H_0 is chosen sufficiently simple such that the corresponding eigenvalue equations

$$H_0\psi_k = E_k\psi_k \tag{7.1}$$

may be solved exactly. The term $\lambda H'$ is the perturbation.

By way of illustration it will be assumed that E_k is non-degenerate and $\lambda H'$ is small enough to ensure that the perturbed energy level $\mathcal{E}_k$ is closer to E_k than to any other unperturbed level. The new eigenvalue problem is

$$H\Psi_k = \mathcal{E}_k\Psi_k$$

where, expanded in powers of λ,

$$\Psi_k = \sum_{n=0}^{\infty} \lambda^n \psi_k^{(n)}$$

$$\mathcal{E}_k = \sum_{n=0}^{\infty} \lambda^n E_k^{(n)}$$

The index n refers to the order of perturbation. When all of these equations are combined one gets

$$(H_0+\lambda H')(\psi_k^{(0)}+\lambda\psi_k^{(1)}+\lambda^2\psi_k^{(2)}+\ldots) = (E_k^0+\lambda E_k^{(1)}+\lambda^2 E_k^{(2)}+\ldots)(\psi_k^{(0)}+\lambda\psi_k^{(1)}+\lambda^2\psi_k^{(2)}\ldots)$$

The correction factors of various order are obtained by equating coefficients of equal powers of λ. Starting with λ^0,

$$H_0\psi_k^{(0)} = E_k^{(0)}\psi_k^{(0)}$$

and $\psi_k^{(0)} = \psi_k$, $E_k^{(0)} = E_k$, as expected. The coefficients of λ give

$$H_0\psi_k^{(1)} + H'\psi_k = E_k\psi_k^{(1)} + E_k^{(1)}\psi_k \tag{7.2}$$

while those of λ^2 yield

$$H_0\psi_k^{(2)} + H'\psi_k^{(1)} = E_k\psi_k^{(2)} + E_k^{(1)}\psi_k^{(1)} + E_k^{(2)}\psi_k$$

and so on.

In order to get the first-order correction to the energy $E_k^{(1)}$, (2) is premultiplied by ψ_k^* and integrated over all space, giving

$$\langle\psi_k|H_0 - E_k|\psi_k^{(1)}\rangle + \langle\psi_k|H' - E_k^{(1)}|\psi_k\rangle = 0 \tag{7.3}$$

From (1) and the Hermitian properties of H_0, it follows that

$$\langle\psi_k|H_0|\psi_k^{(1)}\rangle = \langle H_0\psi_k|\psi_k'\rangle = E_k\langle\psi_k|\psi_k^{(1)}\rangle$$

Equation (3) reduces to the very simple, but important result that

$$E_k^{(1)} = \langle\psi_k|H'|\psi_k\rangle \tag{7.4}$$

Higher order corrections follow in a similar fashion.

To establish the solution $\psi_k^{(1)}$ (or higher orders) the function is expanded in the basis set of the unperturbed eigenfunctions. That is

$$\psi_k^{(1)} = \sum_m a_m^{(1)}\psi_m$$

Substituting back into (2) gives

$$(H_0 - E_k)\sum_m a_m^{(1)}\psi_m + \left(H' - E_k^{(1)}\right)\psi_k = 0$$

This equation is premultiplied by ψ_k^*, and integrated over all space, noting that $H_0\psi_l = E_l\psi_l$ and $\langle\psi_l|\psi_k\rangle = \delta_{kl}$, to give

$$a_l^{(1)}(E_l - E_k) + \langle\psi_l|H'|\psi_k\rangle - E_k^{(1)}\delta_{kl} = 0$$

For $k = l$ this result reduces to (3). For $k \neq l$,

$$a_l^{(1)} = \frac{H'_{lk}}{E_k - E_l} \qquad (H'_{lk} \equiv \langle\psi_l|H'|\psi_k\rangle)$$

Since $a_k^{(1)}$ is not defined by (2), it may be set at $a_k^{(1)} = \langle\psi_k|\psi_k^{(1)}\rangle = 0$, so that

$$\begin{aligned}\psi_k^{(1)} &= \sum_{m\neq k} a_m^{(1)}\psi_m \\ &= \sum_{m\neq k}\frac{H'_{mk}}{E_k - E_m}\psi_m\end{aligned}$$

7.2.1 Time-dependent Perturbation

Approximation methods that lead to an estimate of transition probabilities are of more importance in chemical problems and involve time-dependent perturbations. The total Hamiltonian is split as

$$H = H_0 + \lambda H'(t)$$

where the unperturbed Hamiltonian H_0 is time-independent and $H'(t)$ is a small time-dependent perturbation. The following method [62] is known as Dirac's method of variation of constants. It is supposed that the eigenvalues E_k of H_0 are known, together with the stationary eigenfunctions ψ_k, assumed to be a complete orthonormal set. Since $H_0\psi_k = E_k\psi_k$, the general solution of the time-dependent equation

$$i\hbar\frac{\partial\Psi_0}{\partial t} = H_0\Psi_0$$

is given by

$$\Psi_0 = \sum_k c_k^{(0)}\psi_k e^{-iE_kt/\hbar}$$

where $c_k^{(0)}$ are constants. When the general solution of

$$i\hbar\frac{\partial\Psi}{\partial t} = H\Psi$$

is expanded as

$$\Psi = \sum_k c_k(t)\psi_k e^{-iE_kt/\hbar}$$

the unknown coefficients $c_k(t)$ depend on time. The quantity $|c_k(t)|^2$ is interpreted as the probability of finding the system in state k at time t. When $H'(t) = 0$ the coefficients c_k reduce to the constants $c_k^{(0)}$ which are therefore the initial values of the c_k.

The quantity $|c_k^{(0)}|^2$ gives the probability of finding the system in the stationary state ψ_k before the perturbation is applied. The expansion of Ψ is now substituted into the Schröndiger equation. Since $H_0\psi_k = E_k\psi_k$, it follows that

$$i\hbar\sum_k \dot{c}_k(t)\psi_k e^{-iE_kt/\hbar} = \sum c_k(t)\lambda H'(t)\psi_k e^{-iE_kt/\hbar} \quad \left(\dot{c}_k = \frac{\partial c_k}{\partial t}\right)$$

Taking the scalar product with a particular function ψ_b belonging to the set $\{\psi_k\}$ and using $\langle\psi_b|\psi_k\rangle = \delta_{bk}$ produce the set of coupled equations:

$$\dot{c}_b(t) = (i\hbar)^{-1}\sum_k \lambda H'_{bk}(t)c_k e^{i\omega_{bk}t} \tag{7.5}$$

where

$$H'_{bk}(t) = \langle\psi_b|H'(t)|\psi_k\rangle$$

and the Bohr angular frequency is defined by

$$\omega_{bk} = \frac{E_b - E_k}{\hbar}$$

The system of coupled differential equations is equivalent to the original time-dependent Schrödinger equation, and no approximation has been made. If the perturbation $\lambda H'$ is weak, the coefficients c_k may be expanded in powers of λ as

$$c_k = c_k^{(0)} + \lambda c_k^{(1)} + \lambda^2 c_k^{(2)} + \ldots$$

This expansion is substituted into (5) and coefficients of equal powers are equated to give

$$\dot{c}_b^{(0)} = 0 \quad (7.6)$$

$$\dot{c}_b^{(1)} = (i\hbar)^{-1}\sum_k H'_{bk}(t)e^{i\omega_{bk}t}c_k^{(0)} \quad (7.7)$$

$$\ldots \quad . \quad \ldots\ldots\ldots$$

$$\dot{c}_b^{(s+1)} = (i\hbar)^{-1}\sum_k H'_{bk}(t)e^{i\omega_{bk}t}c_k^{(s)} \quad (7.8)$$

The decoupled equations can be integrated to any given order. Equation (7) simply confirms that the $c_k^{(0)}$ are the time-dependent initial conditions of the problem. The system is assumed to be in an initial well-defined stationary state ψ_a of energy E_a, which implies

$$c_k^{(0)} = \delta_{ka}$$

for discrete states. Substitution into (6) gives

$$\dot{c}_b^{(1)}(t) = (i\hbar)^{-1}H'_{ba}(t')e^{i\omega_{ba}t}$$

which integrates to

$$c_b^{(1)}(t) = (i\hbar)^{-1}\int_{t_0}^{t} H'_{ba}(t')e^{i\omega_{ba}t'}dt'$$

The integration constant has been chosen such that $c_b^{(1)}$ vanishes at $t = t_0$, *i.e.*, before the perturbation is applied. To first order in the perturbation the transition probability for the transition $a \to b$ is then given by

$$P_{ba}(t) = \left|c_b^{(1)}(t)\right|^2$$

If H' is independent of time, except for being turned on at time $t_0 = 0$ and turned off at t, then

$$c_b^{(1)}(t) = \frac{H'_{ba}}{\hbar\omega_{ba}}\left(e^{-i\omega_{ba}t} - 1\right)$$

and the first-order transition probability from state a to state $b \neq a$ is given by

$$\begin{aligned} P_{ba}(t) &= \left|c_b^{(1)}(t)\right|^2 \\ &= \frac{2}{\hbar^2}\left|H'_{ba}\right|^2 F(t,\omega_{ba}) \qquad (7.9) \\ \text{where} \quad F(t,\omega) &= \frac{1-\cos\omega t}{\omega^2} \end{aligned}$$

The function $F(t,\omega)$ has a sharp peak about the value $\omega = 0$.

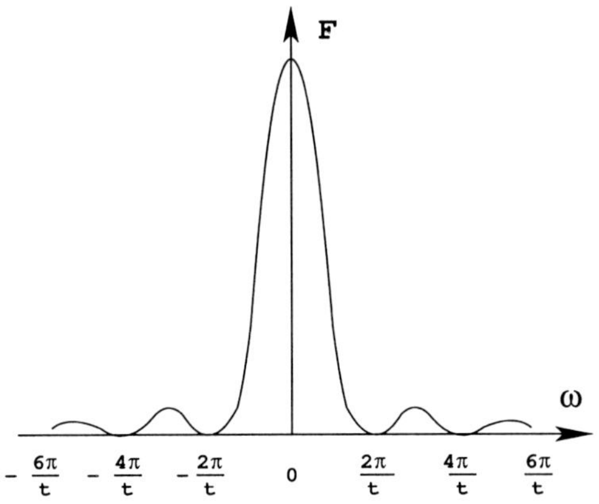

The height of the peak is proportional to $t^2/2$ and the width is approximately $2\pi/t$. Setting $x = \omega t/2$ it is noted[1] that

$$\int_{-\infty}^{\infty} F(t,\omega)d\omega = \frac{t}{2}\int_{-\infty}^{\infty} \frac{1-\cos 2x}{x^2}dx = \pi t \quad ,$$

using a standard integral[2].

[1]For large t the function $F(t,\omega)$ may be written as $F(t,\omega) \simeq \pi t\delta(\omega)$ where

$$\delta(\omega) = \lim_{\beta\to 0}\frac{1-\cos\beta\omega}{\pi\beta\omega^2}$$

is one form of Dirac's delta function.

[2]The integral $\int_0^\infty \frac{1-\cos px}{x^2}dx = \frac{\pi p}{2}$

Since the function F has a peak of width $2\pi/t$ about $\omega_{ba} = 0$, it is clear that transitions to final states b for which ω_{ba} does not deviate from zero by more than $\delta\omega_{ba} \simeq 2\pi/t$ will be strongly favoured. The spread in energy E_b will be in a band

$$\delta E \simeq \hbar\delta\omega = 2\pi\hbar/t$$

about the initial energy E_a, so that the unperturbed energy is conserved to within $2\pi\hbar/t$.

If the transition is such that the unperturbed energy is strictly conserved ($\omega_{ba} = 0$) then from (9) it is seen that

$$P_{ba}(t) = \frac{|H'_{ba}|^2}{\hbar}t^2$$

increases as t^2.

For $\omega_{ba} \neq 0$ the function $F(t, \omega_{ba})$ oscillates between the values 0 and $2/\omega_{ba}^2$ with frequency $\omega_{ba}/2\pi$. As a result, $P_{ba}(t)$ will oscillate with the same frequency about the average value

$$\bar{P}_{ba} = \frac{2\,|H'_{ba}|^2}{\hbar^2\omega_{ba}^2} = \frac{2\,|H'_{ba}|^2}{(E_b - E_a)^2}$$

Instead of transitions to a particular state b, it is often necessary to deal with transitions involving a group of states b' whose energy $E_{b'}$ lies within a given interval $(E_b - \eta, E_b + \eta)$ centred about the value E_b. If the number of states per unit energy is denoted by $\rho_{b'}(E_{b'})$ then the first-order transition probability P_{ba} from the initial state a to the group of final states b' is given by

$$\bar{P}_{ba}(t) = \frac{2}{\hbar^2}\int_{E_b-\eta}^{E_b+\eta} |H'_{b'a}|^2\, F(t, \omega_{b'a})\rho_{b'}(E_{b'})dE_{b'}$$

Assuming that η is small enough so that $H'_{b'a}$ and $\rho_{b'}$ are nearly constant within the integration range, then

$$P_{ba}(t) = \frac{2}{\hbar^2}\,|H'_{ba}|^2\,\rho_b(E_b)\int_{E_b-\eta}^{E_b+\eta} F(t, \omega_{b'a})dE_{b'} \tag{7.10}$$

Next assume that t is large enough so that $\eta >> 2\pi\hbar/t$. It is clear that the overwhelming part of the integral in (10) arises from transitions which conserve the energy within $\delta E = 2\pi\hbar/t$, and can be written as

$$\hbar\int_{-\infty}^{\infty} F(t, \omega_{b'a})d\omega_{b'a} = \hbar\pi t$$

to give the transition probability

$$P_{ba}(t) = \frac{2\pi}{\hbar} |H'_{ba}|^2 \rho_b(E)t$$

with $E = E_a - E_b$. The transition probability per unit time, or transition rate is defined as

$$W_{ba} = \frac{dP_{ba}}{dt}$$

which, to first order in perturbation theory is calculated as

$$W_{ba} = \frac{2\pi}{\hbar} |H'_{ba}|^2 \rho_b(E)$$

This formula, first obtained by Dirac, is commonly known as Fermi's Golden Rule (FGR).

For large t where $F(t, \omega)$ resembles a Dirac delta function, FGR may be written in the state-to-state form

$$\begin{aligned} W_{ba} &= 2\pi|H'_{ba}|^2\delta(\omega) \\ &= \frac{2\pi}{\hbar} |H'_{ba}|^2 \delta(E_b - E_a) \end{aligned}$$

This relationship expresses the result that in the infinite time limit, only transitions which obey energy conservation can be caused by a time independent interaction.

7.2.2 The Variational Method

The variational method is the most widely used approximation method and is particularly useful for obtaining the bound state energies and wave functions of a time-independent Hamiltonian. The method is based on two basic ideas. Firstly, any function can be expanded into a linear combination of other functions. Secondly, if a function is expressed as a linear combination of eigenfunctions of the energy operator, then the average energy associated with the function is a weighted average of the energy eigenvalues. For example, if

$$\phi = \frac{1}{\sqrt{2}}\psi_1 + \frac{1}{\sqrt{2}}\psi_2$$

$$\text{where} \qquad H\psi_1 = E_1\psi_1 \quad , \quad H\psi_2 = E_2\psi_2 \quad , \quad E_1 \neq E_2 \, ,$$

then measurements of the energy states described by ϕ would give the results E_1 and E_2 with the same frequency, to average at $\frac{1}{2}(E_1 + E_2)$.

Now, if some trial wave function consists of a complete orthonormal set of eigenfunctions, the average energy must lie somewhere between the values for the lowest energy and highest energy eigenfunctions respectively. The lowest energy function ψ_0 in the complete set describes, by definition the ground state at energy E_0. Therefore, the average value

$$\bar{E} = \frac{\int \phi^* H \phi \, d\tau}{\int \phi^* \phi \, d\tau} \geq E_0$$

In practice, using trial wave functions with one or more variable parameters, the true ground state may be approached by minimizing the expectation value of the energy with respect to these parameters.

The procedure is illustrated by assuming a hypothetical wave function $\psi = \exp(-cr)$ for the H atom with Hamiltonian

$$-\frac{h^2}{8\pi^2 m}\nabla^2 - \frac{e^2}{4\pi\epsilon_0 r}$$

$$\text{Since} \qquad \nabla^2 = \frac{\partial^2}{\partial r^2} + \frac{2}{r}\frac{\partial}{\partial r} \, , \qquad \text{as before}$$

$$\nabla^2 \psi = \left(c^2 - \frac{2c}{r}\right) \exp(-cr)$$

The average function $\bar{E} = \int \psi H \psi \, d\tau / \int \psi^2 \, d\tau$. All the integrals in this expression contain the standard form

$$\int_0^\infty x^n e^{-ax} \, dx = \frac{n!}{a^{n+1}}$$

to give

$$\bar{E}(c) = \frac{h^2 c^2}{8\pi^2 m} - \frac{e^2 c}{4\pi\epsilon_0}$$

This function has a minimum where

$$\frac{d\bar{E}}{dc} = 0 \quad , \quad i.e. \quad \frac{2ch^2}{8\pi^2 m} - \frac{e^2}{4\pi\epsilon_0} = 0$$

It follows that

$$c = \frac{\pi m e^2}{\epsilon_0 h^2} \quad \text{and} \quad E_{min} = -\frac{me^4}{8\epsilon_0 h^2}$$

which corresponds exactly with the known ground state energy.

Although, in most cases it may not be possible to make as fortunate a choice of variation function as in the previous example, even poor approximations to the actual wave function may yield surprisingly close approximations of the energy.

7.3 Atomic Structure

The initial purpose of pioneer quantum mechanics was to provide the theoretical framework to account for the structure of hydrogen and the nuclear model of atoms in general. The final result, a quantum theory of atomic structure can be discussed in terms of the time-independent Schrödinger equation, in its most general form

$$H\Psi = E\Psi \tag{7.11}$$

In an equivalent classical equation, the variable Ψ cancels to give the Hamiltonian function, which for a single particle of mass m,

$$H = E = \frac{p^2}{2m} + V$$

is equal to the total energy, or the sum of kinetic and potential energies. For a non-classical system the Hamiltonian is an operator that either represents the quantum-mechanical linear momentum by

$$\boldsymbol{p} \rightarrow \frac{h}{2\pi i}\nabla = -\hbar i\nabla$$

or the position variable $\boldsymbol{r}$ by

$$\boldsymbol{r} \rightarrow hi\boldsymbol{\nabla}_p$$

where the x-component of $\boldsymbol{\nabla}_p$ is $\partial/\partial p_x$.

The two Schrödinger equations, in coordinate and momentum space respectively

$$\left\{-\frac{\hbar^2}{2m}\nabla^2 + V\left[\boldsymbol{r},(\hbar/i)\boldsymbol{\nabla}\right]\right\}\Psi(\boldsymbol{r},t) = i\hbar\frac{\partial\Psi}{\partial t}$$

$$\left\{\frac{p^2}{2m} + V\left[(\hbar i\boldsymbol{\nabla}_p),\boldsymbol{p}\right]\right\}\Phi(\boldsymbol{p},t) = i\hbar\frac{\partial\Phi}{\partial t}$$

are equivalent and the functions Φ and Ψ are Fourier transforms of each other:

$$\Phi(\boldsymbol{p},t) = \left(\frac{1}{2\pi\hbar}\right)^{3/2}\int_{-\infty}^{\infty} e^{i(\boldsymbol{p}\cdot\boldsymbol{r})/\hbar}\Psi(\boldsymbol{r},t)dv$$

The important states for atomic systems are those of definite energy E that satisfy the eigenvalue equation (11), which implies that the time-dependence of Ψ for such states is given by

$$i\hbar\frac{\partial\Psi}{\partial t} = E\Psi \quad ; \quad |\Psi|^2 \text{ independent of } t.$$

For such stationary states

$$\Psi(t) = \psi e^{-iEt/\hbar}\,;\ \psi = \Psi(0)$$

For the hydrogen atom

$$H(classical) = \frac{P^2}{2M} + \frac{p^2}{2m} + V$$

which includes kinetic energies of the proton and electron and the potential energy V of interaction,

$$H_{op} = \hat{H} = -\frac{h^2}{8\pi^2}\left[\frac{\nabla^2\Psi}{M} + \frac{\nabla^2\Psi}{m} + V\right]\Psi = E\Psi \tag{7.12}$$

Solution of (12) gives the complete non-relativistic quantum-mechanical description of the hydrogen atom in its stationary states. The wave function is interpreted in terms of

$\Psi^2(xyz) \Longrightarrow$ the joint probability of finding the proton and/or the electron at xyz.

The motions of proton and electron are completely entangled in a holistic sense and the separate densities have no independent meaning.

The mathematics to solve (12) does not exist. In practice a solution is possible by separating the problem into variables describing the centre of mass and relative motion of the electron with respect to the nucleus, respectively. It is assumed that

$$\Psi = \psi'_{com} \cdot \psi_{rm}$$

$$E = E_{com} + E_{rm}$$

$E_{com} \rightarrow$ the translational kinetic energy of the system.

The system is now described by two equations. The COM equation

$$\nabla^2\psi' + \frac{8\pi^2(M+m)}{h^2}E\psi' = 0 \tag{7.13}$$

and the RM equation, for reduced mass μ

$$\nabla^2\psi + \frac{8\pi^2\mu}{h^2}(E - V)\psi = 0 \tag{7.14}$$

The two wave functions are now interpreted as

$(\psi')^2 \rightarrow$ probability density of the proton

$\psi^2 \rightarrow$ probability density of the electron.

It is noted that $(\psi')^2 + \psi^2 \neq \Psi^2$. The symmetry is broken by separating the variables and this eliminates the quantum-mechanical equivalence of proton and electron. Only electrostatic interaction, $V = \frac{e^2}{4\pi\epsilon_o r}$ remains in the electronic wave equation.

Equation (12) can be solved analytically, by separating the variables according to standard procedures. Because of the nature of the Coulomb potential it is necessary to transform to spherical polar coordinates first: *i.e.* $\psi(xyz) \rightarrow \psi(r\theta\varphi)$

To separate the variables it is assumed that

$$\psi(r\theta\varphi) = R(r) \cdot \Theta(\theta) \cdot \Phi(\varphi)$$

For the radial equation, obtained by separating

$$\psi(r\theta\varphi) = \frac{1}{r}P(nl;r) \cdot Y(\theta,\varphi)$$

in appropriate units

$$\left[\frac{d^2}{dr^2} + \frac{2Z}{r} - \frac{l(l+1)}{r^2} - \frac{1}{n^2}\right] P = 0 \qquad (7.15)$$

the boundary conditions are $P(0) = 0$ and $\lim_{r\rightarrow\infty} P = 0$. This equation is valid for hydrogenic atoms with nuclear charge Z.

The solutions of the angular dependent part are the spherical harmonics, Y, known to most chemists as the mathematical expressions describing shapes of (hydrogenic) atomic orbitals. It is noted that Y is defined only in terms of a central field and not for atoms in molecules.

The procedure to solve the R and Θ equations is by expanding the solutions as infinite series according to the Frobenius method. To make physical sense the wave functions must be square-integrable. The infinite series must therefore terminate into a polynomial in order to avoid divergence, and to prevent the wavefunction from going infinite. To achieve this the variables n and l are required to be integers. These integers are known as quantum numbers. A third quantum number, m_l appears in solutions of the Φ equation. It is noted that these quantum numbers are not assumed as in the old quantum theory of Bohr, but emerge quite naturally in the physically acceptable solutions of the Schrödinger equation.

The principal quantum number, $n \rightarrow 1, 2, \cdots, \infty$ describes the quantization of electronic energy

$$E = -\frac{1}{n^2}, \text{ (in atomic units).}$$

$l \to 0, 1, \cdots, (n-1)$, describes the quantization of angular momentum,

$$L^2 = l(l+1)\left(\frac{h}{2\pi}\right)^2.$$

The magnetic quantum number, $m_l = -l, \cdots, +l$, describes the projection of the angular momentum in a specific direction, *i.e.* the direction of an external magnetic field.

Any combination (nl) is colloquially referred to as an "orbital".

For non-hydrogenic atoms with n electrons the electronic Hamiltonian becomes

$$H = -\frac{1}{2}\sum_{i=1}^{n}\left(\nabla_i^2 + \frac{2Z}{r_i}\right) + \sum_{i>j}\frac{1}{r_{ij}} \tag{7.16}$$

(in atomic units) Formal solutions to this problem are not known, not even for He with its two electrons. Approximate perturbation-theory solutions exist, but the closest to an exact solution ever obtained was in terms of the eclipsed model, which considers the two electrons always to stay on opposite sides of the nucleus. Even this relatively simple dynamic system becomes chaotic after a few cycles. These problems are considered next in more detail.

7.3.1 Many-electron Atoms

An intuitive extension of the quantum-mechanical discription to more complicated atoms is by the Thomas-Fermi method [62].

The Thomas-Fermi Model

In the Thomas-Fermi model of the ground-state electronic configuration of complex atoms the N electrons are treated as a Fermi electron gas in the ground state, confined to a region of space by a central potential $V(r)$ that vanishes at infinity. It is assumed that the potential is slowly varying over a distance which is large compared with the de Broglie wavelengths of the electrons, so that enough electrons are present in the volume where $V(r)$ is nearly constant, to justify the use of a statistical approach to study the electron gas. The aim is to calculate the potential $V(r)$ and the electron density $\rho(r)$. To prevent the electron from escaping to infinity the total energy $p^2/2m + V$ cannot be positive. Since the maximum kinetic energy of an electron in a Fermi gas at 0K is E_F, the total classical energy of the most energetic electrons of the system

$$E_{max} = E_F + V(r)$$

Unless E_{max} is independent of r the electrons would migrate to some region where E_{max} has a minimum. Furthermore $E_{max} \leq 0$, as before.

The Fermi energy, and hence the Fermi radius must be a function of r, *i.e.*

$$k_F^2(r) = \frac{2m}{\hbar}\left[E_{max} - V(r)\right]$$

Thus, using (6.8) the electron density

$$\rho(r) = \frac{1}{3\pi^2}\left\{\frac{2m}{\hbar^2}\left[E_{max} - V(r)\right]\right\}^{3/2} \tag{7.17}$$

is seen to vanish when $V = E_{max}$. In the classically forbidden region $V > E_{max}$ it is required that $\rho = 0$, otherwise the maximum kinetic energy E_F goes negative.

The previous expression (15) is readily modified to relate charge density to electrostatic potential,

$$\Psi(r) = -\frac{1}{e}V(r), \quad \text{for electronic charge} \quad -e.$$

A more general formulation, $\Phi(r) = \phi(r) - \phi(0)$, may be obtained by setting $\phi(0) = -E_{max}/e$, a non-negative constant. Hence

$$\begin{aligned} \rho(r) &= \frac{1}{3\pi^2}\left(\frac{2m}{\hbar^2}\right)^{3/2}\left[e\Phi(r)\right]^{3/2} \quad \Phi \geq 0 \\ &= 0 \qquad\qquad\qquad\qquad\qquad\quad \Phi < 0 \end{aligned} \tag{7.18}$$

The equation $\Phi = 0$ (*i.e.* $\phi = \phi_0$ or $V = E_{max}$) may be thought of as determining a boundary $r = r_0$ of an atom (ion) in this model. For a neutral atom ($N = Z$) the electrostatic potential $\phi(r)$ must vanish at the boundary and this requires $\phi_0 = 0$. For an ion $\phi_0 > 0$.

A second relation between $\rho(r)$ and $\Phi(r)$ may be obtained by noting that the sources of $\phi(r)$ are the point charge Ze of the nucleus, located at the origin and the charge distribution due to the N electrons. Treating the charge density $-e\rho(r)$ of the electrons as continuous, Poisson's equation of electrostatics may be used to write

$$\nabla^2\Phi(r) = \frac{1}{r}\frac{d^2}{dr^2}\left[r\Phi(r)\right] = \frac{e}{\epsilon_0}\rho(r) \tag{7.19}$$

Elimination of $\rho(r)$ from the simultaneous equations (18) and (19) for $\Phi \geq 0$, gives

$$\frac{1}{r}\frac{d^2}{dr^2}\left[r\Phi(r)\right] = \frac{e}{3\pi^2\epsilon_0}\left[e\Phi(r)\right]^{3/2}, \ \Phi \geq 0 \tag{7.20}$$

For $\Phi < 0$

$$\frac{d^2}{dr^2}[r\Phi(r)] = 0\,,\ \Phi < 0$$

For $r \to 0$ the leading term of the electrostatic potential must be due to the nucleus, so that the boundary condition at $r = 0$ becomes

$$\lim_{r\to 0} r\Phi(r) = \frac{Ze}{4\pi\epsilon_0}$$

For the N electrons confined to a sphere of radius r_0,

$$4\pi\int_0^{r_0}\rho(r)r^2dr = N$$

The important equation (20) can now be cast into dimensionless form by introducing some function χ, such that

$$r\Phi(r) = \frac{Ze}{4\pi\epsilon_0}\chi$$

and hence

$$\rho(r) = \frac{1}{3\pi^2}\left(\frac{2m}{\hbar^2}\right)^{3/2}\left[\frac{Ze^2\chi}{4\pi\epsilon_0 r}\right]^{3/2}$$

rearranges to

$$\rho = \frac{Z}{4\pi}\left(\frac{2^{7/2}Z^{1/2}}{3\pi a_0^{3/2}r^{3/2}}\right)\chi^{3/2}$$

in terms of the Bohr radius $a_0 = (4\pi\epsilon_0\hbar^2)/me^2$. The expression can be further simplified be defining

$$b = \frac{(3\pi)^{2/3}a_0}{2^{7/3}Z^{1/3}}\,,$$

to give

$$\rho = \frac{Z}{4\pi}\left(\frac{\chi}{br}\right)^{3/2}.$$

The variable x is now introduced, such that $r = bx$ and hence

$$\rho = \frac{Z}{4\pi b^3}\left(\frac{\chi}{x}\right)^{3/2}$$

Equation (20) in dimensionless form follows from

$$\frac{1}{bx}\frac{d^2}{d(bx)^2}\left[\frac{Ze}{4\pi\epsilon_0}\chi\right] = \frac{e}{\epsilon_0}\frac{Z}{4\pi b^3}\left(\frac{\chi}{x}\right)^{3/2}$$

i.e.

$$\frac{d^2\chi}{dx^2} = x\left(\frac{\chi}{x}\right)^{3/2} = x^{-1/2}\chi^{3/2}\,,\ \chi \geq 0 \tag{7.21}$$

known as the Thomas-Fermi equation. For negative χ

$$\frac{d^2\chi}{dx^2} = 0\,,\ \chi < 0 \tag{7.22}$$

The boundary condition at $r = 0$ becomes $\chi(0) = 1$.

Suppose that the atom has a boundary at $r_0 = bx_0$, such that $\chi(x_0) = 0$. Then, by definition $\chi(x) > 0$ for all $x < x_0$ and $\chi(x) < 0$ for all $x > x_0$. The function $\chi(x)$ can clearly not have another zero in the interval $(0, \infty)$.

Equation (22) has the solution

$$\chi = C(x - x_0)$$

where C is a negative constant, and since

$$\chi' = \frac{d\chi}{dx} = C$$

C must be equal to the derivative $\chi'(x_0)$ at the boundary. Knowing the solution for $\chi \geq 0$ therefore determines $\chi(x)$ for all x.

The Thomas-Fermi equation (21) is independent of Z or any other physical constants and is valid for all atoms. The normalization condition in dimensionless form simply becomes

$$\begin{aligned} \mathrm{N} &= Z\int_0^{x_0} x^2\left(\frac{\chi}{x}\right)^{3/2} dx \\ &= Z\int_0^{x_0} x\chi''\,dx \\ &= Z\left[x\chi' - \chi\right]_0^{x_0} \end{aligned}$$

Since $\chi(0) = 1$ and $\chi(x_0) = 0$

$$\frac{N - Z}{Z} = x_0\chi'(x_0) \tag{7.23}$$

For neutral atoms, $N = Z$, (23) requires $\chi'(x_0) = 0$, so that χ' vanishes at the same point as χ. Since this condition cannot be satisfied for a finite value x_0 by non-trivial solutions, the point x_0 must be at infinity. The solution $\chi(x)$ for a neutral atom must hence be asymptotic to the x-axis, $\chi(\infty) = 0$. There is no boundary to the neutral atom in the Thomas-Fermi model.

The universal function $\chi(x)$ obtained by numerical integration and valid for all neutral atoms decreases monotonically. The electron density is similar for all atoms, except for a different length scale, which is determined by the quantity b and proportional to $Z^{-\frac{1}{3}}$. The density is poorly determined at both small and large values of r. However, since most electrons in complex atoms are at intermediate distances from the nucleus the Thomas-Fermi model is useful for calculating quantities that depend on the average electron density, such as the total energy. The Thomas-Fermi model therefore cannot account for the periodic properties of atoms, but provides a good estimate of initial fields used in more elaborate calculations like those to be discussed in the next section.

Solutions $\chi(x)$ that vanish at some finite value $x = x_0$ are such that $N \neq Z$, and hence correspond to positive ions of radius $r_0 = bx_0$. Since the slope of χ is negative at x_0 it means that $Z > N$.

The Hartree-Fock method

The most obvious defect of the Thomas-Fermi model is the neglect of interaction between electrons, but even in the most advanced modern methods this interaction still presents the most difficult problem. The most useful practical procedure to calculate the electronic structure of complex atoms is by means of the Hartree-Fock procedure, which is not by solution of the atomic wave equation, but by iterative numerical procedures, based on the hydrogen model. In this method the exact Hamiltonian is replaced by

$$H = \sum_{i=1}^{N} \left\{ -\frac{1}{2}\nabla_i^2 - \frac{Z}{r_i} + U(r_i) \right\} \tag{7.24}$$

where $U(r_i)$ is a single-particle potential energy, introduced to approximate the inter-electronic interactions.

The first step beyond the statistical model was due to Hartree who derived a wave function for each electron in the average field of the nucleus and all other electrons. This field is continually updated by replacing the initial one-electron wave functions by improved functions as they become available. At each pass the wave functions are optimized by the variation method, until self-consistency is achieved. The angle-dependence of the resulting wave functions are assumed to be the same as for hydrogenic functions and only the radial function (u) needs to be calculated.

The first term of the Hamiltonian for an N-electron system

$$H = -\frac{1}{2}\sum_i \nabla_i^2 - \sum_i \frac{Z}{r_i} + \sum_{i>j} \frac{1}{r_{ij}} \tag{7.25}$$

represents the sum of electronic kinetic energies, the second represents potential energy in the field of the nucleaus and the third, Coulombic interelectronic repulsion.

By grouping one-electron and two-electron terms together the Hamiltonian may be written as

$$H = \sum_i f_i + \sum_{i,j} g_{ij}$$

Because of interelectronic effects this Hamiltonian is not separable. Only when these effects are ignored may the total probability density $\psi^*\psi$ be assumed to be a product of one-electron probability densities and the wave function a product of hydrogenic atomic wave functions

$$\psi(r_1, \theta_1, \phi_1, r_2, \theta_2, \phi_2, \ldots) = \psi_1(r_1, \theta_1, \phi_1) \cdot \psi_2(r_2, \theta_2, \phi_2) \ldots$$

If the given Hamiltonian operates on the product wave function, f_i only operates on $\psi(i)$, *etc.*, and the contribution of this wave function to the expectation value of the energy becomes

$$\int \psi^*(i) f_i \psi(i)\, d\tau_i = \langle i|f|i\rangle$$

The two-electron operator similarly gives a contribution

$$\int \psi^*(i)\psi^*(j) g_{ij} \psi(i)\psi(j)\, d\tau_i d\tau_j = \langle ij|g|ij\rangle$$

Hence, the average energy

$$\bar{E} = \sum_i \langle i|f|i\rangle + \sum_{i>j} \langle ij|g|ij\rangle$$

Numerical methods to evaluate these integrals have been established.

In order to obtain one-electron radial wave functions from the energy expression by the variational method, it is assumed that a set of coefficients ϵ_i exists such that

$$\delta \bar{E}_i = \epsilon_i \cdot \delta \int u_i^2\, d\tau_i$$

for all variations. This assumption ensures that the total energy remains stationary at the minimum.

Variation of a one-electron integral leads to

$$\begin{aligned} \delta I &= \int \delta u \cdot f \cdot u\, dr + \int u \cdot f \cdot \delta u\, dr \\ &= 2 \int \delta u \cdot f \cdot u\, dr \end{aligned}$$

Similarly, variation of a two-electron integral leads to

$$\delta F = 2\int \delta u_i \cdot u_i g_{ij} u_j^2\, dr_i dr_j$$

Also

$$\delta \int u^2\, dr = 2\int \delta u \cdot u\, dr$$

When combined into a single expression

$$2\int \delta u_i (H - \epsilon_i) u_i\, dsr = 0$$

for any value of the variation function δu_i. This means that the second part of the integrand must vanish independently, such that (in atomic units)

$$\left[-\frac{1}{2}\nabla^2 + V(r)\right] u_i = \epsilon_i u_i \tag{7.26}$$

where $V(r)$ represents the average potential field. Equation (26) is known as Hartree's equation.

The average field is calculated as follows:
The potential energy of interaction between point charges q_1 and q_2 is given by

$$V_{12} = \frac{q_1 q_2}{4\pi\epsilon_0 r_{12}}$$

If q_2 is smeared out into a continuous charge distribution over volume v_2 at charge density ρ_2 and at an average distance r_{12} from q_1, then

$$\begin{aligned} V_{12} &= \frac{q_1}{4\pi\epsilon_0}\int \frac{\rho_2}{r_{12}} dv_2 \\ &= \frac{q_1}{4\pi\epsilon_0} e \int \frac{\psi^2}{r_{12}} dv_2 \end{aligned}$$

where ψ is the appropriate one-electron wave function. The total potential energy follows as

$$V(r) = \frac{1}{4\pi\epsilon_0}\sum_{j=2}^{N}\int \frac{e^2\psi^2}{r_{ij}}\, d\tau_j - \frac{Ze^2}{r}$$

where the integration is over all angles.

Since the potential depends on the wave functions, and *vice versa*, the Hartree equation can only be solved by an iterative procedure.

If Hartree's equation is written in the form

$$Hu_i = \epsilon_i u_i$$

the ϵ_i may be assumed to represent individual orbital energies. It is important to note that interelectronic effects need to be included when solving for ϵ_i. In a summation over all ϵ_i these effects would therefore be counted twice and to compensate for that the total energy is expressed as

$$E = \sum_{i=1}^{N} \epsilon_i - \sum_i \sum_{j>i} \frac{e^2\psi^2(i)\psi^2(j)}{r_{ij}} d\tau_i d\tau_j$$

The Self-Consistent-Field (SCF) procedure can be initiated with hydrogenic wave functions and Thomas-Fermi potentials. It leads to a set of solutions $u(r_i)$, each with k nodes between 0 and ∞, with zero nodes for the lowest energy and increasing by one for each higher energy level. The quantum number n can now be defined as $n = l + 1 + k$ to give rise to $1s, 2s, 2p$, *etc.* orbitals.

Spin is not included explicitly in the Hartree calculation and the wave functions are therefore not antisymmetric as required. If anti-symmetrized orbitals are used a set of differential equations

$$H_i f_i = \epsilon_i f_i$$

known as Hartree-Fock equations are obtained. The method is the same as before. In addition to the Coulomb integrals however, the expression now also involves exchange integrals. These extra terms are generated by the use of determinantal rather than the simple product wave functions of the Hartree method. The Hartree-Fock trial functions are Slater determinants. In this case individual wave functions are found to be represented by a set of Schrödinger equations of the form

$$\left[-\frac{1}{2}\nabla_k^2 - \frac{Ze^2}{r_k} + V^C\right]\psi_j - T^{Ex} = E\psi_j(r_k)$$

Such an equation differs from Hartree's equation only by virtue of the extra exchange term, T^{Ex}. Whereas the electronic Coulomb interaction of the Hartree scheme is formulated as

$$V^C = \sum_j \int \frac{e^2|\psi_j(r_j)|^2}{4\pi\epsilon_0|\boldsymbol{r} - \boldsymbol{r}_j|} d\tau_j$$

the exchange term has the form

$$T^{Ex} = V^{Ex}\psi_j(r_k) = \sum_{j=1}^{N} \left[\frac{1}{4\pi\epsilon_0} \int \psi_j^*(r_j) \frac{e^2}{|\boldsymbol{r}_k - \boldsymbol{r}_j|} \psi_k(r_j)\, d\tau_j \right] \psi_j(r_k)$$

where the wave function ψ_k has exchanged roles with the wave function ψ_j by the interchange of electron coordinates $\boldsymbol{r}_k$ and $\boldsymbol{r}_j$. The exchange term is a direct result of the determinant approach, *i.e.* of the antisymmetry of the wave functions. It may be interpreted to mean that the Coulomb interaction energy between electrons with parallel spins is reduced relative to that between electrons with antiparallel spins.

Exchange integrals present the most difficult problem in numerical Hartree-Fock calculations. The simplification introduced by Slater [78], substituting a free-electron gas average for all exchange interactions therefore is a major advance. By analogy with the Coulomb potential in electrostatics which is expressed in terms of the charge density, the exchange potential can be expressed in terms of a fictional exchange charge density. In physical terms, the exchange charge represents the removal, by exclusion interaction, of one electron, from the set of electrons with the same spin, to create a *Fermi* hole. In a free-electron gas, the averaged exchange potential is proportional to the cube root of the electronic charge density. For a total electronic charge density (both spins) of ρ, and assuming equal numbers of up and down spins, the free-electron exchange potential, in atomic units, is given by

$$V^{Ex}(\rho) = -6 \left[\frac{3}{8\pi} |\rho| \right]^{\frac{1}{3}}$$

If the average exchange potential is assumed to depend only on the local electronic charge density, its value at a point $\boldsymbol{r}$ is equal to the $V^{Ex}(\rho)$ for a free-electron gas, and

$$V^{Ex}(\boldsymbol{r}) = -6 \left[\frac{3}{8\pi} |\rho(\boldsymbol{r})| \right]^{\frac{1}{3}}$$

It becomes a function of $\rho(\boldsymbol{r})$, and the problem of calculating exchange integrals is completely avoided.

Numerical Hartree-Fock-Slater Calculations

Herman and Skillman [79] used an HFS algorithm to calculate radial atomic wave functions and energy eigenvalues for all atoms, tabulating all results and the computer software at the same time. They treated all single electronic configurations with one or more open shells on the same basis as configurations having only closed shells. Many-electron wave functions are represented by single-determinantal wave functions, built up from spin orbitals of the type

$$\Phi(n,l,m,s) = R_{nl}(r) \cdot Y_{lm}(\theta,\phi) \cdot \mu_s$$

Multiplet structures are ignored completely and electronic configurations are defined only in terms of the occupation numbers of the various orbitals. Accordingly, the radial HFS wave equations for a free atom or ion are written in the form

$$\left[-\frac{d^2}{dr^2} + \frac{l(l+1)}{r^2} + V(r)\right] P_{nl}(r) = E_{nl}P_{nl}(r)$$

where $P_{nl}(r) = rR_{nl}(r)$, and $V(r)$ is the sum of nuclear Coulomb potential, the total electronic Coulomb potential, and the exchange potential energies. At most values of r, $V(r)$ is equated with

$$V_0(r) = -\frac{2Z}{r} - \frac{2}{r}\int_0^r \sigma(t)\,dt - 2\int_r^\infty \sigma(t)\frac{dt}{t} - 6\left[-\frac{3}{8}\rho(r)\right]^{\frac{1}{3}}$$

$\rho(r) = \sigma(r)/(4\pi r^2)$ is the spherically averaged total electronic charge density (both spins),

$$\sigma(r) = -\sum_{nl} \omega_{nl}\left[P_{nl}(r)\right]^2 ,$$

where ω_{nl} is the occupation number for the orbital (both spins). The $P_{nl}(r)$ are normalized radial wave functions. It is necessary to compensate for the breakdown of the free-electron exchange approximation at large values of r. This is done by defining

$$V(r) = V_0(r) \quad , \quad r < r_0 \quad , \quad V(r) = \frac{-2(Z-N+1)}{r} \quad , \quad r \geq r_0$$

where r_0 is the value of r at which

$$V_0(r) = \frac{-2(Z-N+1)}{r}, \quad i.e. \quad V_0(r_0) = \frac{-2(Z-N+1)}{r_0}$$

The program also provides a facility to correct the calculated values for relativistic effects, starting from the Pauli equation:

$$[H_0(r) + H_m(r) + H_d(r) + H_{so}(r)]\,R(r) = ER(r)$$

in which
$R(r)$ is a two-component spinor;

$H_0(r)$ is the non-relativistic Hamiltonian;
$H_m(r)$ describes the relativistic variation of mass with velocity,

$$H_m(r) = -\frac{\alpha^2}{4}\left[E^0 - V(r)\right]^2$$

where α is the fine-structure constant and E^0 is the non-relativistic energy eigenvalue;
$H_d(r)$ is the relativistic Darwin correction operator

$$H_d(r) = -\frac{\alpha^2}{4}\left[\frac{dV(r)}{dr}\right]\frac{d}{dr}$$

$H_{so}(r)$ is the spin-orbit energy,

$$H_{so}(r) = -\frac{\alpha^2}{4}\left[\frac{-l}{l+1}\right]\frac{1}{r}\left[\frac{dV(r)}{dr}\right]$$

First-order corrected energy, correct to order α^2, is

$$E_{nl}^1 = E_{nl}^0 + \int_0^\infty dr\, r^2 R_{nl}^0(r)\left[H_m(r) + H_d(r) + H_{so}(r)\right] R_{nl}^0(r)$$

Despite many approximations the Herman-Skillman results compare well with experimental values and the program still is a useful tool for calculations related to the electronic configurations of atoms.

7.3.2 Compressed Atoms

The quantum-mechanical description of the hydrogen atom is of central importance as the only example from which derive many generalizations pertaining to chemical systems. It is nevertheless important to note that in many respects it represents real chemical systems rather poorly. Its most glaring defect is perhaps complete neglect of environment as a factor that shapes chemical behaviour. The way in which electronic wavefunctions of free atoms extend to infinity is clearly at variance with chemical reality, especially where atoms find themselves in a crowded environment at the onset of chemical reactions. In a situation like this interatomic contacts largely consist of interacting negative charge clouds. The environment of an atom is therefore approximated well by a uniform electrostatic field that prevents electronic charge density from extending indefinitely. Hence the boundary condition $\psi \to 0$ as $r \to \infty$ might be replaced by $\lim_{r\to r_o}\psi = 0$, where $r_o << \infty$, to obtain a more realistic solution to atomic wave equations for chemically interacting systems.

Although this condition considerably complicates the mathematical atomic model, wave equations for non-hydrogen atoms can be solved numerically with a HF-SCF method under this boundary condition, which forces the wave function to vanish at a finite distance r_o. Hydrogen, as a special case has been examined many times, using various numerical techniques for integration of the wave equation under compression. The single most important finding was an increase of the electronic energy, until the ionization limit is reached on compression to a critical radius [80, 81]. In the Hartree-Fock treatment of non-hydrogen atoms [82] the boundary condition is introduced by multiplying all one-electron wavefunctions before normalization on each iteration by the step function

$$S = \exp\left[-\left(\frac{r}{r_o}\right)^p\right], \text{ for } p >> 1$$

The HFS program [79] was modified especially for this purpose. The general response is the same as for hydrogen, except that all occupied levels are now also affected at the same time. The highest levels are the most sensitive, but even the deepest core levels show an increase. As for hydrogen, the valence level eventually reaches the ionization limit, but this cannot be interpreted directly as an ionization event, as was done for hydrogen.

The quantum numbers, n and l, that quantify the energy levels of different electrons have a precise meaning only at the beginning of the iterative procedure, when they refer to hydrogen-like one-electron wavefunctions. As the potential field changes during the various stages of iteration they loose this meaning and no longer represent sensible quantum numbers. They are simply retained as a book-keeping device, and not to indicate that the independent-electron shell structure used as the starting configuration is preserved in the final self-consistent ground state of the atom. This is clearly reflected in the observation that the ionization energy of a free atom only superficially resembles the HF orbital energy of the most energetic valence electron.

Another complication is that in standard HF calculations all electrons with the same lables (n, l) are assigned similar energies at the average of the multiplet for that level. It is obviously unreasonable to expect all electrons at the valence level to reach the ionization level simultaneously on excitation of the multiplet. Substantial mixing of levels on compression is inevitable and this interaction transfers energy by a process of resonance to a single electron which is thereby promoted to its ionization limit. This process is not unlike ionization in an applied field, often visualized in terms of Fermi's Golden Rule [56]. Excess energy generated at lower levels is likewise transferred to electrons at the highest level and contributes to their excitation. This process is reminiscent of spectral transitions involving groups of states and energies within broad intervals, which is also resolved in terms of FGR that relates transition probabilities to the densities of excited levels. Inverting the argument, but not the logic, for the excess energy in compressed atoms, energy transmission probabilities now get linked to the spacings between calculated HF levels and the valence level. An algorithm based on this argument was built into the HFS software to analyze atomic compressions across the periodic table.

It is assumed that upper and lower levels are separated by a gap of ΔE. If the lower level is occupied by N electrons the transition probability is $P = N/\Delta E$, with a maximum of $P = 1$, which describes fully resonating levels. The probability of resonance within the highest multiplet level is always equal to one. The transferred energy from all affected levels to the topmost level is then calculated as

$$E_T = \sum_{i=1}^{N} \frac{N_i \varepsilon_i}{\Delta E_i}$$

where ε_i is the increase in energy due to compression. For N_n electrons at the highest level compression increases the energy of the most energetic electron by an amount $E_p = E_T + N_n \varepsilon_n$. When E_p matches the binding energy of the valence level, ionization occurs. The ionization limit is therefore reached by compression of each atom to an ionization radius, characteristic of that atom. Atomic ionization radii follow a predictable periodic trend.

The valence-electron wave functions of atoms, compressed beyond their ionization limits are Fourier sums of spherical Bessel functions corresponding to step functions (Compare 6.3.1) of the type

$$f(r) = \begin{cases} \alpha & \text{if} \quad |r| < r_0 \\ 0 & \text{if} \quad |r| > r_0 \end{cases}$$

where r_0 is the ionization radius. For a single electron such a wave function has the form

$$\psi = \left(\frac{3c}{4\pi n}\right)^{\frac{1}{2}} \left(\frac{1}{r_0}\right) \exp\left[-(r/r_0)^p\right] \qquad p >> 1$$

where c is a constant and n is the principal quantum number according to the HF shell structure.

7.4 Molecular Systems

The notion that chemical substances consist of molecules with rigid structure and characteristic shape seems to conflict with quantum theory. So deeply is the dogma embedded in chemical thinking however, and so skillfully has it been grafted on quantum-mechanical roots by now, that the basic idea is rarely disputed any more. The conflict arises from the holistic nature of quantum theory that demands a single state function to describe a pure compound such as liquid water, rather than a composite product function based on individual units[3] of H_2O. To by-pass the problem conventional wisdom diverts attention to the molecular building blocks as they are assumed to exist in a rarefied gaseous state, in the absence of interaction with their neighbours. Still, the structure problem remains intractible without further simplifying assumptions. The most momentous of these assumptions occurs in the Born-Oppenheimer scheme that decouples electronic and nuclear motion.

7.4.1 The Born-Oppenheimer Approximation

A molecule may be seen as a set of nuclei (mass M) and electrons (mass m) with coordinates Q_α and q_i respectively. The Schrödinger equation is

$$H\phi(Q,q) = \left[\sum_\alpha \frac{P_\alpha^2}{2M_\alpha} + \sum_i \frac{p_i^2}{2m} + V(Q,q)\right]\phi(Q,q) = E\phi(Q,q) \qquad (7.27)$$

The potential energy consists of the Coulomb interactions between all pairs of particles:

$$V = \sum_{i,j=1}^{N,E} -\frac{Z_j e^2}{|\boldsymbol{r}_i - \boldsymbol{R}_j|} + \sum_{i>k=1}^{E} \frac{e^2}{|\boldsymbol{r}_i - \boldsymbol{r}_k|} + \sum_{j>l=1}^{N} \frac{Z_j Z_l e^2}{|\boldsymbol{R}_j - \boldsymbol{R}_l|}$$

There is no distinction between electrons and nuclei and no mechanism whereby relative atomic positions can be localized. However, since the lighter electrons may reasonably be assumed to move much faster than the nuclei, the nuclear coordinates may to first approximation be assumed to remain fixed at Q' on the time scale of electronic motion. The electronic wave equation then reduces to

$$\left[\sum_i \frac{p_i^2}{2m} + V(Q',q)\right]\psi(Q',q) = E'(Q')\psi(Q',q) \qquad (7.28)$$

[3]The dilemma is highlighted in the following quotation[83]: "Calculations based on a brute force approach have not contributed anything of significance ... In fact, even a system as $(H_2O)_3$ has so many possible variables in terms of the position of the constituent nuclei that it is not possible to do sufficient calculations to satisfy a statistical mechanism without using grotesque amounts of computer time".

This equation can in principle be solved for all conceivable nuclear configurations. In any particular case it is assumed that $\phi(Q,q) = \psi(Q,q)\nu(Q)$, where $\nu(Q)$ is some vibrational nuclear wave function. Substitution of this form into (27) gives

$$\begin{aligned} H\psi(Q,q)\nu(Q) &= E\psi(Q,q)\nu(Q) \\ &= \left[\sum_\alpha \frac{P_\alpha^2}{2M_\alpha} + \sum_i \frac{p_i^2}{2m} + V(Q,q)\right]\psi(Q,q)\nu(Q) \quad (7.29) \end{aligned}$$

The nuclear kinetic energy is

$$\frac{P^2}{2M}\psi(Q,q)\nu(Q) = -\frac{h^2}{8\pi^2 M}\nabla_\alpha^2\psi(Q,q)\nu(Q)$$

$$= -\frac{\hbar^2}{2M}\left[\psi(Q,q)\nabla_\alpha^2\nu(Q) + 2\nabla_\alpha\psi(Q,q)\cdot\nabla_\alpha\nu(Q) + \nu(Q)\nabla_\alpha^2\psi(Q,q)\right] \quad (7.30)$$

Since the nuclei are massive compared to electrons, $m/M << 1$, they move so slowly in a molecule that the electrons can always maintain their optimal motion at each Q. The electronic wave functions may therefore be assumed to depend on the nuclear positions only and not on their momenta, *i.e.* $\nabla_\alpha\psi(Q,q) \to 0$. The electrons are said to follow the nuclei adiabatically.

Under these assumptions equation (30) reduces to

$$\begin{aligned} \frac{P^2}{2M}\psi(Q,q)\nu(Q) &\simeq -\frac{h^2}{8\pi^2 M}\psi(Q,q)\nabla_\alpha^2\nu(Q) \\ &= \psi(Q,q)\frac{P^2}{2M}\nu(Q) \end{aligned}$$

When this result is substituted into (29)

$$\sum_\alpha \psi(Q,q)\frac{P_\alpha^2}{2M_\alpha} + \left[\sum_i \frac{p_i^2}{2m} + V(Q,q)\right]\psi(q,q)\nu(Q) = E\psi(Q,q)\nu(Q) \quad (7.31)$$

According to (29) the second term on the left becomes $E'\psi(Q,q)\nu(Q)$ and (31) simplifies to

$$\left[\sum_\alpha \frac{P_\alpha^2}{2M_\alpha} + E'(Q)\right]\nu(Q) = E\nu(Q) \quad (7.32)$$

It follows that the wave function for nuclear motion is calculated in an effective potential $E'(Q)$ obtained from the energy eigenvalues of the stationary electronic state.

The *ad hoc* separation of electronic and vibrational states introduced before (6.2.1) is now seen to find its justification in the Born-Oppenheimer approximation. So do most quantum-mechanical analyses of molecular structure. To initiate such an analysis it is necessary to assume a starting set of nuclear coordinates, in terms of which the electronic distribution is calculated. Once the electron density function is known, equation (32) may be used to optimize the structure of the nuclear framework. A quantum-mechanically defined three-dimensional molecular structure seems to emerge almost miraculously during this cycle of computation. Fact remains, somewhere during the mathematical manipulation, molecular geometry, based on chemical intuition or experience, is introduced by hand, if not sleight of hand. Despite convincing claims to the contrary it should be clear that, by definition, no procedure based on the Born-Oppenheimer assumption can generate molecular structure by quantum-mechanical operations. The debilitating step coincides with the neglect of nuclear momentum in order to produce equation (32). Thereby the holistic nature of (27) is destroyed and the whole is reduced to parts. Since one of the parts is classical[4] it cannot remain entangled with the non-classical part. The abstraction that is necessary to establish the classical structure can never be reversed and the quantum-mechanical nature of molecular shape, if any, needs to be established by independent arguments. It is noted in passing that systematic sampling of however many different trial structures, however ingeneously performed, can never reproduce a quantum-mechanical probability density. Ultimately, the only consideration that may decide the quantum nature of a system, is whether or not the electronic and nuclear motion is described by a single wave function. If not, Born-Oppenheimer separation has been introduced along the way, implying that a molecular structure has been assumed *ad hoc*.

7.4.2 The H_2^+ Molecule

The belief that computational chemists obtain molecular structures by solving Schrödinger's equation is often dressed up in so much jargon that the essential arguments are obscured. The general basis of the belief may be examined by considering the simplest possible molecule as a test case[5].

The minimum configuration that defines a molecule is two nuclei and one

[4]Since all atomic coordinates and momenta are fixed the molecule vibrates classically.

[5]The rationale is like settling the issue around perpetual motion machines by the second law of thermodynamics rather than examine each proposed device independently.

electron, *i.e.* H_2^+. The Hamiltonian for this molecule is

$$H\left(r_A, r_B, R\right) = -\frac{\hbar^2}{2}\left[\frac{\nabla_r^2}{m} + \frac{\nabla_A^2}{M} + \frac{\nabla_B^2}{M}\right] - \frac{e^2}{4\pi\epsilon_0}\left[\frac{1}{r_A} + \frac{1}{r_B} - \frac{1}{R}\right]$$

where R is the internuclear separation, and $\boldsymbol{r}_A$, $\boldsymbol{r}_B$ are the position vectors of the electron with respect to protons A and B respectively.

Solutions to the Schrödinger equation $H\phi = E\phi$ are the molecular wave functions ϕ, that describe the entangled motion of the three particles such that $\phi^*\phi$ represents the density of protons and electron as a joint probability without any suggestion of structure. Any other molecular problem, irrespective of complexity can also be developed to this point. No further progress is possible unless electronic and nuclear variables are separated via the adiabatic simplification. In the case of H_2^+ that means clamping the nuclei at a distance R apart to generate a Schrödinger equation for electronic motion only, in atomic units,

$$\left[-\frac{1}{2}\nabla_r^2 - \frac{1}{r_A} - \frac{1}{r_B} + \frac{1}{R} - E\right]\psi = 0$$

This one-particle equation is sufficiently simple so that it is possible to obtain numerical solutions to any degree of accuracy. As first done by Burrau [84] the equation is transformed (eqn. 1.12) into confocal elliptic coordinates (ξ, η, ϕ),

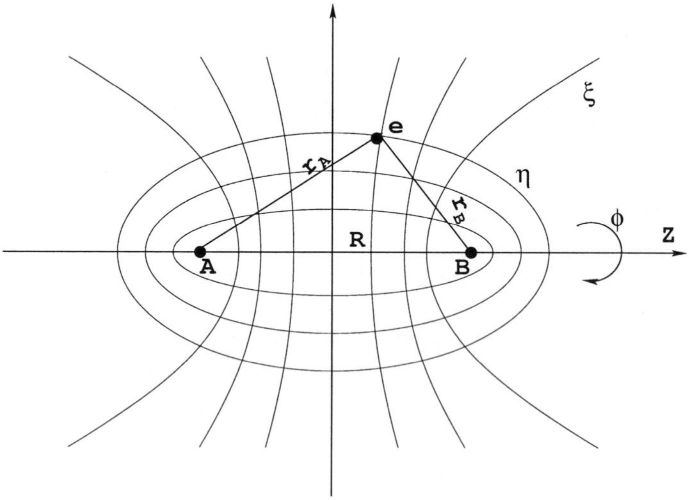

where

$$\xi = \frac{1}{R}\left(r_A + r_B\right)) \quad , \quad 1 \leq \xi \leq \infty$$

$$\eta = \frac{1}{R}(r_A - r_B) \quad , \quad -1 \leq \eta \leq +1$$

$$dV = \frac{R^3}{8}(\xi^2 - \eta^2)d\phi d\eta d\xi \tag{7.33}$$

and ϕ is the azimuthal angle with respect to the Z-axis. The surfaces $\xi =$ constant are confocal ellipsoids of revolution, with the nuclei at the foci, and the surfaces $\eta =$ constant are confocal hyperboloids.

In this coordinate system the Laplacian operator is

$$\nabla^2 = \frac{4}{R^2(\xi^2 - \eta^2)}\left\{\frac{\partial}{\partial \xi}\left[(\xi^2 - 1)\frac{\partial}{\partial \xi}\right] + \frac{\partial}{\partial \eta}\left[(1 - \eta^2)\frac{\partial}{\partial \eta}\right] + \frac{\xi^2 - \eta^2}{(\xi^2 - 1)(1 - \eta^2)}\frac{\partial^2}{\partial \phi^2}\right\}$$

The Schrödinger equation becomes[6]

$$\frac{\partial}{\partial \xi}\left[(\xi^2 - 1)\frac{\partial \psi}{\partial \xi}\right] + \frac{\partial}{\partial \eta}\left[(1 - \eta^2)\frac{\partial \psi}{\partial \eta}\right] + \left(\frac{1}{\xi^2 - 1} + \frac{1}{1 - \eta^2}\right)\frac{\partial^2 \psi}{\partial \phi^2}$$
$$+2R^2\left[\frac{1}{4}\left(E - \frac{1}{R}\right)(\xi^2 - \eta^2) + \frac{1}{R}\xi\right]\psi = 0 \tag{7.34}$$

Since H_2^+ does not have spherical symmetry like the hydrogen atom the angular momentum operator L^2 does not commute with the Hamiltonian, $[L^2, H] \neq 0$. However, H_2^+ does have axial symmetry and therefore L_z commutes with H. The operator $L_z = -i\hbar(\partial/\partial\phi)$ involves only the ϕ coordinate and hence, in order to calculate the commutator, only that part of H that involves ϕ need be considered, *i.e.*

$$\begin{aligned}[L, H] &= \left[-i\frac{\partial}{\partial \phi}, -\frac{2}{R^2(\xi^2 - \eta^2)}\left(\frac{1}{\xi^2 - 1} + \frac{1}{1 - \eta^2}\right)\frac{\partial^2}{\partial \phi^2}\right] \\ &= 0\end{aligned}$$

The electronic wave functions may therefore be chosen to be eigenfunctions of L_z, known from before to be proportional to $\exp(im\phi)$, for $m =$

[6] The derivation requires use of the relationship

$$\frac{1}{r_A} + \frac{1}{r_B} = \frac{2}{r_{AB}(\xi + \eta)} + \frac{2}{r_{AB}(\xi - \eta)} = \frac{4\xi}{r_{AB}(\xi^2 - \eta^2)}$$

$0, \pm 1, \pm 2, \ldots$. The z-component of electronic orbital angular momentum in H_2^+ follows as $m\hbar$ (or m in atomic units). The total electronic orbital angular momentum is not a constant for H_2^+.

The wave equation (34) is found to be separable in terms of the product function

$$\psi(\xi, \eta, \phi) = F(\xi)G(\eta)\Phi(\phi)$$

where $F(\xi)$ and $G(\eta)$ are normalisable solutions of the equations

$$\frac{d}{d\xi}\left[(\xi^2 - 1)\frac{dF}{d\xi}\right] + \left[\frac{R^2}{2}\left(E - \frac{1}{R}\right)\xi^2 + 2R\xi - \frac{m^2}{\xi^2 - 1} + \mu\right]F(\xi) = 0$$

$$\frac{d}{d\eta}\left[(1 - \eta^2)\frac{dG}{d\eta}\right] - \left[\frac{R^2}{2}\left(E - \frac{1}{R}\right)\eta^2 + \frac{m^2}{1 - \eta^2} + \mu\right]G(\eta) = 0$$

and

$$\frac{d^2\Phi}{d\phi^2} = -m^2\Phi$$

μ and m are separation constants. Each electronic term is therefore characterized by three quantum numbers m, n_ξ and n_η. The set of equations can be solved numerically.

Many different procedures have been published, all of them aimed at finding the characteristic values of the parameters m, μ and $\lambda = R^2(E - 1/R)/2$, needed to produce acceptable solutions to the coupled equations. With the allowed values of m known, the procedure consists in finding the relation that must exist between λ and μ to produce an acceptable solution of the η equation, and using this relation to calculate from the ξ equation characteristic values of λ and hence of the energy. The computational details are less important and have often been reduced to reliable computer routines that yield the precise results[85], best represented in terms of binding energy curves, such as those shown below for the ground and first excited states.

The binding energy curve has a minimum at 2.0 bohr = 1.06 Å, showing that the H_2^+ ground state is a stable bound state. The ground-state binding energy $D_e = 0.1026$ hartree = 269 kJ/mol. An entity with these properties can be identified in the spectroscopic excited states of the hydrogen molecule, thought to correspond to an H_2^+ core associated with a highly excited outer electron that has little effect on the potential function of the nuclei. This result shows that the wave-mechanical model provides the correct description of an electron in the field of two clamped nuclei.

The same role that H plays in the theory of complex atoms may be expected for H_2^+ as the prototype from which to generalize electron configurations of complex molecules. The molecular generalization must clearly be subject to the same restrictions that apply to atoms, in addition to the Born-Oppenheimer limitation that rules out an analysis of nuclear coordinates. The most obvious of these limitations are due to the added complexity introduced by interelectronic interactions and the complete neglect of environmental factors. The latter restriction means that the method could only apply to (ideal) gas-phase molecules. This stipulation creates the immediate problem that in this case the Born-Oppenheimer simplification is not necessarily warranted. All molecules such as NH_3 in which the nuclei are known to behave non-classically, would hence be excluded from the scheme and there is no obvious procedure to anticipate this type of quantum behaviour *a priori*. The dilemma created by these arguments is that rigid Born-Oppenheimer structures are more likely to occur in condensed phases where intermolecular interactions could have a decisive influence on electronic distributions. The larger the molecule however, the less important are environmental factors expected to be and the more appropriate the Born-Oppenheimer assumption.

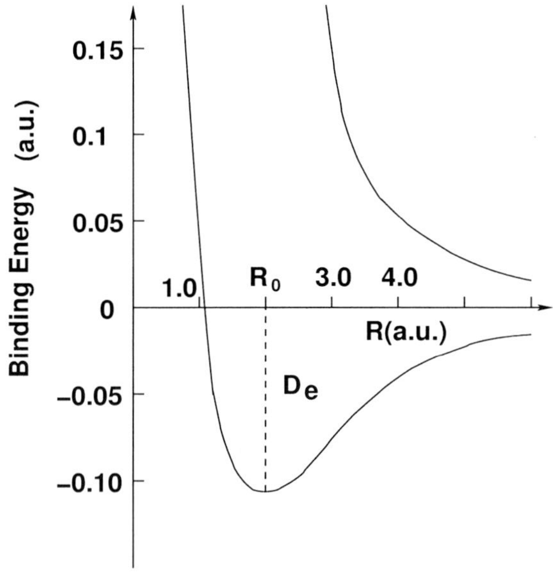

Perhaps the most important general conclusion to be drawn from the one-electron H_2^+ analysis is that chemical bonding intrinsically has nothing to do with electron pairs or electron spin. One remarkable feature that never excites any comment is the bond length of precisely two Bohr radii. Because of that the molecule resembles a system in which an electron simultaneously completes the ground states of two H atoms at the most probable distance of a_0 from each nucleus. This construct is strictly non-classical. In the most precise work [86] the internuclear distance is found to be fractionally less than

$2a_0$ at 1.9972 bohr. This result hints at the fact that the second electron, although far from the nuclei, is not ionized to infinity and still exerts a small perturbation.

These arguments open up the new possibility of synthesizing H_2^+ wave functions from the ground-state hydrogenic functions,

$$(\psi_{1s})_{a,b} = \frac{1}{\sqrt{\pi}} e^{-r_{a,b}} \quad \text{(in atomic units).}$$

The subscripts a, b refer to situations where the electron is more closely associated with either nucleus A, or B. A linear combination such as

$$\phi = \frac{1}{\sqrt{\pi}} \left(c_1 e^{-r_a} + c_2 e^{-r_a}\right) \tag{7.35}$$

where c_1 and c_2 are adjustable parameters would then provide a fair description of H_2^+ for large internuclear distances where the perturbation of an atomic wave function by the second nucleus is unimportant. As the nuclei approach more closely both $1s$ atomic functions will change because of interaction with a second nucleus. As $R \to 0$ the H_2^+ molecule is effectively converted into a hydrogenic He^+ ion with a ground-state wave function proportional to $\exp(-2r)$. In the limit $R \to 0$ the wave function of equation (35) however, reduces to $(1/\sqrt{\pi})(c_1+c_2)\exp(-r)$. Between the two extremes $(0 < R < \infty)$ a more appropriate molecular wave function would therefore be

$$\phi = \frac{1}{\sqrt{\pi}} \left(c_1 e^{-kr_a} + c_2 e^{-kr_b}\right) \quad , \quad 1 < k < 2 \tag{7.36}$$

These propositions may be evaluated by means of variational calculations using (35) and (36) as respective trial functions.

The variational function is of the form

$$\phi = c_1\psi_1 + c_2\psi_2$$

where c_1 and c_2 are parameters to be determined by minimizing the variational integral

$$W = \frac{\int \phi^* H \phi \, d\tau}{\int \phi^* \phi \, d\tau} \geq E_0$$

In the present case the wave functions and parameters are real and hence

$$\begin{aligned} \int \phi^* \phi \, d\tau &= \int \left(c_1^2\psi_1^2 + c_1c_2\psi_1\psi_2 + c_1c_2\psi_2\psi_1 + c_2^2\psi_2^2\right) d\tau \\ &= c_1^2 S_{11} + 2c_1c_2 S_{12} + c_2^2 S_{22} \end{aligned}$$

where, in general, S_{ij} represents an *overlap integral* $S_{ij} = \int \psi_i^* \psi_j d\tau$, and

$$\int \phi^* H \phi \, d\tau = \int \left(c_1^2 \psi_1 H \psi_1 + c_1 c_2 \psi_1 H \psi_2 + c_1 c_2 \psi_2 H \psi_1 + c_2^2 \psi_2 H \psi_2 \right) d\tau$$

Since H is an Hermitian operator the integral

$$\psi_1 H \psi_2 d\tau = \int (\psi_2 H \psi_1)^* \, d\tau = \int \psi_2 H \psi_1 d\tau$$

In terms of the integrals $H_{ij} = \int \psi_i^* H \psi_j d\tau$,

$$\int \phi^* H \phi \, d\tau = c_1^2 H_{11} + 2c_1 c_2 H_{12} + c^2 H_{22}.$$

The variation integral

$$W = \frac{c_1^2 H_{11} + 2c_1 c_2 H_{12} + c^2 H_{22}}{c_1^2 S_{11} + 2c_1 c_2 S_{12} + c_2^2 S_{22}} = \frac{N}{D}$$

is a function of the coefficients c_1 and c_2. A necessary condition for a minimum in W is that its partial derivatives with respect to c_1 and c_2 must vanish at the minimum point. It follows directly that

$$\frac{\partial W}{\partial c_1}(D) + W(2c_1 S_{11} + 2c_2 S_{12}) = 2c_1 H_{11} + 2c_2 H_{12}$$

Setting $\frac{\partial W}{\partial c_1} = 0$ gives

$$c_1(H_{11} - W S_{11}) + c_2(H_{12} - W S_{12}) = 0 \tag{7.37}$$

In the same way the condition $\frac{\partial W}{\partial c_2} = 0$ leads to a second equation

$$c_1(H_{21} - W S_{21}) + c_2(H_{22} - W S_{22}) = 0 \tag{7.38}$$

The simultaneous equations (37) and (38) have non-trivial solutions only if the determinant of coefficients vanishes; giving rise to the *secular equation*

$$\begin{vmatrix} H_{11} - W S_{11} & H_{12} - W S_{12} \\ H_{21} - W S_{21} & H_{22} - W S_{22} \end{vmatrix} = 0$$

In general, for n simultaneous equations, $\det(H_{ij} - W S_{ij}) = 0$.

In the specific case of H_2^+ defined here the secular equation is subscripted by the symbols a and b and the Hamiltonian operator is

$$H = -\frac{1}{2}\nabla^2 - \frac{1}{r_a} - \frac{1}{r_b}$$

Since the nuclei are identical $H_{aa} = H_{bb}$ and since H is hermitian $H_{ab} = H_{ba}$. For normalized $1s$ atomic wave functions the integrals $S_{aa} = S_{bb} = 1$. The secular equation therefore reduces to

$$\begin{vmatrix} H_{aa} - W & H_{ab} - WS_{ab} \\ H_{ba} - WS_{ba} & H_{bb} - W \end{vmatrix} = 0$$

i.e. $H_{aa} - W = \pm(H_{ab} - WS_{ab})$
The two roots

$$W_1 = \frac{H_{aa} + H_{ab}}{1 + S_{ab}} \quad , \quad W_2 = \frac{H_{aa} - H_{ab}}{1 - S_{ab}}$$

are upper bounds to the energies of the ground and first excited electronic states of H_2^+.

To determine the coefficients in the linear combination W_1 is substituted into the secular equation

$$(H_{aa} - W)c_a + (H_{ab} - S_{ab}W)c_b = 0$$

i.e. $c_a/c_b = 1, \quad \phi_1 = c_a(1s_a + 1s_b)$. Normalization requires that

$$|c_a| = \frac{1}{\sqrt{2 + 2S_{ab}}}$$

The normalized trial function corresponding to energy W_1 therefore is

$$\phi_1 = \frac{1}{\sqrt{2}} \frac{1s_a + 1s_b}{\sqrt{1 + S_{ab}}}$$

For the root W_2 it is found that $c_b = -c_a$ and

$$\phi_2 = \frac{1}{\sqrt{2}} \frac{1s_a - 1s_b}{\sqrt{1 - S_{ab}}}$$

To complete the calculation it is necessary to evaluate [7] the various integrals defined symbolically above.

[7] The integrals required in the discussion of the H_2^+ molecule can all be obtained from the basic integral

$$J = \int \frac{e^{-pr_A} e^{-qr_B}}{r_A r_B} d\tau$$

This integral is best evaluated by transformation into confocal elliptical coordinates (33)

To calculate the energy functions W_1 and W_2 it is necessary to evaluate the three integrals H_{aa}, H_{ab} and S_{ab}. It is noted that the variational constant k has the same role as an effective nuclear charge in hydrogenic functions. The normalized functions $1s_A$ and $1s_B$ are therefore of the form

---

so that

$$\begin{aligned} J &= \int \frac{e^{-\frac{R}{2}[(p+q)\xi+(p-q)\eta]}}{\left[\frac{R^2}{4}(\xi^2-\eta^2)\right]}\frac{R^3}{8}(\xi^2-\eta^2)d\phi d\eta d\xi = \frac{R}{2}\int e^{-(a\xi+b\eta)}d\phi d\eta d\xi \\ &= \frac{R}{2}\int_0^{2\pi}d\phi\int_{-1}^{+1}e^{-b\eta}d\eta\int_1^{\infty}e^{-a\xi}d\xi = \frac{R}{2}(2\pi)\left[-\frac{1}{b}\left(e^b-e^{-b}\right)\right]\left(-\frac{1}{a}e^{-a}\right) \\ &= \frac{\pi R}{ab}e^{-a}\left(e^b-e^{-b}\right) = \frac{4\pi}{R}\left(\frac{1}{p^2-q^2}\right)\left(e^{-qR}-e^{-pR}\right) \end{aligned}$$

The other relevant integrals are obtained by differentiating the above result with respect to p and q. That is,

$$\begin{aligned} K &= \int\frac{e^{-pr_A}e^{-qr_B}}{r_A}d\tau = -\frac{\partial}{\partial q}J \\ &= \frac{4\pi}{R}\left[\frac{R}{(p^2-q^2)}e^{-qR}+\frac{2q}{(p^2-q^2)^2}\left(e^{-pR}-e^{-qR}\right)\right] \end{aligned}$$

and

$$\begin{aligned} L &= \int e^{-pr_A}e^{-qr_B}d\tau = -\frac{\partial}{\partial p}K \\ &= \frac{8\pi}{R(p^2-q^2)^2}\left[R\left(pe^{-qR}+qe^{-pR}\right)+\frac{4pq}{(p^2-q^2)}\left(e^{-pR}-e^{-qR}\right)\right] \end{aligned}$$

In the case that $p = q$ the useful integrals are

$$J = \frac{R}{2}\int_0^{2\pi}d\phi\int_{-1}^{1}d\eta\int_1^{\infty}e^{-Rp\xi}d\xi = \left(\frac{R}{2}\right)\left(\frac{4\pi}{Rp}\right)e^{-pR} = \frac{2\pi}{p}e^{-pR}$$

$$K = \frac{\pi}{p^2}(1+pR)e^{-pR}$$

$$L = \frac{\pi}{p^3}\left(1+pR+\frac{1}{3}p^2R^2\right)e^{-pR}$$

The integral $\int[\exp(-2kr_A)]/(r_B)$ transforms into

$$\frac{R^2}{4}\int_0^{2\pi}d\phi\left[\int_{-1}^{+1}e^{-kR\eta}d\eta\int_1^{\infty}\xi e^{-kR\xi}d\xi+\int_{-1}^{+1}\eta e^{-kR\eta}d\eta\int_1^{\infty}e^{-kR\xi}d\xi\right]$$

which integrates to $\dfrac{\pi}{k^3R}\left[1-(kR+1)e^{-2kR}\right]$

$k^{3/2}\pi^{-1/2}\exp(-kr_A)$. It follows immediately that

$$\begin{aligned} S_{ab} &= e^{-kR}\left(1+kR+\frac{1}{3}k^2R^2\right) \\ H_{aa} &= \int \psi_{1s}(r_A)H\psi_{1s}(r_A)d\tau \\ &= \int \psi_{1s}(r_A)\left[-\frac{1}{2}\nabla^2-\frac{1}{r_A}-\frac{1}{r_B}\right]\psi_{1s}(r_A)d\tau \end{aligned}$$

Adding and subtracting k/r_A in the square brackets

$$H_{aa} = \int (1s_a)^*\left[\left(-\frac{1}{2}\nabla^2-\frac{k}{r_A}\right)+\frac{k-1}{r_A}-\frac{1}{r_B}\right](1s_a)\,d\tau$$

Writing $H_a = -(1/2)\nabla^2-(k/r_A)$ for the Hamilton operator of a hydrogenic atom with nuclear charge k,

$$\begin{aligned} H_a(1s_a) &= -\frac{1}{2}k^2(1s_a) \\ H_{aa} &= \int (1s_a)^*\left(H_a+\frac{k-1}{r_A}-\frac{1}{r_B}\right)(1s_a)d\tau \\ &= -\frac{1}{2}k^2+(k-1)\int\frac{1s_a^*1s_a}{r_A}d\tau-\int\frac{1s_a^*1s_a}{r_B}d\tau \end{aligned}$$

In spherical polar coordinates

$$\int\frac{1s_a^2}{r_A}d\tau = \frac{k^3}{\pi}\int_0^\infty\frac{e^{-2kr_A}}{r_A}r_A^2dr_A\int_0^{2\pi}d\phi\int_0^\pi\sin\theta\,d\theta = k$$

H_{aa} follows as

$$\begin{aligned} H_{aa} &= -\frac{1}{2}k^2+k(k-1)-\frac{1}{R}+\left(k+\frac{1}{R}\right)e^{-2kR} \\ &= \frac{1}{2}k^2-k-\frac{1}{R}+\left(k+\frac{1}{R}\right)e^{-2kR} \end{aligned}$$

By the same procedure

$$\begin{aligned} H_{ab} &= \int (1s_a)^*H(1s_b)d\tau \\ &= \int (1s_a)^*\left[\left(-\frac{1}{2}\nabla^2-\frac{k}{r_B}\right)-\frac{1}{r_A}+\frac{k-1}{r_B}\right](1s_b)d\tau \\ &= -\frac{1}{2}k^2S_{ab}-\int\frac{(1s_A)(1s_B)}{r_A}d\tau+(k-1)\int\frac{(1s_a)(1s_b)}{r_B}d\tau \\ &= -\frac{1}{2}k^2S_{ab}+k(k-2)(1+kR)e^{-kR} \end{aligned}$$

Substituting these values of the integrals into the energy expressions, give

$$W_{1,2} = -\frac{1}{2}k^2 + \frac{k^2 - k - \frac{1}{R} + \frac{1}{R}(1+kR)e^{-2kR} \pm k(k-2)(1+kR)e^{-kR}}{1 \pm e^{-kR}\left(1 + kR + \frac{1}{3}k^2R^2\right)}$$

Since only the electronic Hamiltonian has been used, a term $1/R$ must be added to $W_{1,2}$ to account for internuclear repulsion. Finally, the parameter k is varied at fixed values of R to minimize the energy.

The energy expression readily rearranges [87] into the form

$$W = E - \frac{1}{R} = Ak^2 - Bk$$

where the term $1/R$ introduces internuclear repulsion, and where A and B are known functions of the parameter $t = kR$. There is no need to write down the explicit forms of the rather cumbersome expressions for A and B. For any given R the best value of k follows from the variational principle, $(\partial E/\partial k)_R = 0$, *i.e.*

$$2Ak + k^2R\frac{dA}{dt} - B - kR\frac{dB}{dt} = 0$$

so that

$$k = \frac{B + t\frac{dB}{dt}}{2A + t\frac{dA}{dt}}$$

As required by (36), the variational parameter k is calculated to vary between $k = 2$ at $R = 0$ and $k = 1$ at $R > 5a_0$. The parameter k is routinely interpreted as either a screening constant or an effective nuclear charge, as if it had real physical meaning. In fact, it is no more than a mathematical artefact, deliberately introduced to remedy the inadequacy of hydrogenic wave functions as descriptors of electrons in molecular environments. No such parameter occurs within the Burrau [84] scheme.

Setting $k = 1$ generates the expression for the binding energy as determined by a linear combination of simple atomic $1s$ wave functions, *i.e.*

$$(E_{1,2} - E_H) = \frac{1}{R} - \frac{\frac{1}{R} - \left(1 + \frac{1}{R}\right)e^{-2R} \pm (1-R)e^{-R}}{1 \pm e^{-R}\left(1 + R + \frac{1}{3}R^2\right)}$$

where $E_H = -\frac{1}{2}$ hartree is the electronic energy in a normal hydrogen atom.

There are several important conclusions to be drawn from the H_2^+ example. As expected, separation of the electronic wave equation permits exact solution of the one-electron problem. The result provides a benchmark against which to assess the possible use of hydrogenic functions as trial variational functions in more complex molecular problems. The exact results are an equilibrium bond length of 2 bohr (a_0) and the ground-state binding energy of 0.1026 hartree (2.79 eV). Using scaled hydrogenic functions of the type $\exp(-kr)$, the corresponding value for optimal $k = 1.24$ are $r_e = 2$ bohr, $D_e = 2.36$ eV. For $k = 1$, which means using simple H1s trial functions the results are, $r_e = 2.49$ bohr, $D_e = 1.76$ eV. The implications are clear: A linear combination of simple atomic wave functions predicts the existence of a stable molecule qualitatively correctly, but the poor quantitative agreement immediately highlights the inadequacy of atomic wave functions in the simulation of molecular properties. The reason becomes obvious when the molecular wave functions are examined. The charge distribution from the exact solution is closely concentrated about the line between the two nuclei, as shown by the solid curve below.

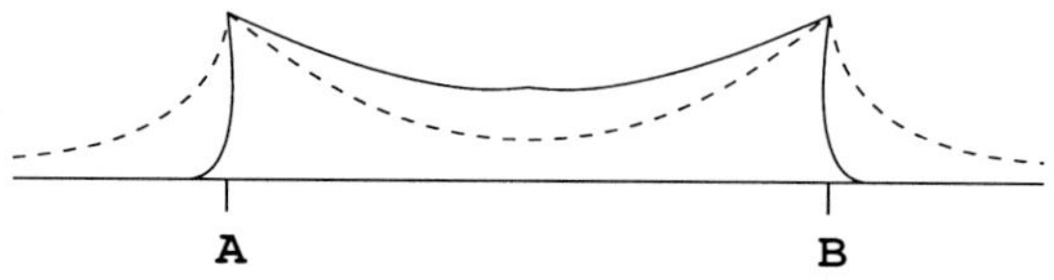

The electron remains in the internuclear region at virtually all times. The amplitude is large and nearly constant at distances midway between the protons. The disadvantage of two repelling protons is more than compensated for by the larger region in which the electron may have negative potential energy in the field of the two protons. Furthermore, the wave function stays smooth over a larger region of space which causes a drop in the kinetic energy. Both of these effects are less pronounced in the wave function constructed by a linear combination of atomic orbitals (LCAO), shown by the stippled curve.

Several workers [88] have managed to improve the LCAO description of H_2^+ by mixing (hybridizing) higher atomic functions, such as $2p_z$ with 1s. This procedure improves the dissociation energy to $D_e = 2.73$ eV. The physical justification of the procedure is hard to rationalize, although explanations such as polarization that happens during bond formation, abound. The real reason for the improvement is probably the mathematical factor that causes more density to concentrate in the internuclear region, with less in the wings. The same effect can be achieved more easily by trial atomic functions that fades at distances closer to the nucleus, such as the wave functions of compressed atoms.

7.4.3 The Hydrogen Molecule

The hydrogen molecule is the simplest molecule that contains an electron-pair bond. The electronic Hamiltonian for H_2, with a term for nuclear repulsion, is

$$H = -\frac{1}{2}\left(\nabla_1^2 + \nabla_2^2\right) - \frac{1}{r_{1a}} - \frac{1}{r_{1b}} - \frac{1}{r_{2a}} - \frac{1}{r_{2b}} + \frac{1}{r_{12}} + \frac{1}{r_{ab}} \tag{7.39}$$

Because of the interelectronic repulsion term $1/r_{12}$, the electronic Hamiltonian is not separable and only approximate solution of the wave equation can be considered. The obvious strategy would be to use H_2^+ wave functions in a variation analysis. Unfortunately, these are not known in functional form and are available only as tables. A successful parameterization, first proposed by James and Coolidge [89] and still the most successful procedure, consists of expressing the Hamiltonian operator in terms of the four elliptical coordinates $\xi_{1,2}$ and $\eta_{1,2}$ of the two electrons and the variable $\rho = 2r_{12}/r_{ab}$. The elliptical coordinates ϕ_1 and ϕ_2, as in the case of H_2^+, do not enter into the ground-state wave function. The starting wave function for the lowest state was therefore taken in the power-series form

$$e^{-\delta(\xi_1+\xi_2)} \sum C_{mnjkp} \left(\xi_1^m \xi_2^n \eta_1^j \eta_2^k \rho^p + \xi_1^n \xi_2^m \eta_1^k \eta_2^j \rho^p\right)$$

where δ and C are parameters that must be determined and m, n, j, k, p are integers. The summation is extended over positive and zero values of the indices, subject to the restriction required by the symmetry of the nuclear positions that $j + k$ must be even. It was pointed out that when the electrons coincide, $r_{12} = 0$ and the term $1/r_{12}$ in the potential energy becomes infinite, a cancelling infinite term arises from the kinetic energy part of the Hamiltonian operator. As a basis for comparison observed values were considered as those given by a Morse curve

$$U(r) = D_e\left[1 - e^{-a(r-r_e)}\right]^2 \quad ; \quad a = \left(2\pi^2 \mu c \omega_e / h D_e\right)^{\frac{1}{2}} ,$$

taking $D_e = 4.73$ eV, $\omega_e = 4375$ cm^{-1}, $r_e = 1.40$ a$_0$ and $\mu = 0.8309 \times 10^{-24}$ g.

In the first approximation, all integers equal to zero, the wave function

$$e^{-\delta(\xi_1+\xi_2)} = e^{-\frac{\delta(r_{1a}+r_{1b})}{r_{ab}}} e^{-\frac{\delta(r_{2a}+r_{2b})}{r_{ab}}}$$

is a product of two one-electron atomic wave functions, and the best value of $\delta = 1.696$ at the observed separation of 1.40 a_0. The binding energy of the molecule is 2.56 eV, which should be compared with the experimental value of 4.73 eV.

In the next approximation all terms that depend on ρ were neglected ($p = 0$ in all cases) and the best energy obtained at $r_{ab} = 1.40$ a$_0$ was 4.27 eV, which differs from the observed by about $\frac{1}{2}$ eV.

The simplest function employed in which the variable ρ occurs is the five-parameter expression

$$e^{-\delta(\xi_1+\xi_2)}\left[a_1\left(\eta_1^2+\eta_2^2\right)+a_2\eta_1\eta_2+a_3\left(\xi_1+\xi_2\right)+a_4\rho\right]$$

(based on the terms 00200, 00110, 10000, 00001).

This expression leads to an energy of 4.507 eV, with $r_{ab} = 1.40$ a$_0$. Now $\delta = 2.3779$.

As a final step James and Coolidge used a thirteen-parameter expression (adding the terms 10200, 10020, 10110, 20000, 00021, 00111, 10001, 00002). The final dissociation energy was within the experimental error of the observed value. Many workers have extended the original work using modern computers which can handle any number of terms. By using hundreds of terms results that are indistinguishable from the most accurate spectroscopically measured values are obtained, but without adding anything conceptually new to the discussion. Functions of such complexity defy physical interpretation and hence there is no obvious way in which the method can be adapted to describe more complicated molecules. In this respect the pioneering treatment of H_2 by Heitler and London (HL) [90] has an intuitive appeal to the chemist who ascribes chemical bonding to localized electron pairs. It was pointed out by James and Coolidge [89] that the use of unchanged individual atomic functions was quite unjustified in a molecule as close-coupled as H_2, but they recognised the advantage of the HL result in which the energy of the molecule is represented as a sum of terms, some of which are just the energy of the separated normal atoms. Derivation of the binding energy then becomes a trivial matter.

The Heitler-London Method

The Hamiltonian for two electrons in the field of two fixed protons is given by (39). For large values of r_{ab} the system reasonably corresponds to two H atoms. The wave functions of the degenerate system are $\psi_1 = u_{1s_a}(1)u_{1s_b}(2)$ and $\psi_2 = u_{1s_b}(1)u_{1s_a}(2)$, where $u_{1s_a}(1)$ is the hydrogenic wave function for electron 1 about nucleus A, *etc.* For smaller values of r_{ab} a linear combination of the two product functions is a reasonable variational trial function, *i.e.* $\psi = \psi_1 + \psi_2$.

The secular equation that corresponds to this linear variational function is

$$\begin{vmatrix} H_{11}-W & H_{12}-WS_{12} \\ H_{12}-WS_{12} & H_{22}-W \end{vmatrix} = 0$$

in which

$$\begin{aligned} H_{11} &= \int\int \psi_1 H \psi_1 d\tau_1\, d\tau_2 \\ H_{12} &= \int\int \psi_1 H \psi_2 d\tau_1\, d\tau_2 \\ S_{12} &= \int\int \psi_1 \psi_2 d\tau_1\, d\tau_2 \\ &= S_{ab}^2 \quad \text{as defined for } \mathrm{H}_2^+ \end{aligned}$$

The secular equation has the same form as for H_2^+, so that the approximate energies and wave functions follow directly as

$$W_{1,2} = \frac{H_{11} \pm H_{12}}{1 \pm S_{12}} \quad \text{and} \quad \phi_{1,2} = \frac{\psi_1 \pm \psi_2}{\sqrt{2 \pm 2S_{12}}}$$

ϕ_1 is symmetric in the positional coordinates of the two electrons and also in the positional coordinates of the two nuclei, whereas ϕ_2 is antisymmetric in both of these sets of coordinates. After some rearrangement the ground-state energy may be written as

$$\begin{aligned} W &= 2E_{1s}(H) + Q + \alpha \quad \text{where} \quad Q = \frac{J}{1+S^2} \quad ; \quad \alpha = \frac{K}{1+S^2} \\ J &= \int \psi_1 \left\{ -\frac{1}{R_{b1}} - \frac{1}{R_{a2}} + \frac{1}{R_{12}} + \frac{1}{R_{ab}} \right\} \psi_1\, d\tau_1\, d\tau_2 \\ &= -2\varepsilon_{aa} + \frac{1}{R_{ab}} + \int \frac{\psi_1^2}{R_{12}} d\tau_1\, d\tau_2 \\ K &= \int \psi_1 \left\{ -\frac{1}{R_{a1}} - \frac{1}{R_{b2}} + \frac{1}{R_{12}} + \frac{1}{R_{ab}} \right\} \psi_2\, d\tau_1\, d\tau_2 \\ &= \frac{S^2}{R_{ab}} - 2S\varepsilon_{ab} + \int \frac{\psi_1\psi_2}{R_{12}} d\tau_1\, d\tau_2 \end{aligned}$$

The integrals ε_{aa}, ε_{ab} and S have all been evaluated in section 7.4.2. The integral

$$\int \frac{\psi_1^2}{r_{12}} = \frac{1}{R} - e^{-2R}\left(\frac{1}{R} + \frac{11}{8} + \frac{3R}{4} + \frac{R^2}{6} \right)$$

was obtained by Heitler and London and the integral $\int \psi_1\psi_2/(r_{12}) d\tau_1\, d\tau_2$ was evaluated by Sigiura [91]. The complicated expression is of academic interest only since in all other applications it will be necessary to resort to numerical integration in any case.

Although the calculated molecular parameters $D_e = 3.15$ eV, $r_e = 1.64$ a_0 do not compare well with experiment the simplicity of the method is the more important consideration. Various workers have, for instance, succeeded to improve on the HL result by modifying the simple $1s$ hydrogenic functions in various ways, and to approach the best results obtained by variational methods of the James and Coolidge type. It can therefore be concluded that the method has the correct symmetry to reproduce the experimental results if atomic wave functions of the correct form and symmetry are used. The most important consideration will be the effect of the environment on free-atom wave functions.

7.4.4 Polyatomic Molecules

Despite the obvious limitation of the LCAO procedure as revealed by the H_2^+ and H_2 problems it still is the most popular scheme used in the theoretical study of polyatomic molecules. There is a bewildering number of approximate methods, commonly distinguished in terms of cryptic acronyms, designated as either *ab initio* or *semi-empirical*, but all of them based on the LCAO construction of *molecular orbitals*. The precise details can be found in many books and reviews. The present summary uses the discussion of Richards and Cooper [92] as a guide.

The method starts from the matrix equation

$$(\boldsymbol{H} - E\boldsymbol{S})\boldsymbol{c} = 0$$

The integrals of the $\boldsymbol{H}$ and $\boldsymbol{S}$ matrices are generally referred to as matrix elements. The condition for a non-trivial solution is that the secular determinant should vanish,

$$|\boldsymbol{H} - E\boldsymbol{S}| = 0$$

When choosing combinations of basis functions which constitute an orthonormal set, $\boldsymbol{S}$ becomes the identity matrix, *i.e.* $S_{ij} = \delta_{ij}$.

In semi-empirical molecular orbital calculations the number of integrals to be computed is greatly reduced by, for example, restricting attention to valence electrons and by ignoring those two-electron integrals which depend on the overlapping of charge densities on different atoms (neglect of diatomic differential overlap) or even by completely neglecting differential overlap arising from the charge densities of different basis functions.

In molecular orbital theory, the wave function for the molecule consists of an anti-symmetrized product of orbitals; one orbital for each individual electron. This is a Slater determinant. Each of the one-electron orbitals is itself a complicated linear combination of atomic orbitals. The molecular wave function is therefore structured as follows: $\Psi = \mathcal{A}\psi$. ($\mathcal{A}$ is an anti-symmetrizing operator). $\psi = \phi_1\phi_2\ldots\phi_n$ (ϕ_i is a one-electron spin orbital, which contains a spin α or β for each orbital and the spatial part of $\phi_i = \sum_p c_{ip}\chi_p$). The χ_p are atomic orbitals.

Examples of molecular orbitals are $1\sigma_g = 1s_A + 1s_B$ and $1\sigma_u = 1s_A - 1s_B$ for H_2^+. The labels g and u distinguish between even and odd symmetries, *i.e.* between molecular orbitals that either remain unchanged or change sign when inverted through the molecular centre of symmetry. For a general heteronuclear diatomic molecule molecular orbitals are built up from atomic functions according to the following scheme:

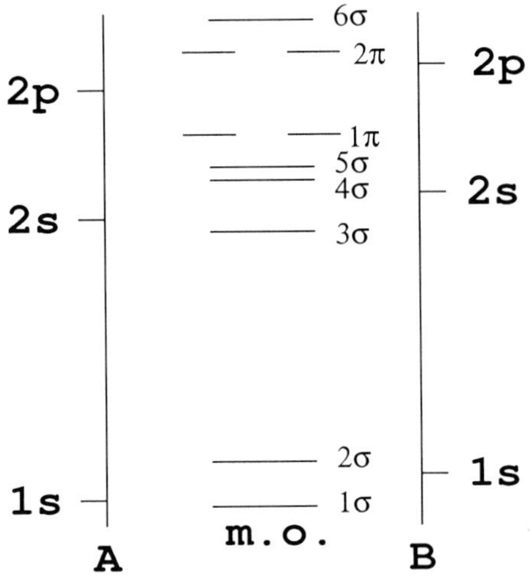

The molecular orbitals are labelled σ and π depending on whether they are symmetrical about the internuclear axis or have a nodal plane passing through the nuclei. The m.o.'s are numbered in sequence of increasing energy, independent of the numbering of the atomic orbitals. This numbering serves to avoid any confusion in cases where atomic orbitals from different levels are combined, as in

$$1\sigma = c_1 1s_A + c_2 1s_B + c_3 2s_A + c_4 2p_A$$

The energy over i electrons and μ nuclei is computed by evaluating the integrals defined by the operation of the general Hamiltonian

$$H = \left\{ \sum_i \left(-\frac{1}{2}\nabla_i^2 \right) - \sum_{i\mu} \frac{Z_\mu}{r_{i\mu}} + \sum_{i<j} \frac{1}{r_{ij}} \right\}$$

on the LCAO m.o. $\phi_i = \sum_p c_{ip}\chi_p$. The only parameters to be optimized are the coefficients c_{ip}. The Hartree self-consistent-field method, first devised for atoms, has been adapted for this purpose.

7.4.5 The LCAO-MO-SCF Method

As for atoms, it is assumed that if the wave function for a molecule is a single product of orbitals, then the energy is the sum of the one-electron energies (kinetic energy and electron-nuclear attractions) and Coulomb interactions between the charge clouds of electrons i and j,

$$E = \sum_i = \varepsilon_i^N + \sum_{i<j} \phi_1^2(1)\frac{1}{r_{12}}\phi_j^2(2)dv_1dv_2$$

The condition that the energy becomes a minimum through a variational procedure, together with the conditions $\int \phi_i\phi_j dv_i dv_j = \delta_{ij}$ gives the Hartree equations

$$\left\{H^N + \sum_j \left(\int \phi_j^2(2)\frac{1}{r_{12}}dv_2\right)\right\}\phi_i(1) = \sum_j \varepsilon_{ij}\phi_j(1)$$

Each ϕ_i is a linear combination of functions χ.

Using an anti-symmetrized, rather than a simple product of one-electron orbitals, the expectation value of the energy becomes a sum of one-electron, Coulomb and exchange terms:

$$\begin{aligned}\bar{E} &= \sum_i \varepsilon_i^N + \sum_{i<j}\int \phi_i^2(1)\frac{1}{r_{12}}\phi_j^2(2)\,dv_1\,dv_2 \\ &- \sum_{i<j}{}' \int \phi_i(1)\phi_j(1)\frac{1}{r_{12}}\phi_i(2)\phi_j(2)\,dv_1\,dv_2\end{aligned}$$

if all orbitals are assumed real. The one-electron core integral

$$\varepsilon_i^N = \int \phi_i^*(1)\left(-\frac{1}{2}\nabla_i^2 - \sum_\mu^N \frac{Z_\mu}{r_{\mu i}}\right)\phi_i(1)\,d\tau$$

In the case of a closed shell molecule, using Dirac notation the expression becomes

$$\bar{E} = 2\sum_k \varepsilon_k^N + 2\sum_{kl}\left[\left\langle\phi_k\phi_l\left|\frac{1}{r}\right|\phi_k\phi_l\right\rangle - \frac{1}{2}\left\langle\phi_k\phi_l\left|\frac{1}{r}\right|\phi_l\phi_k\right\rangle\right]$$

where k and l run over the doubly occupied molecular orbitals. Hence, for closed-shell molecules, where all the electrons are paired with others of opposite spin,

$$\bar{E} = \sum_i 2\varepsilon_i^N + \sum_{i,j}(2J_{ij} - K_{ij})$$

where i and j label the orbitals.

The molecular Hartree-Fock equations are obtained by finding the condition for the energy to be a minimum, $\delta\bar{E} = 0$, and at the same time demanding that the molecular orbitals be orthonormal, $\langle\phi_i|\phi_j\rangle = \delta_{ij}$.

In the variational minimization of $\bar{E}$ it is important to note that only small variations $\delta\phi$ which do not destroy the orthonormality of the m.o.'s are allowed. This means that, at the minimum both $\bar{E}$ and all the m.o. overlap integrals, $S_{ij} = \int \phi_i\phi_j d\tau$, must remain constant for a variation $\delta\phi$. Any linear combination of $\bar{E}$ and S_{ij} must also be constant, and hence

$$c_0\bar{E} + \sum_i\sum_j c_{ij}S_{ij} = \text{constant}$$

Suppose that a particular set of coefficients, ε_{ij} (Lagrangian multipliers) satisfy this equation for any variation. With the factor -2 introduced for later convenience [93] it means that

$$\bar{E} - 2\sum_i\sum_j \varepsilon_{ij}S_{ij} = \text{constant, for any } \delta\phi$$

Hence

$$\delta\bar{E} - 2\delta\sum_i\sum_j \varepsilon_{ij}S_{ij} = 0$$

or, expanding $\bar{E}$

$$2\sum_{i=1}^{n}\delta H_{ii} + \sum_{i=1}^{n}\sum_{j=1}^{n}(2\delta J_{ij} - \delta K_{ij}) - 2\sum_{i=1}^{n}\sum_{j=1}^{n}\varepsilon_{ij}\delta S_{ij} = 0 \qquad (7.40)$$

The variations occur in the ϕ of the m.o.'s, and so

$$\delta S_{ij} = \int \delta\phi_i(1)\phi_j(1)d\tau_1 + \int \phi_i(1)\delta\phi_j(1)d\tau_1$$

$$\delta H_{ii} = \int \delta\phi_i^*(1)H_{(1)}^{core}\phi_i(1)d\tau_1 + \int \phi_i^*(1)H_{(1)}^{core}\delta\phi_i(1)d\tau_1$$

For the two-electron integrals it is convenient to define new operators, *e.g.* a Coulomb operator

$$\hat{J}_i(1) = \int \phi_i^*(2)\left(\frac{1}{r_{12}}\right)\phi_i(2)d\tau_2$$

and similarly for

$$\hat{K}_i(1)\phi_j(1) = \int \phi_i^*(2)\left(\frac{1}{r_{12}}\right)\phi_j(2)d\tau_2\phi_i(1)$$

in terms of which equation (40) becomes

$$\begin{aligned} & 2\sum_i \int \delta\phi_i^*(1)\left[H_{(1)}^{core}\phi_i(1) + \sum_j\left\{2\hat{J}_j(1) - \hat{K}_j(1)\right\}\phi_i(1) - \sum_j \varepsilon_{ij}\phi_j(1)\right] d\tau \\ + \; & 2\sum_i \int \delta\phi_i(1)\left[H_{(1)}^{core}\phi_i^*(1) + \sum_j\left\{2\hat{J}_j^*(1) - \hat{K}_j^*(1)\right\}\phi_i^*(1) - \sum_j \varepsilon_{ij}^*\phi_j^*(1)\right] d\tau \\ = \; & 0 \end{aligned}$$

Since the variations $\delta\phi_i$ and $\delta\phi_i^*$ are independent, each half of the equation must independently equal zero and furthermore each integral must be zero, since the equation holds for any $\delta\phi_i^*$. By the same reasoning it follows that the bracketed term must vanish identically. Thus

$$\left[H_{(1)}^{core} + \sum_j\left\{2\hat{J}_j(1) - \hat{K}_j(1)\right\}\right]\phi_i(1) = \sum_j \varepsilon_{ij}\phi_j(1)$$

for all $i = 1, n$ and for a certain set of constants ε_{ij}.

The collection of operators in brackets is called the *Fock* operator, $\hat{F}$, and hence

$$\hat{F}(1)\phi_i(1) = \sum_j \varepsilon_{ij}\phi_j(1)$$

This looks like an eigenvalue equation, but is not, since instead of regenerating ϕ_i, a sum of functions ϕ_j is obtained. For the complete set of molecular orbitals the equation can be written in matrix notation as

$$\hat{F}\Phi = \Phi\varepsilon$$

This equation can be transformed by a series of matrix manupilations in such a way that the ε matrix is diagonalized, without affecting the meaning of Φ. The procedure simply mixes the m.o.'s which appear in the Slater determinant by internal rearrangements. After transformation

$$\hat{F}'\phi_i' = \varepsilon_i'\phi_i' \quad i = 1, 2 \ldots, n$$

which has the desired eigenvalue form and is usually referred to as the Hartree-Fock equation.

Although the Fock operator appears to be a Hamiltonian there is an important difference, namely the fact that $\hat{F}$ itself is a function of the m.o.'s ϕ and the set of equations must be solved iteratively. The equations are clearly the same as for atoms, but without the simplifying property of spherical symmetry that allows numerical solution.

Since molecular HF equations cannot be solved numerically each m.o. is expanded as an LCAO and the expansion coefficients optimized variationally. The equations in the form

$$H^{SCF}\sum_n c_{in}\chi_n = \varepsilon_i^{SCF}\sum c_{in}\chi_n$$

are also known as *Roothaan* equations. When both sides of the expression is multiplied by χ_m and integrated over all space

$$\sum_n c_{in}\int \chi_m H^{SCF}\chi_n dv = \varepsilon_i^{SCF}\int \chi_m\chi_n dv$$

i.e.

$$\sum c_{in}\left(H_{mn}^{SCF} - \varepsilon_i^{SCF}S_{mn}\right) = 0$$

The secular equation

$$\det\left|H_{mn}^{SCF} - \varepsilon_i^{SCF}S_{mn}\right| = 0$$

can be solved for the eigenvalues once H_{mn} and S_{mn} are known.

In different applications different approximations are introduced during evaluation of the matrix elements H_{mn} and S_{mn}. Procedures in which the matrix elements are evaluated without further approximation, are called *ab initio*, which means from the beginning. The quantities ε_i so obtained represent one-electron energies or *orbital energies*.

By adding up all the ε_i, exchange contributions will be included twice. After correction for this the total electronic energy is obtained as follows:

$$E_e = \sum_{i=1}^{n}\left[2\varepsilon_i - \sum_{j=1}^{n}(2J_{ij} - K_{ij})\right]$$

This expression can also be written as

$$E_e = \sum_{i=1}^{n}\left[2H_{ii} + \sum_{j=i}^{n}(2J_{ij} - K_{ij})\right]$$

or

$$E_e = \sum_{i=1}^{n}(\varepsilon_i + H_{ii})$$

The total energy $E_t = E_e + V_{nn}$, where V_{nn} describes internuclear repulsion.

Basis Sets

Two criteria dictate the choice of basis set:
(i) It should describe the actual wave function well enough to give chemically useful results.
(ii) It should lead to integrals which can readily be evaluated accurately on a computer.
The two most popular basis sets consist of either Slater-type orbitals[8] (STO's) or Gaussian functions. When using STO's one or more are placed on each nucleus - the more the better. The so-called minimal basis set consists of only those STO's which correspond to the occupied a.o.'s in the seperated atom limit. Instead of using Slater's rules to determine orbital exponents they may be varied in order to minimize the energy. Once this optimization has been done for a small molecule the values so established can be used in bigger problems. The basis can be improved by adding additional STO's for various nuclei, *e.g.* with different orbital exponents. If every minimal basis a.o. is represented by two such STO's a "double ζ" set is obtained. The only restriction on the number and type of STO that can be added, seems to be computer time.

Gaussian functions are of the type $\exp(-\alpha r^2)$ and since they produce integrals that are easier to evaluate they are often preferred to STO's. In some applications Gaussian functions are used to approximate STO's, *e.g.* in the STO-3G method, three primitive Gaussians are used to approximate

---

[8]STO's have the normalized form

$$\frac{[2\zeta/a_0]^{n+\frac{1}{2}}}{[(2n)!]^{\frac{1}{2}}} r^{n-1} e^{-\zeta r} Y_l^m(\theta,\phi)$$

To get an accurate representaqtion of a HF orbital an infinite set of STO's is required, but in practice the required accuracy is often obtained by selecting only a few. For He, with ground-state

$$\psi(1)\psi(2)\frac{1}{\sqrt{2}}|\alpha_1\beta_2 - \alpha_2\beta_1|$$

the $1s$ function expressed as

$$\psi = \frac{1}{\sqrt{2}}\sum_{i=1}^{5} c_i \left(\frac{\zeta_i}{a_0}\right)^{3/2} \exp(-\zeta_i r)$$

works well when all c_i and ζ_i are determined variationally. Slater [94] suggests that a.o.'s could be approximated by single STO's with $\zeta = (Z-s)/n$ where Z is the atomic number, n the principal quantum number and s a screening constant which can be calculated with a given set of rules.

each STO.

By improving the basis set lower energies are usually achieved, but there seems to be a limit beyond which a single determinantal wave function cannot go, irrespective of the size of the basis set. This is called the Hartree-Fock limit. One reason why HF energies are too high is because each electron is considered as independent of all others. In fact, their motion is correlated and the difference between the HF energy and the exact energy is known as the correlation energy.

The most popular technique to include correlation effects is known as configuration interaction (CI). Instead of restricting the wave function to be a single determinant, a linear combination of determinants is used. In the case of two determinants, where

$$\psi = c_1 D_1 + c_2 D_2$$

$\bar{E}$ can be minimized as a function of the mixing coefficients,

$$\frac{\partial \bar{E}}{\partial c_i} = 0$$

to give

$$\begin{vmatrix} H_{11} - \bar{E}S_{11} & H_{12} - \bar{E}S_{12} \\ H_{21} - \bar{E}S_{21} & H_{22} - \bar{E}S_{22} \end{vmatrix} = 0$$

$$\text{where, now} \quad H_{ij} = \int D_i H D_j d\tau \quad , \quad S_{ij} = \int D_i D_j d\tau$$

Instead of a.o.'s, as before, configurations (determinantal wave functions) are now interacting. Any number of configurations can be mixed in this way.

Interpretation of *ab initio* results

The computer programs used for *ab initio* calculations may in principle produce vast amounts of information, such as individual atomic or molecular integrals, Fock matrices, overlap matrices, and so on. The quantities of greatest interest to the chemist tend to be energies and coefficients for the m.o.'s and the expectation values of any molecular properties which have been computed. The energies of interest include the electronic energy of the final wave function, the nuclear repulsion energy, and the sum of these - the total energy. The expectation values of the kinetic energy and one-electron eigenvalues (orbital energies) are also available.

Despite the fact that the total electronic energy is not given by the sum of SCF one-electron energies an approximation known as Koopmans' theorem

[95] allows orbital energies to be interpreted as ionization potentials or electron affinities. It is also possible to obtain energies of excited states, contour maps of molecular orbitals and electron densities, molecular constants such as r_e, D_e, ω_e, dipole moments, polarizabilities, coupling constants, diamagnetic susceptibility and shielding, and electric field gradient [96].

The coefficients, particularly for large molecules, are often more useful as the sums of products called the *density matrix*. If $c_{i\lambda p}$ is the coefficient of the pth basis function of symmetry λ in the ith molecular orbital, the element (p, q) of the (partial) density matrix for m.o. i is

$$D_{pq}^{\lambda i} = c_{i\lambda p}c_{i\lambda q}n_i$$

where n_i is the occupancy of orbital i. This matrix can be used, for example, to calculate charges on different nuclei in the highest-occupied (HOMO) and lowest-unoccupied (LUMO) molecular orbitals of organic molecules used to investigate reactivities.

Approximate SCF-MO Methods

Because of the enormous number of integrals to be solved in *ab initio* calculations it is often desirable to look for sensible simplifications when dealing with large molecules. One example is based on the **n**eglect of **d**ifferential **o**verlap between atomic orbitals. DO between two a.o.'s χ_a and χ_b in the volume element dv is $dS = \chi_a(1)\chi_b(1)dv$. **Z**ero DO can be assumed at different levels, *e.g.* for all pairs of a.o.'s in two-electron integrals. This is **c**omplete **n**eglect of DO and it eliminates all two-electron integrals, except

$$\int \chi_a(1)\chi_a(2)\left(\frac{1}{r_{12}}\right)\chi_b(1)\chi_b(2)d\tau$$

and also all overlap integrals.

With the **i**ntermediate NDO method ZDO is not assumed between a.o.'s on the same atom in one-centre electron repulsion integrals. Various other schemes based on different ZDO assumptions together with different schemes of semi-empirical parametrization have been developed. These have become known by their acronyms such as CNDO/1, CNDO/2, INDO, MINDO/3 (m - modified), NDDO (d - diatomic), MNDO *etc.*.

7.4.6 Hückel Molecular-Orbital (HMO) Theory

Hückel Molecular-Orbital theory is one of the more useful theoretical guides of the pratising chemist. The central idea is that in certain molecules it is feasible to distinguish between electrons that are more-or-less localized on atoms and chemical bonds on the one hand, and those electrons delocalized across the entire molecule. The two classes are commonly referred to as σ and π electrons respectively. This distinction is the same as that made before when discussing free electron models in chemistry. The difference is that in HMO theory an attempt is made to include more than a single electron in the simulation. It is vitally important to note however, that by separating π-electrons from core and σ bonding electrons, the HMO approach starts with the same limitation as free electron simulations. The limitation arises from the sectioning of a holistic system to produce parts that lack essential features of the whole. In terms of the algebraic structure of quantum theory according to Primas [67] it means that only one of the parts may be considered non-classical; the other must be a commutative classical system. The price to pay for an elegant quantum-mechanical model for π-electrons is a classical molecular framework. Any attempt to impart quantum properties to the σ-electrons at some later stage would be illogical.

The assumption of ZDO introduces periodicity into an otherwise constant potential free electron scheme, in the same way that the Kronig-Penney potential modifies the simple Sommerfeld model.

The Hückel assumption posits a total wave function of the type

$$\psi(1,2,\ldots,n) = \psi_\pi(1,2,\ldots,k)\psi_\sigma(k+1,\ldots,n)$$

with a Hamiltonian that separates like

$$H(1,2,\ldots,n) = H_\pi(1,2,\ldots,k) + H_\sigma(k+1,\ldots,n)$$

This formulation leads to an expectation energy

$$\begin{aligned}\bar{E} &= \frac{\int \psi_\pi^*\psi_\sigma^*(H_\pi + H_\sigma)\psi_\pi\psi_\sigma d\tau}{\int \psi_\pi^*\psi_\sigma^*\psi_\pi\psi_\sigma d\tau} \\ &= \frac{\int \psi_\pi^* H_\pi \psi_\pi d\tau_\pi}{\int \psi_\pi^*\psi_\pi d\tau_\pi} + \frac{\int \psi_\sigma^* H_\sigma \psi_\sigma d\tau_\sigma}{\int \psi_\sigma^*\psi_\sigma d\tau_\sigma} \\ &= E_\pi + E_\sigma\end{aligned}$$

If these equations were valid one could ignore ψ_σ and minimize E_π by varying ψ_π. Although the equations are not valid and ignore interactions between σ and π electrons, separability is still assumed, since it leads to a number of useful results.

The HMO model further assumes that the wave function ψ_π is a product of one-electron functions and that H_π is the sum of one-electron operators:

$$\begin{aligned}\psi_\pi(1,2,\ldots,n) &= \phi_i(1)\phi_j(2)\ldots\phi_l(n) \\ H_\pi(1,2,\ldots,n) &= H_\pi(1) + H_\pi(2) + \cdots + H_\pi(n)\end{aligned}$$

Hence

$$\frac{\int \phi_i^*(1)H_\pi\phi_i(1)d\tau_1}{\int \phi_i^*(1)\phi_i(1)\,d\tau(1)} = E_i$$

$$E_\pi = E_i + E_j + \cdots + E_l$$

The implied implication that even π-electrons are independent of each other is grossly unjustified. What saves the day is that all wave functions and the overlap function that is being ignored, have the same symmetry. The shape of the molecular orbital remains qualitatively correct.

To compensate for the drastic assumptions an effective Hamiltonian for the system is defined in such a way that it takes into account some of the factors ignored by the model and also factors only known experimentally. The HMO method is therefore referred to as *semi-empirical.* As an example, the Pauli principle is recognized by assigning no more than two electrons to a molecular orbital.

The HMO method is demonstated well by a simple example, such as planar *cis*-butadiene:

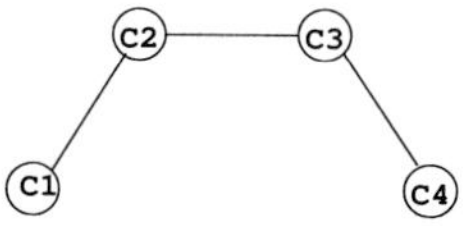

All σ-type a.o.'s are ignored and the four $2p_z$(C) a.o.'s are used as the set of basis functions. The four basis functions ψ_1, ψ_2, ψ_3 and ψ_4 are assumed to be normalized. A variation calculation with this set produces the secular determinant

$$\begin{vmatrix} H_{11}-\varepsilon S_{11} & H_{12}-\varepsilon S_{12} & H_{13}-\varepsilon S_{13} & H_{14}-\varepsilon S_{14} \\ H_{21}-\varepsilon S_{21} & H_{22}-\varepsilon S_{22} & H_{23}-\varepsilon S_{23} & H_{24}-\varepsilon S_{24} \\ H_{31}-\varepsilon S_{31} & H_{32}-\varepsilon S_{32} & H_{33}-\varepsilon S_{33} & H_{34}-\varepsilon S_{34} \\ H_{41}-\varepsilon S_{41} & H_{42}-\varepsilon S_{42} & H_{43}-\varepsilon S_{43} & H_{44}-\varepsilon S_{44} \end{vmatrix} = 0$$

The Hückel method attempts to solve these equations without explicitly writing down an expression for the effective Hamiltonian. It firstly assumes that the integrals H_{ii} along the diagonal

$$H_{ii} = \int \psi_i H_\pi \psi_i d\tau(i)$$

represent the average energy of an electron in the a.o. ψ_i in the field of the entire molecule. Since the dominant part of the potential must arise from interactions with the local atom, all H_{ii} are assumed to be identical for atoms of the same type. The symbol α, also known as the *coulomb* integral, is used to represent the H_{ii}. The *resonance* or *bond integrals* H_{ij} are considered next. They represent the energy of the charge overlap between ψ_i and ψ_j. It is assumed that this integral has a non-zero value of β only for neighbouring atoms. Since the wave functions are normalized all $S_{ii} = 1$. All S_{ij} are assumed to be zero. Although this is a drastic approximation it has little effect on the qualitative nature of the solutions. With these assumptions the secular equation becomes:

$$\begin{vmatrix} \alpha-\varepsilon & \beta & 0 & 0 \\ \beta & \alpha-\varepsilon & \beta & 0 \\ 0 & \beta & \alpha-\varepsilon & \beta \\ 0 & 0 & \beta & \alpha-\varepsilon \end{vmatrix} = 0$$

which is

$$\begin{array}{llllr} c_1(\alpha-\varepsilon) & +c_2\beta & & & =0 \\ c_1\beta & +c_2(\alpha-\varepsilon) & +c_3\beta & & =0 \\ & c_2\beta & +c_3(\alpha-\varepsilon) & +c_4\beta & =0 \\ & & c_3\beta & +c_4\alpha-\varepsilon) & =0 \end{array}$$

In terms of the parameter $m = (\alpha-\varepsilon)/\beta$ the determinant

$$\begin{vmatrix} m & 1 & 0 & 0 \\ 1 & m & 1 & 0 \\ 0 & 1 & m & 1 \\ 0 & 0 & 1 & m \end{vmatrix} = 0$$

expands to

$$m^4 - 3m^2 - 1 = 0$$

If $y = m^2$, the equation reduces to $y^2 - 3y - 1 = 0$, with roots

$$y_1 = \frac{3+\sqrt{5}}{2} = \frac{1+5+2\sqrt{5}}{4} = \left(\frac{1+\sqrt{5}}{2}\right)^2$$

$$y_2 = \left(\frac{1-\sqrt{5}}{2}\right)^2$$

The four roots therefore are:

$$m_1 = -\left(\frac{1+\sqrt{5}}{2}\right) \quad , \quad m_3 = \frac{-1+\sqrt{5}}{2}$$

$$m_2 = \frac{1-\sqrt{5}}{2} \quad , \quad m_4 = \frac{1+\sqrt{5}}{2}$$

and the corresponding energies are:

$$\begin{aligned}\varepsilon_1 &= \alpha + \frac{1+\sqrt{5}}{2}\beta \simeq \alpha + 1.6\beta \\ \varepsilon_2 &\simeq \alpha + 0.6\beta \\ \varepsilon_3 &\simeq \alpha - 0.6\beta \\ \varepsilon_4 &\simeq \alpha - 1.6\beta\end{aligned}$$

Since α is the energy of a π electron in a 2p a.o. and since β refers to an electron in a bond, both of these quantities are negative, and the energies above therefore occur in increasing order. The energy associated with the delocalized bond in the molecular ground state is

$$E = 2\varepsilon_1 + 2\varepsilon_2 = 4\alpha + 4.4\beta$$

The energy associated with the 4 electrons in their atomic orbitals is 4α and so the dissociation energy of the delocalized bond in butadiene is 4.4β. Each orbital contributes an amount of energy $\varepsilon_j = \alpha + m_j\beta$ and it is said to be *bonding, anti-bonding* or *non-bonding* for positive, negative or zero values of m_j.

The coefficients c_{ij} are determined by the normalization conditions:

$$\int \phi_j^2 d\tau = \int \left(\sum_{i=1}^{4} c_{ij}\psi_i\right)^2 d\tau = 1$$

Since all overlap integrals are assumed to be zero, it follows that

$$c_{1j}^2 + c_{2j}^2 + c_{3j}^2 + c_{4j}^2 = 1$$

The equations relating coefficients (and ε) show that

$$c_{21} = \frac{1+\sqrt{5}}{2}c_{11} \quad , \quad c_{31} = \frac{1+\sqrt{5}}{2}c_{41} \quad , \quad c_{11} = c_{41}$$

Hence

$$c_{11}^2 + \left(\frac{1+\sqrt{5}}{2}\right)^2 c_{11}^2 + \left(\frac{1+\sqrt{5}}{2}\right)^2 c_{11}^2 + c_{11}^2 = 1$$

or $c_{11} = c_{41} = 0.37$ and $c_{21} = c_{31} = 0.60$. Thus

$$\phi_1 = 0.37(\psi_1 + \psi_4) + 0.6(\psi_2 + \psi_3$$

Analogous calculations can be done for ϕ_2, ϕ_3 and ϕ_4.

To obtain the π-electron charge density on each atom it is noted that ψ_π was assumed to be a simple product of m.o.'s, *e.g.*

$$\psi_\pi = \phi_1(1)\phi_1(2)\phi_2(3)\phi_2(4)$$

for the ground state. The probability of finding an electron somewhere is the sum of probabilities for each electron and the one-electron density function

$$\rho = 2\phi_1^2 + 2\phi_2^2$$

gives the total charge $\int \rho\, d\tau_\pi = 4$.

To find out how each electron is distributed the m.o.'s are examined separately, *e.g.*

$$\phi_1^2 = c_{11}^2\psi_1^2 + c_{21}^2\psi_2^2 + c_{31}^2\psi_3^2 + c_{41}^2\psi_4^2 + \text{ cross terms, like } 2c_{11}c_{21}\psi_1\psi_2$$

Since all overlap integrals are zero

$$\int \psi_1^2 d\tau = 0.37^2 \int \psi_1^2 d\tau + 0.6^2 \int \psi_2^2 d\tau + 0.6^2 \int \psi_3^2 d\tau + 0.37^2 \int \psi_4^2 d\tau$$

Since all ψ_j are normalized the total charge on any atom follows as $q_i = \sum_j n_j c_{ij}^2$. The occupation number, n_j is the number of electrons in m.o. j. In the present instance all $q_i = 1$.

It is often of interest to compare the π-electron densities in bonds rather than on atoms. Although cross terms above were assumed to vanish because of the HMO zero-overlap assumption, they are actually not zero, especially not between a.o.'s on nearest neighbours. The coefficients of the cross terms can hence be interpreted as a measure of overlap in the bonds. On this basis bond order is defined as $p_{lm} = \sum_j n_j c_{lj} c_{mj}$. For the 1-2 bond in butadiene $p_{12} = 2c_{11}c_{21} + 2c_{21}c_{22} = 0.88 = p_{34}$. $p_{23} = 0.40$. These values show that the delocalized bond contributes more strongly to bind the end pairs than the central pair and gives a simple explanation of the experimental fact that the end bonds are shorter than the central bond.

The obvious limitation that restricts the method to planar hydrocarbons may be partially overcome by redefining modified parameters for conjugated heteroatoms in the form

$$\alpha_X = \alpha_C + h_X \beta_{CC}$$

$$\beta_{XY} = k_{XY} \beta_{CC}$$

and adjusting the new constants empirically.

The unreasonable effectiveness of the HMO method has never been fully explained and presumed refinements have not always stood up to expectation. Although an increasing number of molecules are being examined by *ab initio*-type calculation common concepts such as molecular orbital, bond order, charge density, occupation number, orbital energy, overlap, free valence, reaction index, and many more, in the minds of practising chemists, are the concepts of HMO theory. At one stage it was freely admitted [97] that "in spite of this oversimplified (HMO) approach, important results have been obtained, and therefore the omission of precise quantum mechanical calculations is amply justified". What has probably happened now is that this awareness has been forgotten and replaced by an innocent believe that the popular concepts of chemistry are indeed firmly rooted in quantum theory.

7.4.7 The Extended Hückel Method

Only for a special class of compound with appropriate planar symmetry is it possible to distinguish between (σ) electrons, associated with atomic cores and (π) electrons delocalized over the molecular surface. The Hückel approximation is allowed for this limited class only. Since $\sigma - \pi$ separation is nowhere perfect and always somewhat artificial, there is the temptation to extend the Hückel method also to situations where more pronounced $\sigma - \pi$ interaction is expected. It is immediately obvious that a different partitioning would be required for such an extension. The standard HMO partitioning that operates on symmetry grounds, treats only the π-electrons quantum mechanically and all σ-electrons as part of the classical molecular frame. The alternative is an arbitrary distinction between valence electrons and atomic cores. Schemes have been devised [98, 99] to handle situations where the molecular valence shell consists of either $\sigma + \pi$ or only σ electrons. In either case, the partitioning introduces extra complications. The mathematics of the situation [100] dictates that any abstraction produce disjoint sectors, of which no more than one may be non-classical. In view if the BO approximation already invoked, only the valence sector could be quantum mechanical[9]. In this case the classical remainder is a set of atomic cores in some unspecified excited state, called the *valence state.* One complication that arises is that wave functions of the valence electrons depend parametrically on the valence state.

The procedure followed in an actual **E**xtended **H**ückel analysis formally

[9]The all classical case is considered in the next section.

resembles the HMO process. It starts by fixing the molecular geometry and the basis functions in terms of which to formulate molecular orbitals. The basis set consists of normalized valence-state a.o.'s (f_r), expressed as STO's, one for each valence electron. Specification of the valence-state configuration is an entirely empirical process that requires a measure of precognition. The best known example is the sp^3 valence state assumed for saturated aliphatic carbon atoms. The molecular orbitals are approximated as linear combinations

$$\phi_i = \sum_r c_{ri} f_r$$

and the valence-electron Hamiltonian assumed to be a sum of unspecified one-electron Hamiltonians,

$$H_{val} = \sum_i H(i)$$

In terms of this formulation the problem separates into several one-electron problems [10]

$$H(i)\phi_i = \varepsilon_i \phi_i$$

The variational method then gives the secular equation and equations for the m.o. coefficients

$$\det(H_{rs} - \varepsilon_i S_{rs}) = 0$$

$$\sum_s [H_{rs} - \varepsilon_i S_{rs})c_{si}] = 0$$

Rather than ignore overlap as in HMO all overlap integrals are computed from the basis a.o.'s and the atomic coordinates. Since the $H(i)$ are not specified, evaluation of the H_{rs} presents a computational problem. For $r = s$ the one-electron integral $H_{rr} = \langle f_r|H|f_r\rangle$ looks like the average energy of an electron with wave function f_r centred on atom R in the molecule. The EH method therefore takes H_{rr} as equal to the orbital energy of the a.o. f_r for atom R in its valence state. By Koopman's theorem the orbital energy is next assumed equal to the negative valence-state ionization potential (VSIP) of f_r. In practice simple atomic ionization potentials are used for all types of valence state. Off-diagonal matrix elements are approximated by expressions such as [98]

$$H_{rs} = \frac{1}{2}K(H_{rr} - H_{ss})S_{rs}$$

[10]In HMO theory the assumed separability is partially justified by the symmetry of the Hamiltonian. In the present instance there is no such symmetry and neglect of interaction terms is tantamount to assuming a classical Hamiltonian.

where K is another empirical parameter. The secular equation can now be solved for orbital energies and m.o. coefficients. The total energy is taken as the sum of the one-electron energies.

In one of the simplest applications of EH theory the total energy for methane [93] is calculated as too low by an order of magnitude. In view of the empirical nature of the procedure this is a surprising result and it shows the method to be fatally flawed. As a matter of fact, there is not even agreement on whether or not the calculated energy includes internuclear repulsion. The method fails another simple test. It predicts H_2 to exist as a pair of spatially coincident atoms[101, 93]. Superficially the EH equations appear to have the same form as those for any other LCAO method. However, the necessary assumption that forces separability of the Hamiltonian and coerces a general set of valence electrons into responding like so many independent particles in a symmetrical field, is simply not a realistic proposition. The EH method epitomizes a theory in crisis. It provides something that looks like a respectable argument and seems to link numerous regularities noticed at the chemical bench to the formalisms of quantum theory. In essence it is a classical parametrization of empirical data in terms of ill-defined variables. Pullman [102] considers it "practically disqualified as a valid tool for conformational study" and Jug [103] calls it "obsolete".

7.4.8 Density Functional Theory

The electronic wave function of an n-electron molecule is defined in $3n$-dimensional configuration space, consistent with any conceivable molecular geometry. If the only aim is to characterize a molecule of fixed Born-Oppenheimer geometry the amount of information contained in the molecular wave function is therefore quite excessive. It turns out that the three-dimensional electron density function contains adequate information to uniquely determine the ground-state electronic properties of the molecule, as first demonstrated by Hohenberg and Kohn [104]. The approach is equivalent to the Thomas-Fermi model of an atom applied to molecules.

A collection of electrons in a fixed volume and moving under the influence of an external potential $v(\boldsymbol{r})$ and mutual Coulomb repulsion, has a Hamiltonian of the form

$$H = T + V + U$$

where (in atomic units)

$$\begin{aligned} T &= \frac{1}{2}\int \nabla\psi^*(\boldsymbol{r})\nabla\psi(\boldsymbol{r})d\boldsymbol{r} \\ V &= \int v(\boldsymbol{r})\psi^*(\boldsymbol{r})\psi(\boldsymbol{r})d\boldsymbol{r} \\ U &= \frac{1}{2}\int \frac{1}{|\boldsymbol{r}-\boldsymbol{r}'|}\psi^*(\boldsymbol{r})\psi^*(\boldsymbol{r}')\psi(\boldsymbol{r}')\psi(\boldsymbol{r})d\boldsymbol{r}d\boldsymbol{r}' \end{aligned}$$

For a non-degenerate ground state, Ψ and density operator $\hat{\rho} = \psi^*(\boldsymbol{r})\psi(\boldsymbol{r})$, the electron density is given by

$$\rho(\boldsymbol{r}) = \langle\Psi|\psi^*(\boldsymbol{r})\psi(\boldsymbol{r})|\Psi\rangle$$

which is a *functional*[11] of $v(\boldsymbol{r})$.

Conversely it can be shown that $v(\boldsymbol{r})$ is a unique functional of $\rho(\boldsymbol{r})$, apart from a trivial additive constant. Assume that another potential $v'(\boldsymbol{r})$, with ground state Ψ' gives rise to the same density $\rho(\boldsymbol{r})$. This Ψ' that satisfies a different Schrödinger equation, cannot be equal to Ψ, unless $v'(\boldsymbol{r}) - v(\boldsymbol{r})$ =constant. From the minimal property of the ground state it follows immediately that

$$E' = \langle\Psi'|H'|\Psi'\rangle < \langle\Psi|H'|\Psi\rangle = \langle\Psi|H + V' - V|\Psi\rangle$$

so that

$$E' < E + \int [v'(\boldsymbol{r}) - v(\boldsymbol{r})]\rho(\boldsymbol{r})d\boldsymbol{r}$$

Interchanging primed and unprimed quantities, gives by the same reasoning

$$E < E' + \int [v(\boldsymbol{r}) - v'(\boldsymbol{r})]\rho(\boldsymbol{r})d\boldsymbol{r}$$

Addition of the two previous expressions leads to the inconsistency

$$E + E' < E + E'$$

[11] Recall that a function $f(x)$ is a rule that associates a number with each value of the variable x. A functional $F[f]$ is a rule that associates a number with each function f. For example, the functional $F[f] = \int_{-\infty}^{\infty} f^*(x)f(x)dx$ associates a number, found by integration of $|f|^2$ over all space, with each quadratically integrable function $f(x)$. The variational integral

$$W[\phi] = \frac{\langle\phi|\hat{H}|\phi\rangle}{\langle\phi|\phi\rangle}$$

is a functional of the function ϕ, and gives a number for each well-behaved ϕ.

This absurdity proves that $v(\boldsymbol{r})$ is (to within a constant) a unique functional of $\rho(\boldsymbol{r})$. Also, since $v(\boldsymbol{r})$ fixes H the full many-particle ground state is seen to be a unique functional of $\rho(\boldsymbol{r})$.

Along with Ψ the kinetic and interaction energies must also be functionals of $\rho(\boldsymbol{r})$, such that a universal functional, valid for any number of particles may be defined as

$$F[\rho(\boldsymbol{r})] \equiv \langle\Psi|T+U|\Psi\rangle$$

An energy functional is then defined by

$$E_v[\rho] = \int v(\boldsymbol{r})\rho(\boldsymbol{r})d\boldsymbol{r} + F[\rho]$$

For the correct $\rho(\boldsymbol{r})$, $E_v[\rho]$ equals the ground-state energy E. To identify the correct ground-state density $\rho(\boldsymbol{r})$ consider the energy corresponding to Ψ' for N particles

$$E'_v[\Psi'] \equiv \langle\Psi'|V|\Psi'\rangle + \langle\Psi'|T+U|\Psi'\rangle$$

with the correct ground state Ψ relative to arbitrary variations of Ψ'. In particular, let Ψ' be the ground state associated with a different external potential $v'(\boldsymbol{r})$, such that

$$\begin{aligned} E'_v[\Psi'] &= \int v'(\boldsymbol{r})\rho'(\boldsymbol{r})d\boldsymbol{r} + F[\rho'] \\ &> E'_v[\Psi] = \int v(\boldsymbol{r})\rho(\boldsymbol{r})d\boldsymbol{r} + F[\rho] \end{aligned}$$

Since the number of particles remains constant the energy functional assumes its minimum value for that $\rho(\boldsymbol{r})$ restricted by the condition

$$N[\rho] \equiv \int v(\boldsymbol{r})\rho(\boldsymbol{r})d\boldsymbol{r} = N$$

Because of the long range of the Coulomb interaction, it is convenient to separate out this interaction and write

$$F[\rho] = \frac{1}{2}\int \frac{\rho(\boldsymbol{r})\rho(\boldsymbol{r}')}{|\boldsymbol{r}-\boldsymbol{r}'|}d\boldsymbol{r}d\boldsymbol{r}' + G[\rho]$$

such that $E_v(\rho)$ becomes

$$E_v(\rho) = \int v(\boldsymbol{r})\rho(\boldsymbol{r})d\boldsymbol{r} + \frac{1}{2}\int\int \frac{\rho(\boldsymbol{r})\rho(\boldsymbol{r}')}{|\boldsymbol{r}-\boldsymbol{r}'|}d\boldsymbol{r}d\boldsymbol{r}' + G[\rho]$$

The important conclusion is that in principle, all ground-state molecular properties may be calculated from the ground-state electron density $\rho(x, y, z)$. The challenge is to find the density and use it to calculate energies. A partial solution was found by Kohn and Sham [105].

The universal functional $G[\rho]$ is written as the sum of kinetic and exchange energy functionals

$$G[\rho] = T_s[\rho] + E_{xc}[\rho]$$

such that a variation in ground-state energy

$$\begin{aligned}
\delta E &= \int \delta\rho(\boldsymbol{r})v(\boldsymbol{r})d\boldsymbol{r} + \frac{1}{2}\int \frac{\rho(\boldsymbol{r}')\delta\rho(\boldsymbol{r})}{|\boldsymbol{r}-\boldsymbol{r}'|}d\boldsymbol{r}' + \frac{1}{2}\int \frac{\rho(\boldsymbol{r})\delta\rho(\boldsymbol{r}')}{|\boldsymbol{r}-\boldsymbol{r}'|}d\boldsymbol{r} \\
&+ \delta T_s(\rho) + \delta E_{xc}(\rho) \\
&= \delta\rho(\boldsymbol{r})\left\{\int \left[v(\boldsymbol{r}) + \frac{\rho(\boldsymbol{r}')}{|\boldsymbol{r}'-\boldsymbol{r}|}\right] d\boldsymbol{r} + \frac{\delta T_s[\rho]}{\delta\rho(\boldsymbol{r})} + \frac{\delta E_{xc}[\rho]}{\delta\rho(\boldsymbol{r})}\right\} \\
&= 0
\end{aligned}$$

produces an equation that resembles an eigenvalue equation for a system of non-interacting electrons moving in an external potential

$$\phi = v(\boldsymbol{r}) + \int \frac{\rho(\boldsymbol{r}')}{|\boldsymbol{r}-\boldsymbol{r}'|}d\boldsymbol{r}' + \frac{\delta E_{xc}[\rho]}{\delta\rho(\boldsymbol{r})}$$

In this way the density and energy may be constructed from the self-consistent Schrödinger equation

$$\left[-\frac{1}{2}\nabla^2 + \phi\right]\psi_i(\boldsymbol{r}) = \epsilon_i\psi_i(\boldsymbol{r})$$

by setting

$$\rho(\boldsymbol{r}) = \sum_{i=1}^{N} |\psi_i(\boldsymbol{r})|^2$$

The *Kohn-Sham orbitals* ψ_i have no other physical significance apart from allowing the calculation of the exact ρ. Likewise should Kohn-Sham energies not be confused with molecular-orbital energies. In fact, there is no KS molecular wave function. The major problem with using the KS equation to find ρ and E is to establish the correct functional $E_{xc}[\rho]$ for molecules. In the original use of a DF (Density Functional) method (the HFS method) Slater approximated the exchange energy based on the homogeneous electron-gas model for calculating atomic structure.

Applied to molecules a *local density approximation* is often used to estimate $E_{xc}[\rho]$. If ρ varies slowly with position a reasonable approximation is

$$E_{xc}^{LDA}[\rho] = \int \rho(\boldsymbol{r})\epsilon_{xc}(\rho)d\boldsymbol{r}$$

where $\epsilon_{xc}(\rho)$ is the exchange and correlation energy per electron of a uniform electron gas of density ρ. In a molecule where the positive charges at the nuclei are not uniformly distributed the assumption is an approximation only.

To initiate a DF calculation with E_{xc}^{LDA} a guesstimate of ρ is used to find the functional derivative

$$V_{xc} = \frac{\delta E_{xc}[\rho]}{\delta\rho}$$

which is then used in the KS equation to find an initial set of orbitals ψ_i. These orbitals are used in an itiration to generate improved values of ρ and V_{xc}. At convergence the final values of these variables are used to calculate E_0 and other one-electron properties.

The many applications of DFT have often been reviewed, *e.g.* [106], in detail.

7.4.9 Molecular Geometry

The failure of quantum theory to provide a convincing account of molecular shape is one of the important outstanding scientific problems at the end of the 20th century. On the one hand there is the indisputable fact of millions of molecular structures known from diffraction studies and on the other hand the elementary demonstration that a quantum-mechanical molecule is a holistic unit of electrons and nuclei, characterized by a molecular Hamiltonian (27) of α nuclei and i electrons. The Hamiltonian refers to all situations and permutations consistent with the given partitioning and only a minor fraction of these corresponds to classical stereochemical forms. The common view that the best solution of (27) can be found as the lowest energy form among the alternative classical geometries under Born-Oppenheimer, is therefore unrealistic.

There is no mystery here, only misleading terminology. The so-called "molecular" wave equation in fact describes all possible distributions of the $\alpha + i$ particles and a specific solution can only be the result of an equally specific set of boundary conditions, which depend not only on the physical environment of the system, but also on the history of the particles as a set. It is well known that, without exception, the constituent fragments of a molecule may exist, without change, under exactly the same conditions as the molecule itself. The two systems are described by the same wave function and only have different histories.

The views expressed here have not gained general acceptance and efforts to link molecular structure quantum-mechanically to chemical history are virtually non-existent. This state of affairs is mildly surprising since all theories of synthetic chemistry are historical accounts of interaction sequences, based on statistical models. This stubborn believe that each molecular structure relates to a unique wave function and whose *ab initio* calculation only awaits the next generation of computers can no longer be sustained. The smart alternative is to accept molecular structure for what it is: a fairly rigid nuclear framework that exists over a narrow range of environmental conditions and is stabilized by valence interactions. These interactions are widely known as chemical bonds and most chemists believe that their quantum origins are beyond dispute. The fact remains that chemical bonds only appear once a molecular framework has been fixed. A structured molecule, bonds and all, may thus be identified as a single classical entity. This approach is the alternative to the failed EH procedure that was mentioned before. The alternative has become known as the method of molecular mechanics, which is widely used, albeit apologetically [107], and represents the only successful technique for theoretical optimization of molecular trial structures.

Molecular Mechanics

The philosophy behind molecular mechanics [108] is that quantum interaction between atoms in a molecule can be modelled adequately in terms of empirical classical functions. The molecule is described as an interacting set of atoms. The equilibrium configuration is considered to correspond to a minimum in the molecular potential-energy function (the Born-Oppenheimer surface), which takes into account interactions between all possible pairs of atom in the molecule. The zero point of the potential function is defined to correspond to an arrangement in which each chemical bond, valence angle, bend or torsion has a natural value determined by electronic factors only. In most applications, the method only treats small distortions away from the electronic values. To calculate the strain introduced by distortion, harmonic restoring forces are assumed. The potential strain energy for the deformation of a parameter p from its strain-free value of p_0 is specified in terms of the harmonic restoring force

$$F = -\frac{\partial V}{\partial p} = k_p(p - p_0)$$

where the proportionality constant k_p is known as a harmonic *force constant.* Integration gives the potential energy

$$V_p = \frac{1}{2}k_p(p - p_0)^2$$

since, by definition $V_{p0} = 0$. The total strain in the molecule is obtained as the sum over all interactions

$$V_t = \frac{1}{2}\sum_i^N k_i(\Delta p_i)^2 = V_b + V_\theta + V_\chi + V_\phi + V_r + V_q$$

including bond and angle deformation, bending and torsional terms, non-bonded and Coulombic interactions. In principle anharmonic potentials, such as a Morse function could be used for large distortions that lead to bond rupture, but that is actually beyond the scope of molecular mechanics [110].

It is finally assumed that with all force constants and potential functions correctly specified in terms of the electronic configuration of the molecule, the nuclear arrangement that minimizes the steric strain corresponds to the observed structure of the isolated (gas phase) molecule. In practice however, the adjustable parameters, in virtually all cases, are chosen to reproduce molecular structures observed by solid-state diffraction methods. The parameters are therefore conditioned by the crystal environment and the minimized structure corresponds to neither gas phase nor isolated molecule [109].

The actual calculation consists of minimizing the intramolecular potential energy, or *steric energy*, as a function of the nuclear coordinates. The potential-energy expressions derive from the force-field concept that features in vibrational spectroscopic analysis according to the $\boldsymbol{GF}$-matrix formalism [111]. The $\boldsymbol{G}$-matrix contains as elements atomic masses suitably reduced to match the internal displacement coordinates (matrix $\boldsymbol{D}$) in defining the vibrational kinetic energy T of a molecule:

$$2T = \dot{\boldsymbol{D}}^{\boldsymbol{T}}\boldsymbol{G}\dot{\boldsymbol{D}}$$

In the same notation the potential energy is obtained as

$$2V = \boldsymbol{D}^{\boldsymbol{T}}\boldsymbol{F}\boldsymbol{D}$$

where the matrix $\boldsymbol{F}$ has the quadratic force constants

$$k_{ij} = \frac{\partial^2 V}{\partial D_i \partial D_j}$$

as elements. The classical equations of motion[12] for internal coordinates

$$\frac{d}{dt}\frac{\partial T}{\partial \dot{D}_i} + \frac{\partial V}{\partial D_i} = 0$$

have amplitude solutions of the type

$$D_n = A_n \cos(\alpha t + \varepsilon)$$

provided the secular determinant of coefficients vanishes, *i.e.*

$$|\boldsymbol{F} - \lambda \boldsymbol{G}^{-1}| = 0 \quad \text{or} \quad |\boldsymbol{F}\boldsymbol{G} - \lambda \boldsymbol{E}| = 0$$

The solutions $\lambda_i = \alpha_i^2 = 4\pi^2\nu_i^2$ yield the normal vibration frequencies of the molecule. In principle, force constants can therefore be derived from observed vibrational spectra. However, most force constants used in molecular mechanics are not too well established experimentally.

In molecular mechanics the complete $\boldsymbol{F}$-matrix is known as the *Generalized Valence Force Field* (GVFF) and a simplified form is used in most applications since interaction constants tend to be considerably smaller than the diagonal elements of $\boldsymbol{F}$. Setting all the off-diagonal elements equal to zero produces the Simple VFF, which means that bonding interactions are sufficiently localized to confine normal molecular vibrations to independent neighbourhoods spanning no more than three contiguous covalent bonds. Quantum-mechanically these interactions are described in terms of orthogonal molecular orbitals localized between two, three, or four nuclei. In the classical model they are quantified in terms of bond stretching, angle bending, out-of-plane bending, and torsional force constants. It is generally conceded that this simplification is too drastic, and molecular mechanics requires an intermediate VFF with additional terms representing interaction between localized vibrations and between more remote centres. Mathematically the IVFF defines a potential

$$\begin{aligned} V &= \frac{1}{2}\sum k_b(d-d_0)^2 + \frac{1}{2}\sum k_\theta(\theta-\theta_0)^2 + \frac{1}{2}\sum k_\chi(\chi-\chi_0)^2 \qquad (7.41) \\ &+ \frac{1}{2}\sum k_\phi(\phi-\phi_0)^2 + \sum_{|i-j|\ <3}\sum k_{ij}(p_i-p_i^0)(p_j-p_j^0) + \frac{1}{2}\sum_{|i-j|>2}(\delta_{ij}) \end{aligned}$$

Since the complete GVFF is rarely available for compounds of unknown structure, exact molecular mechanics simulation of small distortions is not

[12]The rate of change of momentum (first term) is equal to the force acting upon the mass (negative of the second term).

possible, and the second best alternative is to use the potential defined by (41) with transferable diagonal elements only. Bond torsions are often too large to justify the harmonic approximation and special potentials are used in molecular mechanics to describe distorsions of this type. The dihedral torsion angle about the central bond of a four-atom fragment is the only characteristic stereochemical parameter in addition to stretching and angle-bending terms.

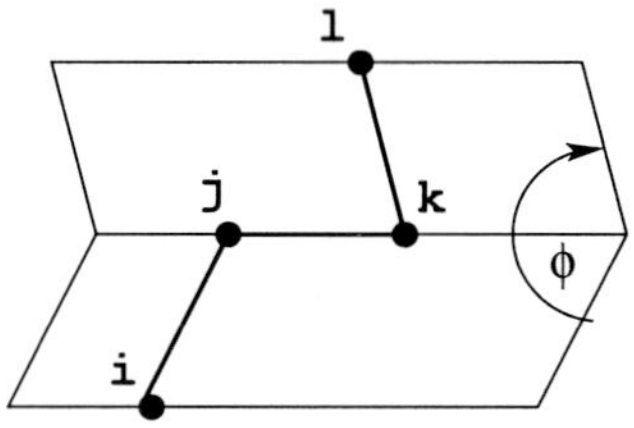

Instead of using harmonic functions the torsional potential is more often described with a small cosine expansion in ϕ

$$V_\phi = k_\phi \sum_{n=0}^{m} C_n \cos n\phi$$

where k_ϕ is the force constant. The coefficients C_n are determined by the observed rotational barrier V_ϕ, the periodicity of the potential, and the natural angle ϕ_0. The potential

$$V_\phi = \frac{1}{2}V(1 + \cos 3\phi)$$

in common use, treats the torsional interaction, assumed to simulate electrostatic repulsion between second-neighbour bonds, as a single term by assuming 3-fold symmetry for aliphatic fragments such as ethane. If the three-fold symmetry is approximate only, the torsional interaction is best described in terms of nine separate contributions as shown in the Newman projection below. In general no more than three of these contributions, identified by crosses, need to be considered if zero torsional strain is defined at $\Delta\phi = |\phi - \phi_0| = 60°$.

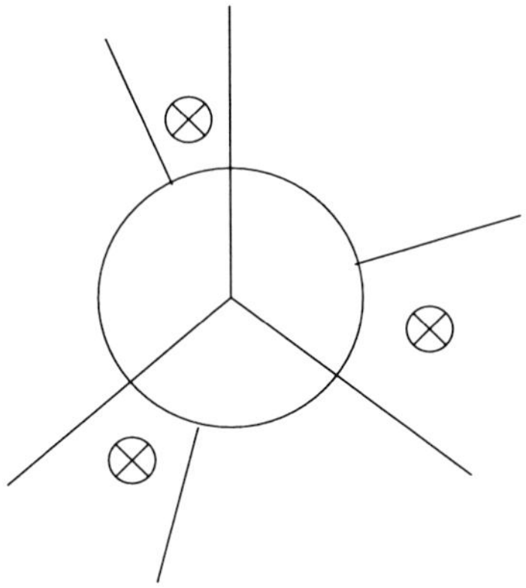

The total torsional strain is then defined by the sum over non-zero contributions

$$V_\phi = \frac{k_\phi}{6}\sum(1+\cos 3\Delta\phi)$$

For delocalized systems that require eclipsing across a double bond the torsion angle free of strain may be defined as $\phi_0 = 0$, now representing an attraction rather than a repulsion between second-neighbour bonds.

The interaction between atoms separated by more than two bonds is described in terms of potentials that represent non-bonded or Van der Waals interaction. A variety of potentials are being used, but all of them correspond to attractive and repulsive components balanced to produce a minimum at an interatomic distance corresponding to the sum of the Van der Waals radii, $V_{nb} = R - A$. The attractive component may be viewed as a dispersive interaction between induced dipoles, $A = c/r_{ij}^6$. The repulsive component is often modelled in terms of either a Lennard-Jones potential, $R = a/r_{ij}^{12}$, or Buckingham potential $R = a\exp(-br_{ij})$.

The final step in the MM analysis is based on the assumption that, with all force constants and potential functions correctly specified in terms of the electronic configuration of the molecule, the nuclear arrangement that minimizes the steric strain corresponds to the observable gas-phase molecular structure. The objective therefore is to minimize the intramolecular potential energy, or steric energy, as a function of the nuclear coordinates. The most popular procedure is by computerized Newton-Raphson minimization. It works on the basis that the vector $\boldsymbol{V_t'}$ with elements $\partial V_t/\partial x_i$, the first partial derivatives with respect to cartesian coordinates, vanishes at a minimum point, *i.e.* $\boldsymbol{V_t'}(\boldsymbol{x_1}) = \boldsymbol{0}$. This condition implies zero net force on each atom in the molecule. If $\boldsymbol{x}$ represents a position vector that describes some trial structure and $\boldsymbol{\delta x}$ is the vector of required corrections, then $\boldsymbol{x_1} = \boldsymbol{x} + \boldsymbol{\delta x}$

and $\boldsymbol{V}_t'(\boldsymbol{x}+\boldsymbol{\delta x}) = \boldsymbol{0}$. In Taylor expansion

$$\boldsymbol{V}_t'(\boldsymbol{x}+\boldsymbol{\delta x}) = \boldsymbol{V}_t'(\boldsymbol{x}) + \boldsymbol{V}_t''\boldsymbol{\delta x} + \cdots = 0$$

For computational simplicity the series is usually truncated after the terms linear in $\boldsymbol{\delta x}$. Then

$$\boldsymbol{V}_t'(\boldsymbol{x}) + \boldsymbol{V}_t''(\boldsymbol{x})\boldsymbol{\delta x}_a = 0$$

where $\boldsymbol{\delta x}_a$ is an approximate correction and $\boldsymbol{V}_t''$ is the Hessian matrix of second derivatives $(\partial^2 \boldsymbol{V}_t)/(\partial x_i \partial x_j)$. The correction vector is then written as

$$\boldsymbol{\delta x}_a = \beta \boldsymbol{V}_t''(\boldsymbol{x})^{-1}\boldsymbol{V}_t(\boldsymbol{x})$$

where β is a damping factor that prevents divergence at the early stages of refinement. $\boldsymbol{V}_t''(\boldsymbol{x})^{-1}$ is the inverse of $\boldsymbol{V}_t''(\boldsymbol{x})$.

Molecular Dynamics

The most serious problem with MM as a method to predict molecular structure is convergence to a false, rather than the global minimum in the Born-Oppenheimer surface. The mathematical problem is essentially still unsolved, but several conformational searching methods for approaching the global minimum, and based on either systematic or random searches have been developed. These searches work well for small to medium-sized molecules. The most popular of these techniques that simulates excitation to surmount potential barriers, has become known as *Molecular Dynamics* [112].

The simplest MD procedure is to keep the total energy constant and allow the molecule to drift through various conformations as the kinetic energy (temperature) fluctuates. This routine, also known as microcanonical MD is useful for simulating internal motion in a molecule.

In *quenched* MD, structures are periodically extracted from the microcanonical MD progression in time and minimized. The configurations found in this set of minimized structures are analyzed for uniqueness and the low-energy subset is said to represent the structure of the molecule. In *annealed* MD, the temperature of the system is incrementally increased and then decreased between low and high temperatures for a number of cycles with the lowest-energy structures being saved for further minimization and analysis.

Dynamical variables are estimated from the gradient information generated during energy minimization. In this way velocities are derived (in one dimension) from the Taylor expansion

$$x(t+\Delta t) = x(t) + \dot{x}(t)\Delta t + \ddot{x}(t)\Delta t^2/2 + \ldots$$

as

$$\dot{x}(t)_+ \approx \frac{x(t) - x(t + \Delta t)}{\Delta t}$$

Alternatively, by considering the interval between $-t$ and present

$$\dot{x}(t)_- \approx \frac{x(t) - x(t - \Delta t)}{\Delta t}$$

A better approximation, in terms of t, Δt and $-\Delta t$ is

$$\dot{x}(t) \approx \frac{1}{2}\left[\dot{x}(t)_+ + \dot{x}(t)_-\right] = \frac{x(t + \Delta t) - x(t - \Delta t)}{2\Delta t}$$

The analogous approximation for second derivatives (forces, $F = m\ddot{x}$) is

$$\ddot{x}(t) \approx \frac{x(t + \Delta t) + x(t - \Delta t) - 2x(t)}{\Delta t^2}$$

or

$$x(t + \Delta t) = 2x(t) - x(t - \Delta t) + \Delta t^2 F(t)$$

To start the process at time $t = 0$ the value of

$$x(-\Delta t) = 2x(0) - x(\Delta t) + \Delta t^2 F(0)$$

is needed. It follows immediately from

$$\ddot{x}(0) \approx \frac{x(\Delta t) - x(-\Delta t)}{2\Delta t}$$

and hence

$$x(\Delta t) = x(0) + \Delta t \dot{x}(0) + \frac{1}{2}\Delta t^2 F(0)$$

Generalized to $3n$ coordinates for n atoms, one has

$$\boldsymbol{r}(t + \Delta t) = 2\boldsymbol{r}(t) - \boldsymbol{r}(t - \Delta t) + \Delta t^2 \boldsymbol{F}(t)$$

$$\boldsymbol{r}(\Delta t) = \boldsymbol{r}(0) + \Delta t \dot{\boldsymbol{r}}(0) + \frac{1}{2}\Delta t^2 \boldsymbol{F}(0)$$

Initial velocities needed to kickstart the simulation is taken as a random distribution. Temperature is calculated from atomic velocities using the ideal-gas relationship

$$\frac{1}{2}\sum_{i=1}^{n} \frac{m_i \dot{\boldsymbol{r}}_i \cdot \dot{\boldsymbol{r}}_i}{n} = \frac{3}{2}kT$$

Chapter 8

Macrosystems

8.1 Introduction

Systems of chemical interest typically contain particles in molar quantity. Mathematical modelling of all interactions in such a system is clearly impossible. Even in a system of non-interacting particles, such as an ideal gas, it would be equally impossible to keep track of all individual trajectories. It is consequently not a trivial matter to extend the mechanical description (either classical or non-classical) of single molecules to macrosystems. It would be required in the first place to define the initial state of each particle in terms of an equation

$$H\psi = E\psi$$

and then to follow the time evolution of that state in terms of an equation

$$ih\frac{\partial\Psi}{\partial t} = E\Psi$$

In other words, it is necessary to define the complete Hilbert space for the entire system.

The classical problem is no less daunting: The initial state is defined in terms of the $3N$ coordinates and $3N$ momenta of the N particles, which together define a $6N$-dimensional *phase space.* Each point in this phase space defines a specific *microstate.* Evolution of the system is described by the trajectory of the starting point through phase space. Not only is it beyond the capabilities of the most powerful computer to keep track of these fluctuations, but should only one of the starting parameters be defined incorrectly, the model fatally diverges very rapidly from the actual situation.

The only known alternatives for the analysis of macrosystems are by the phenomenological approach, known as *thermodynamics*, or by the use of statistics to extrapolate from the known properties of individual particles. This method is called *statistical mechanics*. The common ground between the two approaches is also known as statistical thermodynamics.

8.2 Thermodynamics

Modern thermodynamics has grown from the phenomenological study of matter with a view to establish its general properties without reference to composition. The results are qualitative in nature and represent relationships between apparently unrelated properties of all types of system in macroscopic aggregation.

The typical laboratory instrument used in the study of macrosystems does not respond to microscopic fluctuations at the molecular level and only records responses averaged over time and space coordinates. Only some of the microscopic events that occur in macrosystems produce observable macroscopically averaged variables such as volume, elastic strain or multipole moments, that may be measured by standard procedures. Many others average to zero and remain hidden during macroscopic measurement with mechanical or electromagnetic devices. Special techniques are required to study the macroscopic consequences of these hidden events. The most important of these consequences is the energy associated with hidden atomic motion. This mode of energy transfer occurs in addition to the directly measurable mechanical or electromagnetic modes, known as mechanical or electromagnetic *work*. A typical example of mechanical work may be described in terms of the pressure and volume of a system, $w = -PdV$. Electric work is exemplified by a term $-E_e d\mu$, where E_e is the electric field strength and μ is the electric dipole moment. Energy transfer *via* the hidden atomic modes is called *heat* and the study thereof is thermodynamics.

The observations on which thermodynamics is based refer to macroscopic properties only, and only those features of a system that appear to be temporally independent are therefore recorded. This limitation restricts a thermodynamic analysis to the static states of macrosystems. To facilitate the construction of a theoretical framework for thermodynamics [113] it is sufficient to consider only systems that are macroscopically homogeneous, isotropic, uncharged, and large enough so that surface effects can be neglected, and that are not acted on by electric, magnetic or gravitational fields. The only mechanical parameter to be retained is the volume V. For a mixed system the chemical composition is specified in terms of the *mole numbers* N_i, or the *mole fractions* $[N_k/(\sum_{j=1}^{r} N_j),\ k = 1, 2, \ldots, r]$ of the chemically pure components of the system. The quantity $V/(\sum_{j=i}^{r} N_j)$ is called the *molar volume.* The macroscopic parameters V, N_1, $N_2, \ldots, N_r$ are all *extensive* parameters, which means that their values are additive.

The fundamental assumption of thermodynamics is the conservation of energy, also during its conversion from one form into another. It is cumbersome to account for total energy in all of its forms, while focussing on simple systems that depend on a chosen few. The problem may be avoided by dealing with differences in energy rather than absolute values, and defining a fiducial state of zero energy, convenient for the problem at hand. The energy of a system in any other state, relative to the energy of the system in the fiducial state, is called the thermodynamic *internal energy* U of the system. Like volume and mole number, internal energy is an extensive parameter.

It is noted that all systems in turmoil tend to subside spontaneously to simple states, independent of previous history. It happens when the effects of previously applied external influences damp out and the systems evolve toward states in which their properties are determined by intrinsic factors only. They are called *equilibrium* states. Experience shows that all equilibrium states are macroscopically completely defined by the internal energy U, the volume V, and the mole numbers N_j of all chemical components.

Description of a thermodynamic system requires specification of the way in which it interacts with the environment. An ideal system that exchanges no heat with its environment is said to be protected by an *adiabatic* wall. To change the state of such a system an amount of work equivalent to the difference in internal energy of the two states has to be performed on that system. This requirement means that work done in taking an adiabatically enclosed system between two given states is determined entirely by the states, independent of all external conditions. A wall that allows heat flow is called *diathermal.*

The process described above is *irreversible.* Irreversibility means that, given two states A and B of an adiabatically enclosed system of constant composition, either of the processes $A \to B$ or $B \to A$ may be driven mechanically or electromagnetically, but not both. By such a procedure the energy difference between two states $\Delta U = U_B - U_A$ can always be measured.

If, during a given process the energy difference between final and initial states is not the same as the work performed to drive the process and the mole numbers have remained constant, the difference $\Delta U - w = q$ is interpreted as the heat flux of the system.

The basic problem of thermodynamics is the characterization of the equilibrium state obtained by removal of all internal constraints in a closed, composite system. Internal constraint in this context means any factor that inhibits the flow of energy, volume or matter among the simpler systems that together constitute a composite *closed* system. The removal of such a constraint results in the onset of some spontaneous process and the redistribution of energy or matter between simple systems, ending in a new equilibrium state.

8.2.1 Theoretical Formulation

By analogy with Hamilton's principle of least action, the simplest proposition that could solve the thermodynamic problem is that equilibrium also depends on an extremum principle. In other words, the extensive parameters in the equilibrium state either maximize or minimize some function.

The idea is developed by postulating a function of the extensive parameters that tends to a maximum for any composite system that approaches a state of equilibrium on removal of an internal constraint. This function, to be called the *entropy* S, is defined only for equilibrium states where it assumes definite values. The basic problem of thermodynamics may therefore be considered solved once the entropy is specified in terms of a *fundamental relation* as a function of the extensive parameters[1].

Like the extensive properties of a composite system the entropy also is additive over the constituent subsystems,

$$S = \sum_{\alpha} S^{(\alpha)}$$

where the entropy of each subsystem is a function of the extensive properties of that subsystem alone,

$$S^{(\alpha)} = S^{(\alpha)}\left(U^{(\alpha)}, V^{(\alpha)}, N_j^{(\alpha)}\right)$$

Now suppose that some system in equilibrium is sliced up into N equal segments without disturbing the equilibrium. The extensive properties in each segment will be reduced proportionately, such that

$$S/N = S(U/N, V/N, N_j/N)$$

or

$$S(U, V, N_j) = NS(U/N, V/N, N_j/N)$$

[1]Theoretical formulation of thermodynamics based on the postulates assumed here was pioneered by Callen [113].

This property is described by the statement that entropy is a *homogeneous* first-order function of the extensive parameters. The expression is readily interpreted to define *molar* entropy ($s = S/N$), internal energy (u) and volume (v) for a one-component system, by

$$S(U, V, N) = Ns(u, v, 1)$$

If it is further assumed that the entropy is a continuous, differentiable, monotonically increasing function of the energy, it follows immediately that the partial derivative

$$\left(\frac{\partial S}{\partial U}\right)_{V,N_j} > 0$$

is a positive quantity, later to be related to temperature. It also means that the energy is a single-valued, continuous and differentiable function of S, V, N_j. The fundamental relation $S = S(U, V, N_j)$ can then be solved uniquely for U in the form $U = U(S, V, N_j)$, to give a second equivalent form of the fundamental relation.

The final postulated property of the entropy function is that it vanishes in the state for which $(\partial U/\partial S)_{V,N_j} = 0$, *i.e.* at zero temperature. An immediate implication of this postulate is that S, unlike U has a uniquely defined zero.

Once the total entropy of a composite system has been formulated as a function of the various extensive parameters of the subsystems, the extrema of this total entropy function may in principle be located by direct differentiation and classified as either maxima, minima or inflection points from the sign of the second derivative. Of these extrema, only the maxima represent *stable* equilibria.

8.2.2 Equilibrium Thermodynamics

The first differential of the fundamental equation $U = U(S, V, N_j)$ is

$$dU = \left(\frac{\partial U}{\partial S}\right)_{V,N_j} dS + \left(\frac{\partial U}{\partial V}\right)_{S,N_j} dV + \sum_{j=1}^{r} \left(\frac{\partial U}{\partial N_j}\right)_{S,V,N_{i\neq j}} dN_j$$

The partial derivatives are called *intensive parameters* and given the following symbols

$$\left(\frac{\partial U}{\partial S}\right)_{V,N_j} \equiv T \text{ , the temperature}$$

$$-\left(\frac{\partial U}{\partial V}\right)_{S,N_j} \equiv P \text{ , the pressure}$$

$$\left(\frac{\partial U}{\partial N_j}\right)_{S,V,N_{i\neq j}} \equiv \mu_j \text{ , the (electro)chemical potential of the j th component.}$$

With this notation the differential becomes

$$\begin{aligned} dU &= TdS - PdV + \mu_1 dN_1 + \cdots + \mu_r dN_r \\ &\equiv TdS - \sum_{j=1}^{t} P_j dX_j \end{aligned} \tag{8.1}$$

In the special case of constant mole numbers, $dN_j = 0$,

$$dU = TdS - PdV$$

The term $-PdV$ represents mechanical work, as before and writing internal energy change in terms of heat flux, $dU = w + q$, entropy change may be related to heat flux,

$$dS = q/T$$

Each of the terms TdS, $-PdV$, $\mu_j dN_j$, has the dimensions of energy and since mole number is dimensionless, chemical potential μ, has the same dimensions as energy. The terms $\sum_{j=1}^{r} \mu_j dN_j = w_c$, may be interpreted as chemical work.

Like the entropy expression the fundamental relation as a function of U is also a first-order homogeneous function, such that for constant λ,

$$U(\lambda S, \lambda V, \lambda N_j) \equiv U(\lambda S, \lambda X_j) = \lambda U(S, X_j)$$

When differentiated with respect to λ

$$\frac{\partial U}{\partial(\lambda S)}\frac{\partial(\lambda S)}{\partial \lambda} + \sum_{j=1}^{t} \frac{\partial U}{\partial(\lambda X_j)}\frac{\partial(\lambda X_j)}{\partial \lambda} = U(S, X_j)$$

or

$$\frac{\partial U}{\partial(\lambda S)} S + \sum_{j=1}^{t} \frac{\partial U}{\partial(\lambda X_j)} X_j = U$$

This expression is valid for all λ, and in particular for $\lambda = 1$, in which case

$$\frac{\partial U}{\partial S} S + \sum_{j=1}^{t} \frac{\partial U}{\partial X_j} X_j = U$$

i.e.

$$U = TS + \sum_{j=1}^{t} K_j X_j$$

and the K_j and T are intensive parameters. For simple systems the expression

$$U = TS - PV + \sum_{j=1}^{r} \mu_j N_j \tag{8.2}$$

is referred to as *Euler's* relation. In entropy representation Euler's relation takes the form

$$S = \left(\frac{1}{T}\right) U + \left(\frac{P}{T}\right) V - \sum_{j=1}^{r} \left(\frac{\mu_j}{T}\right) N_j$$

From the definition of temperature, pressure and chemical potential as partial derivatives it follows that these variables may be written as functions of the extensive parameters:

$$T = T(S, V, N_j)$$
$$P = P(S, V, N_j)$$
$$\mu_i = \mu_i(S, V, N_j)$$

These relationships that express intensive parameters in terms of independent extensive parameters, are called *equations of state*. Because of the homogeneous first-order relationship between the extensive parameters it follows that multiplication of each of the independent extensive parameters by a scalar λ, does not affect the equation of state, *e.g*

$$T(\lambda S, \lambda V, \lambda N_j) = T(S, V, N_j)$$

It follows that the temperature of a portion of a system is equal to that of the whole system, in line with experience. The pressure and chemical potential have the same property and together with temperature are said to be *intensive*.

Equivalent relationships are obtained by differentiation of the entropic form of the fundamental relation, formulated as

$$S(U, V, N_j) \equiv S(X_0, X_1, \ldots, N_t)$$

to give

$$dS = \sum_{k=0}^{t} \frac{\partial S}{\partial X_k} dX_k \tag{8.3}$$

The intensive parameters $F_k = \partial S/\partial X_k$ are easily identified by comparison of equations (1) and (3) as

$$F_0 = \frac{1}{T} \quad , \quad F_k = \frac{-P_k}{T} \quad (k = 1, 2, 3 \ldots)$$

The two equivalent, but different choices of independent variable, U or S, are distinguished as *energy* and *entropy representations*, respectively.

Thermal Equilibrium

Consider a closed composite system consisting of two compartments separated by a rigid impermeable diathermal wall. The volumes and mole numbers of the two simple systems are fixed, but the energies $U^{(1)}$ and $U^{(2)}$ may change, subject to the restriction $U^{(1)} + U^{(2)} = \text{constant}$, imposed on the composite closed system. At equilibrium the values of $U^{(1)}$ and $U^{(2)}$ are such as to maximize the entropy.

The total entropy of the composite system

$$S = S^{(1)}\left(U^{(1)}, V^{(1)}, N_j^{(1)}\right) + S^{(2)}\left(U^{(2)}, V^{(2)}, N_j^{(2)}\right)$$

If energy is transferred between the compartments the entropy change is described by

$$dS = \left(\frac{\partial S^{(1)}}{\partial U^{(1)}}\right)_{V^{(1)},N_j^{(1)}} dU + \left(\frac{\partial S^{(2)}}{\partial U^{(2)}}\right)_{V^{(2)},N_j^{(2)}} dU^{(2)}$$

i.e.

$$dS = \frac{1}{T^{(1)}} dU^{(1)} + \frac{1}{T^{(2)}} dU^{(2)}$$

Since the energy of the total system is conserved, $U^{(1)} = -U^{(2)}$, and hence

$$dS = \left(\frac{1}{T^{(1)}} - \frac{1}{T^{(2)}}\right) dU^{(1)}$$

If the amount of heat transferred is infinitesimally small, the extremum principle dictates that the entropy of the system in virtual equilibrium does not change, $dS = 0$. Since $dU^{(1)} \neq 0$, it is implied that, at equilibrium

$$\frac{1}{T^{(1)}} = \frac{1}{T^{(2)}}$$

This result is familiar from the experience that heat flows between two systems in contact until they reach the same temperature. The entropy change between initial and final states may be approximated by

$$\Delta S \simeq \left(\frac{1}{T^{(1)}} - \frac{1}{T^{(2)}}\right) \Delta U^{(1)} > 0$$

Assuming that $T^{(1)} > T^{(2)}$ it follows directly that $\Delta U^{(1)} < 0$, which demonstrates that heat flows spontaneously from a system at the higher temperature to the system at lower temperature.

Mechanical Equilibrium

Consider the same composite system as before, but with the impermeable diathermal wall no longer fixed. Both internal energy as well as the volume $V^{(1)}$ and $V^{(2)}$ may now change, subject to the extra closure condition, $V^{(1)} + V^{(2)} = \text{constant}$.

A differential entropy change in this system

$$\begin{aligned} dS &= \left(\frac{\partial S^{(1)}}{\partial U^{(1)}}\right)_{V^{(1)},N_j^{(1)}} dU^{(1)} + \left(\frac{\partial S^{(1)}}{\partial V^{(1)}}\right)_{U^{(1)},N_j^{(1)}} dV^{(1)} \\ &+ \left(\frac{\partial S^{(2)}}{\partial U^{(2)}}\right)_{V^{(2)},N_j^{(2)}} dU^{(2)} + \left(\frac{\partial S^{(2)}}{\partial V^{(2)}}\right)_{U^{(2)},N_j^{(2)}} dV^{(2)} \end{aligned}$$

reduces to

$$dS = \left(\frac{1}{T^{(1)}} - \frac{1}{T^{(2)}}\right) dU^{(1)} + \left(\frac{P^{(1)}}{T^{(1)}} - \frac{P^{(2)}}{T^{(2)}}\right) dV^{(1)}$$

For a virtual infinitesimal process consisting of heat flow across the wall, or displacement of the wall, at equilibrium there is no change in entropy, $dS = 0$, and hence the conditions for equilibrium are

$$\frac{1}{T^{(1)}} - \frac{1}{T^{(2)}} = 0 \quad \text{and} \quad \frac{P^{(1)}}{T^{(1)}} - \frac{P^{(2)}}{T^{(2)}} = 0$$

i.e.

$$T^{(1)} = T^{(2)} \quad ; \quad P^{(1)} = P^{(2)}$$

The equilization of pressure between the two subsystems in equilibrium agrees with the dictates of mechanics, identifying P as a mechanical pressure.

Chemical Potential and Chemical Equilibrium

Reconsider the equilibrium state of two systems separated by a rigid diathermal wall, but now permeable to one type of material (N_1) and impermeable to all others.

The virtual change in entropy in the appropriate virtual process is

$$dS = \frac{1}{T^{(1)}} dU^{(1)} - \frac{\mu_1^{(1)}}{T^{(1)}} dN_1^{(1)} + \frac{1}{T^{(2)}} dU^{(2)} - \frac{\mu_1^{(2)}}{T^{(2)}} dN_1^{(2)}$$

with the extra closure condition

$$dN_1^{(2)} = -dN_1^{(1)}$$

whence

$$dS = \left(\frac{1}{T^{(1)}} - \frac{1}{T^{(2)}}\right) dU^{(1)} - \left(\frac{\mu_1^{(1)}}{T^{(1)}} - \frac{\mu_1^{(2)}}{T^{(2)}}\right) dN_1^{(1)}$$

As dS must vanish for arbitrary values of both $dU^{(1)}$ and $dN_1^{(1)}$, the equilibrium conditions are

$$\frac{1}{T^{(1)}} = \frac{1}{T^{(2)}} \quad \text{and} \quad \frac{\mu_1^{(1)}}{T^{(1)}} = \frac{\mu_1^{(2)}}{T^{(2)}} \quad \left(i.e. \quad \mu_1^{(1)} = \mu_1^{(2)}\right)$$

In summary, it follows that temperature may be viewed as a *potential* for heat flux, pressure as a potential for volume change and μ as a chemical potential for matter flow.

To establish the direction of matter flow, assume that $T^{(1)} = T^{(2)}$, so that

$$dS = \frac{\mu_1^{(2)} - \mu_1^{(1)}}{T} dN_1^{(1)}$$

For $\mu_1^{(1)} > \mu_1^{(2)}$, and $dS > 0$, it follows that dN_1 is negative. Matter therefore flows from regions of high chemical potential to regions of low chemical potential.

Chemical reaction occurs in systems with changing mole numbers, those of the reactants decreasing and those of the products on the increase. A typical chemical reaction is presented by stoichiometric equations of the type

$$\nu_A \mathrm{A} + \nu_B \mathrm{B} \leftrightharpoons \nu_C \mathrm{C}$$

For a chemical reaction system of r components (reactants and products) the equation can be condensed into

$$\sum_{j=1}^{r} \nu_j \mathrm{A}_j = 0$$

The change in entropy of a virtual chemical process at constant energy and volume

$$dS = -\sum_{j=1}^{r} \frac{\mu_j}{T} dN_j$$

The changes in mole number are proportional to the stoichiometric coefficients

$$dN_j = \mu_j d\gamma$$

where $d\gamma$ is the proportionality factor. The extremum principle then dictates that, at equilibrium

$$\sum_{j=1}^{r} \mu_j \nu_j = 0$$

Chemical reactions, more typically occur in open vessels. The number of variables therefore increases, but the conclusion that the course of reaction is dictated by chemical potentials, remains the same.

The Minimum Energy Principle

The fact that each of the two variables, S and U may be expressed as a function of the other, indicates that the extremum principle could likewise be stated in terms of either entropy or energy. The alternative to the maximum entropy principle is a minimum energy principle, valid at constant entropy, as graphically illustrated for a one-component system in figure 1, below.

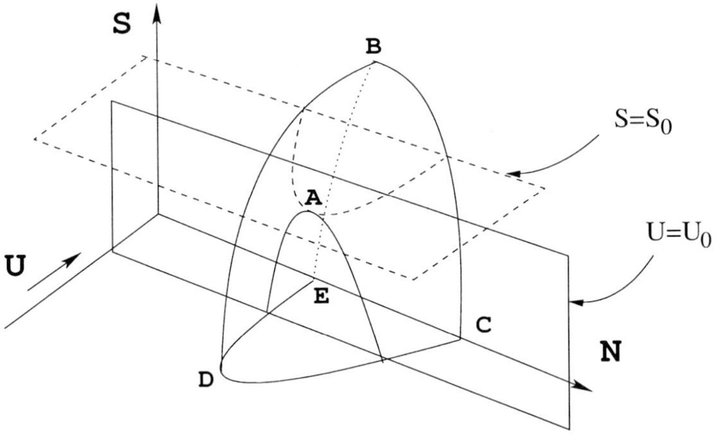

Figure 1: *Diagram to demonstrate that an equilibrium state, identified at point A, may be described as either a maximum entropy or a minimum energy state.*

The surface BCDE represents a segment of the surface defined by the fundamental equation characteristic of a composite system with coordinate axes corresponding to the extensive parameters of all the subsystems. The plane U_0 is a plane of constant internal energy that intersects the fundamental surface to produce a curve with extremum at A, corresponding to maximum entropy. Likewise S_0 is a plane of constant entropy that produces a curve with extremum A that corresponds to minimum energy at equilibrium for the system of constant entropy. This relationship between maximum entropy and minimum energy principles may be likened [113] to a circle that either represents a closed curve of minimum perimeter for given area, or as a curve enclosing maximum area for given perimeter.

A mathematical proof of the inverse relationship starts from the maximum entropy conditions

$$\left(\frac{\partial S}{\partial X}\right)_U = 0 \text{ and } \left(\frac{\partial^2 S}{\partial X^2}\right)_U < 0 .$$

where X represents any external parameter, other than U. All others are kept constant. The proof depends on a relationship between the partial derivatives of a composite function[2]

$$\left(\frac{\partial U}{\partial X}\right)_S = -\frac{(\partial S/\partial X)_U}{(\partial S/\partial U)_X} = -T\left(\frac{\partial S}{\partial X}\right)_U = 0$$

which shows that U has an extremum. The second derivative

$$\left(\frac{\partial^2 U}{\partial X^2}\right)_S = -T\frac{\partial^2 S}{\partial X^2}$$

shows that U has a minimum at $(\partial S/\partial X) = 0$.

8.2.3 Thermodynamic Potentials

In both the entropy and energy representations the extensive parameters are the independent variables, whereas the intensive parameters are derived concepts. This is in contrast to reality in the laboratory where the intensive parameters, like the temperature, are the more easily measurable while a variable such as entropy cannot be measured or controlled directly. It becomes an attractive possibility to replace extensive by intensive variables as

[2] For constant $\psi(x, y, z)$ the first-order differential

$$d\psi = \left(\frac{\partial \psi}{\partial x}\right)_{z,y} dx + \left(\frac{\partial \psi}{\partial y}\right)_{x,z} dy + \left(\frac{\partial \psi}{\partial z}\right)_{x,y} dz = 0$$

If, in addition $dz = 0$, the expression rearranges to

$$\left(\frac{dy}{dx}\right)_{\psi,z} = -\frac{(\partial \psi/\partial x)_{y,z}}{(\partial \psi/\partial y)_{x,z}}$$

mathematically independent. The procedure for switching variables in this way is provided by *Legendre transformation.*

Consider a function of only two variables $f(x, y)$, so that a differential of f has the form

$$df = udx + vdy \tag{8.4}$$

$$\text{where } u = \frac{\partial f}{\partial x} \quad , \quad v = \frac{\partial f}{\partial y} \tag{8.5}$$

It is next attempted to change the basis of description from x, y to a new distinct set of variables u, y (swapping u and x), so that differential quantities are expressed in terms of the differentials du and dy.

Let g be a function of u and y, defined by the equation

$$g = f - ux \tag{8.6}$$

such that

$$\begin{aligned} dg &= df - udx - xdu \\ &= vdy - xdu \text{ , using (4).} \end{aligned} \tag{8.7}$$

The latter relationship has the desired form and the quantities x and v are now functions of the variables u and y,

$$x = -\frac{\partial g}{\partial u} \quad , \quad v = \frac{\partial g}{\partial y}$$

converse to (5).

A familiar example of Legendre transformation is the relationship that exists between the Lagrangian and Hamiltonian functions of classical mechanics [17]. In thermodynamics the simplest application is to the internal energy function for constant mole number $U(S, V)$, with the differentials

$$dU = TdS - PdV$$

$$\frac{\partial U}{\partial S} = T \quad , \quad \frac{\partial U}{\partial V} = -P$$

where it may be useful to replace S as independent variable by T. The Legendre transformation (6) then defines a new function $F = U - TS$, known as the *Helmholtz potential* or Helmholtz free energy. In differential form

$$dF = TdS - PdV - TdS - SdT$$

The complete differential of $F = U(T, V, N_j)$ becomes

$$dF = -SdT - PdV + \sum \mu_j dN_j$$

In similar vein the *enthalpy* $H = U(S, P, N_j)$ is the partial Legendre transform of U that replaces the volume by pressure as an independent variable. Recall that $\partial U/\partial V = -P$, so that the appropriate Legendre transform is

$$H = U + PV$$

$$dH = TdS - PdV + PdV + VdP$$

The complete differential of H is

$$dH = TdS + VdP + \sum \mu_j dN_j$$

The enthalpy is useful in considering isentropic and isobaric processes, but often it becomes necessary to rather deal with isothermal and isobaric processes. In such case one needs a thermodynamic function of T and P alone, defining the *Gibbs* potential $G = U(T, P, N_j)$ as the Legendre transform of U that replaces entropy by temperature and volume by pressure. This transform is equivalent to a partial Legendre transform of the enthalpy,

$$G = H - TS$$

The complete differential of G is

$$dG = -SdT + VdP + \sum \mu_j N_j$$

A Legendre transformation to switch between mole number and chemical potential may also be performed, for instance to produce a potential $U(T, V, \mu)$ for a one-component system as

$$\begin{aligned} U(T, V, \mu) &= U(S, V, N) - TS - \mu N \\ dU &= TdS - PdV + \mu dN - TdS - SdT - \mu dN - Nd\mu \\ &= -SdT - PdV - Nd\mu \end{aligned}$$

This transform is the Grand Canonical Potential of Statistical thermodynamics.

The potential function of intensive variables only, $U(T, P, \mu_j)$ corresponds to the full Legendre transformation

$$\begin{aligned} U(T, P, \mu_j) &= U - TS + PV - \sum \mu_j N_j \\ &= -SdT + VdP - \sum N_j d\mu_j \end{aligned} \tag{8.8}$$

However, by Euler's relation (2)

$$U(S,V,N) = TS - PV + \sum \mu_j N_j$$

and the Legendre transform $U(T,P,\mu_j)$ becomes identically zero. Equation (8) then takes the form

$$\sum_{j=1}^{r} N_j d\mu_j = VdP - SdT$$

known as the Gibbs-Duhem equation.

Legendre transformation does not affect the essential nature of a function and all of the different potentials defined above still describe the internal energy, not only in terms of different independent variables, but also on the basis of different zero levels. In terms of Euler's equation (2) the internal energy consists of three components

$$U = TS - PV + \sum \mu_j N_j$$

or alternatively,

$$\Delta U = q + w + t$$

which defines heat flux, mechanical work and matter flow. By discounting one or more of these terms by performing a partial Legendre transformation, the zero point moves to a new level. In the process, the Euler relations for the various potentials reduce to

$$\begin{aligned} H &= U + PV = TS + \sum \mu_j N_j \\ F &= U - TS = -PV + \sum \mu_j N_j \\ G &= U - TS + PV = \sum \mu_j N_j \\ U(T,V,\mu) &= U - TS - \sum \mu_j N_j = -PV \end{aligned}$$

These formulae explain why the different potentials for one-component closed systems are colloquially known as:

$$\begin{aligned} H &= q_P \text{ , heat content;} \\ F &= w_T \text{ , work function;} \\ G/N &= \mu \text{ , partial molar Gibbs energy.} \end{aligned}$$

$U(T,V,\mu)$ is equivalent to the Helmholtz potential for an open system.

The differential form of the potential expressions show that chemical potential is defined by a first derivative of each potential, *i.e.*

$$\mu = \left(\frac{\partial U}{\partial N}\right)_{S,V} = \left(\frac{\partial H}{\partial N}\right)_{S,P} = \left(\frac{\partial F}{\partial N}\right)_{T,V} = \left(\frac{\partial G}{\partial N}\right)_{T,P}$$

For the open system

$$\left(\frac{\partial U(T,V,\mu)}{\partial \mu}\right)_{T,V} = N$$

As for U it follows without further analysis that the extremum principle for thermodynamic equilibrium at constant entropy should also apply to the other potentials under suitable conditions, *i.e.* constant T for minimum F, constant pressure for minimum H, and constant temperature and pressure for minimum G.

8.2.4 Irreversible Thermodynamics

The usual emphasis on equilibrium thermodynamics is somewhat inappropriate in view of the fact that all chemical and biological processes are rate-dependent and far from equilibrium. The theory of non-equilibrium or irreversible processes is based on Onsager's reciprocity theorem. Formulation of the theory requires the introduction of concepts and parameters related to dynamically variable systems. In particular, parameters that describe a mechanism that drives the process and another parameter that follows the response of the systems. The driving parameter will be referred to as an *affinity* and the response as a *flux*. Such quantities may be defined on the premise that all action ceases once equilibrium is established.

By way of illustration consider a binary composite system characterized by extensive parameters X_k and X'_k in the two subsystems and the closure condition $X_k + X'_k = X^0_k$. The equilibrium values of X_k and X'_k are determined by the vanishing of quantities defined in the sense of equation (3) as

$$\mathcal{F}_k = \left(\frac{\partial S^0}{\partial X_k}\right)_{X^0_k} = \frac{\partial(S - S')}{\partial X_k} = F_k - F'_k$$

Hence, when $\mathcal{F}_k$ is zero the system is in equilibrium, but for non-zero $\mathcal{F}_k$ an irreversible process that takes the system towards equilibrium, occurs. The quantity $\mathcal{F}_k$, which is the difference between intensive parameters in entropy representation, acts as the driving force, or affinity of the non-equilibrium process.

For two systems separated by a diathermal wall and with $X_k = U$, the affinity

$$\mathcal{F}_k = \frac{1}{T} - \frac{1}{T'}$$

No heat flows across the diathermal wall if the difference in inverse temperatures vanishes. Similarly, if X_k is the volume, the affinity $\mathcal{F}_k = [P/T - (P'/T')]$, and if X_k is a mole number the associated affinity is $[\mu'_k/T - (\mu_k/T)]$.

Response to the driving force is defined as the rate of change of the extensive parameter X_k, *i.e.* the flux $J_k \equiv (dX_k/dt)$. The flux therefore stops when the affinity vanishes and non-zero affinity produces flux. The product of affinity and associated flux corresponds to a rate of entropy change and the sum over all k represents the rate of entropy production,

$$\sum_k \mathcal{F}_k J_k = \sum_k \frac{\partial S}{\partial X_k}\frac{dX_k}{dt} = \frac{dS}{dt} \tag{8.9}$$

This equation is useful when extending the definition of affinities to continuous systems.

Consider a three-dimensional system in which flow of energy and matter is driven by applied forces. Flux through the system consists of energy and matter currents and is therefore measurable in terms of current densities. The vectors $\boldsymbol{J}_0$ and $\boldsymbol{J}_k$ in the direction of flow represent the energy and matter current densities for all components k of the system, as the amounts that flow across unit area in unit time.

In order to formulate entropy production, local entropy is assumed to depend on local extensive parameters X_k by the same functional relationship that exists at equilibrium. Thus

$$dS = \sum_k F_k dX_k$$

This definition suggests a reasonable formulation of entropy current density as

$$\boldsymbol{J}_S = \sum F_k \boldsymbol{J}_k$$

The magnitude of the entropy flux is the entropy transported through unit area per unit time, which is the divergence $\nabla \cdot \boldsymbol{J}_S$. It is convenient to define all extensive variables per unit volume (denoted here by lower case symbols) *e.g.*

$$ds = \sum_k F_k dx_k$$

The rate of entropy production per unit volume may then be written as

$$\dot{s} = \frac{\partial s}{\partial t} + \nabla \cdot \boldsymbol{J}_S$$

In the absence of chemical change, all extensive quantities are conserved and hence

$$\frac{\partial x_k}{\partial t} + \nabla \cdot \boldsymbol{J}_k = 0$$

Noting that

$$\frac{\partial s}{\partial t} = \sum_k F_k \frac{\partial x_k}{\partial t}$$

and

$$\nabla \cdot \boldsymbol{J}_s = \nabla \cdot \left(\sum_k F_k \boldsymbol{J}_k \right) = \sum_k \nabla F_k \cdot \boldsymbol{J}_k + \sum_k F_k \nabla \cdot \boldsymbol{J}_k$$

the rate of entropy production reduces to

$$\dot{s} = \sum_k F_k \frac{\partial x_k}{\partial t} + \sum \nabla F_k \cdot \boldsymbol{J}_k + \sum F_k \nabla \cdot \boldsymbol{J}_k$$

and because of the conservation condition, it further reduces to

$$\dot{s} = \sum_k \nabla F_k \cdot \boldsymbol{J}_k \rightarrow \sum_k \mathcal{F}_k \cdot \boldsymbol{J}_k$$

The affinities that define the rate of entropy production in continuous systems are therefore gradients of intensive parameters (in entropy representation) rather than discrete differences. For instance, the affinities associated with the z-components of energy and matter flow for constituent k, in this notation would be

$$\mathcal{F}_0^z = \nabla_z \left(\frac{1}{T} \right) \quad ; \quad \mathcal{F}_k^z = -\nabla_z \left(\frac{\mu_z}{T} \right)$$

Onsager's Theorem

Onsager's theorem deals with reciprocal relations in irreversible resistive processes, in the absence of magnetic fields [114]. The resistive qualifier signifies that the fluxes at a given instant depend only on the instantaneous values of the affinities and local intensive parameters at that instant. For systems of this kind two independent transport processes may be described in terms of the relations

$$X_1 = R_1 J_1 \quad \text{and} \quad X_2 = R_2 J_2$$

which relate each affinity only to the resulting flux, through a characteristic proportionality factor or resistance. However, based on the experience that two transport processes in the same system are likely to interfere with each other, a more appropriate set of phenomenological relations would be

$$X_1 + R_{11}J_1 + R_{12}J_2$$

$$X_2 = R_{21}J_1 + R_{22}J_2$$

Expressed in terms of conductances or *kinetic coefficients*, rather than resistances the relations become

$$J_1 = L_{11}X_1 + L_{12}X_2$$

$$J_2 = L_{21}X_1 + L_{22}X_2$$

where $L_{11} = R_{22}/(R_{11}R_{22} - R_{12}R_{21})$, *etc.*

In general, any linear resistive process is described by the equation

$$J_k = \sum_j L_{jk}\mathcal{F}_j \tag{8.10}$$

As a working example consider the thermoelectric system consisting of two reservoirs of energy and electrons, in which the temperatures are T_1 and T_2 and the electrochemical potentials are μ_1 and μ_2.

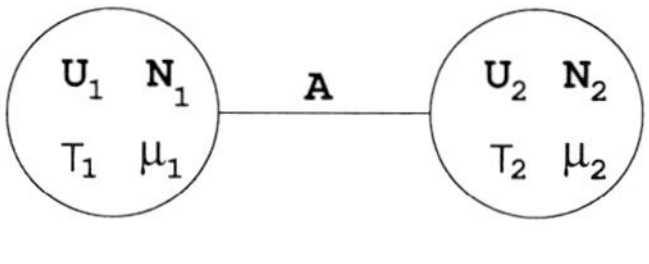

Figure 2: *Configuration of reservoirs of energy and electrons used in the Onsager treatment of thermoelectric effects in wire A.*

When the two reservoirs are connected a steady state is soon established as energy and electrons flow along the wire under the influence of the driving terms $T_2 - T_1$ and $\mu_2 - \mu_1$. As energy and particles flow the joint entropy of the reservoirs increases at a rate given by (9) and (10) as

$$\frac{dS}{dt} = \sum_{k=1}^{2} \mathcal{F}_k J_k = \sum_{j,k=1}^{2} \mathcal{F}_k L_{jk}\mathcal{F}_j$$

or

$$\frac{dS}{dt} = L_{11}\mathcal{F}_1^2 + (L_{12} + L_{21})\mathcal{F}_1\mathcal{F}_2 + L_{22}\mathcal{F}_2^2$$

where the subscripts 1 and 2 refer to energy and electron flow respectively.

Using equation (10) in the form

$$J_k = \frac{d\mathcal{F}_k}{dt} = \sum_j L_{jk}\mathcal{F}_j$$

the variable $\mathcal{F}_2$ above may be replaced by J_2, to give

$$\frac{dS}{dt} = (L_{11} - L_{12}L_{21}/L_{22})\mathcal{F}_1^2 + J_2/L_{22} + [(L_{12} - L_{21})/L_{22}]\mathcal{F}_1 J_2$$

If no current flows $(d\mathcal{F}_2/dt) = J_2 = 0$ and hence the first term is interpreted as the entropy production due to thermal conduction driven by the temperature gradient defined be $\mathcal{F}_1$. In the absence of a temperature gradient however, $\mathcal{F}_1 = 0$ and hence the second term represents entropy production due to resistive heating, which is proportional to the square of the current. Since the sum of these two terms is known experimentally to account for the total entropy production in the process, it is inferred that $L_{12} = L_{21}$.

Onsager's theorem consists of proving that a reciprocal relationship of the type $L_{\alpha\beta} = L_{\beta\alpha}$ between the affinities and fluxes of coupled irreversible processes is universally valid in the absence of magnetic fields.

Onsager found a proof of the general reciprocal relationship by consideration of natural fluctuations that occur in equilibrium systems. It is argued that an imbalance like that which initially exists when two reservoirs are connected as in figure 2, could also arise because of natural fluctuations. When the occasional large fluctuation occurs in an equilibrium system, the subsequent decay of the imbalance would be indistinguishable from the decay that follows deliberate connection of the reservoirs.

Now consider a pair of reservoirs in equilibrium with respect to the extensive parameters X_j and X_k, with instantaneous values of $\hat{X}_j$ and $\hat{X}_k$. Let $\delta\hat{X}_j$ denote a fluctuation from the instantaneous value. The average value of $\delta\hat{X}_j$ is zero, but the average of its square $(\delta\hat{X}_j)^2 \equiv \langle(\delta\hat{X}_j)^2\rangle \neq 0$. Likewise, the average correlation moment $\langle\delta\hat{X}_j\delta\hat{X}_k\rangle \neq 0$.

The crucial step in the Onsager argument concerns the correlation between fluctuations that occur, not simultaneously, but at times τ apart. Such a correlation reflects the effect of one fluctuating parameter on another, of a second kind, that occurs at time τ later. Alternatively it shows the effect on one fluctuating parameter, of a fluctuation of the other kind that occurred at time $-\tau$ earlier.

Assuming internal microscopic time-reversal symmetry for the system of interest means that by reversing the velocity vectors of every particle, the whole system reverses its collective path. This assumption has the important consequence that in a system fluctuating around equilibrium, the fluctuations themselves have time-reversal symmetry. This symmetry means that average values of time-delay correlations must be independent of time direction, such that

$$\langle \delta\hat{X}_j \delta\hat{X}_k(\tau)\rangle = \langle \delta\hat{X}_j \delta\hat{X}_k(-\tau)\rangle$$

This conclusion is not as paradoxical as it sounds. Any configuration in which $\delta\hat{X}_i \neq 0$ is relatively unlikely, and in the majority of cases such a system will not only be closer to equilibrium at the future time $t+\tau$, but also was closer to equilibrium in the past, at time $t-\tau$.

Since only the relative time difference between parameter fluctuations is of significance, the equation for time-delay correlation may be reformulated as

$$\langle \delta\hat{X}_j \delta\hat{X}_k(\tau)\rangle = \langle \delta\hat{X}_j(\tau) \delta\hat{X}_k\rangle$$

Now subtract $\langle \delta\hat{X}_j \delta\hat{X}_k\rangle$ from each side of the equation and divide by τ:

$$\left\langle \delta\hat{X}_j \frac{[\delta\hat{X}_k(\tau) - \delta\hat{X}_k]}{\tau} \right\rangle = \left\langle \frac{[\delta\hat{X}_j(\tau) - \delta\hat{X}_j]}{\tau} \delta\hat{X}_k \right\rangle$$

In the limit $\tau \rightarrow 0$ this equation becomes

$$\left\langle \delta\hat{X}_j \delta\left(\frac{d\hat{X}_k}{dt}\right)\right\rangle = \left\langle \delta\left(\frac{d\hat{X}_j}{dt}\right)\delta\hat{X}_k\right\rangle$$

As argued before, the decay of the fluctuation $\delta(d\hat{X}_k/dt)$ is now assumed to obey the same linear dynamical laws as the macroscopic process, so that

$$\delta\left(\frac{d\hat{X}_k}{dt}\right) = \sum_i L_{ik}\delta\hat{\mathcal{F}}_i$$

and hence

$$\sum_i L_{ik}\langle \delta\hat{X}_j \delta\hat{\mathcal{F}}_i\rangle = \sum_i L_{ij}\langle \delta\hat{\mathcal{F}}_i \delta\hat{X}_k\rangle$$

The final assumption is that each affinity fluctuation is correlated only with fluctuation of its associated external parameters and that cross correlations of the type

$$\begin{aligned}\langle \delta\hat{X}_j \delta\hat{\mathcal{F}}_i\rangle &= 0 \quad \text{for } i \neq j \\ &= k \quad \text{for } i = j\end{aligned}$$

The Onsager relation $L_{ij} = L_{ji}$ then follows directly.

In the presence of a magnetic field time reversal carries the symmetry implication that $\boldsymbol{B} \to -\boldsymbol{B}$ and thus

$$L_{ij}(\boldsymbol{B}) = L_{ji}(-\boldsymbol{B})$$

Some of the well-known effects that have been successfully analyzed in terms of Onsager's theorem include:
The Seebeck Effect: The production of an electromotive force in a thermocouple under conditions of zero electric current. Thermoelectric power is the change in voltage at a thermocouple as a function of temperature.

The Peltier Effect: The evolution of heat caused by the flow of an electric current across an isothermal junction of two materials.

The Thomson Effect: The evolution of heat as an electric current traverses a temperature gradient in a material.

The Hall Effect: In the presence of an orthogonal magnetic field in the z-direction an x-directed electric current produces a y-directed gradient of the electrochemical potential. Similarly an x-directed thermal gradient produces a y-directed gradient of the electrochemical potential, known as the *Nernst effect*.

8.3 Mechanics of Macrosystems

The most important new concept to come from thermodynamics is entropy. Like volume, internal energy and mole number it is an extensive property of a system and together with these, and other variables it defines an elegant self-consistent theory. However, there is one important difference: entropy is the only one of the extensive thermodynamic functions that has no obvious physical interpretation. It is only through statistical integration of the mechanical behaviour of microsystems that a property of the average macrosystem, that resembles the entropy function, emerges.

To put the previous statement into perspective it is necessary to stipulate that any macrosystem with well-defined values of its extensive parameters is made up of myriads of individual particles, each of which may be endowed with an unspecified internal energy, within a wide range consistent with all external constraints. The instantaneous distribution of energy among the constituent particles, adding up to the observed macroscopic energy, defines a microstate. It is clear that any given macrostate could arise as the result of untold different microstates.

A macroscopic measurement senses an average of the properties of many possible microstates. Perpetual interactions between individual particles ensures rapid transition from one microstate to the next. It is therefore reasonable to agree that a macroscopic system samples every permissible microstate with equal probability.

Now suppose that some external constraint on the system, is removed. New microstates, previously inaccessible, become available and transition into these new states may occur. As a consequence the number of microstates among which transitions occur, increases to the maximum permitted by the remaining constraints. This statement is strikingly reminiscent of the entropy postulate of thermodynamics, according to which the entropy increases to the maximum permitted by the imposed constraints. It appears that entropy may be identified with the number of allowed microstates consistent with the imposed macroscopic constraints.

The identification of entropy with available microstates presents one difficulty: Entropy is additive, but the number of microstates is multiplicative[3]. The answer is to identify entropy with the logarithm of the number of available microstates. Thus

$$S = k \ln \Omega$$

where Ω is the number of microstates consistent with the macroscopic constraints. The proportionality factor is chosen to obtain agreement with the temperature scale defined by $T^{-1} = \partial S / \partial U$.

A simple model to illustrate the principles of the statistical approach is a hypothetical two-state system. Each of the N particles is defined to be either in its ground state with zero energy or an excited state with energy ε. If U is the energy of the system then U/ε particles are in the excited state and $(N - U/\varepsilon)$ in the ground state. The number of ways to choose U/ε particles from a total of N, is

$$\Omega = \frac{N!}{(U/\varepsilon)!(N - U/\varepsilon)!}$$

and the entropy is

$$S = k \ln \Omega = k \ln(N!) - k \ln \left(\frac{U}{\varepsilon}!\right) - k \ln \left[\left(N - \frac{U}{\varepsilon}\right)!\right]$$

By Stirling's approximation the logarithm of the factorial of a large number

$$\ln(n!) \simeq n \ln n - n$$

[3]The number of microstates available to two rolling dice is $6 \times 6 = 36$.

the expression for the entropy reduces to

$$S=\left(\frac{U}{\varepsilon}-N\right)k\ln\left(1-\frac{U}{N\varepsilon}\right)-\frac{U}{\varepsilon}k\ln\left(\frac{U}{N\varepsilon}\right)$$

The thermal equation of state is obtained from

$$\frac{dS}{dU}=\frac{1}{T}=\frac{k}{\varepsilon}\ln\left(\frac{N\varepsilon}{U}-1\right)$$

Noting that $U<N\varepsilon$, the temperature is observed to be a strictly positive quantity. Solving the equation of state for the energy gives

$$U=\frac{N\varepsilon}{1+e^{\varepsilon/kT}}$$

The statistical analysis required for real systems is no different in conception from the treatment of the hypothetical two-state system. The elementary particles from which the properties of macroscopic aggregates may be derived by mechanical simulation, could be chemical atoms or molecules, or they may be electrons and atomic nuclei. Depending on the nature of the particles their behaviour could best be described in terms of either classical or quantum mechanics. The statistical mechanics of classical and quantum systems may have many features in common, but equally pronounced differences exist. The two schemes are therefore discussed separately here, starting with the simpler classical sytems.

8.3.1 Classical Systems

Problems in statistical mechanics of classical systems are nearly always treated by means of Hamilton's equations of motion.

A single classical particle is described by a second-order differential equation

$$F=m\frac{d^2q}{dt^2}=\frac{dp}{dt} \tag{8.11}$$

For a large number of particles in an isolated system the motion is dictated by the three conservation laws of

$$\text{linear momentum}:\quad\sum_{i=1}^{N}p_i=\text{constant}$$

$$\text{angular momentum}:\quad\sum_{i=1}^{N}p_i\times q_i=\text{constant}$$

$$\text{energy}: \quad \sum_i \epsilon_i = \sum_{i=1}^{N} \frac{p_i^2}{2m_i} + \phi(q_1, q_2, \ldots, q_N) = \text{constant}$$

in terms of which (11) is readily reduced into Hamilton's canonical first-order equations of motion:

$$\frac{dp}{dt} = F = -\left(\frac{\partial \phi}{\partial q}\right)_p = -\left(\frac{\partial E}{\partial q}\right)_p \tag{8.12}$$

$$\frac{dq}{dt} = \frac{p}{m} = \frac{1}{2m}\left(\frac{\partial p^2}{\partial p}\right)_q = \left[\frac{\partial}{\partial p}\left(\frac{p^2}{2m}\right)\right]_q = \left(\frac{\partial E}{\partial p}\right)_q \tag{8.13}$$

For N particles in a system there are $2N$ of these first-order equations. For given initial conditions the state of the system is uniquely specified by the solutions of these equations. In a conservative system F is a function of q. If q and p are known at time t_0, the changes in q and p can therefore be determined at all future times by the integration of (12) and (13). The states of a particle may then be traced in the coordinate system defined by $p(t)$ and $q(t)$, called a *phase space.* An example of such a phase space for one-dimensional motion is shown in figure 3.

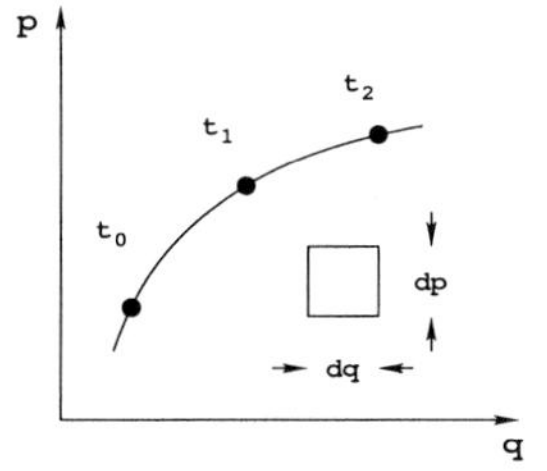

Figure 3: *Phase space for one-dimensional motion. An elemental area (dqdp) in this phase space has the dimensions of time-energy, units of joule-sec.*

An important example of one-dimensional motion is provided by a simple harmonic oscillator. The equation of motion is

$$\ddot{q} = -\left(\frac{k}{m}\right) q$$

with a solution containing two arbitrary constants,

$$q = A\cos(\omega t + \alpha)$$

in which ω is an angular velocity. The second derivative

$$\ddot{q} = -\omega^2 A\cos(\omega t + \alpha) = -\omega^2 q$$

confirms the solution to be acceptable provided $\omega = \sqrt{k/m}$.

The conjugate momentum

$$p = m\dot{q} = -m\omega A \sin(\omega t + \alpha)$$

The orbit described by the representative point in the phase plane (pq) is given by the relationship between q and p. From

$$q^2 = A^2 \cos^2(\omega t + \alpha)$$

$$p^2 = m^2\omega^2 A^2 \sin^2(\omega t + \alpha)$$

follows that

$$\frac{q^2}{A^2} + \frac{p^2}{m^2\omega^2 A^2} = \cos^2(\omega t + \alpha) + \sin^2(\omega t + \alpha) = 1 \tag{8.14}$$

This equation defines an ellipse with semiaxes A and $m\omega A$, determined by the initial values of p and q:

$$A^2 = q_0^2 + \frac{p_0^2}{m^2\omega^2}$$

The value of the phase constant α does not affect the size or shape of the orbit, but it does affect the position of the orbit as a function of time. If at time $t = t_0$ a number of different identical oscillators have $q = 0$ but values of $\dot{q}$ lying between v and $-v$ the representative points will lie on a straight line coincident with the p axis. In time, these points will describe their individual orbits but remain collinear.

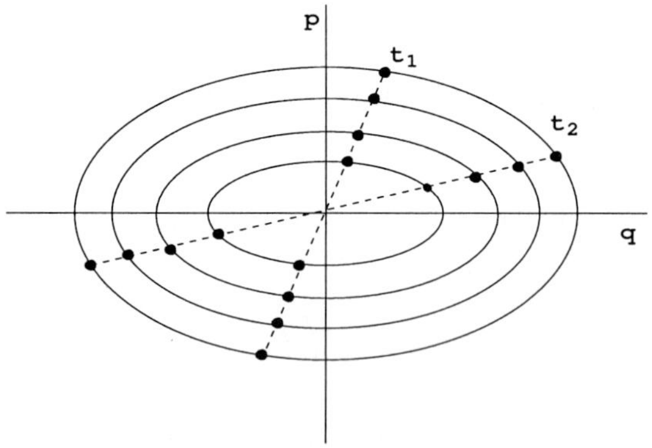

Figure 4: *Orbits in the phase space of points representing simple oscillators.*

When equation (14) is rearranged into the form

$$\frac{mq^2\omega^2}{2} + \frac{p^2}{2m} = \frac{m\omega^2 A^2}{2} \tag{8.15}$$

the l.h.s. represents the total energy, equal to a constant. Elliptic orbits in the phase space therefore represent states of constant energy. These energy curves never intersect. One-dimensional motion is rare in systems of chemical interest. More generally, for systems with f degrees of freedom points in phase space travel on a surface of constant energy, momentum and angular momentum, in a $2f$-dimensional vector space.

The phase space for three-dimensional motion of a single particle is defined in terms of three cartesian position coordinates and the three conjugate momentum coordinates. A point in this six-dimensional space defines the instantaneous position and momentum and hence the *state* of the particle. An elemental hypothetical volume in six-dimensional phase space $dp_xdp_ydp_zdq_xdq_ydq_z$, is called an *element*, in units of (joule-sec)^3. For a system of N such particles, the instantaneous states of all the particles, and hence the state of the system of particles, can be represented by N points in the six-dimensional space. This so-called μ-space, provides a convenient description of a particle system with weak interaction. If the particles of a system are all distinguishable it is possible to construct a $6N$-dimensional phase space (3N position coordinates and 3N conjugate momenta). This type of phase space is called a Γ-space. A single point in this space defines the instantaneous state of the system of particles. For f degrees of freedom there are $2f$ coordinates in μ-space and $2Nf$ coordinates in the Γ space.

Liouville's Theorem

Even had it been possible to measure all state variables of a system at a given instant, provision must still be made for some variation in the accuracy of the measurement. The instantaneous state of the system should therefore be viewed as described by an element, rather than a point of phase space. The one-dimensional oscillator may be considered again to examine the effect of this refinement.

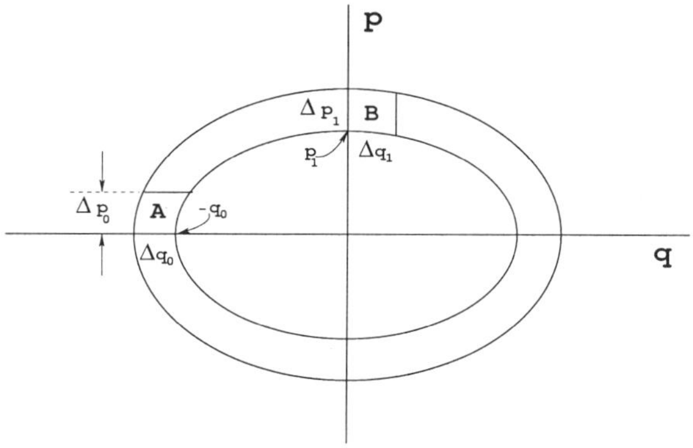

Figure 5: *Trajectory of an element representing a simple harmonic oscillator in phase space.*

Suppose that the position and momentum are measured at time t_0 when $p = 0$ and the potential energy has reached a maximum. At this point the constant $A = q_0$, $E = m\omega^2 q_0^2/2$ and hence

$$-\left(\frac{\partial E}{\partial q}\right)_p = m\omega^2 q \quad , \quad \left[-\left(\frac{\partial E}{\partial q}\right)_p\right]_{-q_0} = m\omega^2 q_0 \tag{8.16}$$

If the accuracy of the measurement is Δp_0 in p and Δq_0 in q, the accuracy of the calculated energy at t_0 follows from (15) as

$$\Delta\epsilon = \frac{p}{2m}\Delta p + \frac{1}{2}mq\omega^2\Delta q = \frac{1}{2}m\omega^2 q_0 \Delta q_0$$

This energy difference may be interpreted in terms of two elliptical trajectories separated by $\Delta\epsilon$ and with a phase lag between the leading and following edges of the element, $\Delta p\Delta q$ that moves along the fuzzy trajectory. The two edges remain separated in time by a fixed amount Δ and define the elements A and B at $-q_0$ and p_1 respectively.

The time Δt_0 required by the oscillator to traverse the element A is

$$\Delta t_0 = \Delta p_0 \Bigg/ \left[\left(\frac{dp}{dt}\right)\right]_{t_0}$$

and it follows from (12) and (16) that

$$\Delta t_0 = \Delta p_0 \Bigg/ \left[-\left(\frac{\partial E}{\partial q}\right)_p\right]_{-q_0} = \frac{2\Delta p_0}{m\omega^2 q_0} = \frac{\Delta p_0 \Delta q_0}{\Delta\epsilon}$$

A quarter cycle later each point on the boundary of A at $p = 0$ has moved to a new position at $q = 0$ and $p = p_1$, now defining a differently shaped element B, as shown in figure 5. At this new position the time required to traverse element B is

$$\Delta t_1 = \Delta q_1 \Bigg/ \left(\frac{dq}{dt}\right)_{p_1}$$

In terms of (13),

$$\left(\frac{dq}{dt}\right)_{p_1} = \left[\left(\frac{\partial E}{\partial p}\right)_q\right]_{p_1}$$

the previous expression reduces to

$$\Delta t_1 = \Delta q_1 \bigg/ \left(\frac{\Delta\epsilon}{\Delta p_1}\right)$$

i.e.

$$\begin{aligned}\Delta t_1 \Delta\epsilon &= \Delta q_1 \Delta p_1 = \text{constant} \\ &= \Delta t_0 \Delta\epsilon = \Delta q_0 \Delta p_0\end{aligned}$$

The elements A and B therefore have equal areas. Liouville's theorem states that an element in phase space is conserved, which means that the element within which a system can be found is constant. Further, if the range $\Delta\epsilon$ in the phase space is divided into equal elements, the system spends equal times passing through these elements, or *cells*.

8.3.2 Statistical Mechanics

Because it is physically impossible to know the starting coordinates and momenta of all particles in a macroscopic system, statistical mechanics may be used to predict the state and time evolution of the system only when the initial state is known from the measurement of bulk properties. This is tantamount to starting from an initial state represented, not by a single system, but by a large collection (ensemble) of identical systems, made up from the same number and types of molecules in the same sort of enclosure. Together they represent the macroscopic system in various dynamical states, resulting from different initial conditions. Each point in phase space then represents a separate system, and each trajectory is independent of all others. Instead of studying the motion of a single system, one looks at the streaming of the whole ensemble of points.

To show [115] that Liouville's theorem holds in any number of phase-space dimensions it is useful to restate some special features of Hamilton's equations,

$$\dot{q}_i = \frac{\partial H}{\partial p_i} \qquad \dot{p}_i = -\frac{\partial H}{\partial q_i}$$

A transformation $\boldsymbol{q}, \boldsymbol{p} \rightarrow \boldsymbol{q}', \boldsymbol{p}'$ possessing the property that the canonical equations of motion also hold for the new coordinates and momenta, is called a canonical transformation.

If J and K are two single-valued and differentiable functions of the generalized coordinates and momenta, the expression

$$\{J, K\} = \sum_{i=1}^{n} \left(\frac{\partial J}{\partial q_i}\frac{\partial K}{\partial p_i} - \frac{\partial K}{\partial q_i}\frac{\partial J}{\partial p_i}\right)$$

is called the *Poisson bracket* of J and K. It becomes useful, for example when expressing the total time derivative of a dynamic quantity $A(q, p, t)$:

$$\begin{aligned}\frac{dA}{dt} &= \frac{\partial A}{\partial t} + \sum \frac{\partial A}{\partial q_i} \cdot \frac{\partial q_i}{\partial t} + \sum \frac{\partial A}{\partial p_i} \cdot \frac{\partial p_i}{\partial t} \\ &= \frac{\partial A}{\partial t} + \sum \left(\frac{\partial A}{\partial q_i} \cdot \frac{\partial H}{\partial p_i} - \frac{\partial A}{\partial p_i} \cdot \frac{\partial H}{\partial q_i} \right) \\ &= \frac{\partial A}{\partial t} + \{A, H\} \qquad (8.17)\end{aligned}$$

The phase space (Γ space) of the system is the Euclidean space spanned by the $2n$ rectangular Cartesian coordinates q_i and p_i. Every possible mechanical state of the system is represented by exactly one point in phase space (and conversely each point in phase space represents exactly one mechanical state).

The objects of statistical mechanics however, are not single systems, but *ensembles* of many, say ν, physically similar systems, *i.e.* systems with Hamiltonians $H(q_i, p_i, t)$. The distribution of such a *virtual ensemble* over the phase space will be described by means of the density function $\varrho(q_i, p_i, t)$, called the *phase density*. It is a real non-negative quantity and will always be normalized to unity, *i.e.*

$$\varrho = \varrho(q, p, t) \geq 0 \qquad \int_\Gamma \varrho(q, p, t) d\Omega = 1$$

where $d\Omega$ denotes the "volume" element of phase space. The phase space density should be interpreted as a *probability density* (probability of phase) rather than a simple density D, such that $\varrho = D/N$, where N is the total number of systems in the ensemble. Although both D and N may be thought of as infinite in order to get a continuous distribution, ϱ must be finite and normalized.

The phase average (or ensemble average) of a dynamical quantity $A(q, p)$ is defined as

$$\bar{A} = \int_\Gamma A(q, p) \varrho(q, p, t) d\Omega$$

Motion of the virtual ensemble in phase space may be likened to fluid flow by introducing a $2n$-dimensional vector of phase velocity $\boldsymbol{v}$ with components $q_i, p_i (i = 1, 2, \ldots, n)$. Since the systems of the virtual ensemble can neither be created nor destroyed in the flow, the equation of continuity

$$\frac{\partial \varrho}{\partial t} + \mathrm{div}(\varrho \boldsymbol{v}) = 0$$

must hold. By definition

$$\begin{aligned} \operatorname{div} \boldsymbol{v} &= \sum_i \frac{\partial}{\partial q_i}(\dot{q}_i) + \sum_i \frac{\partial}{\partial p_i}(\dot{p}_i) \\ &= \frac{\partial}{\partial q_i}\left(\frac{\partial H}{\partial p_i}\right) - \frac{\partial}{\partial p_i}\left(\frac{\partial H}{\partial q_i}\right) = 0 \end{aligned}$$

The "fluid" formed from the phase points is thus incompressible. Furthermore,

$$\operatorname{div}(\varrho \boldsymbol{v}) = \sum \dot{q}_i \frac{\partial \varrho}{\partial q_i} + \varrho \sum \frac{\partial \dot{q}_i}{\partial q_i} + \sum \dot{p}_i \frac{\partial \varrho}{\partial p_i} + \varrho \sum \frac{\partial \dot{p}_i}{\partial p_i}$$

Hence

$$\frac{\partial \varrho}{\partial t} = -\{\varrho, H\}$$

This relationship is known as Liouville's equation. It is the fundamental theorem of statistical mechanics and is valid only for the phase space of generalized coordinates and momenta, whereas it is not, for example valid for the phase space of generalized coordinates and velocities. This is the basic reason that one chooses the Hamiltonian equations of motion in statistical mechanics.

Liouville's equation may also be written in the form

$$\frac{D\varrho}{Dt} = \frac{\partial \varrho}{\partial t} + \{\varrho, H\} = 0$$

where D/Dt indicates a *substantial derivative* or a derivative that follows the motion. Vanishing of the substantial derivative indicates that, as any point moves through Γ space, there is no change with time of the density of points in the immediate neighbourhood of the point. This corollary to Liouville's theorem is known as the conservation of density in phase. A second, already familiar corollary is the conservation of *extension* of phase. This corollary states that, following the motion, there is no change with time in the volume occupied by a given set of points, although the shape is continually changing.

Consider a small region in phase space, $\Delta\tau$, sufficiently small that the density is essentially constant within this region. The number of system points $\Delta n = \varrho\Delta\tau$. If the surface of the element is always determined by the same system points originally in the surface, the behaviour of the element can be followed as a function of time. None of the points within the surface may escape and there are no sources or sinks of system points. Consequently

$$\frac{D(\Delta n)}{Dt} = \frac{D\varrho}{Dt} + \varrho\frac{D(\Delta\tau)}{Dt} = 0$$

The factor $D\varrho/Dt$ vanishes as before, so that

$$\frac{D}{Dt}(\Delta\tau) = 0$$

That the latter equation also holds for an extension of any size may be established by summing the results for a number of elements.

8.3.3 Equilibrium Systems

Statistical mechanics is used to analyze both equilibrium and non-equilibrium systems. Only the former, simpler application will be considered here.

In order for an ensemble to represent a system in equilibrium, the density of phase must remain constant in time. This means that Liouville's equation is satisfied, which requires that ϱ is constant with respect to the coordinates $\boldsymbol{q}$ and $\boldsymbol{p}$, or that $\varrho = \varrho(\alpha)$, where $\alpha = \alpha(\boldsymbol{q}, \boldsymbol{p})$ is a constant of the motion. Constants of motion are properties of an isolated system which do not change with time. Familiar examples are energy, linear momentum and angular momentum. For constants of motion $\{H, \alpha\} = 0$. Hence, if $\varrho = \varrho(\alpha)$ and α is a constant of motion, then

$$\{H, \varrho\} = \frac{d\varrho}{d\alpha}\{H, \alpha\} = 0$$

and by Liouville's equation this condition implies that

$$\frac{\partial \varrho}{\partial t} = 0$$

Therefore, if $\varrho = \varrho(\alpha)$, the ensemble represents a steady state or equilibrium distribution. The two most important steady-state distributions are known as *microcanonical* and *canonical* ensembles.

The classical microcanonical ensemble

The microcanonical, or *energy shell* ensemble is a one-parameter set of distributions used in the study of isolated systems when the volume and energy of the system are known. The distribution function is defined by distributing system points evenly in the region between two neighbouring energy surfaces in phase space; no points being distributed outside this region. Mathematically the ensemble is described by

$$\begin{aligned} \varrho &= \varrho_0 \quad \text{(a constant) for energy } H(q,p) \text{ between } E \text{ and } E + \Delta E \\ &= \Omega^{-1} \\ \varrho &= 0 \text{ , elsewhere} \end{aligned}$$

By letting $\Delta E \to 0$ one gets a *surface ensemble* in which all systems have exactly the energy E. $\varrho(E)$ is in statistical equilibrium because E is a constant of motion.

Since the phase density $\varrho = \Omega^{-1}$ is constant in the energy interval between E and $E + \Delta E$, the probability that a particular state is occupied reaches a maximum when $\Delta E \to 0$, and in general

$$p \propto \frac{\Omega}{\Delta E}$$

As before, it is convenient to define instead, an additive quantity related to the probability, by

$$\Phi = \ln \frac{\Omega}{\Delta E} \quad \text{such that} \quad \varrho = e^{-\Phi}(\Delta E)^{-1}$$

Since ΔE is negligibly small compared to Ω, the physical interpretation of Φ follows from

$$\begin{aligned} \Phi &= \ln \Omega - \ln \Delta E \\ &\simeq \ln \Omega = \frac{S}{k} \end{aligned}$$

as the entropy. The exponential form $\exp \Phi = \Omega/\Delta E$ defines the set of states accessible to the system and thus represents the same function for the microcanonical ensemble which for other ensembles is denoted as the partition function.

If Ω is defined to be the phase volume enclosed by the microcanonical energy surface, the exponential form, in the limit $\Delta E \to 0$, becomes

$$e^{\Phi} = \frac{d\Omega}{dE}$$

known as the Gibbs energy function. It follows that

$$\frac{e^{\Phi}}{\Omega} = \frac{d \ln \Omega}{dE} \equiv k^{-1} \frac{dS}{dE} = \frac{1}{kT}$$

This formulation provides the microcanonical statistical analogue $e^{-\Phi}\Omega$ of the empirical temperature. The width of the microcanonical energy distribution

$$\Delta E = e^{-\Phi}\Omega = kT$$

as expected.

There is a fundamental problem associated with the definition of any constant energy ensemble. Since the momenta have an infinite range the integration

$$\int \varrho d\Omega = \int \varrho dq_i\, dp_i = 1, i = 1, \ldots, n$$

for constant $\varrho(E)$ cannot be performed unless $\varrho = 0$ at all points, or if the distribution of points is restricted to one special region of phase space. The microcanonical ensemble may appear to meet the latter requirement. However, real isolated systems are impossible since any system must exchange energy with the surroundings. Although the microcanonical ensemble may therefore be used to describe large hypothetical isolated systems such as an ideal gas, a more realistic ensemble is needed to include possible variation of energy.

The classical canonical ensemble

Consider a gas whose phase density in Γ space is represented by a microcanonical ensemble. Let it consist of molecules with μ-spaces μ_i with probability distributions ϱ_i. Denote the element of extension in μ_i by ϕ_i. Since energy exchanges may occur between the molecules, μ_i cannot be represented by a microcanonical distribution. There must be a finite density corresponding to points of the ensemble that do not satisfy the requirement of constant energy. Nevertheless, the simultaneous probability that molecule 1 be within element $d\phi_1$ of its μ-space, molecule 2 within $d\phi_2$ of its μ-space, *etc.*, equals the probability that the whole gas be in the element

$$d\phi = d\phi_1 d\phi_2 \ldots d\phi_\nu \text{ of } \mu \text{ space.}$$

Hence

$$\varrho_1(H_1)d\phi_1\varrho_2(H_2)d\phi_2 ... \varrho_\nu(H_\nu)d\phi_\nu = \varrho(H)d\phi \tag{8.18}$$

so that

$$\varrho_1(H_1)\varrho_2(H_2)\ldots\varrho_\nu(H_\nu) = \varrho(H_1 + H_2 + \cdots + H_\nu)$$

It is necessary that this functional equation should be satisfied for every value of the total energy $H = \sum H_i$, although the constant $\varrho(H)$ for any given H is still described by the microcanonical distribution.

Equation (18) holds for any ν. If the ensemble consisted of only two points,

$$\varrho_1(H_1) \times \varrho_2(H_2) = \varrho(H_1 + H_2)$$

It follows that $\varrho_i(0) = 1$ for every i. Since the logarithmic function relates products to sums it is logical to denote $\ln \varrho_i$ by f_i, such that

$$f_i(H_i) + f_2(H_2) = f(H_1 + H_2) \tag{8.19}$$

On putting $H_2 = 0$, equation (19) reads

$$f_1(H_1) + f_2(0) = f(H_1)$$

and since $f_2(0) = 0$, $f_1 = f$. Thus all f_i are seen to be the same function, f. Equation (19) is therefore equivalent to

$$f(x) + f(y) = f(x + y)$$

For $y = x$, one has $2f(x) = f(2x)$, and so by induction,

$$f(nx) = nf(x) \tag{8.20}$$

for any integer n. Writing (20) in the form

$$f\left(n\left[\frac{x}{n}\right]\right) = nf\left(\frac{x}{n}\right)$$

immediately allows the alternatives

$$f\left(\frac{x}{n}\right) = \frac{1}{n}f(x) \quad \text{and} \quad f\left(\frac{m}{n}x\right) = \frac{m}{n}f(x)$$

where m is another integer. This last relationship shows that (20) is generally valid for all rational numbers, as well as integers. It is therefore possible to write

$$f(x) = f(x \times 1) = xf(1) = x \times (\text{constant})$$

This final result identifies the only function that satisfies (19), as $f_i = cH$, whence

$$\varrho_i(H_i) = e^{cH_i}$$

Since ϱ_i is a probability it must remain finite for every H_i, a quantity which may tend to infinity, but not $\to -\infty$. Hence c is a negative constant, formulated by Gibbs as $-1/\theta$.

An ensemble that satisfies all previous conditions and $d\varrho/dt = 0$ for a conservative system is therefore defined by

$$\varrho = e^{-H/\theta}$$

This *canonical ensemble* is also a one-parameter distribution function for various values of θ, independent of $\boldsymbol{q}$ and $\boldsymbol{p}$.

A more general formulation, often used, writes

$$\varrho = e^{(\psi - H)/\theta}$$

where both distribution parameters are independent of q and p. The relationship between the parameters follows from the normalization integral

$$\begin{aligned}\int \varrho d\Omega &= 1\\ &= \int e^{(\psi-H)/\theta}d\Omega\\ &= e^{\psi/\theta}\int e^{-H/\theta}d\Omega\end{aligned}$$

Hence

$$e^{-\psi/\theta} = \int e^{-H/\theta}d\Omega = Q \tag{8.21}$$

where Q is known as either the *phase integral*, the *partition function* or the *sum of states*. In terms of Q, the parameter

$$\psi = -\theta \ln Q \tag{8.22}$$

According to equation (21) the systems of the canonical ensemble are conservative. Each system moves independently of all others and the phase integral exists for each of them. Each system therefore moves on a surface of constant energy and corresponds to a microcanonical ensemble. In this sense the canonical ensemble is built up from a multitude of microcanonical ensembles. Quantities defined for the microcanonical ensemble may therefore be averaged over the canonical ensemble. The original system which is represented by the canonical ensemble however, cannot be described, even approximately, as conservative. It is customary to denote the Hamiltonian of the systems of the canonical ensemble as the Hamiltonian of the original system, which is not justified.

The canonical ensemble is often stated to describe a system in contact with a thermal reservoir. States of all energies, from zero to arbitrarily large values are available to the system, but all states no longer have equal probabilities. The system does not spend the same fraction of time in each state. To determine the probability distribution among the available microstates it is important to understand that the system plus reservoir constitute a closed system, to which the principle of equal probability applies once more.

The total probability p_j of the system being in state j is equal to the fraction of the total number of states (system plus reservoir) in which the subsystem is in state j, with energy E_j, *i.e.*

$$p_j = \frac{\Omega_{res}(E_{tot} - E_j)}{\Omega_{tot}(E_{tot})}$$

where Ω_{tot} is the total number of states, system plus reservoir.

The classical grand canonical ensemble

The microcanonical ensemble is used to describe isolated systems and the canonical ensemble describes closed systems. In the first case there is no exchange of either energy or matter, whereas energy exchange is allowed in the second instance. The most general situation is an open system which exchanges both matter and energy with its environment and between systems. This definition implies an ensemble made up of canonical ensembles with the number of particles varying from zero to infinity, and designated as a *grand canonical ensemble.* A probability that allows a system to have variable particle number can clearly no longer be defined as a phase density. The classical grand canonical ensemble can therefore not be represented as a distribution in Γ space and in fact falls outside the domain of classical statistical mechanics.

In quantum statistical mechanics where a *density operator* replaces the classical phase density the statistics of the grand canonical ensemble becomes feasible. The problem with the classical formulation is not entirely unexpected in view of the fact that even the classical canonical ensemble that predicts equipartitioning of molecular energies, is not supported by observation.

Before extending the discussion to quantum systems it is instructive to use classical statistics to investigate the relationship between statistical mechanics and thermodynamics.

8.3.4 Statistics and Thermodynamics

The most important objective of statistical mechanics is an explanation of thermodynamic properties in terms of the mechanical behaviour of molecules that constitute the system of interest. For this purpose it is necessary to know, or to assume, some connection between the statistical ensemble and the thermodynamic system.

There are two fundamental assumptions of statistical mechanics as applied to thermodynamics. The first axiom was stated by Tolman [116] as the hypothesis of *equal a priori probabilities.* It postulates that equal probabilities for different regions in the phase space correspond to elements of the same magnitude. A phase point in a system with energy in the range E to $E + \delta E$ is therefore equally likely to be found in any of the different equal elements H_i within the range, and this element is conserved. This postulate is analogous to making the assumption as to the probabilities for heads or tails when flipping a coin.

The second fundamental assumption is the *ergodic* hypothesis. According to this hypothesis, each representative point describes a trajectory that eventually brings it into every position of the phase space consistent with its energy. The argument can be explained in terms of the microcanonical ensemble.

For an isolated system based on the microcanonical ensemble with phase points equally distributed in a narrow energy range, all systems exhibit the same long-term behaviour and spend the same fraction w_i of time in different elementary regions. For a total number N of systems at time t,

$$N = \sum_i N_i$$

Over a long time τ, the number of phase points traversed is

$$N(w_i\tau) = \int_0^\tau N_i dt$$

Since N_i is independent of the time for a mocrocanonical ensemble in statistical equilibrium, one finds

$$w_i = \frac{N_i}{N}$$

In other words, the time average of any quantity pertaining to any single system agrees with the ensemble average. Dynamical systems that obey this equivalence are said to be *ergodic*. It means that any system, left to itself for long enough, will take on every possible configuration.

In its extreme form the ergodic hypothesis is clearly untenable. Only probability statements can be made in statistical mechanics and these have nothing to do with sequences in time [117]. Not surprisingly, a totally convincing proof of the ergodic theorem in its many guises has not been formulated. The current concensus still is that an axiomatic basis, completely independent of ergodic theory should be chosen [115] for the construction of statistical mechanics.

In practical application [117] a problem in statistical mechanics is approached by setting up the proper ensemble to represent the available information about the state of the system. The average values of various quantities can then be determined by averaging over the ensemble.

The ensemble average of a quantity u, which is a function of the canonical variables p_i and q_i, is given by the integral

$$\bar{u} = \int u e^{(\psi - H)/\theta} dq_1 \ldots dq_n \, dp_1 \ldots dp_n$$

Suppose that the system is described by n normal coordinates q_i. The Hamiltonian function, or energy, is given by

$$H = \sum_{i=1}^{n} \left(a_i p_i^2 + b_i q_i^2\right)$$

and the average energy is

$$\bar{H} = \int \sum_{i=1}^{n} \left(a_i p_i^2 + b_i q_i^2\right) e^{\frac{\psi - \Sigma\left(a_i p_i^2 + b_i q_i^2\right)}{\theta}} dq_1 \ldots dp_n$$

To evaluate this integral the average value of $a_1 p_1^2$ is considered first.

$$\begin{aligned}
\overline{a_1 p_1^2} &= e^{\psi/\theta} \int_{-\infty}^{\infty} a_1 p_1^2 e^{-a_1 p_1^2/\theta} dp_1 \int_{-\infty}^{\infty} e^{-a_2 p_2^2/\theta} dp_2 \ldots \int_{-\infty}^{\infty} e^{-b_n q_n^2/\theta} dq_n \\
&= e^{\psi/\theta} \left(\frac{\pi\theta^3}{4a_1}\right)^{\frac{1}{2}} \prod_{i=2}^{n} \left(\frac{\pi\theta}{a_i}\right)^{\frac{1}{2}} \prod_{i=1}^{n} \left(\frac{\pi\theta}{b_i}\right)^{\frac{1}{2}}
\end{aligned}$$

using a standard integral[4]. To satisfy equation (21) it is required that

$$e^{\psi/\theta} = \prod_{i=1}^{n} \left(\frac{\pi\theta}{a_i}\right)^{\frac{1}{2}} \left(\frac{\pi\theta}{b_i}\right)^{\frac{1}{2}}$$

so that

$$\overline{a_1 p_1^2} = \frac{\theta}{2}$$

The result is independent of the coefficient a_1 and is the same for all coordinates and momenta. Hence $\bar{H} = n\theta$. This expression resembles the equipartition theorem according to which each degree of freedom has the average energy kT, half of it kinetic and half potential, and suggests that the *distribution modulus* θ be identified with temperature.

Statistical definition of temperature

To establish the meaning of θ it is noted that equation (19) provides an approximate description of two equilibrium systems brought into thermal

---

4

$$\int_0^{\infty} x^{2n} e^{-ax^2} dx = \frac{1.3.5 \ldots (2n-1)}{2^{n+1} a^n} \sqrt{\frac{\pi}{a}}$$

contact without disturbing the equilibrium[5]. A more precise formulation would require

$$f_1(H_1) + f_2(H_2) = f(H_1 + H_2 + V_{12})$$

where the term V_{12} represents the energy of interaction between the two systems. By expansion of both systems in size, the interaction term V_{12} can be made arbitrarily small and be neglected. If both systems are represented by canonical ensembles before thermal contact is established, then

$$f_1 = \frac{\psi_1 - H_1}{\theta_1} \quad \text{and} \quad f_2 = \frac{\psi_2 - H_2}{\theta_2}$$

The condition that preserves equilibrium may then be written

$$\frac{\psi_1 - H_1}{\theta_1} + \frac{\psi_2 - H_2}{\theta_2} = \frac{\psi_{12} - H_{12}}{\theta_{12}}$$

where the r.h.s. represents the combined system. The necessary and sufficient condition satisfying this relationship is that $\theta_1 = \theta_2 = \theta_{12}$, *i.e.* from (19)

$$(\psi_1 + \psi_2) - (H_1 + H_2) = \psi_{12} - (H_1 + H_2)$$

and from (22) $\psi_{12} = \psi_1 + \psi_2$.

The distribution modulus, θ of the canonical ensemble thus possesses the property of an empirical temperature $\theta = kT$. The proportionality constant, by analogy with ideal gas kinetic theory, is the same as Boltzmann's constant.

The considerations above show that temperature is definable only for macroscopic (in the strict sense, infinitely large) systems, whose definition implies neglect of the disturbance during first thermal contact. The canonical ensemble is now seen to be characterized by its temperature and typically describes a system in equilibrium with a constant temperature bath.

The Maxwell-Boltzmann distribution

For a system with many degrees of freedom the canonical distribution of energy closely approximates the microcanonical distribution about the most

[5] If two bodies have the same temperature and are put into contact, the composite system is in thermodynamic equilibrium at the same temperature unless there is a significant interaction between the two bodies. If there is a chemical reaction, for example, the resulting system will eventually come to equilibrium at a lower or higher temperature than that of the individual systems before they were put into contact. If the interaction is negligible, the whole system will be in equilibrium at the original temperature.

probable value of the energy. That means that the energy distribution represented by a canonical ensemble has a very sharp maximum and the density of points in other energy ranges is negligible. To demonstrate this proposition it is necessary to calculate the probability that particle i has an energy H_i corresponding to p_i, q_i. Regardless of the states of all other particles this probability is given by the integral $\int \varrho d\Omega$ over the momenta and coordinates of all particles, except i itself, *i.e.*

$$\varrho_i dp_i dq_i = \Lambda e^{-H_i/\theta} d\Omega_i \tag{8.23}$$

where the new normalizing constant Λ follows from

$$\int \Lambda e^{-H_i/\theta} d\Omega_i = 1$$

For an ideal gas, in the absence of external potentials

$$\begin{aligned} \frac{1}{\Lambda} &= \int \cdots \int e^{-(\sum p_i^2)/2m\theta} dp_i dq_i \\ &= \tau \int \cdots \int e^{-(\sum p_i^2)/2m\theta} dp_i \end{aligned}$$

where τ is the volume of the gas. Thus[6]

$$\frac{1}{\Lambda} = \tau \left[\int_{-\infty}^{\infty} e^{-u^2/2m\theta} du \right]^3 = \tau (2\pi m\theta)^{3/2}$$

Substituted into (23)

$$\varrho_i dp_i = \frac{1}{\tau} (2\pi m\theta)^{-3/2} e^{-H_i/\theta} d\Omega_i$$

Under the assumption of isotropy the element in momentum space may be replaced by a spherical shell, in polar coordinates $4\pi p^2 dp$. Substituting $p^2 = 2mH$ this factor becomes

$$4\pi p^2 dp = 4\pi (2mH) \left(\sqrt{\frac{m}{2H}} \right) dH$$

[6] Using

$$\int_0^{\infty} e^{-a^2 x^2} dx = \frac{1}{2a} \sqrt{\pi}$$

The probability distribution in μ-space becomes

$$d\varrho_i = 2\pi(\pi\theta)^{-3/2}\sqrt{H}e^{-H/\theta}dH \tag{8.24}$$

In Γ space the elementary shell on the hypersphere is defined by the radial coordinate

$$R = \sqrt{\sum_i (p_i^2/2m)} \quad ; \quad (H = R^2)$$

and $3N-1$ angular coordinates. The number of systems in the shell of thickness dR, *i.e.*

$$\begin{aligned} 4\pi R^{3N-1}dR &= 2\pi R^{3N-2}(2RdR) \\ &= 2\pi H^{(3N-2)/2}dH \ , \end{aligned}$$

defines the energy distribution

$$d\varrho = \frac{2\pi}{(\pi\theta)^{3/2}}H^{(3N-2)/2}e^{-H/\theta}dH$$

Since the factor $H^{(3N-2)/2}$ increases rapidly with H and the exponential factor falls off rapidly, the distribution has a sharp maximum.

The maximum occurs at the most probable value of the energy when $dP/dH = 0$, for the distribution function P, which has the form

$$P = \frac{d\varrho}{dH} = AH^n e^{-H/\theta}$$

Hence

$$nAH^{n-1}e^{-H/\theta} - \frac{1}{\theta}e^{-H/\theta}AH^n = 0$$

i.e.

$$AH^n e^{-H/\theta}\left\{\frac{n}{H} - \frac{1}{\theta}\right\} = 0$$

such that[7]

$$H_p = n\theta$$

---

[7]In μ-space the most probable kinetic energy $\epsilon_p = \frac{1}{2}mc_p^2 = kT$, which differs from the average energy according to the equipartition principle,

$$\bar{\epsilon} = \frac{1}{2}m\bar{c}^2 = \frac{3}{2}kT$$

For this value of the energy the exponential factor becomes a constant and the distribution a function of H only, like the microcanonical ensemble. As a matter of fact, as the number of systems in the ensemble approaches infinity, the canonical distribution becomes increasingly sharp, thus approaching a microcanonical surface ensemble.

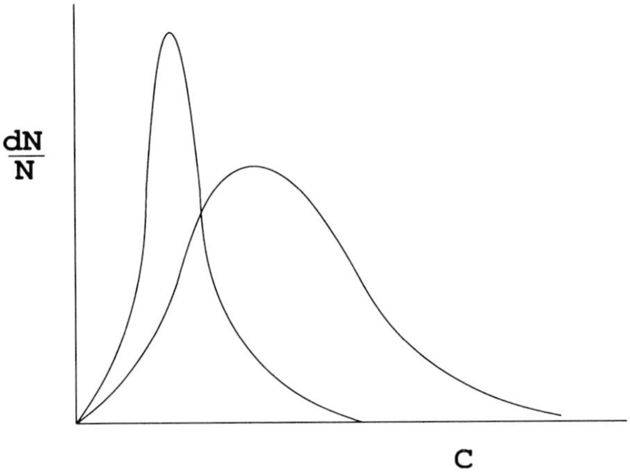

Figure 6: *The Maxwell-Boltzmann distribution of molecular velocities at two different temperatures.*

The distribution function (24) for an ideal gas, shown in figure 6 is known as the Maxwell-Boltzmann distribution and is specified more commonly [118] in terms of molecular speed, as

$$\frac{dN}{N} = 4\pi \left(\frac{m}{2\pi kT}\right)^{3/2} C^2 e^{-mC^2/2kT} dC$$

obtained by putting $H = \frac{1}{2}mC^2$, $\theta = kT$, in (24). The number of particles with speeds between C and $C + dC$ is dN, and N is the total number of particles.

Thermodynamic Functions

Another way of looking at the Boltzmann distribution assumes that the energy spectrum consists of closely spaced, but fixed energy levels, ϵ_ν. The probability that level ν is populated is specified in terms of the canonical distribution

$$\varrho_\nu \propto e^{-\epsilon_\nu/kT}$$

The proportionality constant is independent of the specific state of the system and is determined by the normalization requirement $\sum_\nu \varrho_\nu = 1$. Writing

$$\varrho_\nu = \frac{e^{-\epsilon_\nu/kT}}{Q}$$

it follows that

$$\sum_\nu \varrho_\nu = \frac{1}{Q}\sum_\nu e^{-\epsilon_\nu/kT} = 1$$

defining Q as equivalent to the sum over states of equation (21), *i.e.*

$$Q = \sum_\nu e^{-\epsilon_\nu/kT}$$

The partition function provides a direct method to estimate thermodynamic functions statistically.

In view of the ergodic hypothesis the average value of an observable property may be regarded as the quantity measured under specified conditions. In this way the internal energy of a system corresponds to the average energy of the canonical distribution:

$$\begin{aligned} U = \bar{\epsilon}_\nu &= \sum_\nu \varrho_\nu \epsilon_\nu \\ &= \frac{\sum_\nu \epsilon_\nu e^{-\epsilon_\nu/kT}}{Q} \\ &= \frac{\frac{\partial Q}{\partial T}\cdot kT^2}{Q} \end{aligned}$$

Hence

$$U = kT^2\frac{\partial \ln Q}{\partial T}$$

To establish [5] the thermodynamic meaning of the *index of probability* $\eta = (\psi - H)/\theta$, it is assumed that the distribution $\varrho = e^\eta$ changes as the condition of the system changes, but always subject to the normalization condition $\int e^\eta d\Omega = 1$. It follows that the derivative $d\int e^\eta d\Omega = 0$. It is assumed that both θ and ψ, as well as extensive parameters ξ_i may be altered, such that

$$\begin{aligned} d\int e^\eta d\Omega &= \int e^\eta\left[\frac{d\psi}{\theta} - \frac{\eta d\theta}{\theta} - \frac{1}{\theta}\sum_i \frac{\partial H}{\partial \xi_i} d\xi_i\right] d\Omega = 0 \\ &= \frac{d\psi}{\theta} - \frac{\overline{\ln\varrho}}{\theta} d\theta - \frac{1}{\theta}\sum_i \bar{f}_i d\xi_i \end{aligned} \quad (8.25)$$

where horizontal bars indicate average values, *i.e.* $\bar{R} = \int \varrho R d\Omega$, and $f_i = -\frac{\partial H}{\partial \xi_i}$ represents the force created by changing the i-th extensive parameter. Since $\overline{\ln\varrho} = (\psi - \bar{H})/\theta$, it follows that

$$\psi = \theta\cdot\overline{\ln\varrho} + \bar{H}$$

$$d\psi = d\bar{H} + \theta d(\overline{\ln \varrho}) + \overline{\ln \varrho d\theta}$$

When this result is substituted into (25) and rearranged, one gets

$$d\bar{H} = -\theta d(\overline{\ln \varrho}) - \sum_i \bar{f}_i d\xi_i \tag{8.26}$$

As before, $d\bar{H}$ is interpreted as the increase in total internal energy of the thermodynamic system, $\theta = kT$ and $\bar{f}_i$ must represent actual forces acting on the real system. Equation (26) is then seen to be the exact analogue of the basic equation (1) of chemical thermodynamics [118]

$$dU = TdS - \sum_j P_j dX_j \quad [e.g.\ dU = TdS - PdV]$$

provided that $-(\ln \varrho)k = S$, the entropy of the system.

Equation (25) can be written in the form

$$-d\psi = SdT + \sum_j P_j X_j$$

that immediately identifies ψ with the Helmholtz free energy function, F.

8.4 Quantum Statistics

Despite the fact that classical statistics provides an appealing theoretical justification of the basic equations of thermodynamics, there are some annoying contradictions, of which the principle of equipartitioning of energy is the most glaring. The resolution to all of the associated problems was found through the use of quantum mechanics, rather than Hamiltonian mechanics as the basis of statistical mechanics. The definition of statistical ensembles and the derivation of thermodynamic analogues will be seen to remain the same. The interesting fact is that quantum mechanics is itself a statistical theory that applies to virtual ensembles. The important question to be settled, in view of this, relates to the appropriate space to use in exact statistical quantum mechanics.

Classical systems evolve from initial mechanical states that may, in principle, be known completely. The systems of classical ensembles have well-defined mechanical states and the initial system is always found in one of these states, although it is not known which one. In quantum mechanics however, the state of a system, including its initial state, is represented by a

Ψ function which carries maximum information about that system. Definition of the Ψ-function itself, depends on a probability aggregate or quantum-mechanical ensemble. The mechanical state of the systems of this ensemble cannot be defined more precisely than by stating the Ψ-function. It follows that the same Ψ-function and hence the same mechanical state must be assumed for all systems of the quantum-mechanical ensemble. A second major difference between classical and quantum states is that the Ψ-function that describes the quantum-mechanical ensemble is not a probability density, but a probability amplitude. By comparison the probability density for coordinates $\boldsymbol{q}$ is

$$W(\boldsymbol{q}, t) = \Psi^* \Psi$$

From this relationship arises the possible interference of probabilities, a concept which is totally foreign to classical theory.

The previous discussion only applies when a Ψ-function for a system exists and this situation is described as a *pure ensemble.* It is a holistic ensemble that cannot be generated by a combination of other distinct ensembles. It is much more common to deal with systems for which maximum information about the initial state is not available in the form of a Ψ-function. As in the classical case it then becomes necessary to represent the initial state by means of a *mixed ensemble* of systems with distinct Ψ-functions, and hence in distinct quantum-mechanical states.

The topic that is commonly referred to as statistical quantum mechanics deals with mixed ensembles only, although pure ensembles may be represented in the same formalism. There is an interesting difference with classical statistics arising here: In classical mechanics maximum information about all subsystems is obtained as soon as maximum information about the total system is available. This statement is no longer valid in quantum mechanics. It may happen that the total system is represented by a pure ensemble and a subsystem thereof by a mixed ensemble.

8.4.1 The Quantum Formalism

The quantum-mechanical state is represented in abstract Hilbert space on the basis of eigenfunctions of the position operator, by $\Psi(\boldsymbol{q}, t)$. If the eigenvectors of an abstract quantum-mechanical operator are used as a basis, the operator itself is represented by a diagonal square matrix. In wave-mechanical formalism the position and momentum matrices reduce to multiplication by q_i and $(h/2\pi i)(\partial/\partial q_i)$ respectively. The corresponding expectation values are

$$\bar{q}_i = \int \Psi^*(\boldsymbol{q}, t) q_i \Psi(\boldsymbol{q}, t) d\boldsymbol{q}$$

$$\bar{p}_i = \frac{h}{2\pi i}\int \Psi^*(\boldsymbol{q},t)\frac{\partial}{\partial q_i}\Psi(\boldsymbol{q},t)d\boldsymbol{q}$$

For systems with classical analogues every observable quantity is an algebraic function of q_i and p_i. The wave-mechanical operator acting on the Ψ-function is

$$\boldsymbol{A} = A\left(\boldsymbol{q}, \frac{h}{2\pi i}\frac{\partial}{\partial \boldsymbol{q}}\right)$$

In particular, the wave-mechanical Hamiltonian

$$\boldsymbol{H} = \sum_{i=1}^{N}\left(-\frac{h^2}{8\pi^2 m_i}\nabla_i^2\right) + V(\boldsymbol{q})$$

The evolution of the system in time is described by the Schrödinger equation

$$\boldsymbol{H}\Psi = \frac{h}{2\pi i}\frac{\partial\Psi}{\partial t} = 0 \tag{8.27}$$

The choice of representation is arbitrary and one basis can be mapped into another by unitary transformation. Thus, let $\psi(n,\boldsymbol{q})$ and $\varphi(l,\boldsymbol{q})$ be two countable sets of basis vectors, such that

$$\varphi(l,\boldsymbol{q}) = \sum_n \psi(n,\boldsymbol{q})U_{nl} \quad ; \quad \psi(n,\boldsymbol{q}) = \sum_l \varphi(l,\boldsymbol{q})U_{ln}^{-1}$$

The unitary matrix $\boldsymbol{U}$ is defined by the relationships

$$U_{ml} = \int \psi^*(m,\boldsymbol{q})d\boldsymbol{q} \quad ; \quad U_{kn}^{-1} = U_{nk}^* = \int \varphi^*(k,\boldsymbol{q})\psi(n,\boldsymbol{q})d\boldsymbol{q}$$

$$\boldsymbol{U}\boldsymbol{U}^\dagger = \boldsymbol{U}^\dagger\boldsymbol{U} = 1$$

Abstract vectors defined in the bases $\psi(n,\boldsymbol{q})$ and $\varphi(l,\boldsymbol{q})$ as the orthonormal sets

$$\Psi_a(\boldsymbol{q},t) = \sum_n a(n,t)\psi(n,\boldsymbol{q}) \quad \text{and} \quad \Psi_b = \sum_l b(l,t)\varphi(l,\boldsymbol{q})$$

transform like

$$\boldsymbol{\Psi}_b = \boldsymbol{U}^{-1}\boldsymbol{\Psi}_a = \boldsymbol{U}^\dagger\boldsymbol{\Psi}_a$$

$$\boldsymbol{\Psi}_a = \boldsymbol{U}\boldsymbol{\Psi}_b = (\boldsymbol{U}^{-1})^\dagger\boldsymbol{\Psi}_b$$

The matrices that represent abstract operators in the two bases are related by the transformations

$$\boldsymbol{A}_b = \boldsymbol{U}^{-1}\boldsymbol{A}_a\boldsymbol{U} \quad ; \quad \boldsymbol{A}_a = \boldsymbol{U}\boldsymbol{A}_b\boldsymbol{U}^{-1}$$

The foregoing unitary transformations may be interpreted as the analogues of canonical transformations in classical mechanics.

The probability of finding the system in the state characterized by the value n, is

$$W(n,t) = a^*(n,t)a(n,t) \quad \text{with} \quad \sum_n a^*(n,t)a(n,t) = 1$$

The expectation value determined by the probability amplitude $a(n,t)$ is

$$\bar{A} = \sum_m \sum_n a^*(m,t) A_{mn} a(n,t) \tag{8.28}$$

Other formulations that are analogous to wave mechanics in an arbitrary basis, are obtained by defining the operator $\boldsymbol{A}^{(a)}$ according to

$$\boldsymbol{A}^{(a)} a(m,t) = \sum_n A_{mn} a(n,t)$$

In terms of this operator

$$\bar{A} = \sum_m a^*(m,t) \boldsymbol{A}^{(a)} a(m,t)$$

exactly as for the position and momentum operators. The transformed Hamiltonian defines a transformed Schrödinger equation in terms of the probability amplitude $a(m,t)$,

$$\boldsymbol{H}^{(a)} a(m,t) + \frac{h}{2\pi i} \frac{\partial a(m,t)}{\partial t} = 0 \tag{8.29}$$

In the Schrödinger picture operators in the case of a closed system do not depend explicitly on the time, but the state vector is time dependent. However, the expectation values are generally functions of the time. The commutator of the Hamiltonian operator $\boldsymbol{H} = -(h/2\pi i)(\partial/\partial t)$ and another operator $\boldsymbol{A}$, is defined by

$$\begin{aligned} [\boldsymbol{A}, \boldsymbol{H}] &= \boldsymbol{AH} - \boldsymbol{HA} \\ &= -\frac{h}{2\pi i} \left\{ \boldsymbol{A} \frac{\partial \Psi}{\partial t} - \frac{\partial (\boldsymbol{A}\Psi)}{\partial t} \right\} \end{aligned}$$

$$\Psi^*[\boldsymbol{A}, \boldsymbol{H}]\Psi = \frac{h}{2\pi i} \left\{ \Psi^* \frac{\partial \boldsymbol{A}}{\partial t} \Psi \right\}$$

Now

$$\bar{\boldsymbol{A}} = \int \Psi^* \boldsymbol{A} \Psi d\boldsymbol{q}$$

Hence

$$\begin{aligned}\frac{\partial\bar{\boldsymbol{A}}}{\partial t} &= \int \Psi^* \frac{\partial \boldsymbol{A}}{\partial t}\Psi d\boldsymbol{q} \\ &= \frac{2\pi i}{h}\int \Psi^*[\boldsymbol{A},\boldsymbol{H}]\Psi d\boldsymbol{q} \\ &= \frac{2\pi i}{h}\overline{[\boldsymbol{A},\boldsymbol{H}]} \qquad (8.30)\end{aligned}$$

If $\boldsymbol{A}$ is the Hamiltonian, it follows that

$$\frac{d\bar{\boldsymbol{H}}}{dt} = 0$$

For a closed system, therefore, the expectation value of the energy is constant in time, which is the energy theorem of quantum mechanics.

The time evolution in the Schrödinger representation may be described in terms of a development operator $\boldsymbol{T}(t,t_0)$ by the equation

$$\frac{d\boldsymbol{T}}{dt} = \frac{2\pi i}{h}\boldsymbol{HT}$$

with solutions

$$\boldsymbol{T}(t,t_0) = e^{-\frac{2\pi i}{h}\boldsymbol{H}(t-t_0)}$$

For an arbitrary representation it follows that

$$a(n,t) = e^{-(2\pi i/h)\boldsymbol{H}^{(a)}(t-t_0)}a(n,t_0)$$

In the Heisenberg picture the operators themselves depend explicitly on the time and the time evolution of the system is determined by a differential equation for the operators. The time-dependent Heisenberg operator $\boldsymbol{A}_H(t)$ is obtained from the corresponding Schrödinger operator $\boldsymbol{A}_S$ by the unitary transformation

$$\boldsymbol{A}_H(t) = \boldsymbol{T}^\dagger \boldsymbol{A}_S \boldsymbol{T}$$

Taking derivatives of this equation with respect to time, gives

$$\begin{aligned}\frac{d\boldsymbol{A}_H}{dt} &= \boldsymbol{T}^\dagger\frac{\partial \boldsymbol{A}_S}{\partial t}\boldsymbol{T} + \frac{\partial \boldsymbol{T}^\dagger}{\partial t}\boldsymbol{A}_S\boldsymbol{T} + \boldsymbol{T}^\dagger\boldsymbol{A}_S\frac{\partial \boldsymbol{T}}{\partial t} \\ &\equiv \frac{\partial \boldsymbol{A}_H}{\partial t} + \frac{2\pi i}{h}\left[-\boldsymbol{H}\boldsymbol{T}^\dagger(\boldsymbol{A}_S\boldsymbol{T}) + \boldsymbol{T}^\dagger\boldsymbol{A}_S\boldsymbol{H}\boldsymbol{T}\right] \\ &= \frac{\partial \boldsymbol{A}_H}{\partial t} + \frac{2\pi}{ih}[\boldsymbol{A}_{\boldsymbol{H}},\boldsymbol{H}_{\boldsymbol{H}}] \qquad (8.31)\end{aligned}$$

Equation (31) is known as Heisenberg's equation of motion and is the quantum-mechanical analogue of the classical equation (17). The commutator of two quantum-mechanical operators multiplied by $(2\pi i/h)$ is the analogue of the classical Poisson bracket. In quantum mechanics a dynamical quantity whose operator commutes with the Hamiltonian, $[A, H] = 0$, is a constant of the motion.

Second quantization

Statistical mechanics deals explicitly with the motion of particles. The common quantization procedure that provides a quantum description of classical particles by the introduction of operators, such as the momentum operator, $p \to -\hbar i\frac{\partial}{\partial q}$ however, replaces the classical particle description by a wave field. The basis of quantum statistics is therefore not obvious. On the other hand, quantization of a classical wave field produces a particle description of the field. The best-known example is the appearance of photons in the quantized electromagnetic field.

In order to obtain the particle description required for quantum statistics, it may therefore be necessary to quantize the quantum-mechanical wave field a second time. This procedure, known as *second quantization*, starts from the wave field once quantized:

$$-\frac{\hbar^2}{2m}\nabla^2\psi + V(\boldsymbol{q})\psi = i\hbar\dot{\psi}$$

$$-\frac{\hbar^2}{2m}\nabla^2\psi^* + V(\boldsymbol{q})\psi^* = -i\hbar\dot{\psi}^*$$

treating ψ and ψ^* as independent variables.

The two field equations may be generated from the Lagrangian density

$$L = \int \psi^* \left\{ i\hbar\dot{\psi} - V(\boldsymbol{q})\psi + \frac{\hbar^2}{2m}\nabla^2\psi \right\} d^3q \tag{8.32}$$

noting that the Lagrangian equations

$$\frac{d}{dt}\frac{\delta L}{\delta\dot{\psi}^*} - \frac{\delta L}{\delta\psi^*} = -\left\{ i\hbar\dot{\psi} - V(\boldsymbol{q})\psi + \frac{\hbar^2}{2m}\nabla^2\psi \right\} = 0$$

immediately lead to the first of the field equations again.

A canonically conjugate momentum may be obtained from (32) in the usual way as the derivative of the Lagrangian,

$$\pi = \frac{\delta L}{\delta\dot{\psi}} = i\hbar\psi^*$$

The Hamiltonian is derived from the Lagrangian by applying the rule

$$\begin{aligned} H &= \int (\pi\dot{\psi} - L)d^3q \\ &= \int \left\{ i\hbar\dot{\psi}\psi^* - i\hbar\psi^*\dot{\psi} - \psi^*\frac{\hbar^2}{2m}\nabla^2\psi + \psi^* V\psi \right\} \psi(\boldsymbol{q})d^3q \end{aligned}$$

substituting from (32). The first two terms cancel and hence

$$H = \int \psi^*(\boldsymbol{q}) \left\{ -\frac{\hbar^2}{2m}\nabla^2 + V(\boldsymbol{q}) \right\} \psi(\boldsymbol{q})d^3q \tag{8.33}$$

which is of the same form as the expectation value of energy in Schrödinger wave mechanics. It is important to note that this Hamiltonian (33) is not an operator and should be treated like a classical variable. In the same way, the functions ψ and ψ^* are treated as classical fields.

Since these field amplitudes are known to be solutions of Schrödinger's equation, the solutions of

$$\left(-\frac{\hbar^2}{2m}\nabla^2 + V \right) \psi_\mu = i\hbar\dot{\psi}_\mu$$

in the form

$$\psi_\mu = e^{-iE_\mu t/\hbar}\varphi(\boldsymbol{q})$$

may be used as the basis for their expansion. If the time factor is included with the coefficients of the expansion

$$b_\mu(t) = \sum_\mu b(0)e^{-iE_\mu t/\hbar}$$

the wave functions are

$$\begin{aligned} \psi(\boldsymbol{q}) &= \sum_\mu b_\mu \varphi_\mu(\boldsymbol{q}) \\ \psi^*(\boldsymbol{q}) &= \sum_\mu b_\mu^\dagger \varphi_\mu(\boldsymbol{q}) \end{aligned}$$

When the field is quantized, each term in the expansion may be considered as representing a particle. Substituting for ψ and ψ^* into (33) gives a new type of Schrödinger equation:

$$H\Phi = \left(\sum_\mu b_\mu^\dagger b_\mu E_\mu \right) \Phi = E\Phi$$

using the orthonormal property of the φ_μ. The product $b_\mu^\dagger b_\mu = n_\mu$, known as the *number operator*, defines the number of energy quanta E_μ.

Starting from the standard Heisenberg commutator $[p,q] = -\hbar i$, it is readily shown [119] that the coefficients satisfy the following commutation rules

$$[b_\mu^\dagger, b_\nu] = \delta_{\mu\nu} \quad , \quad [b_\mu, b_\nu] = 0 \quad , \quad [b_\mu^\dagger, b_\nu^\dagger] = 0 \tag{8.34}$$

A single particle may now be described by a Schrödinger equation

$$\left(b^\dagger b E\right)_\mu \psi = E\psi$$

This equation is now multiplied on the left by $b^\dagger$, *i.e.*

$$b^\dagger(b^\dagger b E_\mu)\psi = E b^\dagger \psi$$

Using (34):

$$b^\dagger b = (bb^\dagger - 1) \quad , i.e.$$
$$(bb^\dagger - 1)b^\dagger E_\mu \psi = E b^\dagger \psi$$
$$bb^\dagger E_\mu(b^\dagger \psi) = (E + E_\mu)(b^\dagger \psi)$$

Hence $b^\dagger\psi$ is a new eigenfunction with eigenvalue $(E + E_\mu)$. By applying the operator $b^\dagger$ n times, starting from the vacuum state (ψ_0, E_0), the nth excited eigenfunction

$$\psi_n = (b^\dagger)^n \psi_0$$

is reached. In normalized form

$$\psi_n = N_n (b^\dagger)^n \psi_0$$

It will be shown by induction that the normalization factor $N_n = 1/\sqrt{n!}$. The procedure assumes that, if the condition is valid for some $n = n_0$ it can also be shown to be valid for $n = n_0 + 1$. Since the condition is clearly valid for $n = 0$, $(0! = 1)$ it would then be valid for all n.

To complete the proof, put

$$\psi_{n_0+1} = c_{n_0+1} b^\dagger \psi_{n_0}$$

where c_{n_0} is a constant to be determined, and that ψ_{n_0} is thereby normalized.

$$\begin{aligned}\langle \psi_{n_0+1} | \psi_{n_0+1} \rangle &= |c_{n_0+1}|^2 \langle b^\dagger \psi_{n_0} | b^\dagger \psi_{n_0} \rangle = 1 \\ &= |c_{n_0+1}|^2 \langle \psi_{n_0} | bb^\dagger \psi_{n_0} \rangle\end{aligned}$$

From (34) it then follows that

$$|c_{n_0+1}|^2 \langle \psi_{n_0+1} | (b^\dagger b + 1) \psi_{n_0+1} \rangle = 1$$

Finally, since ψ_{n_0} is an eigenfunction of the operator $E_{n_0}b^\dagger b$, it follows, as before, that

$$|c_{n_0+1}|^2(n_0+1)\langle\psi_{n_0}|\psi_{n_0}\rangle = 1$$

and

$$c_{n_0+1} = \frac{1}{\sqrt{n_0+1}}$$

Substituting back for $n = n_0$ one has

$$\psi_{n_0+1} = \frac{1}{\sqrt{(n_0+1)!}}(b^\dagger)^{n_0+1}\psi_0$$

to complete the proof.

By subjecting the amplitudes $b^\dagger, b$ to the commutation relationships (34) the classical Hamiltonian

$$H = \sum_\mu E_\mu b_\mu^\dagger b_\mu$$

is thus transformed into a Hamiltonian operator

$$H = \int \psi^*(\boldsymbol{q})\left\{-\frac{\hbar^2}{2m}\nabla^2 + V(\boldsymbol{q})\right\}\psi(\boldsymbol{q})d^3q$$

that acts on an eigenfunction

$$\Phi = \prod_\mu \frac{1}{\sqrt{n_\mu!}}(b^\dagger)^{n_\mu}\Phi_0$$

producing a Schrödinger equation $H\Phi = E\Phi$, with $E = \sum_\mu E_\mu n_\mu$. The *vacuum* state is defined by $b_\mu\Phi_0 = 0$, for all μ.

The operator $b^\dagger$ applied n times is seen to generate the nth excited eigenfunction from the ground state,

$$\psi_n = \frac{1}{\sqrt{n!}}(b^\dagger)^n\psi_0$$

and to multiply the energy n-fold. This may be interpreted to mean that n energy quanta were generated and $b^\dagger$ is called a *creation operator*. When b is applied to ψ a quantum is destroyed, so that b is an *annihilation operator*.

In summary, a series of energy levels E_μ is determined by the original Schrödinger equation. These energy levels may be occupied by the n_μ quanta, or *particles*. This conclusion shows that field quantization guarantees the corpuscular character even in the case of the Schrödinger wave field.

Second quantization transforms the Schrödinger particle density into a *particle density operator*,

$$\varrho(\boldsymbol{q}) = \psi^\dagger(\boldsymbol{q})\psi(\boldsymbol{q})$$

The corresponding expectation value for the particle density is then given by

$$\overline{\varrho(\boldsymbol{q})} = \langle\Phi|\psi^\dagger(\boldsymbol{q})\psi(\boldsymbol{q})|\Phi\rangle$$

In the case of a single particle this expression reduces to the usual Schrödinger density.

The fact that every state may be occupied by several particles shows that the second quantization particles are bosons. However, in terms of different commutation relations an equivalent scheme may be obtained for fermions. To achieve this objective the wave functions are written in decomposed form as before:

$$\psi(\boldsymbol{q}) = \sum_\mu a_\mu \varphi_\mu(\boldsymbol{q})$$

$$\psi^\dagger(\boldsymbol{q}) = \sum_\mu a_\mu^\dagger \varphi_\mu^*(\boldsymbol{q})$$

in terms of the eigenfunctions $\varphi_\mu(\boldsymbol{q})$ in first quantization.

To establish the appropriate commutation rules the new creation operator $a^\dagger$ is applied twice to the vacuum state Φ_0 in order to create two particles in the same state μ by forming $a_\mu^\dagger a_\mu^\dagger \Phi_0$. Since this two-particle state is not allowed, it is required that

$$a_\mu^\dagger a_\mu^\dagger \Phi_0 = 0$$

This condition applies not only to the vacuum state, but to any arbitrary state Φ which can only mean that $a_\mu^\dagger a_\mu^\dagger = 0$. To incorporate this condition into the previous set of commutation rules for Bose fields, it is sufficient to change the negative into positive signs, such that

$$\begin{aligned} a_\mu^\dagger a_\nu^\dagger + a_\nu^\dagger a_\mu^\dagger &= 0 \\ a_\mu^\dagger a_\nu + a_\nu a_\mu^\dagger &= \delta_{\mu\nu} \\ a_\mu a_\nu + a_\nu a_\mu &= 0 \end{aligned}$$

Commutation relations of this type are written with + signs, *i.e.*

$$[A, B]_+ = AB + BA$$

Particles whose creation and annihilation operators satisfy these relationships are called *fermions*. It is found that [119] these commutation relations lead to wave functions in space that are antisymmetric.

The Hamiltonian operator is once more given by

$$\begin{aligned} H &= \int \psi^\dagger(\boldsymbol{q})\left\{-\frac{\hbar^2}{2m}\nabla^2 + V(\boldsymbol{q})\right\}\psi(\boldsymbol{q})d^3q \\ &\equiv \sum_\mu E_\mu a_\mu^\dagger a_\mu \\ H\Phi &= E\Phi \end{aligned}$$

For a ground state with the assumed property $a_\mu \Phi_0 = 0$ for all μ, the eigenfunctions are

$$\Phi_{\{n\}} = \prod_\mu (a^\dagger)^{n_\mu}\Phi_0$$

Since $(a^\dagger)^2 = 0$, it follows that the *occupation number* n may only be 0 or 1.

8.4.2 The Density Matrix

The quantum-mechanical equivalent of phase density is known as the *density matrix* or *density operator*. It is best understood in the case of a mixed ensemble whose systems are not all in the same quantum state, as for a pure ensemble.

The density operator (matrix) is Hermitian and for an arbitrary countable basis may be represented by a square matrix, that may be infinite, and with elements

$$\varrho_{mn} = \frac{1}{\nu}\sum_{i=1}^{\nu} a_i^*(m,t)a_i(n,t) = \overline{a^* a_n} \tag{8.35}$$

The diagonal element of the density matrix, $W(n) = \overline{a_n^* a_n}$ is the probability that a system chosen at random from the ensemble occurs in the state characterized by n, and implies the normalization

$$\sum_n \overline{a_n^* a_n} = \text{trace } \boldsymbol{\varrho} = 1$$

The expectation value of the operator $\boldsymbol{A}$ is defined by

$$\overline{A} = \sum_{m,n} A_{mn}\overline{a_m^* a_n}$$

or in matrix notation

$$\overline{A} = \text{trace } \boldsymbol{A\varrho} = \text{trace } \boldsymbol{\varrho A} \tag{8.36}$$

Integration of the phase density over classical phase space corresponds to finding the trace of the density matrix in quantum mechanics. Transition to a new basis is achieved by unitary transformation

$$\begin{aligned}\boldsymbol{\varrho}_b &= \boldsymbol{U}^{-1}\boldsymbol{\varrho}_a\boldsymbol{U}\\ \boldsymbol{F}_b\boldsymbol{\varrho}_b &= \boldsymbol{U}^{-1}\boldsymbol{F}_a\boldsymbol{\varrho}_a\boldsymbol{U}\end{aligned}$$

Invariance of the trace of a matrix under unitary transformation corresponds to the invariance of phase density under canonical transformation in classical theory.

The density matrix is defined in second quantization as an operator by an equation

$$\varrho_{mn} = \int \psi^*(m,\boldsymbol{q})\boldsymbol{\varrho}\psi(n,\boldsymbol{q})d\boldsymbol{q}$$

For a pure ensemble the elements of the density matrix are

$$\varrho_{mn} = a_m^* a_n$$

and hence

$$\sum_k \varrho_{mk}\varrho_{kn} = \sum_k a_m^* a_k a_k^* a_n = a^* a_n$$

or

$$\boldsymbol{\varrho}\boldsymbol{\varrho} = \boldsymbol{\varrho}$$

Suppose that the density matrix has been brought to diagonal form, whereby it is immediately apparent that all its eigenvalues should be either zero or one. More precisely, since tr$\boldsymbol{\varrho}$=1, only one of the eigenvalues can be equal to one and all the others are zero.

As defined by equation (35) the density matrix in Schrödinger presentation depends explicitly on the time. The time derivatives of the matrix elements are

$$\frac{\partial \varrho_{mn}}{\partial t} = \frac{1}{\nu}\sum_{i=1}^{\nu}\left[\frac{\partial a_i^*(m,t)}{\partial t}a_i(n,t) + \frac{\partial a_i(n,t)}{\partial t}a_i^*(m,t)\right]$$

Using equation (29) this equation becomes

$$\begin{aligned}\frac{\partial \varrho_{mn}}{\partial t} &= \frac{1}{\nu}\left[\frac{2\pi i}{h}H_{mi}a_i^*(m,t)\cdot a_i(n,t) - \frac{2\pi i}{h}H_{ni}a_i(n,t)\cdot a_i^*(m,t)\right]\\ &= \sum_{i=1}^{\nu}\frac{2\pi i}{h}\left[H_{mi}\varrho_{in} - \varrho_{mi}H_{in}\right]\end{aligned}$$

In terms of the matrix, or operator, $\boldsymbol{\varrho}$,

$$\frac{\partial \boldsymbol{\varrho}}{\partial t} = -\frac{2\pi}{h}[\boldsymbol{\varrho}, \boldsymbol{H}] \tag{8.37}$$

Equation (37) is the quantum statistical analogue of Liouville's equation. To find the quantum analogue of the classical principle of conservation of phase density the solution to (37) is written in the form

$$\boldsymbol{\varrho}_S(t) = T\boldsymbol{\varrho}_S(t_0)T^\dagger$$

In the Heisenberg picture, obtained by unitary transformation

$$\boldsymbol{\varrho}_H(t) = T^\dagger\boldsymbol{\varrho}_S(t)T = \boldsymbol{\varrho}_S(t_0)$$

and thus

$$\frac{d\boldsymbol{\varrho}_H}{dt} = 0$$

If the averages defined by (36) are independent of the time, the ensemble is in statistical equilibrium and $\partial\boldsymbol{\varrho}/\partial t = 0$. As a sufficient condition for equilibrium it is, according to (37) therefore necessary that $[\boldsymbol{\varrho}, \boldsymbol{H}] = 0$. In general therefore, an ensemble is in statistical equilibrium if the density operator commutes with the Hamiltonian.

For conservative systems with time-independent Hamiltonian the density operator may be defined as a function of one or more quantum-mechanical operators $\boldsymbol{A}$, *i.e.* $\boldsymbol{\varrho} = \varphi(\boldsymbol{A})$. This definition implies that for statistical equilibrium of an ensemble of conservative systems, the density operator depends only on constants of the motion. The most important case is $\boldsymbol{\varrho} = \varphi(\boldsymbol{H})$. If the eigenvectors of the Hamiltonian are chosen as a basis, then

$$H_{mn} = E_n\delta_{mn}$$

where the E_n are the eigenvalues. It follows that $\varrho_{mn} = \varphi(E_n)\delta_{mn}$ and the density matrix will be diagonal in the energy presentation.

8.4.3 Macroscopic Variables

As for classical systems, measurement of the properties of macroscopic quantum systems is subject to experimental error that exceeds the quantum-mechanical uncertainty. For two measurable quantities F and G the inequality is defined as $\Delta F\Delta G >> \delta F\delta G$.The state vector of a completely closed system described by a time-independent Hamiltonian $\boldsymbol{H}$, with eigenvalues E_n and eigenfunctions ψ_n, is represented by

$$\Psi(\boldsymbol{q}, t) = \sum_n a_n\psi_n e^{-2\pi iEt/h}$$

where $\sum_n |a_n|^2 = 1$ and the ψ_n are either symmetric or anti-symmetric. The eigenvalues will be ordered in the sequence $0 < E_1 < E_2 < \ldots$. For many degrees of freedom the spectrum is a dense quasi-continuum and many eigenvalues may be grouped together into groups $[\Delta E]_K$ covering domains of values within the experimental uncertainty. Such a group of eigenvalues may be likened to classical energy shells of the type shown in figure 5.

It should be possible to distinguish between eigenvalues of different energy shells, but not between those of the same shell, by a macroscopic measurement that corresponds to a diagonal matrix whose elements for all ψ_n of a shell $[\Delta E]_K$ are equal to an average value E_K. The associated operator $\{H\}$ is defined with the same accuracy with which the measurement can be carried out.

The previous argument is valid for all observables, each represented by a characteristic operator $\{X\}$ with experimental uncertainty ΔX. The problem is to identify an elementary cell within the energy shell, to be consistent with the macroscopic operators. This cell would constitute a linear sub-space over the Hilbert space in which all operators commute with the Hamiltonian. In principle each operator may be diagonalized by unitary transformation and only those elements within a narrow range along the diagonal that represents the minimum uncertainties would differ perceptibly from zero.

If the small regions of this narrow strip that fall outside the shells $[\Delta E]_N$ near the junctions are ignored, $\{X\}$ is seen to share all its eigenvectors with $\{H\}$.

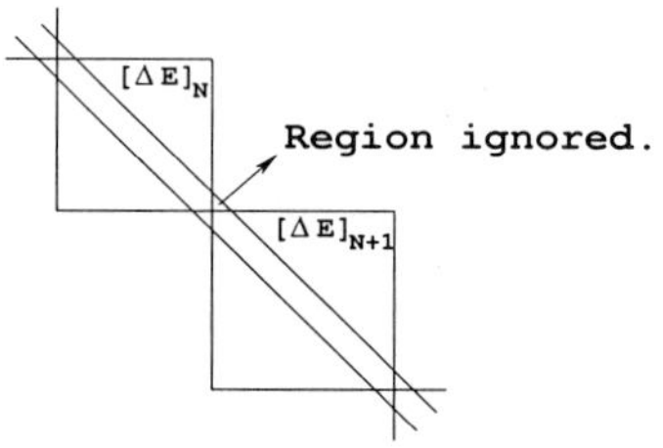

Figure 7: *Diagram to explain the definition of a macroscopic operator.*

Several macroscopic operators $\{H\}$, $\{F\}$, $\{G\}$, corresponding to different macroscopic variables would automatically decompose the energy shells into maximal subspaces or *cells* of dimension ω_ν, that represent eigenvalues for all macroscopic operators. For M cells per shell it follows that

$$\sum_{\nu=1}^{M} \omega_\nu = \Omega$$

Choosing as orthonormal basis $\psi_{i\nu} (i = 1, \ldots, \omega_\nu; \nu = 1, \ldots, M)$ in the energy shell with

$$\langle \psi_{i\nu} | \psi_{j\mu} \rangle = \delta_{ij} \delta_{\nu\mu}$$

ensures that the $\psi_{i,\nu}$ exactly span the cell, being eigenfunctions of all macroscopic operators.

In the cells thus defined, all macroscopic operators would simultaneously have sharply defined values and hence provide the quantum analogue of the classical cells in Γ-space.

8.4.4 Axioms of the Theory

The implimentation of quantum statistical ensemble theory applied to physically real systems presents the same problems as in the classical case. The fundamental questions of how to define macroscopic equilibrium and how to construct the density matrix remain. The ergodic theory and the hypothesis of equal *a priori* probabilities again serve to forge some link between the theory and working models.

According to ergodic theory macroscopic equilibrium properties are identical to the time average of the corresponding observables, ideally extended over infinite times. In principle the average could be calculated by solving the Schrödinger equation for the system, which is not practically feasible. The quantum ergodic hypothesis [120] states that every state inside the energy shell $[\Delta E]_K$ of a closed system can be reached from any other state in the same energy shell, either directly or through other states. The necessary perturbation of the Hamiltonian to allow the transition to take place may be effected through either internal interactions or through the electromagnetic field.

Transition between states of probabilities p_i and p_j is assumed to obey Fermi's *master equation*

$$\frac{dp_i}{dt} = \sum_j w_{ij}(p_j - p_i) \tag{8.38}$$

with transition probabilities $w_{ij} = w_{ji}$.

The stationary solution of the equation is

$$p_i = \Omega^{-1}$$

Thus, in equilibrium the system spends on the average equal times in each of the Ω states. The calculation of the time average can therefore be replaced by averaging over the quantum statistical microcanonical ensemble. However, as in classical theory, equating time average with microcanonical ensemble average remains conjectural.

Construction of the density operator can also not be achieved without assumption of an additional axiom: All quantum states of a system compatible with the knowledge revealed by macroscopic measurement have equal *a priori* probabilities and random *a priori* phases. This axiom implies that for a system as defined above all diagonal elements of the density matrix ϱ belonging to the ith cell must be equal. Hence

$$\overline{|b_{ij}|^2} = \varrho_{jj} = \varrho_{kk} = \overline{|b_{ik}|^2} \quad (j, k = 1, \ldots, \omega_i)$$

For the off-diagonal elements in this cell it is generally valid that

$$\varrho_{jk} = \frac{1}{\nu} \sum b_{ji} b_{ik} = \frac{1}{\nu} \sum r_j r_k e^{i(\alpha_j - \alpha_k)} \quad j \neq k$$

where r_j is the modulus and α_j the phase of b_{ij}. The assumption of random *a priori* phases then states that the phases are so distributed over the system of the ensemble that the average of the off-diagonal density matrix vanishes.

In the case of an equilibrium system the Hamiltonian is the same as that of an ensemble of conservative systems in statistical equilibrium. If the energy of the system is measured to lie between E_K and $E_K + \Delta E$, then the representative ensemble is also restricted to the energy shell $[\Delta E]_K$. From the hypotheses of equal *a priori* probabilities and random *a priori* phases it then follows that the diagonal elements of the density matrix lying inside $[\Delta E]_K$ are all equal and that all others vanish. The density matrix of the quantum statistical microcanonical ensemble is thereby determined as

$$\begin{aligned} \varrho_{nm} &= \Omega^{-1}\delta_{nm} \quad \text{for } E_K \leq E_n \leq E_K + \Delta E \\ \varrho_{nm} &= 0 \quad \text{in all other cases.} \end{aligned}$$

8.4.5 Symmetry Considerations

The major difference between classical and quantum mechanical ensembles arises from the symmetry properties of wave functions which is not an issue in classical systems.

The principles involved are illustrated well by a system consisting of two particles. In the first instance it may be assumed that the particles are so weakly interacting that their respective wave functions do not overlap significantly. The Hamiltonian may then be assumed to consist of two parts

$$H = H_1 + H_2$$

and the wave equation $H\psi = E\psi$ may be solved by separation of variables. The total energy $E = \epsilon_1 + \epsilon_2$, and the eigenfunction solution $\psi_1(\epsilon_1)$ will have large values at coordinates where $\psi_2(\epsilon_2)$ is small, and *vice versa.* The state of energy E, which may be represented by either $\psi_a = \psi_1(\epsilon_1)\psi_2(\epsilon_2)$ or $\psi_b = \psi_1(\epsilon_2)\psi_2(\epsilon_1)$, is therefore degenerate because of *exchange degeneracy.* According to this description the particles are distinguishable in the classical sense. Examples are atoms on a crystal lattice or molecules in a dilute gas.

In the more general situation individual quantum-mechanical particles cannot be distinguished in terms of locality and the appropriate solutions are best defined as linear combinations, either of the symmetric type

$$\psi_S = \psi_a + \psi_b = \frac{1}{\sqrt{2}}\left[\psi_1(\epsilon_1)\psi_2(\epsilon_2) + \psi_2(\epsilon_1)\psi_1(\epsilon_2)\right]$$

or the antisymmetric type

$$\psi_A = \psi_a - \psi_b = \frac{1}{\sqrt{2}}\begin{vmatrix} \psi_1(\epsilon_1) & \psi_2(\epsilon_2) \\ \psi_1(\epsilon_2) & \psi_2(\epsilon_2) \end{vmatrix}$$

Since the Hamiltonian is symmetric in space coordinates the time-dependent Schrödinger equation prevents a system of identical particles in a symmetric state from passing into an anti-symmetric state. The symmetry character of the eigenfunctions therefore is a property of the particles themselves. Only one eigenfunction corresponds to each eigenfunction and hence there is no exchange degeneracy.

ψ_A and ψ_B cannot simultaneously be valid functions for a system of particles. Had they been, their linear combination would also satisfy the wave equation, and that would produce once more, the wave functions ψ_a or ψ_b, characteristic of distinguishable particles. Hence for indistinguishable particles, as occur in wave mechanics, the wave function must be either symmetrical or anti-symmetrical. It is well known that (elementary or composite) particles with integral spin have symmetric wave functions and those with half-integral spin have anti-symmetric wave functions.

It is noted that if $\epsilon_1 = \epsilon_2$ the anti-symmetric wave function vanishes, $\psi_a = 0$. Two identical particles with half-spin can hence not be in the same non-degenerate energy state. This conclusion reflects Pauli's principle. Particles with integral spin are not subject to the exclusion principle ($\psi_S \neq 0$) and two or more particles may occur in the same energy state.

The argument may be extended to degenerate energy levels. Suppose there are three distinct wave functions corresponding to particle I at energy ϵ_1, *i.e.*

$$\psi_I^{(1)}(\epsilon_1) \quad , \quad \psi_I^{(2)}(\epsilon_1) \quad , \quad \psi_I^{(3)}(\epsilon_1)$$

and for particle II,

$$\psi_{II}^{(1)}(\epsilon_1) \quad , \quad \psi_{II}^{(2)}(\epsilon_1) \quad , \quad \psi_{II}^{(3)}(\epsilon_1)$$

Three independent anti-symmetric wave functions can be written for these two particles:

$$\psi_a = \begin{vmatrix} \psi_I^{(1)} & \psi_{II}^{(1)} \\ \psi_I^{(2)} & \psi_{II}^{(2)} \end{vmatrix}$$

$$\psi_b = \begin{vmatrix} \psi_I^{(2)} & \psi_{II}^{(2)} \\ \psi_I^{(3)} & \psi_{II}^{(3)} \end{vmatrix}$$

$$\psi_c = \begin{vmatrix} \psi_I^{(3)} & \psi_{II}^{(3)} \\ \psi_I^{(1)} & \psi_{II}^{(1)} \end{vmatrix}$$

The total number of independent wave functions for the two-particle system is therefore given by the number of combinations of the two upper suffixes from the total of three.

In a more general case there may be N_i particles at the energy level ϵ_i, corresponding to g_i independent wave functions. A typical wave function for the system is represented by a determinant of order N_i,

$$\Psi(\boldsymbol{q}) = \begin{vmatrix} \psi_I^{(1)} & \psi_{II}^{(1)} & \dots & \psi_{N_i}^{(1)} \\ \psi_I^{(2)} & \psi_{II}^{(2)} & \dots & \psi_{N_i}^{(2)} \\ \dots & \dots & \dots & \dots \\ \psi_I^{(N_i)} & \psi_{II}^{(N_i)} & \dots & \psi_{N_i}^{(N_i)} \end{vmatrix} = \Psi(\epsilon_i)$$

According to Pauli's principle the number of upper suffixes $(1), \dots, (N_i)$ must be less than g_i. The number of independent wave functions that may be formulated in this way must be equal to the number of ways of choosing N_i different upper suffixes from a total of g_i.

If the energy levels of a large system, consisting of N particles, are denoted by

$$\epsilon_0, \epsilon_1, \dots, \epsilon_i, \dots$$

the statistical state of the system is specified by a set of positive integers $N_0, N_1, \dots, N_i, \dots,$ denoting the number of particles at each energy level. Conservation principles demand that

$$N = \sum_i N_i \quad , \quad E = \sum E_i = \sum_i N_i \epsilon_i$$

Let each elemental particle (i) be described by a wave function $\psi_j(\boldsymbol{q})_i$ $(j = k, s; i = 1, n)$. Each microstate of the system has a wave function

$$\Psi_\mu(\boldsymbol{q}_1, \boldsymbol{q}_2, \dots, \boldsymbol{q}_n)$$

with allowed energy level $E_\mu = \sum_j E_j$. The different energy states within the energy level have different wave functions. The statistical description of a system must therefore depend on the number of particles allowed in each state, and this is dictated by symmetry.

The simplest situation, known as the quantized Maxwell-Boltzmann distribution occurs with a system of N distinguishable particles, with eigenfunction

$$\Psi_\mu = \prod_{j,i} \psi_j(\boldsymbol{q}_i)$$

Any interchange of particles (i) among the energy levels (j) leads to a new Ψ for the whole system. With g_i states available to energy level ϵ_i each of the N_i molecules has a choice of g_i states. The number of eigenfunctions corresponding to the eigenvalue ϵ_i is then given by [118]

$$\Omega_{MB} = N_i! \prod_i \frac{g_i^{N_i}}{N_i!}$$

The distribution is valid for weakly interacting heavy molecules at high temperatures.

For non-localized indistinguishable particles the eigenfunctions are generalizations of the linear combinations ψ_S and ψ_A. For particles with integral spins

$$\Psi_\mu = \sum_P \boldsymbol{P}\psi_j(\boldsymbol{q}_i)$$

where P denotes permutation. The number of possible arrangements may be visualized as follows.

At the energy level ϵ_i there are g_i ways to start any sequence - one particle for each of the g_i states. In each of these sequences the remaining $N_i - 1$ particles can be arranged in any order, or permutation of $(g_i + N_i - 1)$ objects. The number of ways in which N_i particles can be arranged is $g_i(g_i + N_i - 1)!$. Since g_i states can be arranged in any order, the previous result must be divided by $g_i!$. The particles are indistinguishable and hence no other new arrangements are possible. The number of possible symmetric eigenfunctions correspondind to ϵ_i are thus given by the Bose-Einstein distribution

$$\Omega_{BE} = \prod_i \frac{(g_i + N_i - 1)!}{(g_i - 1)!N!}$$

For indistinguishable particles that obey Pauli's priciple, the wave function $\Psi_\mu(\boldsymbol{q})$ takes the form

$$\Psi_\mu = \sum_P \mp P\psi_j(\boldsymbol{q})$$

where - and + stand for odd and even permutations respectively. The wave function vanishes identically whenever two rows or columns of the determinantal wave function are equal. Hence, for g_i states at energy level ϵ_i there are $g_i!$ possible arrangements, of which $g_i - N_i$ are empty. The number of anti-symmetric eigenfunctions for all energy levels is given by the Fermi-Dirac distribution

$$\Omega_{FD} = \prod_i \frac{g_i!}{N_i!(g_i - N_i)!}$$

It should be noted that in case of $g_i >> N_i$, both Ω_{BE} and Ω_{FD} reduce to Ω_{MB}. It can therefore be concluded that classical statistics is a limiting case of quantum statistics.

To give thermodynamic meaning to the distribution functions it is necessary to establish the distribution for which the number of possible arrangements, interpreted as thermodynamic probability, is a maximum: Assuming that all of the variables g_i, N_i, $g_i - N_i >> 1$, the application of Stirling's formula to the probabilities defined above, leads to

$$\begin{aligned}
\ln \Omega_{MB} &= \sum_i N_i \left[\ln \frac{g_i}{N_i} + 1 \right] \\
\ln \Omega_{BE} &= \sum_i \left[(N_i + g_i) \ln \left(\frac{g_i}{N_i} + 1 \right) - g_i \ln \frac{g_i}{N_i} \right] \\
\ln \Omega_{FD} &= \sum_i \left[(N_i - g_i) \ln \left(\frac{g_i}{N_i} - 1 \right) + g_i \ln \frac{g_i}{N_i} \right]
\end{aligned}$$

For given E and N the maxima are obtained by setting

$$\delta \ln \Omega_{MB} = \sum_i \ln \frac{g_i}{N_i} \delta N_i = 0$$

$$\delta \ln \Omega_{BE} = \sum_i \ln \left(\frac{g_i}{N_i} + 1 \right) \delta N_i = 0$$

$$\delta \ln \Omega_{FD} = \sum_i \ln \left(\frac{g_i}{N_i} - 1 \right) \delta N_i = 0$$

subject to

$$\delta N = \sum_i \delta N_i = 0$$

$$\delta E = \sum_i \epsilon_i \delta N_i = 0$$

By the method of Lagrangian multipliers (α and β in the following) it is found in all cases for arbitrary δN_i that

$$\delta \ln \Omega + \ln \alpha \sum_i \delta N_i - \beta \sum_i \epsilon_i \delta N_i = 0$$

This relationship defines the following three equilibrium quantum distributions

$$\text{Maxwell-Boltzmann} \quad : \quad N_i = g_i \alpha e^{-\beta \epsilon_i}$$

$$\text{Bose-Einstein} \quad : \quad N_i = \frac{g_i \alpha e^{-\beta \epsilon_i}}{1 - \alpha e^{-\beta \epsilon_i}}$$

$$\text{Fermi-Dirac} \quad : \quad N_i = \frac{g_i \alpha e^{-\beta \epsilon_i}}{1 + \alpha e^{-\beta \epsilon_i}}$$

8.4.6 The Microcanonical Ensemble

The microcanonical ensemble in quantum statistics describes a macroscopically closed system in a state of thermodynamic equilibrium. It is assumed that the energy, number of particles and the extensive parameters are known. The Hamiltonian may be defined as

$$\boldsymbol{H} = \boldsymbol{H}_0 + \boldsymbol{H}'$$

where $\boldsymbol{H}'$ is a perturbation operator to compensate for the fact that the system is not closed in the quantum-mechanical sense. The macroscopic energy condition is interpreted to mean that all matrix elements of the perturbation operator that lie outside the interval E to $E + \Delta E$ must vanish. Such a system will be represented by a microcanonical ensemble with density operator of elements

$$\varrho_{nm} = \begin{cases} \Omega^{-1}\delta_{nm} & \text{for } E \leq E_n \leq E + \Delta E \\ 0 & \text{in all other cases.} \end{cases}$$

In this interpretation Ω is the number of linearly independent eigenfunctions of the unperturbed Hamiltonian in the interval ΔE. From (36) the microcanonical average of an observable represented by the operator $\boldsymbol{A}$ in an arbitrary basis, is

$$\overline{A} = \Omega^{-1} \text{trace } \boldsymbol{A} \tag{8.39}$$

This equation may be used to derive the quantum mechanical virial theorem. For this purpose it is necessary to define the kinetic operator

$$\boldsymbol{E}_{kin} = \sum_{i=1}^{N} \frac{p^2}{2m_i} = -\frac{h^2}{8\pi^2 m_i} \nabla_i^2$$

and the virial operator

$$[\boldsymbol{V}] = \frac{1}{2}\sum_{i=1}^{N} \boldsymbol{q}_i \cdot \frac{\partial U}{\partial \boldsymbol{q}_i}$$

where U denotes potential energy. The virial theorem states that

$$\bar{E}_{kin} = \overline{[V]}$$

or, because of (39)

$$\text{trace } \boldsymbol{E}_{kin} = \text{trace}[\boldsymbol{V}]$$

The proof is based on the fact that the diagonal elements of the commutator $[\boldsymbol{A}, \boldsymbol{H}]$ vanish in the basis of energy eigenfunctions. Because of the Hermitian properties of $\boldsymbol{A}$

$$\int \psi_m^*[\boldsymbol{A}, \boldsymbol{H}]\psi_n d\boldsymbol{q} = (E_m - E_n)\int \psi_m^* \boldsymbol{A}\psi_n d\boldsymbol{q}$$

As a special case

$$\int \psi_m[\boldsymbol{q_i} \cdot \boldsymbol{p_i}, \boldsymbol{H}]\psi_m d\boldsymbol{q} = 0$$

which is simplified by the following commutator identities [121]

$$\begin{aligned}
[\boldsymbol{q_i} \cdot \boldsymbol{p_i}, \boldsymbol{H}] &= [\boldsymbol{q_i}, \boldsymbol{H}]\boldsymbol{p_i} + \boldsymbol{q_i}[\boldsymbol{p_i}, \boldsymbol{H}] \\
[\boldsymbol{q_i}, \boldsymbol{H}] &= \frac{ih}{2\pi m_i}\boldsymbol{p}_i \\
[\boldsymbol{p_i}, \boldsymbol{H}] &= -\frac{ih}{2\pi}\frac{\partial U}{\partial \boldsymbol{q}_i}
\end{aligned}$$

where U is the total potential energy operator of the system.

From these identities it follows that

$$\int \psi_m^*[\boldsymbol{q_i} \cdot \boldsymbol{p_i}, \boldsymbol{H}]\psi_m d\boldsymbol{q} = \frac{ih}{\pi}\int \psi_m^* \left[\frac{\boldsymbol{p}_i^2}{2m_i} - \frac{1}{2}\boldsymbol{q}_i \frac{\partial U}{\partial \boldsymbol{q}_i}\right]\psi_m d\boldsymbol{q} = 0$$

The term in square brackets implies that

$$(E_{kin})_{mm} = [V]_{mm}$$

to prove the theorem. The result is seen to depend on the properties of the quantum mechanical operators only, independent of statistics. The quantity $-\partial U/\partial q_i$ represents a generalized force component in the direction of the ith coordinate, such as pressure in a gas, which may therefore be calculated directly from the virial condition.

Any generalized force may be represented by an operator

$$\boldsymbol{X}_j = \frac{\partial \boldsymbol{H}}{\partial x_i}$$

where x_i is an extensive parameter. To relate the force to the microcanonical ensemble it is assumed that Ω may be expressed as a continuous differential function of the energy. The energy interval is defined macroscopically and hence the eigenvalues must be so densely distributed as to constitute a continuum. The derivative of the Hamiltonian is defined in terms of the expression

$$\frac{\partial \boldsymbol{H}}{\partial x_j} dx_j = \boldsymbol{H}(x_j + dx_j) - \boldsymbol{H}(x_j)$$

According to (36) the microcanonical average of the generalized force is

$$\overline{X}_j = \Omega^{-1} \text{trace } \boldsymbol{X}_j$$

Hence

$$\overline{X}_j = \Omega^{-1} \text{trace } \boldsymbol{X}_j = (\Omega dx_j)^{-1} [\text{trace } \boldsymbol{H}(x_j + dx_j) - \text{trace } \boldsymbol{H}(x_j)]$$

When $\boldsymbol{H}$ is represented in terms of its eigenfunctions, the trace becomes a sum over eigenvalues, such that

$$\overline{X}_j = \Omega^{-1} \sum_m \frac{\partial E_m}{\partial x_j}$$

or in macroscopic formulation

$$\overline{X}_j = \left(\frac{\partial E}{\partial x_j} \right)_\Omega \tag{8.40}$$

Next, consider an arbitrary change of state under which E and x_j are independent variables and the variable $\Phi = \Phi(E, x_j)$, such that

$$d\Phi = \frac{\partial \Phi}{\partial E} dE + \sum_j \frac{\partial \Phi}{\partial x_j} dx_j$$

Furthermore

$$\sum_j \frac{\partial \Phi}{\partial x_j} = \frac{\partial \Phi}{\partial E} \sum_j \frac{\partial E}{\partial x_j}$$

Using (37) it follows that

$$d\Phi = \frac{\partial \Phi}{\partial E} dE + \frac{\partial \Phi}{\partial E} \sum_j \overline{X}_j dx_j$$

This equation resembles the general equation (3) that describes entropy in terms of the intensive parameters F_k, provided that

$$\frac{\partial\Phi}{\partial E} = F_0 = \frac{1}{T} \quad \text{and} \quad \left(\frac{\partial\Phi}{\partial E}\right)\overline{X}_j = F_k \quad (j = 1, 2, \ldots)$$

Setting $\Phi = \ln\Omega$ produces the statistical definition of entropy,

$$dS \propto d\ln\Omega = \frac{1}{T}dU - \sum_{k=0}^{t} F_k dX_k$$

To match the thermodynamic temperature scale, $S = k\ln\Omega$.

8.4.7 The canonical ensemble

The canonical ensemble was developed as the appropriate description of a system in thermal equilibrium with its surroundings by free exchange of energy. Following the discussion of classical systems the density operator of the canonical ensemble is introduced axiomatically as

$$\varrho = e^{(\psi - H)/\theta}$$

or, more explicitly [115],

$$\varrho = e^{\psi/\theta} \sum_{l=0}^{\infty} \frac{1}{l!}\left(\frac{H}{\theta}\right)^l$$

where all symbols have the same meaning as before. From the normalization of (35), trace $\varrho = 1$, it follows from

$$\varrho = e^{\psi/\theta} \cdot e^{-H/\theta}$$

that

$$\varrho e^{-\psi/\theta} = e^{-H/\theta}$$

and by taking the trace, that

$$e^{-\psi/\theta} = \text{trace } e^{-H/\theta}$$

In the basis of energy eigenfunctions

$$\varrho_{nm} = e^{(\psi - E_n)/\theta}\delta_{nm}$$

The probability of finding a randomly chosen system in an eigenstate characterized by n is

$$W(n) = \varrho_{nn} = e^{(\psi - E_n)/\theta} = 1$$

or

$$e^{-\psi/\theta} = \sum_n e^{-E_n/\theta} = Q = \text{trace } e^{-H/\theta}$$

The last expression is commonly written in the form

$$Q = \text{trace } e^{-\beta H} \tag{8.41}$$

For energy levels E_l with degeneracy Ω_l

$$W(E_l) = \Omega_l e^{(\psi - E_l)/\theta} \quad \text{and} \quad e^{-\psi/\theta} = \sum_l \Omega_l e^{E_l/\theta}$$

As indicated before (8.3.3.2) this formulation demonstrates that the canonical ensemble is made up of a large number of microcanonical ensembles. When calculating properties like the internal energy from the canonical ensemble, the values so obtained may reasonably be expected to be the same as those found from the microcanonical ensemble. Indeed, the two ensembles can be shown to be equivalent for large systems, in which case the sum

$$Q = \sum_l \Omega_l e^{-\beta E_l}$$

may be replaced in the continuum limit by the integral

$$Q = \int_0^\infty \overline{\Omega}(E) e^{-\beta E} dE$$

where $\overline{\Omega}(E)$ is the *density of states.* This integral is the *Laplace transform*[8] of the microcanonical $\Omega(E)$. Since Laplace transforms are unique, the two functions contain identical information.

The function Q is the canonical partition function or sum over states. The eigenvalues E_n are difficult, if not impossible, to obtain. It is significant,

[8] The Laplace transform of a function $f(x)$ is defined by the integral

$$\int_0^\infty e^{-px} f(x) dx$$

if it exists, in which case it is unique [6].

therefore that a canonical ensemble calculation can be carried out independent of the exact solutions of Schrödinger's equation. This fact is understood from the way that traces are independent of the representation of a matrix[9]. Thus, once $\boldsymbol{H}$ is known any complete set of wave functions may be used to compute Q. In other words, $Q = \exp(-\beta A)$ may be calculated without actually solving Schrödinger's equation with the Hamiltonian $\boldsymbol{H}$.

According to (39) the canonical average of an observable $\boldsymbol{A}$ is

$$\overline{A} = \text{trace } \boldsymbol{A}e^{(\psi - H)/\theta}$$

and in the basis of energy eigenfunctions

$$\overline{A} = \sum_n A_{nn}e^{(\psi - E_n)/\theta} = \sum_l A_l\Omega_l e^{(\psi - E_l)/\theta}$$

where A_l is the average over the Ω_l eigenfunctions of level E_l. The canonical average of the energy

$$\overline{E} = \sum_l E_l\Omega_l e^{(\psi - E_l)/\theta}$$

Differentiation of (41), using the formalism defined for the microcanonical ensemble, in terms of a generalized force operator, yields

$$0 = \text{trace}\left[\left(\frac{d(\psi/\theta)}{dx_j} - \frac{1}{\theta}\frac{\partial \boldsymbol{H}}{\partial x_j}\right)e^{(\psi - H)/\theta}\right]$$

or

$$\overline{X}_j = \overline{\left(\frac{\partial H}{\partial x_j}\right)} = \frac{\partial \psi}{\partial x_j}$$

In terms of the total differential of (41)

$$-d(\psi/\theta) = \text{trace}\left[\left(-Hd\left(\frac{1}{\theta}\right) - \frac{1}{\theta}\sum_j \frac{\partial H}{\partial x_j}dx_j\right)e^{(\psi - H)/\theta}\right]$$

which, in terms of $\overline{E}$ and $\overline{X}_j$ reduces to

$$d\psi = \frac{\psi}{\theta}d\theta - \frac{\overline{E}}{\theta} - \sum_j \overline{X}_j dx_j.$$

[9] trace $A =$ trace $SS^{-1}A =$ trace $S^{-1}AS$

Finally, in terms of the canonical average of the probability exponent

$$\overline{\eta} = \frac{\psi - \overline{E}}{\theta}$$

one has

$$d\psi = \overline{\eta} d\theta - \sum_j \overline{X}_j dx_j$$

These equations have the same form as the familiar thermodynamic expressions

$$-S = \frac{F - U}{T}$$

$$dF = -SdT - \sum_j X_j dx_j$$

As in the classical expression (25) the quantity ψ can be inferred directly as representing the statistical analogue of the Helmholtz free energy. The average behaviour of the canonical ensemble thus obeys the laws of thermodynamics.

From the normalization condition the relationship between Helmholtz free energy and the canonical partition function

$$Q = \text{trace } e^{-\beta H}$$

follows immediately as

$$F = -kT \ln Q$$

leading to the expressions for

$$\text{entropy}: \; S = k\left(\ln Q + T\frac{\partial \ln Q}{\partial T}\right)$$

$$\text{internal energy}: \; U = -kT^2\frac{\partial \ln Q}{\partial T}$$

$$\text{and pressure}: \; P = kT\frac{\partial \ln Q}{\partial V}$$

The statistical analogue of entropy for the canonical ensemble follows as $S = -k\overline{\eta}$. Since $\overline{\eta} = \text{trace } \boldsymbol{\eta} e^{\eta}$, the entropy expression becomes

$$S = -k \text{ trace } (\boldsymbol{\varrho} \ln \boldsymbol{\varrho})$$

which is the most general statistical definition that relates the entropy directly to the density matrix.

8.4.8 The grand canonical ensemble

Limitation to ensembles that allow exchange of energy, but not of matter, with their environment is unnecessarily restrictive and unrealistic. What is required is an ensemble for which the particle numbers, N_j also appear as random variables. As pointed out before, the probability that a system has variable particle numbers N_i and occurs in a mechanical state $(\boldsymbol{p}, \boldsymbol{q})$ can not be interpreted as a classical phase density. In quantum statistics the situation is different. Because of second quantization the grand canonical ensemble, like the microcanonical and canonical ensembles, can be represented by means of a density operator in Hilbert space.

In terms of the creation operator $b^\dagger$ of second quantization each energy level has an eigenfunction

$$\psi_n = \frac{1}{\sqrt{n!}}\left(b^\dagger\right)^n \psi_{no}$$

that represents n particles at a given level. The effect of the creation operator is to increase the number of particles in state j by unity, *i.e.*

$$\psi_{N_j} = \frac{1}{\sqrt{N_j}} b^\dagger \psi_{N_j-1} \quad \text{or} \quad \left(b^\dagger\right)^{N_j}_{N_j-1} = \sqrt{N_j} \tag{8.42}$$

In terms of particle states

$$\psi_{no} = \sqrt{\frac{N_j!}{N_j}}\psi_j$$

the state of the system, in terms of the eigenfunctions of the individual particles may then be formulated as

$$\psi(\boldsymbol{q}) = \left(\frac{\prod N_j!}{N!}\right)^{1/2} \sum \boldsymbol{P}\left[\prod \psi_j(\boldsymbol{q}_i)\right] \tag{8.43}$$

where $\boldsymbol{P}$ is again the sequence of permutation operators that interchange the $\boldsymbol{q}_i$.

For a system of fermions the occupation numbers N_j are restricted to values of 0 and 1 and the eigenfunctions are the Slater determinants

$$\psi(\boldsymbol{q}) = (N!)^{-1/2} \sum_P (-1)^{|P|} P \prod \psi_j(\boldsymbol{q}_i) \tag{8.44}$$

The definition of a grand canonical density operator requires that the system Hamiltonian be expressed in terms of creation and annihilation operators.

For a system of N identical particles that interact by two-body central forces

$$\begin{aligned} H &= -\sum_{i=1}^{N}\left[\frac{h^2}{8\pi^2 m}\nabla_i^2 - u(\boldsymbol{q})\right] + \sum_{i>j} u(\boldsymbol{q}_i, \boldsymbol{q}_j) \\ &= H_i^{(1)} + \sum_{i>j} U^{(2)} \end{aligned}$$

where $u(\boldsymbol{q}_i)$ is the potential energy of particle i in an external field.

The operator $H^{(1)}$ may be written as the sum of one-particle operators that act on $\psi(\boldsymbol{q}_i)$ and are symmetric with respect to all particles, $H^{(1)} = \sum_i H_i^{(1)}$. The matrix elements

$$H_{j,k}^{(1)} = \int \psi_j(\boldsymbol{q}_i) H^{(1)} \psi_k(\boldsymbol{q}_i) d\boldsymbol{q}_i$$

in the basis of (43) will be different from zero only for transitions that either leave the particle numbers N_j unchanged, or increase one occupation number by unity as the other decreases by one. Following commutation rules a general term $H^{(1)} = H_{j,k}^{(1)}\sqrt{N_j N_k}$, may therefore be written as $H^{(1)} = H_{j,k}^{(1)} b_j^\dagger b_k$, in terms of creation and annihilation operators.

The interaction term is formulated in terms of an operator which acts simultaneously on functions of two sets of coordinates, $\boldsymbol{q}_i$ and $\boldsymbol{q}_l$,

$$U^{(2)} = \sum_{i>l} f_{il}^{(2)} = \frac{1}{2} \sum_{j,k,m,n} \left[f^{(2)}\right]_{mn}^{jk} b_k^\dagger b_j^\dagger b_m b_n$$

where

$$\left[f^{(2)}\right]_{mn}^{jk} = \int\int \psi_j^*(\boldsymbol{q}_1)\psi_k^*(\boldsymbol{q}_2) f^{(2)} \psi_m(\boldsymbol{q}_1)\psi_n(\boldsymbol{q}_2) d\boldsymbol{q}_1 d\boldsymbol{q}_2$$

Finally

$$\boldsymbol{H} = \sum_{j,k} H_{jk}^{(1)} b_j^\dagger b_k + \frac{1}{2} \sum_{j,k,m,n} \left[U^{(2)}\right]_{mn}^{jk} b_k^\dagger b_j^\dagger b_m b_n$$

As a logical extension, particle number operators are next defined such that $\boldsymbol{N} = \sum_j N_j = \sum_j b_j^\dagger b_j$. The commutation rules require that

$$[\boldsymbol{H}, \boldsymbol{N}] = 0 \tag{8.45}$$

so that the particle number represents a constant of the motion. From here onwards a subscript to N will be used to differentiate between different types of particle.

Incorporation of particle number operators defines the density operator of the grand canonical ensemble as[10]

$$\varrho = e^{(\Omega+\sum_{i=1}^{\sigma}\mu_i N_i - H)/\theta} \tag{8.46}$$

The discussion that follows closely parallels the same presented before for the canonical ensemble. Thus from normalization of (35) follows that

$$e^{\Omega/\theta} = \text{trace } e^{(\mu N - H)/\theta} \tag{8.47}$$

where $\mu N = \sum_{i=1}^{\sigma} \mu_i N_i$.

On account of (45) the density matrix may be diagonalized in the basis of energy eigenfunctions, such that

$$\varrho_{nm} = e^{(\Omega+\mu N - E_l)/\theta}\delta_{nm}$$

is the probability that a randomly chosen system of the ensemble has the set of particle numbers N and is found in the quantum state with eigenvalue E_l. In terms of this representation

$$\sum_n \varrho_{nn} = \sum_{N_1=0}^{\infty} \cdots \sum_{N_\sigma=0}^{\infty} e^{(\Omega+\sum_{i=1}^{\sigma}\mu_i N_i)/\theta}\left[\sum_l e^{-E_l/\theta}\right] = 0$$

or

$$e^{-\Omega/\theta} = \sum_N e^{\mu N/\theta} Q^{(N)} \tag{8.48}$$

where $Q^{(N)}$ is the canonical partition function of a system with the set N of particle numbers. Equation (48) shows that the grand canonical ensemble is made up of canonical ensembles, which in turn had been shown to be combinations of microcanonical ensembles. This argument confirms that the grand canonical ensemble is in statistical equilibrium.

According to (39) the average of an observable over the grand canonical ensemble is

$$\overline{A} = \text{trace } \boldsymbol{A} e^{(\Omega+\mu N - H)/\theta} \tag{8.49}$$

In terms of energy eigenfunctions

$$\overline{A} = \sum_N \left[\sum_n A_{nn}^{(N)} e^{(\Omega+\mu N - E_n)/\theta}\right] \tag{8.50}$$

---

[10]Following tradition the symbol Ω is used here with different meaning, but in the context there should be no confusion with phase volume.

so that the average energy is

$$\overline{E} = \sum_N \left[\sum_n E_n e^{(\Omega+\mu N - E_n)/\theta} \right]$$

and the average number of particles in the ith component is

$$\overline{N_i} = \sum_{N_1=0}^{\infty} \cdots \sum_{N_\sigma=0}^{\infty} N_i e^{(\Omega+\sum_{i=1}^{\infty} \mu_i N_i)/\theta} Q^{(N)}$$

A generalized force operator is again introduced as before, by

$$X_j = \frac{\partial H}{\partial x_j}$$

Differentiation of the normalization condition (48) with respect to x_j yields

$$0 = \text{trace} \left[\left(\frac{\partial(\Omega/\theta)}{\partial x_j} - \frac{1}{\theta}\frac{\partial H}{\partial x_j} \right) e^{(\Omega+\mu N - H)/\theta} \right]$$

or

$$\overline{X_j} = \frac{\partial \overline{H}}{\partial x_j} = \frac{\partial \Omega}{\partial x_j}$$

The total derivative of (47)

$$-d(\Omega/\theta) = \text{trace} \left[\left(\sum_{i=1}^{\sigma} N_i d(\mu_i/\theta) - H d(1/\theta) - \frac{1}{\theta} \sum_j \frac{\partial H}{\partial x_j} dx_j \right) e^{(\Omega+\mu N - H)/\theta} \right]$$

Using (50) and substituting for $\overline{X_j}$ the expression reduces to

$$\begin{aligned} d\Omega &= \left(\frac{\Omega + \sum_{i=1}^{\sigma} \mu_i \overline{N_i} - \overline{E}}{\theta} \right) d\theta - \sum_{i=1}^{\sigma} \overline{N_i} d\mu_i - \sum_j \overline{X_j} dx_j \\ &\equiv \overline{H} d\theta - \sum_{i=1}^{\sigma} \overline{N_i} d\mu_i - \sum_j \overline{X_j} dx_j \end{aligned} \tag{8.51}$$

Equation (51) again has the form of a thermodynamic differential equation for a characteristic function and may be used to identify the thermodynamic analogues of the grand canonical ensemble.

As before, the necessary and sufficient condition that thermal equilibrium depends on equal temperatures identifies $\theta = kT$. By similar arguments it follows that equilibrium across a semi-permeable wall requires that $(\mu_i/\theta)_1 =$

$(\mu_i/\theta)_2$, so that material equilibrium requires $(\mu_i)_1 = (\mu_i)_2$. The statistical parameters μ_i thus possess the essential property of chemical potentials.

In order to identify further parameters equation (51) is compared with the thermodynamic equation (8.2.3)

$$d(PV) = SdT + PdV + \sum_{i=1}^{\sigma} N_i d\mu_i$$

to yield

$$\Omega \to -PV\,,\ \overline{H} \to -S/k\,,\ \theta \to kT\,,\ \overline{N_i} = N_i\,,\ \mu_i \to \mu_i\,,\ \overline{X_j} \to P$$

In analogy with (41) a grand partition function is defined by

$$\Xi = \text{trace } e^{\alpha N - \beta H}$$

where $\alpha = \frac{\mu}{kT}$, $\beta = \frac{1}{kT}$.
As demonstrated for the canonical ensemble, the form

$$\ln \Xi = \beta(\mu N - F)$$

rearranges to an expression for the grand canonical potential

$$PV = kT \ln \Xi$$

From this relationship follow expressions for

$$\text{entropy} : S = k\left(\ln \Xi + T\frac{\partial \ln \Xi}{\partial T}\right)$$

$$\text{number of molecules} : \overline{N_i} = kT\frac{\partial \ln \Xi}{\partial \mu_i}$$

$$\text{pressure} : \overline{P} = kT\frac{\partial \ln \Xi}{\partial V}$$

$$\text{internal energy} : U = kT^2\frac{\partial \ln \Xi}{\partial T}$$

In the basis of energy eigenfunctions the grand partition function is

$$\Xi = \sum_N e^{\frac{\mu N}{kT}}\left[\sum_n e^{-E_n/kT}\right]$$

The definition of entropy in the grand canonical ensemble follows from the definition of $-S/k$ according to (50), combined with (46) and (49), as

$$S = -k \text{ trace}(\boldsymbol{\varrho} \ln \boldsymbol{\varrho})$$

which is identical to the functional relation between entropy and density operator of the *petit* canonical ensembles.

8.4.9 Generalized Ensembles

As in energy representation the fundamental thermodynamic equation in entropy representation (3) may also be subjected to Legendre transformation to generate a series of characteristic functions designated as Massieu-Planck (MP) functions, Φ_m. The index m denotes the number of intensive parameters introduced as independent variables, *i.e.*

$$\Phi_m = S - \sum_{i=1}^{m} F_i X_i$$

The intensive parameter F_i, conjugate to the extensive parameter X_i is defined by

$$F_i = \left(\frac{\partial S}{\partial X_i}\right)_{X_j} \qquad (i = 1, \ldots, r; j \neq i)$$

and $S = \sum_{i=1}^{r} F_i X_i$. The MP functions must therefore satisfy the differential equation

$$d\Phi_m = -\sum_{i=1}^{m} X_i dF_i + \sum_{j=m+1}^{r} F_j dX_j$$

or

$$\Phi_m = \Phi_l - \sum_{i=l+1}^{m} F_i X_i \quad (l < m)$$

Well-known examples of MP functions are $-F/T$ and $-G/T$.

The relations involving canonical partition functions

$$\begin{aligned} S &= k \ln \Omega \\ -\frac{F}{T} &= k \ln Q \\ \frac{PV}{T} &= k \ln \Xi \end{aligned}$$

are all related in the same way to an MP function. It is therefore clear that a general relationship exists such that

$$\Phi_m = k \ln \Xi_m$$

where Ξ_m is the partition function of a generalized ensemble that depends on m intensive parameters.

Because of the relationship between the different canonical ensembles the generalized partition function is best defined as the generalization of the microcanonical definition of entropy in terms of the Gibbs energy function $e^{\Phi} = \Omega$, *i.e.*

$$\Xi_m = e^{\Phi_m} = F(\Omega^*)$$

where Ω^* is a cumulative partition function over all extensive parameters

$$\Omega^* = \int_0^{X_r} dX_r \ldots \int_0^{X_2} dX_2 \int_0^{X_1} \Omega dX_1$$

where X_1 represents energy, X_r represents volume and all intermediate X_i represent particle numbers. Thus

$$e^{\Phi} = \frac{\partial^r \Omega^*}{\partial X_1 \ldots \partial X_r}$$

The partition function of an ensemble with m intensive parameters becomes

$$\Xi_m = e^{\Phi_m} = \int_0^{\infty} \ldots \int_0^{\infty} e^{-\sum_{i=1}^m F_i X_i} d_{1 \ldots m} \left[\Omega^* \left(X_{1 \ldots r}\right)\right]$$

If $\Omega^*(E)$ is differentiable in the ordinary sense the partition function of a generalized ensemble with m intensive parameters is the m-fold Laplace transform of the microcanonical partition function e^{Φ},

$$\Xi_m \equiv e^{\Phi_k} = \int_0^{\infty} \ldots \int_0^{\infty} e^{-\sum_{i=1}^m F_i X_i} e^{\Phi_m} \prod_{i=1}^{m} dX_i$$

For $m = 1$

$$\Xi_1 = \int_0^{\infty} e^{-\beta E} e^{\Phi} dE$$

the canonical partition function Q appears as the Laplace transform of e^{Φ}.

For $m = 2$ one instance of interest refers to the transformation $E \to \beta$, $N \to -\alpha$, where $\alpha = \mu/kT$. If Ω^* is differentiable with respect to particle number the grand partition function

$$\Xi = \int_0^{\infty} \int_0^{\infty} e^{-\alpha N - \beta E} e^{\Phi} dN dE$$

appears as the two-fold Laplace transform of the function e^{Φ}.

The statistical partition functions are seen to be related by Laplace transformation in the same way that thermodynamic potentials are related by Legendre transformation. It is conjectured that the Laplace transformation of the statistical partition functions reduces asymptotically to the Legendre transformation of MP in the limit of infinitely large systems.

8.5 Non-equilibrium Statistical Mechanics

The difference between equilibrium and non-equilibrium systems exists in the time-dependence of the latter. An example of a non-equilibrium property is the rate at which a system absorbs energy when stirred by an external influence. If the stirring is done monochromatically over a range of frequencies an absorption spectrum is generated. Another example of a non-equilibrium property is the relaxation rate by which a system reaches equilibrium from a prepared non-equilibrium state.

For systems close to equilibrium the non-equilibrium behaviour of macroscopic systems is described by *linear response theory*, which is based on the *fluctuation-dissipation theorem.* This theorem defines a relationship between rates of relaxation and absorption and the correlation of fluctuations that occur spontaneously at different times in equilibrium systems.

The term *fluctuation* is used to describe the continual random transitions that occur among the microstates of a thermodynamic system. If the system consists of a subsystem in diathermal contact with a thermal reservoir, the subsystem and reservoir together experience these transitions among their joint microstates. The transitions may lead to states of high subsystem energy and sometimes to states of low subsystem energy, as the constant total energy is shared in different proportions between subsystem and reservoir. The subsystem energy thereby fluctuates around its equilibrium value.

Alternatively, the subsystem may simply be part of a larger system that constitutes a reservoir. In such a case the fluctuations are said to be *local* within a nominally homogeneous system.

Deviations from an average value $\overline{x}$ is itself a fluctuating variable

$$\delta x = x - \overline{x}$$

with an average value of $\overline{\delta x} = \overline{x} - \overline{x} = 0$.

The mean square deviation $\langle (x - \overline{x})^2 \rangle$ however, measures the extent of the fluctuations and is non-zero.

$$\begin{aligned}
\overline{(\delta x)^2} &= \overline{(x - \overline{x})^2} = \overline{x^2 - 2x\overline{x} + \overline{x}^2} \\
&= \overline{x^2} - 2\overline{x}\overline{x} + \overline{x}^2 = \overline{x^2} - \overline{x}^2
\end{aligned}$$

For the canonical ensemble, the sum over energy states is

$$Q = \sum_{\nu} e^{\beta \epsilon_{\nu}}$$

and the average values

$$
\begin{aligned}
\bar{\epsilon} &\equiv U = \sum_{\nu} \epsilon_{\nu} e^{\beta\epsilon_{\nu}} / \sum_{\nu} e^{\beta\epsilon_{\nu}} \\
\overline{\epsilon^2} &= \sum_{\nu} \epsilon_{\nu}^2 e^{\beta\epsilon_{\nu}} / \sum_{\nu} e^{\beta\epsilon_{\nu}}
\end{aligned}
$$

It follows that

$$
\begin{aligned}
\frac{\partial Q}{\partial \beta} &= \sum_{\nu} \epsilon_{\nu} e^{\beta\epsilon_{\nu}} = \bar{\epsilon} Q \\
\frac{\partial^2 Q}{\partial \beta^2} &= \sum_{\nu} \epsilon_{\nu}^2 e^{\beta\epsilon_{\nu}} = \overline{\epsilon^2} Q
\end{aligned}
$$

Hence

$$
\begin{aligned}
\overline{(\delta\epsilon)^2} &= \overline{\epsilon^2} - \bar{\epsilon}^2 \\
&= \frac{1}{Q}\frac{\partial^2 Q}{\partial \beta^2} - \frac{1}{Q^2}\left(\frac{\partial Q}{\partial \beta}\right)^2 = \frac{\partial}{\partial \beta}\left(\frac{1}{Q}\frac{\partial Q}{\partial \beta}\right) \frac{\partial \bar{\epsilon}}{\partial \beta}
\end{aligned}
$$

Recall that $\beta = -1/kT$ and hence

$$
\frac{\partial \beta}{\partial T} = -\frac{1}{k}\frac{\partial}{\partial T} T^{-1} = k^{-1} T^{-2}
$$

to give

$$
\overline{(\delta\epsilon)^2} = \frac{\partial \bar{\epsilon}}{\partial T}\frac{\partial T}{\partial \beta} = kT^2 \frac{\partial U}{\partial T} = NkT^2 c_v
$$

where $C_V = Nc_v$ is the heat capacity of the N particle system.

This informative result shows that the mean square energy fluctuations are proportional to the size of the system and to the rate at which energy changes with temperature. The relative root mean square dispersion of the energy

$$
\frac{\sqrt{\overline{(\delta E)^2}}}{\bar{\epsilon}} = \sqrt{\frac{kT^2 C_V}{U^2}} \propto \frac{1}{\sqrt{N}}
$$

For large systems ($N \to \infty$) the fluctuation amplitudes become negligible relative to the mean values, and thermodynamics becomes precise.

For systems with large C_V the fluctuations in energy are correspondingly large. Furthermore the energy fluctuations in all systems become very small at low temperatures, where $C_V \to 0$.

A familiar example of *dissipation* is friction. A particle pushed through a medium experiences frictional drag or force, proportional to the velocity

$$f_{drag} = -\eta v$$

where η is the frictional constant. The presence of the velocity-dependent force implies that a fluid will absorb energy as a particle is dragged through the fluid. In other words, as a result of friction, energy is dissipated and the medium heats up.

The relationship between fluctuation and dissipation is reminiscent of the reciprocal Onsager relations that link affinity to flux. The two relationships become identical under Onsager's regression hypothesis which states that the decay of a spontaneous fluctuation in an equilibrium system is indistinguishable from the approach of an undisturbed non-equilibrium system to equilibrium. The conclusion important for statistics, is that the relaxation of macroscopic non-equilibrium disturbances is governed by the same (linear) laws as the regression of spontaneous microscopic fluctuations of an equilibrium system. In the specific example discussed above, the energy fluctuations of a system in contact with a heat bath at temperature T,

$$\overline{\Delta E^2} = kT^2C$$

relate to the thermodynamic response function $C_V = dE/dT$. that measures the rate of change as the controlling restriction is varied.

Whenever a mechanism dissipates mechanical energy as heat there exists an inverse mechanism by which heat can generate a mechanical fluctuation. There is always a connection between fluctuation and dissipation.

8.5.1 The fluctuation-dissipation theorem

The general nexus between fluctuation and dissipation was examined by Callen and Welton [122] in terms of the fluctuations of appropriate generalized forces and the *impedance* in a linear dissipative system. A system is considered to be dissipative if capable of absorbing energy when subjected to a time-periodic perturbation, *e.g.* an electrical resistor that absorbs energy from an impressed periodic potential.

For a linear dissipative system, an impedance may be defined, and the proportionality constant between power and the square of the perturbation amplitude is simply related to the impedance. In the electrical case

$$\text{Power} = (\text{voltage})^2 R/|Z|^2$$

The applied perturbation is treated quantum-mechanically, relating power dissipation to certain matrix elements of the perturbation operator. It shows that for small perturbations, a system with densely distributed energy levels is dissipative and linear.

The Hamiltonian H of the system is defined as

$$H = H_0(q_k, p_k) + VQ(q_k, p_k)$$

where H_0 is the unperturbed Hamiltonian, Q is a function of coordinates and momenta and V a function of time, that measures the instantaneous magnitude of the perturbation. As an example V may be the impressed voltage and $Q = \sum e_i x_i / L$, where e_i is particle charge, x_i the distance from one end of the resistor and L its total length. If the applied perturbation varies sinusoidally,

$$V = V_0 \sin \omega t$$

The power dissipation is computed by standard time-dependent perturbation theory (7.2.1).

Let $\psi_1, \psi_2, \ldots \psi_n, \ldots$ be the set of eigenfunctions of the unperturbed Hamiltonian, so that

$$H_0 \psi_n = E_n \psi_n$$

The true eigenfunction ψ, may be expanded in terms of ψ_n,

$$\psi = \sum_n a_n(t) \psi_n$$

Substitution into Schrödinger's equation

$$H_0 \psi + (V_0 \sin \omega t) Q \psi = i\hbar \frac{\partial \psi}{\partial t}$$

generates a set of first-order equations for the coefficients $a_n(t)$. If the energy levels of the system are densely distributed, the total induced transition probability of the system initially in state ψ_n follows as

$$\frac{\pi V_0^2}{2\hbar} \left\{ |\langle E_n + \hbar\omega | Q | E_n \rangle|^2 \rho(E_n + \hbar\omega) + |\langle E_n - \hbar\omega | Q | E_n \rangle|^2 \rho(E_n - \hbar\omega) \right\}$$

The symbol $\rho(E)$ indicates the density-in-energy of the quantum states in the neighbourhood of E, so that the number of states between E and $E + \delta E$ is $\rho(E)\delta E$.

Each transition from the state ψ_n to the state with eigenvalue $E + \hbar\omega$ is accompanied by the absorption of energy $\hbar\omega$, and each transition from ψ_n to the state with eigenvalue $E - \hbar\omega$ by the emission of energy $\hbar\omega$. The rate of absorption of energy by a system initially in the nth state is

$$\frac{1}{2}\pi V_0^2\omega \left\{|\langle E_n + \hbar\omega|Q|E_n\rangle|^2\rho(E_n + \hbar\omega) - |\langle E_n - \hbar\omega|Q|E_n\rangle|^2\rho(E_n - \hbar\omega)\right\}$$

To predict the behaviour of real thermodynamic systems, it is necessary to average over all initial states, weighting each according to the Boltzmann factor $\exp(-E_n/kT)$. For weighting factor $f(E_n)$,

$$\frac{f(E_n + \hbar\omega)}{f(E_n} = \frac{f(E_n)}{f(E_n - \hbar\omega)} = \exp\left(-\frac{\hbar\omega}{kT}\right) \tag{8.52}$$

The power dissipation follows as

$$\text{Power} = \frac{1}{2}\pi V_0^2\omega \sum_n \left\{|\langle E_n + \hbar\omega|Q|E_n\rangle|^2\rho(E_n + \hbar\omega)\right.$$

$$\left. -|\langle E_n - \hbar\omega|Q|E_n\rangle|^2\rho(E_n - \hbar\omega)\right\} f(E_n)$$

The summation over n may be replaced by integration over energy

$$E_n(\quad) \rightarrow \int_0^\infty (\quad)\rho(E)dE$$

whence

$$\text{Power} = \frac{1}{2}\pi V_0^2\omega \int_0^\infty \rho(E)f(E)\left\{|\langle E + \hbar\omega|Q|E\rangle|^2\rho(E + \hbar\omega)\right.$$

$$\left. -|\langle E - \hbar\omega|Q|E\rangle|^2\rho(E_n - \hbar\omega)\right\} dE$$

It follows that a small periodic perturbation applied to a system, the eigenstates of which are densely distributed in energy, leads to a power dissipation quadratic in the perturbation. For such a linear system it is possible to define an impedance $Z(\omega)$, the ratio of the force V to the response $\dot{Q}$, where all quantities are now assumed to be in standard complex notation, $V = Z(\omega)\dot{Q}$. The instantaneous power is $V\dot{Q}R(\omega)/|Z(\omega)|^2$, where $R(\omega)$, the resistance, is the real part of $Z(\omega)$.

If the applied perturbation is not sinusoidal, but some general function of time $V(t)$, and if $v(\omega)$ and $\dot{q}(\omega)$ are the Fourier transforms of $V(t)$ and $\dot{Q}(t)$, the impedance is defined in terms of the Fourier transforms

$$v(\omega) = Z(\omega)\dot{q}(\omega) \tag{8.53}$$

In this notation the general linear dissipative system is described by

$$\frac{R}{|Z|^2} = \pi\omega \int_0^\infty \rho(E)f(E)\left\{|\langle E+\hbar\omega|Q|E\rangle|^2\rho(E_n+\hbar\omega)\right.$$

$$\left.-|\langle E_n-\hbar\omega|Q|E\rangle|^2\rho(E-\hbar\omega)\right\}dE \tag{8.54}$$

To establish the relationship between dissipation and fluctuation, consider a system left in thermal equilibrium, without any applied force. It is expected that spontaneous fluctuation $\dot{Q}$ is associated with a spontaneously fluctuating force.

Let $\langle V^2\rangle$ be the mean square value of the fluctuating force, and $\langle\dot{Q}^2\rangle$ the mean square value of $\dot{Q}$. Suppose that the system is known to be in the nth eigenstate. Since H_0 is hermitian the expectation value of $\dot{Q}$, $\langle E_n|\dot{Q}|E_n\rangle$ vanishes and the mean square fluctuation of $\dot{Q}$ is given by the expectation value of $\dot{Q}^2$, *i.e.*

$$\begin{aligned}
\langle E_n|\dot{Q}^2|E_n\rangle &= \sum_m \langle E_n|\dot{Q}|E_m\rangle\langle E_m|\dot{Q}|E_n\rangle \\
&= \hbar^{-2}\sum_m \langle E_n|H_0Q-QH_0|E_m\rangle\langle E_m|H_0Q-QH_0|E_n\rangle \\
&= \hbar^{-2}\sum_m (E_n-E_m)^2|\langle E_m|Q|E_n\rangle|^2
\end{aligned}$$

Introducing a frequency ω by $\hbar\omega = |E_n-E_m|$, the summation over m may be replaced by two integrals over ω (one for $E_n<E_m$ and one for $E_n>E_m$):

$$\begin{aligned}
\langle E_n|\dot{Q}^2|E_m\rangle &= \hbar^{-2}\int_0^\infty (\hbar\omega)^2|\langle E_n+\hbar\omega|Q|E_n\rangle|^2\rho(E_n+\hbar\omega)\hbar d\omega \\
&+ \hbar^{-2}\int_0^\infty (\hbar\omega)^2|\langle E_n-\hbar\omega|Q|E_n\rangle|^2\rho(E_n-\hbar\omega)\hbar d\omega \\
&= \int_0^\infty \hbar\omega^2\{\langle E_n+\hbar\omega|Q|E_n\rangle|^2\rho(E_n+\hbar\omega) \\
&+ |\langle E_n-\hbar\omega|Q|E_n\rangle|^2\rho(E_n-\hbar\omega)\}d\omega
\end{aligned}$$

The thermodynamic fluctuation is obtained by multiplying the fluctuation in the nth state by the weighting factor $f(E_n)$, and summing

$$\langle\dot{Q}^2\rangle = \sum_n f(E_n)\int_0^\infty \hbar\omega^2\{|\langle E_n+\hbar\omega|Q|E_n\rangle|^2\rho(E_n+\hbar\omega)$$

$$+|\langle E_n-\hbar\omega|Q|E_n\rangle|^2\rho(E_n-\hbar\omega)\}d\omega$$

As before, the summation over n is replaced by integration over the energy spectrum and introduction of the density factor $\rho(E)$. Thus, finally

$$\begin{aligned}\langle\dot{Q}^2\rangle &= \int_0^\infty \hbar\omega^2 \Big[\int_0^\infty \rho(E)f(E)\{|\langle E+\hbar\omega|Q|E\rangle|^2\rho(E+\hbar\omega) \\ &+ |\langle E-\hbar\omega|Q|E\rangle|^2\rho(E-\hbar\omega)\}dE\Big]\,d\omega \end{aligned} \tag{8.55}$$

or, using (53) to define impedance,

$$\begin{aligned}\langle V^2\rangle &= \int_0^\infty |Z|^2\hbar\omega^2 \Big[\int_0^\infty \rho(E)f(E)\{|\langle E+\hbar\omega|Q|E\rangle|^2\rho(E+\hbar\omega) \\ &+ |\langle E-\hbar\omega|Q|E\rangle|^2\rho(E-\hbar\omega)\}dE\Big]\,d\omega \end{aligned} \tag{8.56}$$

The expressions for both dissipation and fluctuation involve the constructs

$$\int_0^\infty \rho(E)f(E)\{|\langle E+\hbar\omega|Q|E\rangle|^2\rho(E+\hbar\omega) \pm |\langle E-\hbar\omega|Q|E\rangle|^2\rho(E-\hbar\omega)\}dE \tag{8.57}$$

The negative sign is associated with $R/|Z|^2$ and the positive sign with $\langle V^2\rangle$. It will be shown that the two values of (57) are simply related, and thus establish the desired relation between $\langle V^2\rangle$ and $R(\omega)$.

Consider first the quantity $C(-)$ obtained by selecting the negative sign.

$$C(-) = \int_0^\infty f(E)|\langle E+\hbar\omega|Q|E\rangle|^2\rho(E+\hbar\omega)\rho(E)dE$$

$$-\int_0^\infty f(E)|\langle E-\hbar\omega|Q|E\rangle|^2\rho(E)\rho(E-\hbar\omega)dE$$

In the second integral the quantity $\langle E-\hbar\omega|Q|E\rangle$ vanishes for $E<\hbar\omega$ and by changing the integration variable $E\to E+\hbar\omega$ in this part, the expression reduces to

$$C(-) = \int_0^\infty |\langle E+\hbar\omega|Q|E\rangle|^2\rho(E+\hbar\omega)\rho(E)f(E)\{1-f(E+\hbar\omega/f(E)\}dE$$

Using (52) it further reduces to

$$C(-) = \{1-\exp(-\hbar\omega/kT)\}\int_0^\infty |\langle E+\hbar\omega|Q|E\rangle|^2\rho(E+\hbar\omega)\rho(E)f(E)dE$$

If $C(+)$ denotes the value of (57) corresponding to the positive sign, then, in identical fashion, one finds

$$C(+) = \{1+\exp(-\hbar\omega/kT\}\int_0^\infty |\langle E+\hbar\omega|Q|E\rangle|^2\rho(E+\hbar\omega)\rho(E)f(E)dE$$

With these alternative expressions for (57) it is now possible to write from (54)

$$\frac{R(\omega)}{|Z(\omega)|^2} = \pi\omega\left\{1 - \exp\left(-\frac{\hbar}{kT}\right)\right\}\int_0^\infty |\langle E+\hbar\omega|Q|E\rangle|^2\rho(E+\hbar\omega)\rho(E)f(E)dE$$

and from (56)

$$\langle V^2\rangle = \int_0^\infty |Z|^2\hbar\omega^2\left\{1 + \exp\left(-\frac{\hbar\omega}{kT}\right)\right\}$$
$$\times\int_0^\infty |\langle E+\hbar\omega|Q|E\rangle|^2\rho(E+\hbar\omega)\rho(E)f(E)dE$$

Comparison of these equations directly yields the fundamental theorem

$$\langle V^2\rangle = \frac{2}{\pi}\int_0^\infty R(\omega)E(\omega,T)d\omega \tag{8.58}$$

where

$$E(\omega,T) = \frac{1}{2}\hbar\omega + \hbar\omega\left[\exp\left(\frac{\hbar\omega}{kt}\right) - 1\right]^{-1}$$

It may be recognized that $E(\omega,T)$ is, formally, the expression for the mean energy of an oscillator of natural frequency ω, at temperature T.

At high temperatures, $E(\omega,T)$ takes its equipartition value of $E(\omega,T) \simeq kT$, $(kT >> \hbar\omega)$ to yield the so-called Nyquist formula that describes voltage fluctuations (Nyquist noise) generated in a linear electrical system, *i.e.*

$$\langle V^2\rangle = \frac{2}{\pi}kT\int R(\omega)d\omega \tag{8.59}$$

To reiterate, a system with generalized resistance $R(\omega)$ exhibits, in equilibrium, a fluctuating force given by (58), or at high temperature, by (59).

Application of the F-D theorem produced [122] several significant results. Apart from the Nyquist formula these include the correct formulation of Brownian motion, electric dipole and acoustic radiation resistance, and a rationalization of spontaneous transition probabilities for an isolated excited atom.

8.5.2 Chemical Reaction

Irreversible statistical mechanics could in principle be used to develop a fundamental theory of reaction rates by treating chemical reaction as the response to chemical potential fluctuations against a reaction barrier, A.

$$\langle \mu^2\rangle \simeq \int A(\omega)\phi(\omega,T)d\omega$$

where ϕ represents reactant energy and ω relates to collision frequency.

Although the formulation of such a theory has never been achieved, Eyring's absolute reaction rate model [123] has several features in common with such theory.

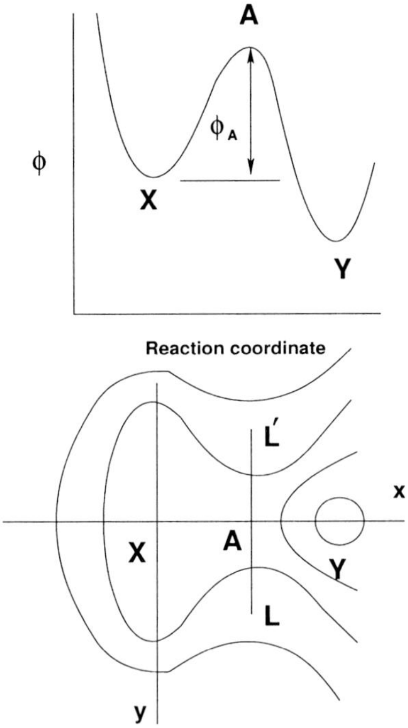

Figure 8: *Top: Potential energy as a function of position for a one-dimensional system. Bottom: Potential-energy contours for an atom moving in two dimensions.*

It is assumed that the reaction X $\rightarrow$ Y that proceeds *via* an activated state at A starts in a local minimum near configuration X, but is able to move through A to a final configuration Y at lower energy, as in figure 8. The amount ϕ_A is known as the *activation energy*, and typically $\phi_A >> kT$. If the system is kept at fixed temperature, its thermal fluctuations will explore the configurations around X and may eventually reach the activated configuration, from where it descends to Y. The system thus has a definite probability per unit time, R, of escaping from its *metastable state* near X. The rate of transfer increases with temperature.

The mechanism whereby an atom, adsorbed on a solid surface at X, moves towards a site of lower energy at Y, is also shown in figure 8. The contours of potential energy show that the route of lowest energy from X to Y passes through the saddle point at A.

To determine the rate of transfer a large set of systems, each with an atom either to the left or near the saddle point is considered. Once an atom has moved to the right of LL′ the process is regarded as having occurred. The rate R is then defined as the net flux of atoms across LL′, subject to several assumptions, introduced to ensure forward reaction. For n atoms in equilibrium near A, moving with mean Maxwell velocity $\overline{v} = \sqrt{2kT/\pi m}$ towards the right, the required flux density is $(1/2)\overline{v}n$, and the total flux is

$$F = \frac{1}{2}\overline{v}n_x e^{-\phi_A/kT} \int_{LL'} e^{-\delta\phi/kT} d\xi$$

where n is represented by the Boltzmann distribution $n = n_x \exp(-\phi/kT)$, ϕ_X defines the potential zero and $\delta\phi = \phi - \phi_A$. The total number of atoms concentrated near X may be written as

$$C = n_x \int_A e^{-\phi/kT} dxdy$$

where the integral is taken over the area of the plane near X.

For small values of ξ it is assumed that $\delta\phi$ varies quadratically with ξ along LL′, so that the flux integral may be written as $\sqrt{2\pi}\sigma_\xi$, where σ_ξ is the rms value of ξ, averaged over the Boltzmann distribution. The variable σ_ξ is interpreted as a measure of the window through which escaping molecules pass at A, and is temperature-dependent. In the same way the concentration integral is rewritten as $2\pi\sigma_x\sigma_y$ in terms of rms deviations in x and y for atoms vibrating in the potential minimum at X. In terms of these new quantities the rate for the activated process calculated as $R = F/C$ is given by

$$R = \left(\frac{1}{2\pi}\right)\left(\frac{kT}{m}\right)^{1/2}\left(\frac{\sigma_\xi}{\sigma_x\sigma_y}\right)e^{-\phi_A/kT}$$

Introducing the force constant for vibration in the x-direction at X, as

$$\alpha_x = \frac{kT}{\sigma_x^2}$$

and the corresponding vibration frequency

$$\nu_x = \frac{1}{2\pi}\left(\frac{\alpha_x}{m}\right)^2$$

the rate expression reduces to

$$R = \nu_x\left(\frac{\sigma_\xi}{\sigma_y}\right)e^{-\phi_A/kT} \tag{8.60}$$

The frequency ν_x is interpreted as an *attempt frequency* that describes the number of times per second that the atom tries to escape. The rest of the expression represents the probability of escape at any given attempt. The term σ_ξ/σ_y specifies the ratio of the window sizes at A and X through which the atom has to pass. The final term is a temperature-dependent Boltzmann factor that specifies the ratio of equilibrium number densities at A and X.

Different forms of (60) derived for a variety of systems have the exponential temperature dependence in common. In practice ϕ_A is invariably much larger than kT. The Boltzmann factor therefore increases rapidly with temperature and the remainder of the expression may safely be treated as constant over a small temperature range. The activation energy may be determined experimentally from an Arrhenius plot of $\ln R$ vs $1/T$, which should be a straight line of slope $-\phi_A/k$.

Chapter 9

Chemical Change

9.1 Introduction

The inconclusive quest for a fundamental theory of chemistry, consistent with quantum mechanics and relativity, has been described in the preceeding three chapters. To put the problem into perspective once more, the place of chemistry within the sciences needs to be restated.

All evidence suggests that the entire universe exists in a state of constant dynamic change. Those changes external to the solar system have been the focus of theoretical physics and many of the leading exponents claim that, on that front the end is in sight. The formulation of a theory of everything is thought to be imminent.

Within the solar system the observable changes are of a different kind, best described as chemical change. The most striking common feature of those chemical reactions driven by solar energy is their cyclic nature, linked to planetary motion. All phenomena, collectively known as life, or growth, are of this type. Their essential characteristic is a state far from equilibrium. For a life process, equilibrium is synonymous with death and chemical change after death is a rapid slide towards equilibrium. The most advanced chemical theories deal with these posthumous effects and related reactions only, albeit rather superficially. A fundamental theory to predict conditions for the onset of elementary chemical change is not available.

The most important clue to explain the essence of chemical reaction has probably been the discovery of *catalysis*, without which a chemical industry would not exist. Catalysis is not understood in detail but it clearly dictates the course of a reaction by re-inforcing the spontaneous fluctuations that occur in a reaction medium. The most dramatic effects are produced in autocatalysis where a reaction product acts as a catalyst. Such feedback leads to the creation of apparently chaotic structures, like turbulence that appears when laminar flow in a fluid system breaks down.

To deal with runaway sytems that result from positive feedback, the linear response mechanism assumed in near-equilibrium irreversible thermodynamics, has to be abandoned. Non-linear thermodynamic analysis of systems far from equilibrium was pioneered by Progogine [124] based on the realization that, far from equilibrium, systems may still evolve to a steady state, but in general this state can no longer be characterized in terms of some suitably chosen potential, such as entropy production for near-equilibrium states. Thus, while turbulent motion may appear as irregular and chaotic on the macroscopic scale, it is, on the contrary highly organized on the microscopic scale [125]. The argument goes that large-scale fluctuations result in the formation of dissipative structures. They occur when fluctuations threaten existing structures and the system reaches a critical *bifurcation* point. At this stage it is no longer possible to predict the next state of the system, which has to choose among new available paths which it follows to the next bifurcation. A dissipative structure created at bifurcation is due to self-organization and represents a form of supramolecular organization.

Many examples of post-critical organization in the form of dissipative structures have been recognized, but a detailed analysis of transition through a critical point has never been achieved. A theory of chemical reaction far from equilibrium, therefore is a long way off, but some progress has been made towards the understanding of critical phenomena associated with phase transformation.

Phase transitions, although rarely described in these terms represent the second example of chemical change in the local environment. The most familiar example is transformation between the phases of water, which is responsible for all observed climatic and weather changes on the planet. Volcanic activity is another dramatic manifestation of phase transformations.

9.2 Phase Change

The best-known examples of phase transition are the liquid-vapour transition (evaporation), the solid-liquid transition (melting) and the solid-vapour transition (sublimation). The relationships between the phases, expressed as a function of P, V and T consitute an equation of state that may be represented graphically in the form of a *phase diagram*. An idealized example, shown in figure 1, is based on the phase relationships of argon [126].

The diagram on the left shows a series of *isotherms* and the $P-V$ conditions that determine phase stabilities at each temperature. At the highest temperatue, T_1, and low pressure argon is an ideal gas, which means that PV is constant and the isotherm is a straight line. The high-pressure deviation from linearity indicates decreased compressibility. The isotherm T_3 lies below the Boyle temperature [1] and the gas becomes more compressible because of intermolecular attractions.

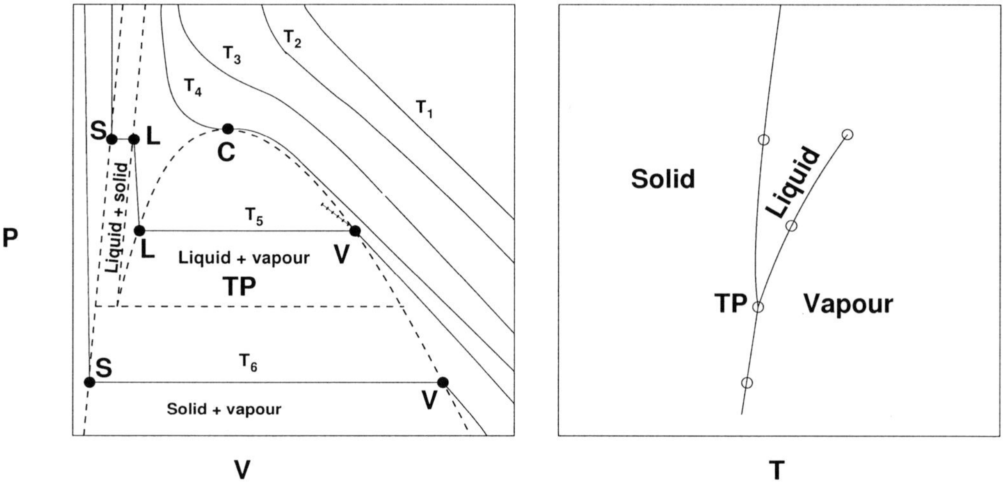

Figure 1 *Typical isotherms (on logarithmic scale) and phase diagram based on the argon system, after [126].*

The isotherm T_4 passes through the critical point C, where compressibility becomes infinite.

At lower temperatures phase separation occurs. The isotherm T_5 enters a mixed-phase region at point V where the pressure corresponds to the *liquid vapour pressure* at that temperature. Liquid droplets condense from the vapour and this causes a decrease in volume. The pressure remains constant and condensation continues until the entire system is converted into liquid, at point L. Within the mixed-phase region a sample of uniform density is *unstable.* Any fluctuation in density promotes growth of the high-density phase at the expense of the low-density phase, and the system breaks up into two phases. Continuing along T_5 the liquid shows highly reduced compressibility, but when it reaches the *melting pressure* at L it starts to solidify along $L - S$. Finally, isotherm T_6 shows that when the vapour reaches the *sublimation pressure* at V it starts converting into solid until only solid remains at S.

[1]The lowest temperature at which, for one mole of gas, $PV/RT = 1$ as $P \to 0$.

On the $P-T$ diagram, at right, each of the mixed-phase regions is represented by a single *co-existence line*. The value of the chemical potential must be the same for the two phases at any point on a co-existence line. The lines meet at the *triple point* where all three phases are in equilibrium. The triple-point line on the $P-V$ diagram is marked TP.

A given phase may persist beyond the point at which transition to another phase should properly occur. On the T_5 isotherm, for instance, it is possible to compress the vapour in clean conditions beyond point V, as shown by the dots, without condensation occurring. This vapour is *supercooled*, and $\mu_V > \mu_L$. When disturbed, the supercooled vapour condenses at once. In the same way, clean liquid may be superheated without boiling, in which case $\mu_L > \mu_V$. Supercooled or superheated phases are *metastable*. They appear to be stable, but are thermodynamically unstable, since another state of lower chemical potential exists.

During each phase transition of the type illustrated here, both of the intensive parameters P and T remain constant. Because of the difference in density however, when a certain mass of liquid is converted into vapour, the total volume (extensive parameter) expands. From the Gibbs-Duhem equation (8.8) for one mole of each pure phase,

$$d\mu = vdP - sdT$$

it follows that

$$\left(\frac{\partial\mu}{\partial T}\right)_P = -s \quad , \quad \left(\frac{\partial\mu}{\partial P}\right)_T = v$$

Although at equilibrium $\mu_V = \mu_L$, it therefore follows that the first derivatives of μ_V and μ_L are different. In particular,

$$\left[\frac{\partial\left(\mu_V - \mu_L\right)}{\partial T}\right]_P = -\left(s_V - s_L\right) < 0$$

$$\left[\frac{\partial\left(\mu_V - \mu_L\right)}{\partial P}\right]_T = v_V - v_L > 0$$

A phase transition of this type is therefore known as a *first-order transition.*

It may happen in a phase transition that $s_2 - s_1 = 0$ and $v_2 - v_1 = 0$ which means that the first derivatives of the chemical potentials are continuous across the transition point. This latter type is indiscriminately commonly referred to as *second-order transitions.*

Phase transitions are not restricted to $P-V-T$ systems and may occur in other types of system. Well-known transitions occur in magnetic systems, where the extensive variable corresponding to the volume is magnetic moment and the intensive variable corresponding to pressure is the applied magnetic field.

In general then, a first-order phase transition occurs when a co-existence line is crossed and the system passes through a two-phase region, the two phases being separated in space. The two phases are substantially different and some of their extensive functions of state take on different values. Typically there is a change in entropy, magnetization, enthalpy or volume. If a route that passes through a critical point is followed however, a second-order transition occurs. There is no separation into spatially distinct phases. At the critical point the two phases are identical, so there can be no supercooling and superheating, and the whole system passes into the new state imperceptibly and uniformly. There is no discontinuity in any function of state, although derivatives of the functions of state such as compressibility, magnetizability and heat capacity may become infinite at the critical point.

A critical point may often be regarded as the onset of an ordering process: when a ferromagnet is cooled through its *Curie point* magnetization appears spontaneously; when a superconductor is cooled through the critical temperature in zero field superconducting electron pairs appear. This type of ordering process is no different from the appearance of dissipative structures when microscopic fluctuations reach critical proportions. A close parallel between disorder phenomena and the understanding of phase relationships and their connection with chemical processes is indicated.

9.3 Disorder

Disorder is an intuitive concept, best appreciated in terms of deviations from an ordered array. The highest degree of order in the physical world occurs in a crystal, made up of identical atoms or molecules, uniformly packed in regular rows and planes to fill all available space. Any displacement from the regular array clearly lowers the symmetry of the arrangement. Any change in the degree of disorder is in fact accompanied by a change of symmetry, but not necessarily a lowering of symmetry. It is therefore important not to confuse the concepts of disorder and symmetry and to clearly distinguish between *short-range* and *long-range order.* To avoid the specification of symmetry, an *order parameter* is often introduced to quantify the progress of a phase transition in terms of a measurable extensive variable.

Models of disorder are not in short supply [127]. For the simplest, ideal systems, it is assumed that no interaction occurs between particles or degrees of freedom and disorder is purely statistical. Interaction between the particles of real systems inevitably leads to correlations and instabilities, the actual cause of phase transitions.

A powerful, though simple model postulates a distribution of interacting points on a regular lattice. In the first instance interaction may be considered as confined to occur between neighbouring sites only. The total energy, summed over pairs of nearest-neighbour sites is given by

$$E = -J \sum_{ij} s_i s_j$$

where J is a coupling constant. This simple, *Ising* or *spin* model may be applied directly to magnetic systems such as a ferromagnet with *up* and *down* spin states, where s_i is a spin index, $s_i = \pm 1$.

In an applied magnetic field $\boldsymbol{B}$ the Hamiltonian takes the form

$$H = -\sum_i s_i \mu \boldsymbol{B} - J \sum_{ij} s_i s_j$$

with the first sum taken over all spins. When $J > 0$, it will be energetically favourable for neighbouring spins to be aligned. Hence, it is anticipated that for low enough temperature, this stabilization will lead to a cooperative phenomenon called spontaneous magnetization. That is, through nearest-neighbour interaction, a given magnetic moment can influence the alignment of spins at a macroscopic distance away. These long-range correlations between spins are associated with long-range order in which net magnetization may occur even in the absence of a magnetic field.

The value of the Ising model lies therein that it is the only model of disorder to have produced valid theoretical predictions of spontaneous phase changes. To understand the role of symmetry it is noted that spontaneous magnetization, starting from a random distribution of spins, amounts to a process of ordering that destroys an existing isotropic arrangement.

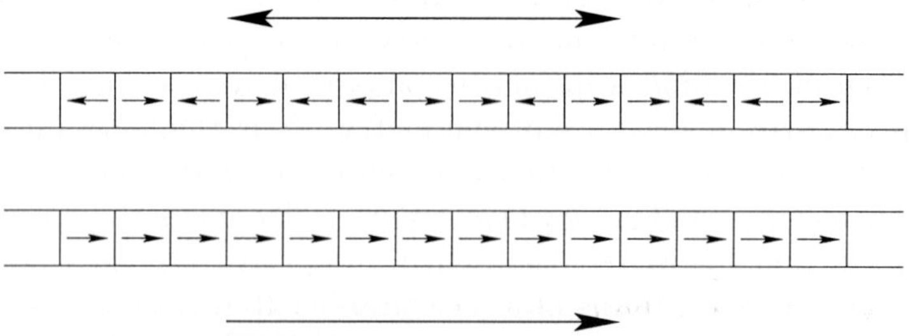

Figure 2: *Fragments of a randomly disordered and an ordered arrangement of one-dimensional Ising spins. The long arrows identify equivalent symmetry*

directions.

Of the two one-dimensional arrangements shown in figure 2, the disordered state has the higher symmetry since left and right directions are equivalent.

9.3.1 Thermodynamics of Disorder

Before a statistical description of disorder is further explored, it is instructive to examine a classical thermodynamic model, known as the *Landau theory.* This theory first introduced the concept of an order parameter, so defined that it is zero in the high-temperature phase and non-zero in the low-temperature phase. Using the ferromagnetic transition as an example, the order parameter is defined as the magnetic moment, which measures the cooperative alignment of the spins. When an order parameter changes by $d\phi$, an amount of work, $dW = y\,d\phi$ is done on the system. The parameter y is called the *conjugate field.* In the case of the ferromagnetic transition it is the external magnetic field, if present. For the vapour-liquid critical point the order parameter is defined as the volume difference of the co-existing phases and the conjugate field is defined by $dW = PdV$, as the applied pressure.

The order-disorder transition of a binary alloy (*e.g.* CuZn) provides another instructive example. The body-centred lattice of this material may be described as two interpenetrating lattices, A and B. In the disordered high-temperature phase each of the sub-lattices is equally populated by Zn and Cu atoms, in that each lattice point is equally likely to be occupied by either a Zn or a Cu atom. At zero temperature each of the sub-lattices is entirely occupied by either Zn or Cu atoms. In terms of fractional occupation numbers for A sites, an appropriate order parameter may be defined as

$$\phi = \frac{N^A_{\text{Zn}} - N^A_{\text{Cu}}}{N^A}$$

Above the transition temperature $\phi = 0$ and below the critical temperature $\phi > 0$. At $T = 0$, $\phi = \pm 1$. Any order parameter defined to have a magnitude of unity at zero temperature is said to be *normalized.* Criteria for testing the theory include a number of *critical exponents*, which are accurately known from experimental measurement.

Although transition across a critical point may proceed without any first-order discontinuity, the fact that there is a change of symmetry implies that the two phases must be described by different functions of the thermodynamic variables, which cannot be continued analytically across the critical point. The order parameter serves to compensate for the reduction in symmetry. Although it is a regular function of temperature it can develop a discontinuous derivative at the critical temperature. Likewise, several measurable quantities show divergent behaviour at a critical point. The nature of these singularities is described by the critical exponents that may be formulated in terms of the parameter

$$t = \frac{T - T_c}{T_c}$$

where T_c denotes critical temperature. Some of the critical exponents, denoted by Greek characters, describe the divergent behaviour of,

Heat capacity	:	$C \sim |t|^{-\alpha}$
Order parameter	:	$\phi \sim |t|^{\beta}$
Susceptibility	:	$\chi \sim |t|^{-\gamma}$
Field on critical isotherm	:	$\phi \sim y^{-1/\delta}$

9.3.2 Landau Theory

The Landau theory applies in the vicinity of a critical point where the order parameter is small and assumed continuous. The Gibbs function

$$G = G(T, P, \phi, n_i)$$

may then be approximated by a series expansion in the order parameter ϕ

$$G = G_0 + G_1\phi + G_2\phi^2 + G_3\phi^3 + \ldots$$

where the G_j are functions of T, P, n_i. Because of symmetry the Gibbs function should be even in ϕ, (*e.g.* there is no intrinsic difference between sets of up and down spins.) and odd powers of ϕ disappear from the expansion.

If the continuous order parameter is written in terms of an *order parameter density* $m(x)$, $\phi = \int m(x)dx$ and the Gibbs function

$$G = \int \psi[m(x)]dx$$

where x is a single parameter that represents all spatial dimensions and $\psi(m)$ is called the *Landau free energy*.

The expansion of ψ in terms of m is usually written in the form

$$\psi(m) = \psi_0 + \frac{1}{2}r_0 m^2 + u_0 m^4 + \ldots \tag{9.1}$$

It is customary [129] to define u_0 independent of T, whilst $r_0 = a_0 t$. In thermodynamic equilibrium the free energy must be a minimum as a function of m and m is assumed to have the *mean-field* value of $\overline{m} = \phi/V$. By this assumption the internal field acting at each point in the domain is the same, and equal to the mean internal field for the whole domain. This approximation is valid if the fluctuations in order are small, or if the interaction producing the internal field is long range and averaged over the states of many sites. In most phase transitions, however, the real interaction has a very short range and the fluctuations become very large near the critical point. The approximation ignores all of these fluctuations in the vicinity of the critical point.

In the absence of external fields ($\psi_0 = 0$) the behaviour of $\psi(m)$, as specified by eqn (1), is shown in figure 3 for $t > 0$ and $t < 0$. It illustrates how a transition from a regime with $\overline{m} = 0$ to one with $\overline{m} \neq 0$ can be described by an underlying continuous function ψ.

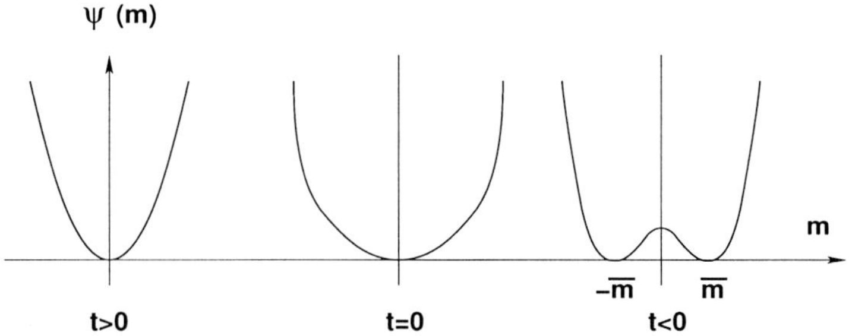

Figure 3: *Landau free energy at different temperatures. Spontaneous symmetry breaking occurs for t < 0, giving rise to a second-order phase transition at t=0.*

The Landau free energy is invariant under $m \to -m$, but this symmetry is spontaneously broken[2] when $t < 0$. The transition is second-order because $\overline{m}$ is continuous at $t = 0$. The condition that determines $\overline{m}$ is

$$\left.\frac{\partial \psi}{\partial m}\right|_{m=\overline{m}} = 0 = r_0\overline{m} + 4u_0\overline{m}^3$$

[2]When a continuous symmetry is spontaneously broken there is a non-countable infinity of ground states(with energy taken as zero) orthogonal to one another. The system must choose one of these. In a quantum system there is no immediately apparent reason why the system cannot exist in a linear combination of all the ground states. In the thermodynamic limit when the number of particles tends to infinity however, there is no operator that could simultaneously change the states of all particles from one ground state to another. In this limit, therefore, the different ground states generate separate Hilbert spaces, and transitions among them are forbidden. A *superselection rule* [128] is said to insulate one ground state from another. A superselection rule operates between subspaces if there are neither spontaneous transitions between their state vectors and if, in addition, there are no measurable quantities with finite matrix elements between their state vectors.

The only real solution is

$$\overline{m} = \left\{ \begin{array}{ll} 0 & (t > 0) \\ \pm(a_0/4u_0)^{1/2}|t|^{1/2} & (t < 0) \end{array} \right\}$$

which immediately implies $\beta = 1/2$.
It is noted that

$$\left.\frac{\partial^2 \psi}{\partial m^2}\right|_{m=\overline{m}} \geq 0$$

All critical exponents calculated according to the mean-field assumption are close to, but different from their experimental values. For example, experimental values of β range between 0.32 and 0.39. Although mean-field theory is not qualitatively reliable it shows how a perfectly regular function such as the Landau free energy can lead to a singular Gibbs energy. Mean field theories that often work well away from the critical point clearly fails near T_c. Typical failures, as measured by calculated critical exponents are shown in figure 4.

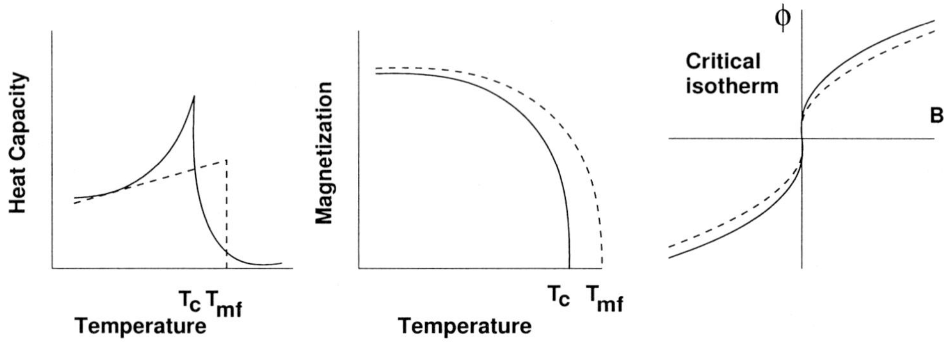

Figure 4: *Deviations of mean field theory (dashed lines) from experimental measurements described by critical exponents (solid lines) near the critical point. T_{mf} is the critical temperature predicted by mean field theory.*

Mean field theory fails because fluctuations are always important near the critical point. Consider, for example, a ferromagnet just above T_c in zero field. Although the mean internal field is zero, the actual field acting on the nearest neighbour of a given site with known up spin, may be quite large. Just above T_c the magnetizability is large and the local internal field due to the given spin, is almost certain to line up the spins at neighbouring sites. By the same argument these neighbouring sites should line up the second nearest sites, and so on. The probability of being spin up will die away only slowly with distance. The system is said to have a large *coherence* or *correlation length*, ξ. Although the mean magnetization may be zero, the *local* magnetization is not. There will be regions of size ξ with substantial magnetization in a definite direction, although the boundaries and direction of these magnetized domains will change in the course of thermal fluctuations. The existence of such domains explains the large heat capacity that persists above T_c, even though the *mean* magnetization is zero. The ferromagnetic fluctuations of spin up and spin down regions also exist below T_c except that one type now predominates.

A similar phenomenon occurs at the liquid-vapour critical point. The system contains relatively large regions of liquid-like and vapour-like density, which fluctuate continuously[3].

9.3.3 The Van der Waals Equation

The Van der Waals equation of state is perhaps the best-known example of a mean-field theory. It was first proposed in the form

$$\left(P + a(n/V)^2\right)(V - nb) = nRT$$

to correct the ideal-gas equation of state, by taking into account the finite volume of a gas molecule (parameter b) and intermolecular attraction (a). In terms of molar volume ($v = V/n$)

$$P = \frac{RT}{v - b} - \frac{a}{v^2} \tag{9.2}$$

Each molecule is imagined to be a hard elastic sphere surrounded by an attractive force field.

Some isotherms corresponding to the Van der Waals equation are shown in figure 5. At a certain critical temperature T_c the isotherm has an intermediate form that goes through an inflection point at critical pressure and volume of P_c and V_c. To satisfy this requirement the first and second derivatives must vanish,

$$\frac{\partial P}{\partial v} = \frac{RT}{(v-b)^2} + \frac{2a}{v_3} = 0$$

$$\frac{\partial^2 P}{\partial v^2} = \frac{2RT}{(v-b)^3} - \frac{6a}{v^4} = 0$$

[3]These regions scatter light, and the system has a milky appearance, known as *critical opalescence*.

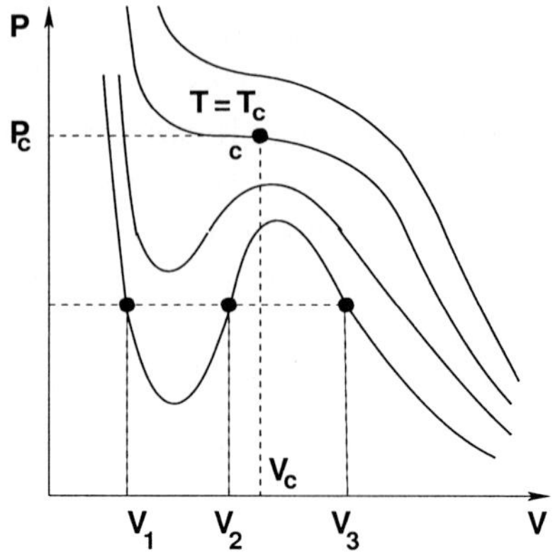

Figure 5: *$P - V$ diagram of the Van der Waals equation of state.*
The solutions to these simultaneous equations are

$$V_c = 3b \qquad P_c = \frac{a}{27b^2} \qquad T_c = \frac{8a}{27Rb}$$

In terms of these quantities *reduced* variables are defined as

$$P_r = \frac{P}{P_c} \qquad V_r = \frac{v}{V_c} \qquad T_r = \frac{T}{T_c}$$

By substituting $P = P_r P_c$ and similar substitutions for v and T, after dropping subscripts r, equation (2) transforms into

$$PP_c + \frac{RTT_c}{VV_c - b} - \frac{a}{V_2 V_c^2}$$

i.e.

$$\frac{aP}{27b^2} = \frac{8aT}{27b(3bV - b)} - \frac{a}{9b^2V^2}$$

which simplifies into

$$\left(P + \frac{3}{V^2}\right)\left(V - \frac{1}{3}\right) = \frac{8}{3}T \tag{9.3}$$

where P, V, and T, respectively denote pressure, volume and temperature in units of the corresponding critical constants. For example, $t = T - 1$.

The equation does not contain the constants a and b, characteristic of the substance, and therefore applies to all substances. Hence it is known as the *law of corresponding states.*

Part of each isotherm below T_c corresponds to a physically impossible negative compressibility. This apparent problem may arise if the system is not homogeneous in this region because of the co-existence of two phases. Two such phases can be in equilibrium if they are at the same P and T and co-exist if the free energy for the mixed system is less than that of the corresponding homogeneous phase.

The free energy is calculated along an isotherm, at fixed total volume,

$$A(T,V) = -\int P\,dV$$

Graphical integration may proceed as shown in figure 6.

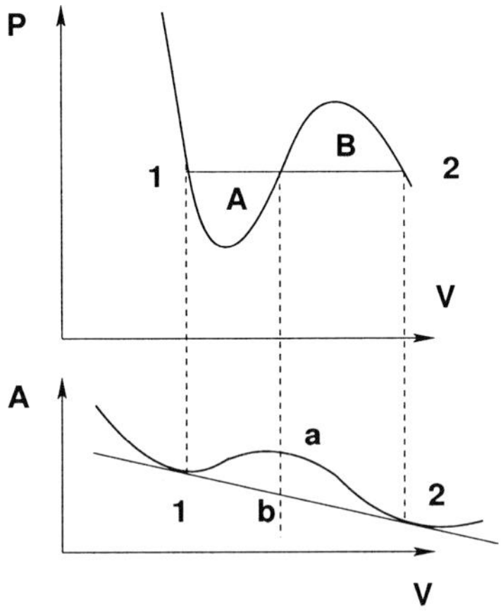

Figure 6: *The Maxwell construction.*

The diagram shows that states 1 and 2 may coexist because they have the same P and T. At point b on the common tangent to the integration curve minima, the free energy is a linear combination of the free energies of states 1 and 2. The free energy at b is lower than the free energy a of the homogeneous sytem at the same T and V. Between points 1 and 2 the system is therefore predicted to separate into two phases in equilibrium at the same pressure, subject to the conditions of equal pressure,

$$-\frac{\partial A}{\partial V_1} = \frac{\partial A}{\partial V_2}$$

and common tangent,

$$\frac{A_2 - A_1}{V_2 - V_1} = \frac{\partial A}{\partial V_2}$$

These conditions combine to give

$$\left(\frac{\partial A}{\partial V_1}\right)(V_2 - V_1) = -(A_2 - A_1)$$

or

$$P_1 (V_2 - V_1) = \int_{V_1}^{V_2} P\, dV$$

Phase equilibrium requires that $A_2 = A_1$ and hence that the integral vanish. All conditions are satisfied if the points 1 and 2 are located such that the areas $A = B$. This geometry defines the *Maxwell construction.* It shows that stable liquid and vapour states correspond to minima in free energy and that $A_L = A_V$ when the external pressure line cuts off equal areas in the loops of the Van der Waals isotherm. At this pressure that corresponds to the saturated vapour pressure, a first-order phase transition occurs.

The Maxwell construction produces a Van der Waals phase diagram that resembles the experimental results of argon, shown in figure 1, in all respects.

To demonstrate that equation (3) describes a mean-field theory of gas-liquid transitions it will be shown how it can be obtained by minimizing a Landau free-energy function. This objective is achieved by working backwards.

Define the order parameter V with conjugate field P and suppose that (3) is obtained by minimization of the Landau free energy $\psi(V, P, T)$, *i.e.* such that (3) is the equation that corresponds to $\partial\psi/\partial V = 0$. Thus by integration of (3) at constant P and T,

$$\begin{aligned} \psi &= \int \left[P + \frac{3}{V_2} - \frac{8}{3}T\left(\frac{1}{V - \frac{1}{3}}\right)\right] dV \\ &= PV - \frac{3}{V} - \frac{8}{3}T \ln\left(V - \frac{1}{3}\right) + C \end{aligned} \qquad (9.4)$$

To work in the opposite direction, minimization of the Landau free energy with respect to the order parameter, $\partial\psi/\partial V = 0$, clearly yields

$$P = \frac{8}{3}\left(\frac{T}{V - \frac{1}{3}}\right) - \frac{3}{V^2}$$

which is Van der Waals' equation (3).

A schematic sketch [129] of ψ, according to (4) is shown in figure (7).

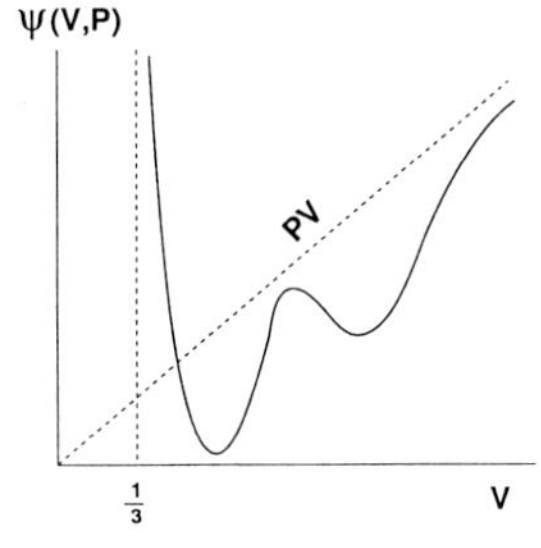

Figure 7: *Landau free energy that leads to the Van der Waals equation of state.*

In the region of a first-order transition ψ has equal minima at volumes V_1 and V_2, in line with the Maxwell construction. The mixed phase is the preferred state in the volume range between V_1 and V_2. It follows that the transition from vapour to liquid does not occur by an unlikely fluctuation in which the system contracts from vapour to liquid at uniform density, as would be required by the maximum in the Van der Waals function. Maxwell construction allows the *nucleation* of a liquid droplet by local fluctuation within the vapour, and subsequent growth of the liquid phase.

9.4 The Scaling Hypothesis

Failure of mean-field theories near the critical point has stimulated the development of several alternative theories. One of the more promising is that parameters near a critical point obey *scaling laws.*

The phenomenological magnetic equation of state, based on the Brillouin formula, has a form related to a scaling law:

$$x = \tanh^{-1} y - \left(\frac{T_c}{T}\right) y$$

where $x = \mu_B \boldsymbol{B}/kT$, $\boldsymbol{B}$ is the effective field and $y = \overline{m}$, is the mean-field magnetization. For small y the inverse function expanded to third power is approximated by

$$\tanh^{-1} y = y + \frac{1}{3} y^3$$

The equation, $x = ty + (1/3)y^3$, assumed valid near the critical point has a scaling property in x, t and y. For any value of μ it may be rewritten in the form

$$\mu^3 x = (\mu^2 t)(\mu y) + \frac{1}{3}(\mu y)^3$$

or $$x' = t'y' + \frac{1}{3}(y')^3$$

where $$x' = \mu^3 \quad , \quad t' = \mu^2 t \quad , \quad y' = \mu y$$

In other words, if x, t and y are replaced by their scaled values x', t' and y' the equation still holds.

Scaling laws apply in any situation where the physical description is not changed when the scale of the system changes. The reason why this idea may gain importance near the critical point is because of the expected growth of long-range order. Minor fluctuations become magnified into large ones. Stated more succinctly, the coherence length increases and becomes infinite at the critical point. Thus there is nothing to characterize the scale of the fluctuations. The essential scaling assumption is that, near the critical point, the correlation length ξ is the only characteristic length of the system, in terms of which other lengths must be measured. It was pointed out by Kadanoff that the appearance of fluctuations at the critical point is unaffected by changing the length scale, provided other relevant parameters are scaled accordingly.

Free energy in the critical region is assumed to split into regular and singular parts (G_r and G_s respectivlely), only the latter of which obeys a scaling law,

$$\lambda G_s = f(\lambda^n t, \lambda^m y) \tag{9.5}$$

for any λ. It can be shown that the singular part of the critical exponents must be related to n and m. For example, the singular part of the heat capacity, calculated at zero field,

$$C_s = -T_c \left(\frac{\partial^2 G_s}{\partial T^2} \right)_{y=0}$$

From

$$G_s = \frac{1}{\lambda} f(\lambda^n t, 0)$$

and

$$\frac{\partial(\lambda^n t)}{\partial T} = \frac{\partial}{\partial T}[\lambda^n(1 - T/T_c)] = -\frac{\lambda^n}{T_c}$$

it follows that

$$\begin{aligned}\frac{\partial G}{\partial T} &= \frac{\partial}{\partial(\lambda^n t)}\left[\frac{1}{\lambda}f(\lambda^n t)\right]\frac{\partial(\lambda^n t)}{\partial T}\\ &= \frac{1}{\lambda}f'\left(-\frac{\lambda^n}{T_c}\right)\\ \frac{\partial}{\partial T}\left(\frac{\partial G}{\partial T}\right) &= \frac{\partial}{\partial(\lambda^n t)}\left[-\frac{1}{T_c}f'\lambda^{n-1}\right]\frac{\partial(\lambda^n t)}{\partial T}\\ &= \frac{1}{T_c}f''\lambda^{n-1}\frac{\lambda^n}{T_c}\end{aligned}$$

Hence

$$-T_c\frac{\partial^2 G}{\partial T^2} = -\frac{1}{T_c}f''\lambda^{2n-1}$$

Since the value of λ is arbitrary it may be chosen such that $\lambda^n|t|$ remains constant when T varies. It means that f'' remains constant, and thus

$$C \propto \lambda^{2n-1} \propto |t|^{\frac{1}{n}-2} \text{ which implies } \alpha = 2 - \frac{1}{n}$$

Similar methods produce the following values of the critical exponents:

$$\begin{aligned}\beta &= \frac{1-m}{n}\\ \gamma &= \frac{2m-1}{n}\\ \delta &= \frac{m}{1-m}\end{aligned} \tag{9.6}$$

Elimination of m and n between the four equations gives the following relations between critical exponents:

$$\begin{aligned}\alpha + 2\beta + \gamma &= 2\\ \alpha + \beta(\delta+1) &= 2\\ \gamma(\delta+1) &= (2-\alpha)(\delta-1)\\ \beta\delta &= \beta+\gamma\end{aligned}$$

Known experimental values [126] are found to satisfy these relations within experimental error. The scaling hypothesis seems to be successful.

9.5 Renormalization Group

Scaling laws provide an improved estimate of critical exponents without a scheme for calculating their absolute values or elucidating the physical changes that occur in the critical region.

Critical phenomena belong to a class of problem that involves a large number of degrees of freedom. In contrast, most theoretical methods work only when there is a single independent variable, *i.e.* one degree of freedom. Schrödinger's equation is a typical example. Under normal circumstances the 10^{23} or so, degrees of freedom for a macroscopic system can be reduced enormously, as happens in the theories of hydrodynamics. The properties of a macroscopic system may be reconstructed from those of a microscopic sample only. Thus a liquid of only 1000 atoms, say, would probably have approximately the same energy per unit volume and density as the same liquid with a mole of atoms at the same temperature and pressure.

The minimum size to which a sample can be reduced without qualitatively changing its properties corresponds to the correlation length. If the correlation length is small the properties of the system can be calculated by a variety of methods, for instance Hartree-Fock. The assumption is that the properties of matter in the bulk can be related to the properties of a small cluster of atoms, noting that even a cluster of three has too many degrees of freedom to be solved without considerable simplification.

In the critical region that marks the onset of a phase transition the correlation length becomes much larger than atomic spacings and the previous arguments no longer apply. At the critical point ξ is infinite and near the critical point ξ is large. Computational methods to deal with this problem must be designed so as to reduce the *effective* correlation length. One method that has been used with some success, known as *block transformation* leads to a scaling law, but the reasons for the appearance of scaling near T_c remains conjectural.

During magnetic transition the problem is to take into account all the spins within the rapidly increasing volume defined by the coherence length. Because of the large coherence length near the critical point neighbouring Ising spins are likely to be correlated, as shown for the two-dimensional case in figure 8.

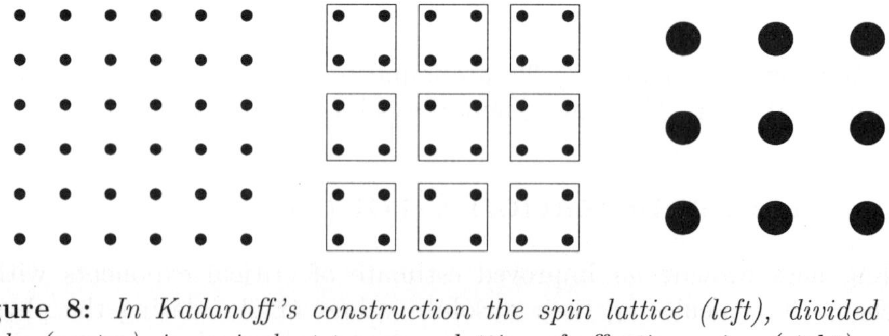

Figure 8: *In Kadanoff's construction the spin lattice (left), divided into blocks (centre), is equivalent to a new lattice of effective spins (right).*

In Kadanoff's [130, 131] two-dimensional block-spin model four neighbouring spins are assumed to have identical spins, either up or down, near the critical point. The block of four then acts like a single effective spin. The lattice constant of the effective new lattice is double the original lattice constant. The coherence length measured in units of the new lattice constant will hence be at half of its original measure. Repetition of this procedure allows further reduction in ξ by factors of two, until finally one has an effective theory with $\xi = 1$. At each step it is convenient to define *renormalized* block spins such that their magnitude is ± 1 instead of ± 4. The energy of such blocked spins is

$$E_1 = \sum_{n,i} K_1 s_n^{(1)} s_{n+i}^{(1)}$$

where K_1 is a constant, $s_n^{(1)}$ the block spin variable and n labels sites on the effective new lattice with unit vectors i. The only difference between E_1 and the original interaction

$$E_o = \sum_{n,j} K s_n s_{n+j}$$

is the change in constant from K to K_1. Such an operation of coarse-graining, followed by rescaling is called a *renormalization group* (RG) transformation. It is so named because it renormalizes the coupling constants, which have a group-like property: if $R_1(K)$ and $R_2(K)$ are RG transformations, so is $R_1R_2(K)$. However, they do not really form a group. Blocked spins cannot be "unblocked", and hence there is no inverse transformation.

There are two spin-spin interactions of the original type that couple a set of block spins to a neighbouring block. If the blocks are parallel the coupling energy is $2K$, or $-2K$ if they are anti-parallel. Hence $K_1 = 2K$ and $\xi(K_1) = \frac{1}{2}\xi(K)$.

It is actually an oversimplication to assume that the spins in the new block must all have the same orientation. It is sufficient that the new block has two possible states. If an effective Ising interaction exists between the blocks, the relationship between the coupling constants is more complicated, but analytic

$$K_1 = f(K) \qquad \text{and} \qquad \xi[f(K)] = \frac{1}{2}\xi(K) \tag{9.7}$$

even for $K = K_c$.

To explore the implications of this model for critical behaviour it is noted that $K_c = f(K_c)$ so that $\xi[f(K_c)]$ is infinite. Secondly, suppose that K is near K_c, such that approximately

$$f(K) = f(K) + \lambda(K - K_c)$$

where $\lambda = df/dK$ for $K = K_c$. This means that

$$f(K) - K_c = \lambda(K - K_c) \tag{9.8}$$

If it is now assumed that $\xi(K)$ behaves as $(K - K_c)^{-\nu}$ for K near K_c, one has

$$\frac{\xi[f(K)]}{\xi(K)} = \left\{\frac{f(K) - K_c}{K - K_c}\right\}^{-\nu}$$

which by (7) and (8) reduces to $\frac{1}{2} = [\lambda]^{-\nu}$. If $f'(K_c)$ could be determined, ν may be calculated from $\nu = \ln 2/\ln \lambda$.

To determine $\xi(K)$ it may be assumed known for $K > 2K_c$, where ξ should not be very large and relatively easy to calculate. Next suppose that $K_c < K < 2K_c$ and construct a sequence of effective constants K_1, K_2, *etc.* to satisfy $K_{n+1} = f(K_n)$, with $K_1 = f(K)$. Kadanoff's formula then implies

$$\xi(K) = 2^n \xi(K_n) \tag{9.9}$$

This formula (9) shows that no matter what value K takes $(K > K_c)$, n can be chosen such that $\xi(K_n)$ is small enough to ensure $K_n > 2K_c$. Then $\xi(K_n)$ would be known and $\xi(K)$ is determined by (9). This calculation is only possible if the function $f(K)$ is known.

The importance of this analysis is the idea that, starting from an analytic function $f(K)$ one generates non-analytic behaviour for $\xi(K)$ at the point K_c for which $f(K_c) = K_c$.

It was shown by Wilson [131] that the Kadanoff procedure, combined with the Landau model, may be used to identify the critical point, verify the scaling law and determine the critical exponents without obtaining an exact solution, or specifying the nature of fluctuations near the critical point. The Hamiltonian for a set of Ising spins is written in suitable units, as before

$$H_0 = -J_0 \sum_{ij} \sigma_i \sigma_j - \sum_i \sigma_i B_0 \tag{9.10}$$

The corresponding free energy is expressed by the relation

$$e^{-G/kT} = \sum_{C_0} e^{-H_0/kT} \cdot e^{-G_{r0}/kT} \tag{9.11}$$

where G_{r0} is the non-singular part of the free energy which is of non-magnetic origin, and the sum is taken over all possible configurations C_0.

When the lattice is doubled by blocking, equation (11) serves to describe a sum over the C_1 blocked-spin configurations. The spins that are blocked out act as a medium which transmits an effective interaction and contribute to the effective external field over the blocked spins. This procedure amounts to the separation of the sum over all configurations C_0 into a double sum over the C_1 blocked configurations and the configurations r of the blocked out part, *i.e.*

$$e^{-G/kT} = \sum_{C_1} \left[\sum_r e^{-H_0/kT} \right] e^{G_{r0}/kT}$$

The two types of interaction in the block-spin system contribute to the Hamiltonian in the form

$$H_0 = H_1 + R_0$$

where H_1 is the effective Hamiltonian for blocked spins and R_0 is constant. Thus

$$e^{-G/kT} = \sum_{C_1} e^{-H_1/kT} \times e^{-G_{r1}/kT}$$

where the new effective regular part of the free energy is defined by

$$G_{r1} = G_{r0} + R_0$$

and now includes terms of magnetic origin.

In a much simplified approach H_1 is assumed to have the same form as (10). The only difference is that the new Hamiltonian refers to block spins only with a new effective coupling constant J_1 and a new effective field B_1. In terms of reduced variables $K = J/kT$, $b = B/kT$, the new coupling constant and field, written as functions of the old, are

$$K_1 = f(K_0) \qquad b_1 = g(b_0) \tag{9.12}$$

Once the functions f and g have been established the blocking procedure may be repeated many times to produce a sequence of effective coupling constants K_0, K_1, $K_2 \ldots$, and fields b_0, b_1, $b_2 \ldots$, together with regular parts of the free energy G_{r0}, G_{r1}, $G_{r2} \ldots$. The operation of finding the new values of K, b and G_r from the old is renormalization and the set of them all constitute the RG.

Once the so-called, fixed point $K_c = f(K_c)$ has been reached, changing the length scale by another iteration does not change the effective Hamiltonian, and fluctuations appear unchanged on the new scale. In practice such limiting behaviour occurs for $B = 0$ and a given value of J_0 at some definite temperature, which may be identified as the critical temperature. Calculations of this type are extremely complicated and sensitive to the model chosen to represent interactions. Values in good agreement with experiment have been obtained.

To predict the critical exponents it follows by taking derivatives of (12) that, near the critical point

$$\delta K_n \simeq f'\delta K_{n-1} \quad , \quad b_n \simeq g'b_{n-1}$$

where $\delta K_n = K_c - K_n$. After n iterations

$$K_n = K_c - \delta K_n \simeq (J_0/kT_c)\left[1 - (f')^n t\right]$$

and

$$b_n \simeq (g')^n b_0 \simeq (g')^n B_0/kT_c$$

The singular part of the free energy after n iterations G_{sn} has a value per spin which is a function of K_n and b_n. Since J_0 and T_c are constants the value per spin may be regarded as a function of $(f')^n t$ and $(g')^n B_0$. After n blockings the number of spins is $n/(2^d)^n$, where d is the number of dimensions. Thus

$$G_{sn}2^{dn} = G\left[(f')^n t, (g')^n B_0\right]$$

Writing $2^{dn} = \lambda$, and approximating f' and g' from (7) as $(2^d)^p$ and $(2^d)^q$ respectively, there follows

$$\lambda G = G(\lambda^p t, \lambda^q B_0)$$

This expression has the form of the scaling hypothesis (5) and the critical exponents may be written down in terms of f' and g' using (6).

It is concluded that RG calculations provide a means of calculating T_c, a justification of the scaling hypothesis and a method of finding the critical exponents.

9.6 Chemical Reaction

Phase transition has many aspects in common with chemical interaction, but the latter is exceedingly more complicated. Models of phase transformation can therefore not provide much more than a crude indication of how chemical reactions occur.

Chemical reactions, like phase transitions are driven by interparticle interactions that give rise to fluctuations and cooperative effects. The closest parallel of a phase A transition is with a polymerization reaction, such as

$$n\mathrm{X} \rightarrow \mathrm{X}_n$$

Formulated as a reversible reaction

$$n\mathrm{X} \rightleftharpoons \mathrm{X}_n \tag{9.13}$$

the formal similarity is complete. The critical point of the phase transition becomes the transition state of the chemical reaction. The common type of reaction

$$\mathrm{A} + \mathrm{B} \rightleftharpoons \mathrm{AB}$$

may be viewed as a special case of (13). This cryptic, but common formulation of a chemical reaction is highly misleading since it masks all interactions, except the A-B chemical interaction which is recognized as causing the chemical change.

Before a chemical bond is established weaker interactions lead to the formation of clusters, such as

$$\mathrm{A}_n + \mathrm{B}_m \rightleftharpoons (\mathrm{AB})_k$$

because of long-range effects. Co-existence of these clusters constitutes a critical region characterized by three correlation lengths, ξ_A, ξ_B and ξ_{AB}, rather than the single parameter of the phase transition. The complexity of the situation grows exponentially when A and B represent molecules rather than atoms. This critical mixture does not lead to chemical interaction without some activation process that enables the formation of chemical bonds in the $(AB)_k$ clusters. An entirely new factor that enters the equation at this point therefore discriminates between a critical point and a transition (or activated) state; between chemical reaction and phase transformation.

The prospect of elucidating the course and mechanism of chemical reactions by some procedure such as RG looks rather bleak. What has been suggested [132] is that RG could be used to identify an effective interaction to describe molecules at the bond level and replace *ab initio* computations that start at the individual electron level. However, the formal resemblance of chemical reactions to phase transitions may clarify some properties of the former.

The progress of a first-order phase transition is summarized in figure 9. Each of the circled diagrams show the $G-V$ relationship at particular points on or near to a co-existence line, or its extrapolation into the critical region.

Along the first-order co-existence line the two phases are in equilibrium, and this equilibrium may shift in either direction with slight changes in energy and pressure. Chemical reactions are not defined in one-component systems and a parallel situation therefore cannot occur. Proceeding down the extension of the co-existence line through the critical region sees a flattening of the free-energy curve over a larger volume until it *bifurcates* at the critical point. The phase symmetry in the co-existence region has a parallel in chemical reactions of the type

$$\mathrm{D} \rightleftharpoons \mathrm{L}$$

that inverts chirality by passing over a racemic transition state at a critical point.

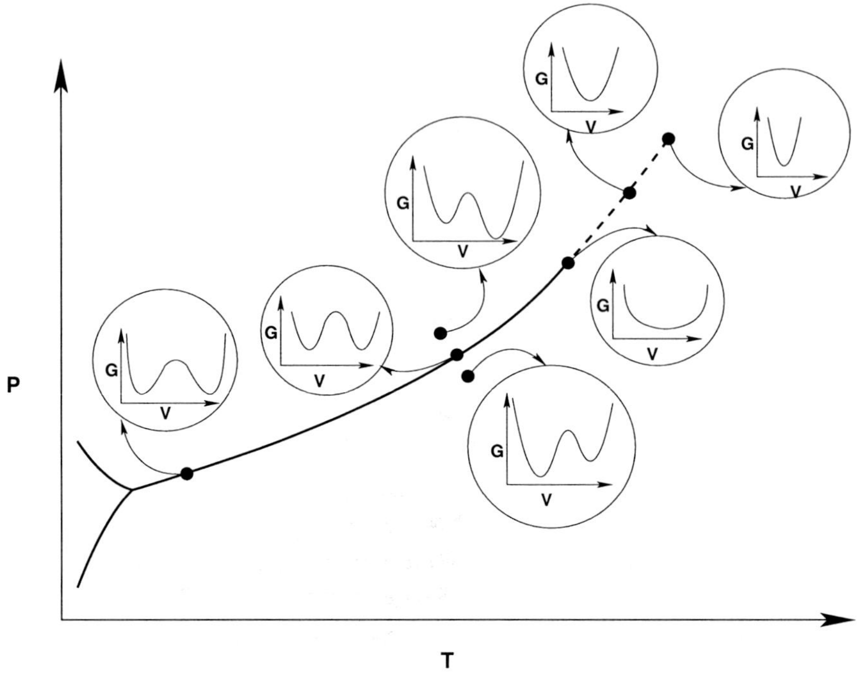

Figure 9: *Minima in the Gibbs potential in the vicinity of a co-existence curve curve and the critical point for a phase transition in a one-component system.*

The infinite potential barrier, shown schematically in figure 10 corresponds to a superselection rule that operates below the critical temperature [133]. Above the critical temperature the quantum-mechanical superposition principle applies, but below that temperature the system behaves classically. The system bifurcates spontaneously at the critical point. The bifurcation, like second-order phase transformation is caused by some interaction that becomes dominant at that point. In the case of chemical reactions the interaction leads to the rearrangement of chemical bonds. The essential difference between chemical reaction and second-order phase transition is therefore epitomized by the formation of chemically different species rather than different states of aggregation, when the symmetry is spontaneously broken at a critical point.

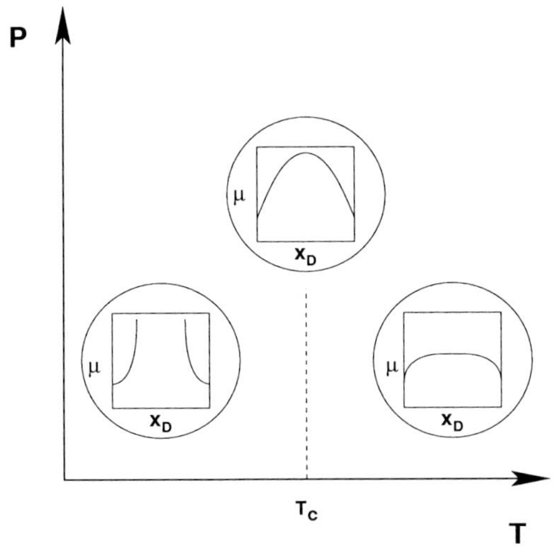

Figure 10: *Spontaneous resolution of enantiomers below a critical point, shown diagrammatically in terms of the variation of chemical potential with mole fraction, near the critical point.*

The qualitative picture of chemical change is clear. The reactant system, in an otherwise fixed environment, approaches an activated, or *valence* state, at a critical temperature. In addition to the appearance of normal critical phenomena, the chemical system is further prepared for reaction by long-range quantum-mechanical activation. This feature falls outside the scope of statistical thermodynamics and needs elucidation in terms of molecular quantum fields.

9.7 Conclusion

The development of theoretical chemistry ceased at about 1930. The last significant contributions came from the first of the modern theoretical physicists, who have long since lost interest in the subject. It is not uncommon today, to hear prominent chemists explain how chemistry is an experimental science, adequately practiced without any need of quantum mechanics or the theories of relativity. Chemical thermodynamics is routinely rehashed in the terminology and concepts of the late nineteenth century. The formulation of chemical reaction and kinetic theories take scant account of statistical mechanics and non-equilibrium thermodynamics. Theories of molecular structure are entirely classical and molecular cohesion is commonly analyzed in terms of isolated bonds. Holistic effects and emergent properties that could distinguish chemistry from physics are generally ignored.

The theories that should feature prominently in the understanding of chemical effects have been summarized in this volume, without demonstrating their application. The way forward has been indicated by Primas [67] and in the second volume of this work the practical use of modern concepts such as spontaneous symmetry breaking, non-local interaction, bohmian mechanics, number theories and space-time topology, to elucidate chemical effects will be explored. The aim is to stimulate renewed theoretical interest in chemistry.

Bibliography

[1] G.N. Lewis, *J. Am. Chem. Soc.*, 1916, **38**, 762.

[2] R.W. Gurney, *Proc. Roy. Soc. (London)*, 1932, **134A**, 137.

[3] J.O'M. Bockris and A.K.N. Reddy, *Modern Electrochemistry*, 1973, 2nd. ed., Wiley, N.Y.

[4] P.W. Atkins, *Physical Chemistry*, 1998, 6th ed., University Press, Oxford.

[5] H.Margenau and G.M. Murphy, *The Mathematics of Physics and Chemistry*, 1943, Van Nostrand, N.Y.

[6] M.M. Nicholson, *Fundamentals and Techniques of Mathematics for Scientists*, 1961, Longmans, Green and Co., London.

[7] G. Stephenson, *Mathematical Methods for Science Students*, 2nd ed., 1973, Longman, London.

[8] R.G. Mortimer, *Mathematics for Physical Chemistry*, 1981, Macmillan, N.Y.

[9] *International Tables for X-ray Crystallography*, Vol. II, 1972, Kynoch, Birmingham.

[10] G.P. Harnwell, *Principles of Electricity and Electromagnetism*, 2nd ed., 1949, Mc Graw-Hill, N.Y.

[11] M. Hamermesh, *Group Theory*, 1962, Addison-Wesley,M Reading.M

[12] A.W. Joshi, *Elements of Group Theory for Physicists*,M 1973, Halsted Press, N.Y.M

[13] F.A. Cotton, *Chemical Applications of Group Theory*,M 1971, 2nd. ed., Wiley, N.Y.M

[14] H.F. Jones, *Groups, Representations and Physics*, 1990,M Hilger, Bristol. M

[15] Wu-Ki Tung, *Group Theory in Physics*, 1985, WorldM Scientific, Singapore.M

[16] W. H. Fegan, *Introduction to Compact Lie Groups,* 1991, World Scientific, Singapore.

[17] H. Goldstein, *Classical Mechanics*, 1950, Addison-Wesley, Reading.

[18] A.O. Barut, *Electrodynamics and Classical Theory of Fields and Particles*, 1964, Macmillan, New York.

[19] M. Friedman, *Foundation of Space-Time Theories*, 1983, University Press, Princeton, NJ.

[20] S. Nettel, *Wave Physics*, 1992, Springer-Verlag, Heidelberg.

[21] C.W. Wong, *Introduction to Mathematical Physics*, 1991, University Press, Oxford.

[22] *Handbook of Chemistry and Physics*, Chemical Rubber Company, any edition.

[23] C.A. Coulson, *Waves*, 7th edition, 1955, Oliver and Boyd.

[24] P.G. Drazin and R.S. Johnson, *Solitons: an introduction*, University Press, Cambridge, 1989.

[25] R.B. Lindsay, *Mechanical Radiation*,1960, McGraw-Hill, N.Y.

[26] A.D. Sakharov, *Soviet Physics - Doklady*, 1968, **12**, 1040.

[27] F. Winterberg, *Z. Naturforsch.*, 1988, **43a**, 1131.

[28] A.P. Levich (ed.) *On the way to understanding the time phenomenon: The constructions of time in natural science*, 1995, World Scientific, Singapore.

[29] R. Penrose, *The emperor's new mind*, 1989, University Press, Oxford.

[30] A.S. Eddington, *The nature of the physical world*, 1948, University Press, Cambridge.

[31] J.A Stratton, *Electromagnetic Theory*, 1941, Mc Graw-Hill, New York.

[32] R.P. Bauman, *Absorption Spectroscopy*, 1962, Wiley, N.Y.

[33] P. Debye, *Polar Molecules*, 1929, Dover Publications, N.Y.

[34] E. Charney, *The Molecular Basis of Optical Activity*, 1979, Wiley, N.Y.

[35] M. Born, *Die Relativitätstheorie Einsteins*, 2nd. ed., 1921, Springer, Berlin.

[36] H.A. Lorentz, in *Fortschr. Math. Wiss., Heft 2, Das Relativitätsprinzip*, 1920, Teubner, Leipzig.

[37] D.J. Griffiths, *Introduction to Electrodynamics*, 1981, Prentice-Hall, Englewood Cliffs.

[38] J.J. Sakurai, *Advanced Quantum Mechanics*, 1967, Addison-Wesley, Reading, Mass.

[39] P.M. Morse and H. Feshbach, *Methods of Theoretical Physics*, 1953, McGraw-Hill, NY.

[40] W.G. Dixon, *Special Relativity*, 1978, University Press, Cambridge.

[41] R. Adler, M. Bazin and M. Schiffer, *Introduction to General Relativity*, 1965, McGraw-Hill, NY.

[42] H. Weyl, *Sitzber. Preuss. Akad. Wiss. Berlin*, 1918, 465.

[43] E. Schrödinger, *Z. Phys.*,1922, **12**, 13.

[44] F. London, *Z. Phys.*, 1927,**42**, 373.
[45] H. Weyl, *Z. Phys.*, 1929, **56**, 330.
[46] C. Hong-Mo and T.S. Tsun, *Some Elementary Gauge Theory Concepts*, 1993, World Scientific, Singapore.
[47] Y. Nambu, *Phys. Rev.*, 1960, **117**, 648.
[48] J. Goldstone, *Nuovo Cimento*, 1961, **19**, 154.
[49] L.H. Ryder, *Quantum Field Theory*, 1985, University Press, Cambridge.
[50] P.W. Higgs, *Phys. Rev.*, 1966, **145**, 1156.
[51] H. Genz, *Die Entdeckung des Nichts*, 1994, Carl Hauser Verlag, München.
[52] J.C.A. Boeyens, *S. Afr. J. Sci.*, 1995, **91**, 220.
[53] N.H. Frank, *Introduction to Electricity and Optics*, 2nd ed., 1950, McGraw-Hill, N.Y.
[54] J.J. Sakurai, *Modern Quantum Mechanics*, 1985, Addison-Wesley, Redwood City.
[55] C.W. Kilmister, (ed) *SCHRÖDINGER Centenary celebration of a polymath*, 1987, University Press, Cambridge.
[56] J.L. Martin, *Basic quantum mechanics*, 1981, Clarendon Press, Oxford.
[57] D. Bohm, *Quantum Theory*, 1959, Dover Publications., N.Y., 1989.
[58] J.H. Jeans, *Report on Radiation and the Quantum-theory*, 1914, The Physical Society, London.
[59] H. Haken and H.C. Wolf, *The Physics of Atoms and Quanta*, translated by W.D. Brewer, 1994, Springer-Verlag, Heidelberg, 4th edition.
[60] I.R. Levine, *Quantum Chemistry*, 4th edition, 1991, Prentice-Hall, Englewood Cliffs.
[61] P. Roman, *Advanced Quantum Theory*, 1965, Addison-Wesley, Reading, Mass.
[62] B.H. Bransden and C.J. Joachain, *Physics of Atoms and Molecules*, 1983, Longman, London.
[63] L.I. Schiff, *Quantum Mechanics*, 1955, Mc Graw-Hill, N.Y.
[64] A. Sommerfeld, *Ann. Phys.*, 1916, **51**, 1.
[65] A. I. Akhiezer and V.B. Berestetskii, *Quantum Electrodynamics*, Translated from 2nd Russian ed. by G.M. Volkoff, 1965, Interscience, New York.
[66] K. Gottfried, *Quantum Mechanics. Volume I. Fundamentals*, 1966, Addison-Wesley.
[67] H. Primas, *Chemistry, Quantum Mechanics and Reductionism*, 1983, Springer, Berlin.
[68] S.F.A. Kettle *Symmetry and Structure*, 2nd ed., 1995, Wiley, N.Y.
[69] J.C.A. Boeyens, *S. Afr. J. Chem.*, 2000, **53**, 49.

[70] C. Djerassi, *Optical Rotatory Dispersion*, 1960, McGraw-Hill, N.Y.

[71] C.S. Johnson and L.G. Pederson, *Problems and Solutions in Quantum Chemistry and Physics*, 1974, Addison-Wesley, Reading.

[72] A. Amann, *Can Quantum mechanics account for Chemical Structures?* in W. Gans, A. Amann and J.C.A. Boeyens, *Fundamental Principles of Molecular Modeling*, 1996, Plenum, New York.

[73] C. Kittel, *Introduction to Solid State Physics*, 5th ed., 1976, Wiley, N.Y.

[74] E. Heilbronner and H. Bock, *The HMO Model and its Application*, translated by E. Martin and A.J. Rackstraw, 1976, Verlag Chemie, Weinheim.

[75] C.J.H. Schutte, *Wave Mechanics of Atoms, Molecules and Ions*,1968, Arnold, London.

[76] S.D. Travlos and J.C.A. Boeyens, *J. Cryst. Spectr. Res.*, 1990, **20**, 433.

[77] J.C.A. Boeyens, *J. Phys. Chem*, 1967, **71**, 2969.

[78] J.C. Slater, *Phys. Rev.*, 1951, **81**, 385.

[79] F. Herman and S. Skillman, *Atomic Structure Calculations*, 1963, Prentice-Hall, Englewood Cliffs.

[80] A. Michels, J. de Boer and A. Bijl, *Physica*, 1937, **4**, 981.

[81] A. Sommerfeld and H. Welker, *Ann. Phys.*, 1938, **32**, 56.

[82] J.C.A. Boeyens, *J. Chem. Soc. Faraday Trans.*, 1994, **90**, 3377.

[83] W.G. Richards in F. Franks, *Water. A Comprehensive Treatise.* Vol. VI, 1979, Plenum Press, N.Y.

[84] Ø. Burrau, *Kgl. Danske Videnskab. Selskab.*, 1927, **7**, 1.

[85] H. Wind, *J. Chem. Phys.*, 1965, **42**, 2371.

[86] L.J. Schaad and W.V. Hicks, *J. Chem. Phys.*, 1970, **53**, 851.

[87] C.A. Coulson, *Trans. Faraday Soc.*, 1937, **33**, 1479.

[88] F. Weinhold, *J. Chem. Phys.*, 1971, **54**, 530.

[89] H.M. James and A.S. Coolidge, *J. Chem Phys.*, 1933, **1**, 825.

[90] W.Heitler and F. London, *Z. Physik*, 1927, **44**, 455.

[91] Y. Sigiura, *Z. Physik*, 1927, **45**, 1.

[92] W.G. Richards and D.L. Cooper, *Ab initio molecular orbital calculations for chemists,* 2nd ed., 1983, Clarendon Press, Oxford.

[93] J.P. Lowe, *Quantum Chemistry*, 2nd ed., 1993, AcademicM Press, San Diego.

[94] J.C. Slater, *Quantum theory of atomic structure*, 1960,M McGraw-Hill, N.Y.M

[95] T. Koopmans, *Physica*, 1933, **1**, 104.

[96] A. Hinchliffe, *Ab Initio determination of molecularM properties*, 1987, Hilger, Bristol.

[97] K. Higasi, H. Baba and A. Rembaum, *Quantum OrganicM Chemistry*, 1965, Wiley, N.Y.

[98] M. Wolfsberg and L. Helmholz, *J.Chem. Phys.*,1952, **20**, 837.

[99] R. Hoffmann, *J. Chem. Phys.* 1963, **39**, 1397.

[100] G.A. Raggio, *States and composite systems in W* algebraic quantum mechanics*, 1981, Dissertation ETH No. 6824, Zürich.

[101] G.H. Wagnière, *Introduction to Elementary Molecular Orbital Theory and to Semiempirical Methods*, 1976, Springer, Berlin.

[102] B. Pullman, *Advances in Quantum Theory*, 1977, **10**,251.

[103] K. Jug, *Theor. Chim. Acta*, 1980, **54**, 263.

[104] P. Hohenberg and W. Kohn, *Phys. Rev. B*, 1964, **136**, 864.

[105] W. Kohn and L.J. Sham, *Phys. Rev. A*, 1965, **140**, 1133

[106] R.O. Jones and O. Gunnarson, *Rev. Mod. Phys.*, 1989, **61**, 689.

[107] N. Allinger in S.J. Formoshino, I.G. Csizmadia and L.G. Arnaut (eds.), *Theoretical and Computational Models for Organic Chemistry*, 1991, Nato ASI Series C339, Kluwer, Dordrecht.

[108] J.C.A. Boeyens, *Structure and Bonding*, 1985, **63**, 65.

[109] J.C.A. Boeyens, in W.Gans, J.C.A. Boeyens and A. Amann, *Fundamental Principles of Molecular Modeling*, 1996, Plenum, N.Y., p. 99.

[110] J.C.A. Boeyens and P. Comba, *Coord. Chem. Revs*, 2001, **212**, 3.

[111] E.B. Wilson, J.C. Decius and R.C. Cross, *Molecular Vibrations*, 1955, McGraw-Hill, N.Y.

[112] A.K. Rappè and C.A. Casewit, *Molecular mechanics across chemistry*, 1997, University Science Books, Sausalito.

[113] H.B. Callen, *Thermodynamics and an Introduction to Thermostatistics*, 1985, 2nd ed., Wiley, N.Y.

[114] L. Onsager, *Phys. Rev.*, 1932, **37**, 405.

[115] A. Münster, *Statistical Thermodynamics*, 1969, Springer, Berlin.

[116] R.C. Tolman, *The principles of statistical mechanics*, 1938, Clarendon Press, Oxford.

[117] W.V. Houston, *Principles of mathematical physics*, 1948, 2nd ed., McGraw-Hill, N.Y.

[118] S.L. Soo, *Analytical Thermodynamics*, 1962, Prentice-Hall, Englewood Cliffs.

[119] H. Haken, *Quantum Field Theory of Solids*, 1976, North Holland, Amsterdam.

[120] J.E. Mayer and M. Goeppert-Mayer, *Statistical mechanics*, 1940, N.Y.

[121] W.L. Clinton, *J. Chem. Phys.*, 1960, **33**, 1603.

[122] H.B. Callen and T.A. Welton, *Phys. Rev.*, 1951, **83**,34.

[123] J.R. Waldram, *The theory of thermodynamics*, 1985, University Press, Cambridge.

[124] P. Glansdorff and I. Progogine, *Thermodynamic theory of Structure, Stability and Fluctuations,* 1971, Wiley, New York.

[125] G. Nicolis and I. Prigogine, *Self-organization in Non-equilibrium systems*, 1977, Wiley, New York.

[126] J.R. Waldram, *The theory of thermodynamics*, 1985, University Press, Cambridge.

[127] J.M. Ziman, *Models of disorder*, 1979, University Press, Cambridge,

[128] G.C. Wick, A.S. Wightman and E.P. Wigner, *Phys. Rev.*, 1952, **88**, 101.

[129] K. Huang, *Statistical Mechanics*, 1987, 2nd ed., Wiley, N.Y.

[130] L.P. Kadanoff, *Physics*, 1965, **2**, 263.

[131] K.G. Wilson and J. Kogut, *Phys. Reports, C*, 1974, **12**, 75.

[132] K.G. Wilson, *Rev. Mod. Phys.* 1983, **55**, 583.

[133] A. Amann, in W. Gans *et al.*, (eds.) *Fundamental Principles of Molecular Modeling*, 1996, Plenum, N.Y.

Index